Rapid Guide to
Hazardous Air Pollutants

HOWARD J. BEIM
JENNIFER SPERO
LOUIS THEODORE

VAN NOSTRAND REINHOLD
I(T)P® an International Thomson Company

New York • Albany • Bonn • Boston • Detroit • London • Madrid • Melbourne
Mexico City • Paris • San Francisco • Singapore • Tokyo • Toronto

Copyright © 1998 by Van Nostrand Reinhold

I(T)P®   An International Thomson Publishing Company
The ITP logo is a registered trademark used herein under license

Printed in the United States of America

For more information, contact:

Van Nostrand Reinhold   Chapman & Hall GmbH
115 Fifth Avenue   Pappelallee 3
New York, NY 10003   69469 Weinheim
                               Germany

Chapman & Hall   International Thomson Publishing Asia
2-6 Boundary Row   221 Henderson Road #05-10
London   Henderson Building
SE1 8HN   Singapore 0315
United Kingdom

Thomas Nelson Australia   International Thomson Publishing Japan
102 Dodds Street   Hirakawacho Kyowa Building, 3F
South Melbourne, 3205   2-2-1 Hirakawacho
Victoria, Australia   Chiyoda-ku, 102 Tokyo
                               Japan

Nelson Canada   International Thomson Editores
1120 Birchmount Road   Seneca 53
Scarborough, Ontario   Col. Polanco
Canada MIK 5G4   11560 Mexico D.F. Mexico

1 2 3 4 5 6 7 8 9 10 RDCV 01 00 99 98

Library of Congress Cataloging-in-Publication Data

Beim, Howard.
    Rapid guide to hazardous air pollutants / Howard Beim, Jennifer
Spero, Louis Theodore.
        p. cm.
    Includes bibliographical references and index.
    ISBN 0-442-02515-7 (pbk.)
    1. Air—Pollution—Handbooks, manuals, etc.   2. Pollutants—
Handbooks, manuals, etc.   I. Spero, Jennifer.   II. Theodore,
Louis.   III. Title.
TD883.1.B37   1997
363.738—dc21   97-26495
                                                           CIP

http://www.vnr.com
product discounts • free email newsletters
software demos • online resources

email:info@vnr.com

A service of I(T)P®

To my wife, Lucy, for her loving encouragement

and to

Luigi, Vito, and Lou, who showed me the way.

Howard Beim

To my friend Daniel J. Franco for his help and patience.

Jennifer Spero

To my friends Howie and Jen.

Louis Theodore

CONTENTS

Reasonable care has been taken to assure the accuracy of the information contained in the *Rapid Guide to Hazardous Air Pollutants*. However, the authors and the publisher cannot be responsible for errors or omissions in the information presented or for any consequences arising from the use of the information published in the *Rapid Guide to Hazardous Air Pollutants*. Accordingly, reference to original sources is encouraged. The authors and the publisher strongly encourage all readers, and users of chemicals, to follow the manufacturers' or suppliers' current instructions, technical bulletins, and material safety data sheets (MSDSs) for specific use, handling, and storage of all chemical materials. Reporting of any errors or omissions is solicited in order to assure that appropriate changes may be made in future editions.

PREFACE

The 1990 Amendments to the Clean Air Act established a list of 189 hazardous air pollutants (HAPs). Most of these had not been previously regulated. The requirement to treat a new long list of chemicals in an environmentally responsible manner fuels a need for a quick source of information on those chemicals. This need will impact on a wide variety of people. Engineers and scientists will be joined in this need by consultants, lawyers, insurance professionals, plant managers, small business operators, students, and others, even though some may have little engineering, chemical, or science background. As regulations pertaining to the 189 hazardous air pollutants increase, the need for this information will intensify.

To date, the authors feel that no one source covers all of the 189 hazardous air pollutants in the manner presented in this guide. It is hoped that this book will serve to fill the growing information need.

For additional and more detailed information on the Hazardous Air Pollutants, the reader is directed to the list of references found under the heading "Sources" in the section entitled "About This Book."

The authors strongly recommend that the reader review the "About This Book" section before proceeding to use this book. The information it contains about the organization and content of this Rapid Guide will speed locating the hazardous air pollutants, assist in interpreting the data presented, and aid in quickly extracting precisely the information needed.

Howard J. Beim
Jennifer Spero
Louis Theodore
New York

ACKNOWLEDGMENTS

Approximately one-half of the entries in the body of the Rapid Guide have two or three letters in parentheses following the name of the hazardous air pollutant. These letters are the initials of a contributor. The authors wish to thank the contributors for all of their valuable assistance. The names of the contributors and their initials are listed below.

Elena R. Capone (ec)
Richard Carbonaro (rc)
Robert T. Ciotti (rtc)
Keith J. Colacioppo (kjc)
Sean Culliney (sc)
Gianni Del Duca (gd)
Bella Devito (bd)
Saul Ferguson (sf)
Paul F. Fernandez (pf)
Yolanda Foley (yf)
Christine D. Hellwege (ch)

Bruce Howie (bh)
James H. Love (jl)
Sameet Master (sm)
James Morrissey (jm)
Robert Pettenato (rp)
Christopher Pilek (cp)
Nikki Sanders (ns)
Richard C. Shaw (rs)
Andrew H. Shelofsky (ahs)
Deyanshi Trivedi (dt)
Michael Zeolla (mz)

INTRODUCTION

The Clean Air Act has been evolving since it was first passed by Congress in 1967 as the Air Quality Act. It was in the 1970 amendments that Congress began to address the problems caused by "hazardous air pollutants" or "air toxics." Implementation of the regulations issued under the 1970 amendments, however, led to much litigation. Largely in response to that litigation, Congress, in 1977, again amended the Clean Air Act to clarify clean air goals and to set timetables for the U.S. Environmental Protection Agency (EPA) to develop the specific standards needed to achieve those goals. Congress also directed the EPA to consider regulating a number of specific hazardous air pollutants such as arsenic, polycyclic organic matter, and radionuclides. Another result of the 1977 amendments was the development of national emission standards for hazardous air pollutants (NESHAPs).

Hazardous air pollutants (HAPs) were addressed in Section 112 of the 1977 amendments. That portion of the act called on the EPA to create a list of hazardous air pollutants that would include any chemical substance that could produce serious health effects if present in the nation's air. The 1977 amendments further required that all hazardous air pollutants on the list be regulated so as to reduce their concentration in the air to a level that would protect the health of the public with "an ample margin of safety."

Implementation of these 1977 hazardous air pollutant provisions resulted in numerous lawsuits and counterlawsuits in the 1980s. Citing studies that estimated the risk of cancer in heavily polluted areas to be as high as 1 in 10,000 annually, public interest and environmental groups insisted that the EPA had to list all harmful chemicals, with special emphasis on known and suspected carcinogens. Those groups also contended that the Clean Air Act required the EPA to regulate all sources of those listed hazardous air pollutants in a manner that eliminated all risk to the health of the public without regard to cost or the availability of the technology needed to implement those regulations. Other groups countered that chemicals had to be clearly proven to be a serious health problem at the concentrations found in the air before they could be listed and regulated. Such proof would require many scientific studies that could take decades to complete and cost huge amounts of money. Even if these studies were undertaken, their scientific validity was certain to be challenged upon completion. Lack of adequate facilities, trained personnel, and the money to conduct the number of such studies needed would, in the opinion of the pro-regulation groups, almost completely block implementation of the hazardous air pollutant program and leave the public unprotected. On the other side of the argument, industry claimed that the cost and other burdens of regulating many substances that might not be harmful would strangle the nation's economy. By 1990, thirty-three substances had been listed. Others were under consideration.

The 1990 amendments to the Clean Air Act sought to break the litigation deadlock by changing the approach to regulating hazardous air pollutants. Section 112 of the 1990 amendments created a list of 189 hazardous air pollutants, thus ending the debate over which chemicals should be on the list. In selecting these substances, Congress made significant use of the Toxic Release Inventory (TRI) compiled by the EPA from reports required by the Emergency Planning and Community Right-to-Know Act of 1986. Substances can be added to or deleted from the list by the EPA. Petitions for such changes may be submitted to the EPA. Substances that affect only the environment and not public health can also be listed.

The 1990 amendments also changed the basis of regulating hazardous air pollutants in another important way. Previously, the EPA had to analyze the health risk posed by each hazardous air pollutant and then determine the level of restrictions on emissions that would adequately protect the health of the public from that chemical. Under the 1990 amendments, the sources that emitted any of the 189 listed hazardous air pollutants (HAPs) would be grouped together into categories and their emissions restricted to what was technologically feasible rather than to what was needed to protect the health of the public. This regulatory standard was called the "maximum achievable control technology" (MACT) standard. The MACT standard would use the best technology currently available. It does not mandate the development of new or better technology. The development of MACT standards does take cost into account, thus settling another major area of disputes generated by the 1977 amendments.

As part of the process of implementing the changes in the approach to regulating hazardous air pollutants mandated by the 1990 amendments, the EPA was directed to create a list of these categories of sources of hazardous air pollutants. Examples of categories to be included on such a list would be chemical manufacturing plants and oil refineries. A "major source" would be one that emitted 10 tons/year of any one of the hazardous air pollutants on the list or a total of 25 tons/year of any combination of listed hazardous air pollutants. Major sources would be regulated by the EPA to the MACT standard through a permitting process. A list of source categories was published by the EPA in July 1992. The EPA was also mandated to study other, smaller, sources of hazardous air pollutants such as gas stations and dry cleaners. These sources are called "area sources."

MACT standards for 112 hazardous air pollutants were issued as an EPA final rule in March 1994. These MACTs covered approximately 370 synthetic and nonsynthetic organic chemical manufacturers and 940 manufacturing processes.

After the MACT standard has been put in place and allowed to operate for a number of years, the EPA must determine if there is any "residual risk" to health from that hazardous air pollutant. Residual risk exists for known and suspected carcinogens if studies show that the emissions would produce a risk of cancer greater than 1 in 1,000,000 annually for the most exposed individual. A finding of residual risk would require the EPA to propose additional standards to protect public health with an "ample margin of safety."

ABOUT THIS BOOK

The Beim, Spero, and Theodore *Rapid Guide to Hazardous Air Pollutants* has been written for both academic and professional use. In addition, it has not only been written for technical individuals who work in the environmental area or in environmentally related fields, but also for nontechnical (in an environmental sense) individuals, such as office workers, secretaries, doctors, lawyers, etc., and last but not least, the consumer. In effect, it is a quick reference guide that may be used whenever and wherever information about hazardous air pollutants is likely to be sought.

The Guide consists of thirteen fields of information on each of the 189 hazardous air pollutants (HAPs) listed in section 112 of the 1990 Amendments to the Clean Air Act. Every attempt has been made to present as much information as possible in a 2–3 page setting. Information in some of the fields was not available for a few of the HAPs. This was indicated by the notation "NOT AVAILABLE." In addition, seventeen of the HAPs are categories or groups of compounds rather than single substances. Examples of these HAPs are Antimony Compounds, Coke Oven Emissions, Polycyclic Organic Matter, and Radionuclides. For the HAPs that are the compounds of a specific element (such as Antimony Compounds), the properties of a few representative compounds as well as properties characteristic of the entire group are given. For the HAPs that are groups of compounds with very different properties (such as Coke Oven Emissions), every effort was made to provide as much specific information as possible.

ALPHABETICAL ARRANGEMENT and HAZARDOUS AIR POLLUTANT NAMES

The hazardous air pollutant entries are listed by the names and in the alphabetical order used in section 112 of the 1990 Amendments to the Clean Air Act—with one exception. The seventeen Hazardous Air Pollutants, such as Antimony Compounds and Fine Mineral Fibers, that are categories or groups of chemicals for which no CAS number is available have been integrated into the alphabetical order rather than being grouped together at the end of the list as they are in the 1990 Amendments to the Clean Air Act. Also, numbers, letters (such as N or n), Greek letters, and prefixes (except for bis) such as sym, tris, ortho, meta, and para that precede the name do not affect the alphabetical order.

It is important to check the alphabetical order carefully as the use of two words for a name rather than one word could affect its location. Names consisting of two or more words come before one word names. For example, Methylene diphenyl diisocyanate comes before 4,4'-Methylenedianiline. The most common one word/two word variations of the names of the hazardous air pollutants, as well as many of their most common synonyms, may be found in Appendix I. If the name of the chemical being sought is not one of the entries and cannot be found in Appendix I, it may be one of the many other possible synonyms. In that case, the CAS Number Cross-Reference List (Appendix II) may be used to help locate the chemical. If the reader is not aware of the CAS number, two good sources for locating synonyms and their corresponding CAS numbers are:

1. *Sax's Dangerous Properties of Industrial Materials*, Richard J. Lewis Sr., Van Nostrand Reinhold, New York, 9th edition, 1996 (in 3 volumes), and
2. *Suspect Chemicals Sourcebook*, Roytech Publications, Bethesda, Maryland, 1996 edition.

CAS NUMBER CROSS-REFERENCE LIST

This list is found in Appendix II. It gives the names of the hazardous air pollutant entries in CAS number order. For more about the names of the entries, see the preceding section on alphabetical arrangement. The CAS number follows the form xxxx-xx-x. The field on the left has a variable number of digits. The middle field always has two digits and the field at the right, one digit. The numbers are ordered first by the field at the left, second by the middle field and last by the field, at the right. For more about CAS numbers, see the explanation of the CAS # field that follows.

FIELDS OF INFORMATION

The thirteen fields of information provided in the body of the guide for each HAP entry are listed and explained in the following descriptions. Special detail is provided for the Health Risk field.

CAS

Chemical Abstracts Service registry number. The CAS registry number is a unique number assigned to each substance by the Chemical Abstract Service of the American Chemical Society. It definitively identifies the substance regardless of how it is named or how its formula is written. No two substances have the same CAS #. HAPs such as Nickel Compounds and Fine Mineral Fibers, which are groups of substances, do not have a specific CAS #. An example of a HAP with a specific CAS # is acrolein; acrolein's CAS # is 107-02-8.

MOLECULAR FORMULA

The number of atoms of each element in one molecule of the substance. For example, C_6H_7N is the molecular formula of aniline. Subscripts of one are understood to be present without being written. HAPs such as Nickel Compounds and Fine Mineral Fibers, which are groups of substances, do not have a specific molecular formula.

FORMULA WEIGHT

The sum of the atomic weights of the atoms in the molecular formula. For example, 153.81 is the sum of the atomic weights of the one carbon and four chlorines in the HAP CCl_4, carbon tetrachloride. HAPs such as Nickel Compounds and Fine Mineral Fibers, which are groups of substances, do not have a specific formula weight.

SYNONYMS

Other names by which the chemical is known. Many of these names are trade names. The most commonly used synonyms are given. Synonyms may also be found in Appendix I. The following are examples of synonyms by which some HAPs are known: toluene is also known as methyl benzene; methyl ethyl ketone is often called MEK;

and polychlorinated biphenyls are referred to as aroclors. An exhaustive treatment of synonyms is not possible in a rapid guide format.

USES

Special emphasis was given to two categories of use.

Manufacture of other chemicals
A list of the materials and other chemicals manufactured with the use of this chemical is provided. Generally, the HAP is chemically changed by these manufacturing processes.

Used as a component in the manufacturing process
A list of the ways the HAP is used in which it usually retains its chemical properties. The end product of which it may be a part will likely have properties that are significantly different from those of the HAP.

PHYSICAL PROPERTIES

Properties that can be measured or observed without converting the HAP into a different chemical substance. For example, boiling point temperature, density, specific heat, vapor pressures, and color may be determined without changing the HAP into a different substance.

CHEMICAL PROPERTIES

Properties whose measurement or observation requires converting the HAP into a different chemical substance. Some examples are as follows: reactivity with air and/or water; reactivity with skin or metal; flash point temperature; and heat of combustion.

HEALTH RISK

Health Risk is subdivided into five categories.

- ◆ Toxic Exposure Guidelines
- ◆ Toxicity Data
- ◆ Other Toxicity Indicators
- ◆ Acute Risks
- ◆ Chronic Risks

Details on each of these Health Risk categories are provided in the following.

◆ Toxic Exposure Guidelines
These guidelines are intended to provide the limits of exposure below which those who work with the chemical will not suffer any adverse health effects. It is expected that the worker is, otherwise, in good health. These numbers add meaningfulness to toxicity data.

Note: Use of a respirator to reduce exposure to a HAP to below the Toxic Exposure Guideline requires professional expertise to determine the choice and effectiveness of the respirator. The use of a respirator can place a strain on respiratory and cardiac systems. A medical exam by a specially trained health professional is needed prior to the use of a respirator to determine if the use of a respirator poses a risk to the health of the wearer.

The Toxic Exposure Guidelines cited are as follows.

ACGIH TLV TWA
American Conference of Governmental Industrial Hygienists (ACGIH) threshold limit value (TLV) time-weighted average (TWA). The ACGIH, a private organization, issues its TLV TWAs annually for approximately 600 airborne substances. The TLV is the concentration below which a healthy worker could be exposed for 8 hours/day, 5 days/week, for 20 years without developing any disease. Time-weighted average means that the TLV could be exceeded for a time if it is balanced by a time of lower exposure. For each TLV TWA, a ceiling value places an upper limit on the concentration to which the worker can be exposed regardless of how much time is spent at concentrations below the TLV TWA. A short-term exposure limit (STEL) is a 15 minute TWA exposure that should not be exceeded at any time during a workday. For example, the TLV TWA for the HAP carbon tetrachloride is 5 parts per million (ppm) with a STEL of 30 ppm.

OSHA PEL TWA
Occupational Safety and Health Administration (OSHA) permissible exposure limits (PEL) Time Weighted Aver-

age (TWA). This is similar to the ACGIH TLV TWA but issued as a regulation by OSHA to protect workers. PELs are published in Title 29 of the Code of Federal Regulations (CFR) Part 1900 Section 1000.

NIOSH REL TWA
National Institute of Occupational Safety and Health (NIOSH) recommended exposure limits (REL) time-weighted average (TWA). NIOSH is a government agency that conducts research and makes recommendations on occupational safety and health but does not issue regulations. The NIOSH REL TWA is similar to the ACGIH TLV TWA.

IDLH
Immediately dangerous to life or health. IDLH values are issued by NIOSH and reflect a concentration of a chemical that is likely to cause death or immediate or delayed permanent adverse health effects. The IDLH is also the concentration of the chemical that could incapacitate a worker and thus prevent escape from a contaminated area. As a margin of safety, IDLH values are based on the effects that might occur as a consequence of a 30 minute exposure. NIOSH cautions, however, that the worker exposed to the IDLH should not continue to work until the 30 minutes is up. Every effort should be made to exit immediately.

◆ Toxicity Data
These data usually reflect the results of animal testing. The table of relative acute toxicity criteria given on the facing page was published by the National Institute for Occupational Safety and Health (NIOSH) in the Registry of the Toxic Effects of Chemical Substances (RTECS) in 1976. It is widely used to interpret animal toxicity data. As the table indicates, for animal toxicity data, the lower the number, the greater the toxicity. The measures of toxicity used in the table, LD_{50} and LC_{50}, (and others) are explained in the section that follows the table.

Rating	Key Words	LD$_{50}$ Single Oral Dose* (mg/kg)	LC$_{50}$ Inhalation Vapor Exposure* (ppm)	LD$_{50}$ Skin** (mg/kg)
4	Extremely hazardous	1	10	5
3	Highly hazardous	50	100	43
2	Moderately hazardous	500	1000	340
1	Slightly hazardous	5000	10,000	2800
0	No significant hazard	>5000	>10,000	>2800

*Rats
**Rabbits

Animal toxicity data usually provide the route of entry into the body (oral ingestion, inhalation, absorption through the skin, etc.) first, followed by the test subject (mouse, rat, human, etc.), followed by the measure of toxicity. The most common measures of toxicity are as follows.

LD$_{50}$

Lethal Dose$_{50\%}$. The LD$_{50}$ is the dose of the chemical that killed 50% of the test animals when administered by a route of entry other than inhalation. The dose of the chemical (usually solids or liquids) is given as mg/kg, which represents milligrams of chemical per kilogram of body weight of the test animal. The LD$_{50}$ is expressed in this manner because more chemical is needed to kill a larger animal. The oral rat LD$_{50}$ for the HAP calcium cyanamide, for example, is 158 mg/kg.

LC$_{50}$

Lethal Concentration$_{50\%}$. Same as LD$_{50}$ except that the route of entry is inhalation. The concentration of the inhaled chemical (usually gases) is expressed as parts per million (ppm) or milligrams per cubic meter (mg/m^3). Either unit may be converted into the other by use of the following equation.

$$mg/m^3 = (ppm)(\text{formula weight of the substance})(0.0409)$$

Utilizing information found in the entry for the HAP 2-nitropropane, the following example illustrates the use of the equation.

2-nitropropane has an inhalation rat LC_{50} of 400 ppm. The formula for 2-nitropropane is $C_3H_7NO_2$. Its formula weight is 89.11. Using these data in the equation gives $(400 \text{ ppm})(89.11)(0.0409) = 1458 \text{ mg/m}^3$.

LDLo
Lethal Dose Low. The lowest dose that killed any of the animals in the study when administered by a route of entry other than inhalation. For example, a study cited in the entry for the HAP bromoform reported a subcutaneous rabbit LDLo of 410 mg/kg.

LCLo
Lethal Concentration Low. Same as LDLo except that the route of entry is inhalation. The HAP aniline, for example, has an inhalation rat LCLo of 250 ppm.

TDLo
Toxic Dose Low. The lowest dose used in the study that caused any toxic effect (not just death) when administered by a route of entry other than inhalation. The entry for the HAP β-propiolactone, for example, shows that benign tumors were produced by a subcutaneous mouse TDLo of 69 mg/kg.

TCLo
Toxic Concentration Low. Same as TDLo except that the route of entry is inhalation. An inhalation rat TCLo of 100 parts per billion (ppb) reported for the HAP bis(chloromethyl)ether caused malignant tumors.

◆ Other Toxicity Indicators
The indicator most frequently cited here is the EPA estimate of the per million lifetime risk of cancer to the most exposed individual.

◆ Acute Risks
The risks associated with short exposures to high concentrations.

◆ Chronic Risks
The risks associated with long-term exposures to low concentrations.

HAZARD RISK

All risks other than health risks. These risks are most often accidents, fires, explosions, corrosive hazards, and chemicals that are incompatible with the HAP. The National Fire Protection Association (NFPA) ratings are given for those HAPs that have been rated by that private organization. The NFPA ratings use a scale of 0 to 4 where 0 represents no hazard and 4 represents extreme hazard. The NFPA ratings are primarily intended as a cautionary guide to firefighters dealing with a fire in an area where the rated chemical is stored.

EXPOSURE ROUTES

In addition to the routes of entry into the body, the activity, environment, source, or occupation that results in exposure to the HAP is given where available.

REGULATORY STATUS/INFORMATION

This field of information lists the primary laws and citations of the Code of Federal Regulations (CFR) where the chemical is regulated. Citations of the CFR have the following form: Title # CFR Part #.Section #. For example, 40CFR372.65 means Title 40 of the CFR Part 372 Section 65. As all titles of the CFR are revised annually, these references are subject to change.

ADDITIONAL/OTHER COMMENTS

This field gives a collection of any information that is especially notable but not found in the other fields. The information in this field will vary with the properties of the HAP. Examples are as follows: symptoms, first aid, firefighting methods, personal protective equipment, and safe methods of storage.

SOURCES

Sources of more detailed information on the HAPs are given here. The following is a list of the most frequently cited sources of additional information.

ACGIH Threshold Limit Values (TLVs) and Biological Exposure Indices (BEIs) for 1996. ACGIH. Cincinnati, Ohio.

CRC Handbook of Chemistry and Physics. David R. Lide, editor in chief. CRC Press, Inc. Boca Raton, Florida. 77th edition, 1996–1997.

CRC Handbook of Thermophysical and Thermochemical Data. David R. Lide and Henry V. Kehiaian. CRC Press, Inc. Boca Raton, Florida. 1994.

Environmental Contaminant Reference Databook, volume 1. Jan C. Prager. Van Nostrand Reinhold. New York. 1995.

Environmental Law Index to Chemicals. C. C. Lee, editor. Government Institutes, Inc. Rockville, Maryland. 1996 edition.

Fire Protection Guide to Hazardous Materials. National Fire Protection Association. Quincy, Massachusetts. 11th edition, 1994.

Hawley's Condensed Chemical Dictionary. Richard J. Lewis, Sr. Van Nostrand Reinhold. New York. 13th edition, 1997.

Health Effects Notebook for Hazardous Air Pollutants. U.S. Environmental Protection Agency. December 1994.

Material Safety Data Sheets (MSDS). MSDS are produced by the companies that manufacture and sell a particular chemical.

Merck Index. Susan Budavari, editor. Merck and Co. Whitehouse Station, New Jersey. 12th edition, 1996.

NIOSH Pocket Guide to Chemical Hazards. Centers for Disease Control and Prevention. Cincinnati, Ohio. June 1994.

Sax's Dangerous Properties of Industrial Materials. Richard J. Lewis, Sr. Van Nostrand Reinhold. New York. 9th edition, 1996 (in 3 volumes).

Suspect Chemicals Sourcebook. Roytech Publications. Bethesda, Maryland. 1996 edition.

ACETALDEHYDE (ec)

CAS #: 75-07-0
MOLECULAR FORMULA: C_2H_4O
FORMULA WEIGHT: 44.06
SYNONYMS: acetaldehyd (German), acetic aldehyde, aldehyde acetique (French), aldeide acetica (Italian), ethanal, ethyl aldehyde, octowy aldehyd (Polish)

USES
- ◆ The predominant use is as an intermediate in the synthesis of other chemicals.
- ◆ synthetic flavor and adjuvant
- ◆ manufacture of synthetic resins, dyes, paraldehyde, perfumes, aniline dyes, plastics, synthetic rubber, silvering mirrors, hardening gelatin fibers
- ◆ chemical intermediate for acetic acid, peracetic acid, pyridine, pyridine bases, chloral and glyoxal monomer for polyacetaldehyde and comonomer for copolymers, and ester production
- ◆ alcohol denaturant
- ◆ oxidation promoter in manufacture of terephthalic acid

PHYSICAL PROPERTIES
- ◆ a colorless, mobile liquid
- ◆ pungent, suffocating odor, but at dilute concentrations it has a fruity and pleasant odor
- ◆ leafy, green taste
- ◆ boiling point 21°C
- ◆ melting point −123.5°C
- ◆ vapor pressure 740 mm Hg at 20°C
- ◆ vapor density 1.52
- ◆ miscible with water, alcohol, ether, benzene, gasoline, toluene, xylene, turpentine, and acetone
- ◆ heat of vaporization latent 136 cal/g
- ◆ surface tension 29.04 dynes/cm at 20°C
- ◆ odor threshold, recognition in air $= 2.1 \times 10^{-1}$ ppm
- ◆ viscosity 0.02456 mPa at 20°C
- ◆ density 0.788 at 16°C
- ◆ surface tension 21.2 mN/m at 20°C

CHEMICAL PROPERTIES
- ◆ reacts with oxidizers, halogens, amines, strong bases, and acids
- ◆ heat of combustion -246.4×10^5 J/kg
- ◆ flash point −8°C (−36°F)
- ◆ autoignition temperature 365°C
- ◆ flammability limits 4.0% lower and 60% in air upper
- ◆ specific heat 0.650

HEALTH RISK
- ◆ Toxic Exposure Guidelines
  - • OSHA 8-hr time-weighted average 200 ppm (360 mg/m³)
  - • ACGIH TLV 180 mg/m³

- EPA Cancer Risk Level (1 in a million excess lifetime risk) 5.0×10^{-4} mg/m^3
  - RfC 0.009 mg/m^3
- ◆ Toxicity Data
  - eye-human 50 ppm/15M
  - inhalation-human TCLo 134 ppm/30M
  - oral-rat LD$_{50}$ 661 mg/kg
  - inhalation-rat LC$_{50}$ 37 g/m^3/30M
  - inhalation-rat TCLo 735 ppm/6H/2Y
  - intraperitoneal-rat LDLo 500 mg/kg
  - LOAEL rats 728 mg/m^3
  - NOAEL rats 273 mg/m^3
- ◆ Acute Risks
  - The primary acute (short-term) effect of inhalation exposure to acetaldehyde is irritation of the eyes, skin, and respiratory tract in humans. Erythema, coughing, pulmonary edema, and necrosis may also occur and, at extremely high concentrations, respiratory paralysis and death.
  - Acute inhalation resulted in a depressed respiratory rate and elevated blood pressure in experimental animals.
  - Tests involving acute exposure of animals have demonstrated acetaldehyde to have low acute toxicity from inhalation and moderate acute toxicity from oral or dermal exposure.
  - target organs: liver, kidneys, lungs, blood, central nervous system
- ◆ Chronic Risks (Noncancer)
  - In hamsters, chronic (long-term) inhalation exposure to acetaldehyde has produced changes in the nasal mucosa and trachea, growth retardation, slight anemia, and increased kidney weight.
  - Symptoms of chronic intoxication of acetaldehyde in humans resemble those of alcoholism.
  - The RfC for acetaldehyde is based on degeneration of olfactory epithelium in rats.
- ◆ Reproductive/Developmental Effects
  - Acetaldehyde has been shown, in animals, to cross the placenta to the fetus.
  - Data from animal studies suggest that acetaldehyde may be a potential developmental toxin.
- ◆ Cancer Risk
  - Human data regarding the carcinogenic effects of acetaldehyde are inadequate.
  - An increased incidence of nasal tumors in rats and laryngeal tumors in hamsters has been observed following inhalation exposure to acetaldehyde.
  - EPA has classified acetaldehyde as a Group B2 probable human carcinogen of low carcinogenic hazard.

HAZARD RISK
- ◆ Acute Hazard Level

2

- Acute hazard stems from disaster potential and inhalation-irritation problems: extreme exposures can cause respiratory paralysis.
◆ Degree of Hazard to Public Health
  - extreme danger of inhalation poisoning and irritation
◆ highly flammable liquid

EXPOSURE ROUTES
◆ Acetaldehyde is ubiquitous in the ambient environment. It is an intermediate product of higher plant respiration and formed as a product of incomplete wood combustion in fireplaces and woodstoves, coffee roasting, burning of tobacco, vehicle exhaust fumes, and coal refining and waste processing. Hence, many individuals are exposed to it by breathing ambient air.
◆ It is formed in the body from the breakdown of ethanol; this would be a source among those who consume alcoholic beverages.
◆ direct contact with eyes and mucous membranes
◆ intratracheal and intravenous
◆ accidental swallowing

REGULATORY STATUS/INFORMATION
◆ NIOSH usually recommends that occupational exposure to carcinogens be limited to the lowest feasible concentration.
◆ RCRA Requirements
  - When acetonitrile, as a commercial chemical product or manufacturing chemical intermediate, becomes a waste, it must be managed according to federal or state hazardous waste regulations.
◆ designated as a hazardous substance under section 311(b)(2)(A) of the federal Water Pollution Control Act and further regulated by the Clean Water Act Amendments of 1977 and 1978
◆ FIFRA Requirements
  - Acetonitrile is exempted from the requirement of a tolerance for residues on apples and strawberries.
◆ EPA's Office of Air Quality Planning and Standards, for hazard ranking under section 112(g) of the Clean Air Act Amendments, has ranked acetaldehyde in the nonthreshold category.

ADDITIONAL/OTHER COMMENTS
◆ Acetaldehyde is oxidized rapidly and exothermally by air to acetic acid.
◆ First Aid
  - flush skin with water
  - if inhaled, remove to fresh air; if not breathing, give artificial respiration
  - remove and wash contaminated clothing promptly
  - if swallowed, wash out mouth with water provided person is conscious
◆ Fire fighting: carbon dioxide, dry chemical powder, or appropriate foam

- ◆ dangerous to aquatic life in low concentrations
- ◆ storage: store in a cool, dry, well-ventilated location separate from other reactive hazards

SOURCES
- ◆ *Environmental Contaminant Reference Databook*, volume 1, 1995
- ◆ *Fire Protection Guide to Hazardous Materials*, 11th edition, 1994
- ◆ *Health Effects Notebook for Hazardous Air Pollutants*, U.S. Environmental Protection Agency, December 1994
- ◆ *Material Safety Data Sheets (MSDS)*
- ◆ *Merck Index*, 12th edition, 1996
- ◆ *NIOSH Pocket Guide to Chemical Hazards*, June 1994
- ◆ *Sax's Dangerous Properties of Industrial Materials*, 9th edition, 1996

ACETAMIDE (ec)

CAS #: 60-35-5
MOLECULAR FORMULA: C_2H_5NO
FORMULA WEIGHT: 59.07
SYNONYMS: acetic acid amide, acetimidic acid, amid kyseliny octove (Polish), ethanamide, methanecarboxamide

USES
- ◆ Direct
  - • solvent and wetting agent for many organic compounds
  - • alcohol denaturant
  - • plasticizer
  - • stabilizer
  - • cryoscopic agent
- ◆ Indirect
  - • intermediate for melamine, thioacetamide, pesticides, pharmaceuticals, rubber chemicals, and plastics

PHYSICAL PROPERTIES
- ◆ odorless when pure, but frequently has a mousy odor
- ◆ deliquescent hexagonal crystals
- ◆ colorless crystals
- ◆ boiling point 221°C at 760 mm Hg
- ◆ melting point 82.3°C at 760 mm Hg
- ◆ vapor pressure 1 mm Hg at 65°C
- ◆ density .9986 g/cm³ at 85°C
- ◆ log octanol/water partition coefficient −1.26
- ◆ odor threshold 140–160 mg/m³

CHEMICAL PROPERTIES
- ◆ decomposes in hot water
- ◆ One gram dissolves in 0.5 mL water, 2 mL alcohol, 6 mL pyridine
- ◆ soluble in chloroform, glycerol, hot benzene

HEALTH RISK
- ◆ Toxicity Data

- subcutaneous-rat LD_{50} 10 g/kg
- oral-rat LD_{50} 7000 mg/kg
- oral-rat TDLO 431 g/kg/1Y-C
- oral-mouse LD_{50} 12,900 mg/kg
- intraperitoneal-rat LD_{50} 10,300 mg/kg
- intraperitoneal-mouse LD_{50} 1000 mg/kg

◆ Acute Risks
  - Tests involving acute exposure of animals have shown acetamide to have low to moderate acute toxicity from oral exposure.
  - Material is irritating to mucous membranes and upper respiratory tract.
  - Acetamide causes mild skin irritation in humans from acute (short-term) exposure.

◆ Chronic Risks
  - carcinogen
  - may alter genetic material
  - target organs: liver
  - an experimental teratogen
  - may cause cancer, heritable genetic damage

HAZARD RISK

◆ incompatibilities: strong oxidizing agents, strong acids, strong bases, strong reducing agents
◆ toxic fumes of carbon monoxide, carbon dioxide, nitrogen oxides

EXPOSURE ROUTES

◆ Occupational exposure to acetamide may occur for those workers in the plastics and chemical industries.
◆ No information is available on the assessment of personal exposure to acetamide.
◆ moderately toxic by intraperitoneal and possibly other routes
◆ contact with eyes

REGULATORY STATUS/INFORMATION

◆ IARC Cancer Review Group 2B IMEMDT 7,56,87
◆ Animal Sufficient Evidence IMEMDT 7,389,87
◆ Community Right-to-Know List
◆ reported in EPA TSCA Section 8(B) chemical inventory

ADDITIONAL/OTHER COMMENTS

◆ prepared by fractional distillation of ammonium acetate
◆ First Aid
  - flush skin and eyes with copious amounts of water
  - if inhaled, remove to fresh air; if not breathing, give artificial respiration
  - remove and wash contaminated clothing promptly
  - if swallowed, wash out mouth with water provided person is conscious
◆ Fire Fighting
  - extinguishing media: water spray, carbon dioxide, dry chemical powder, or appropriate foam

- special procedures: wear self-contained breathing apparatus and protective clothing to prevent contact with eyes and skin
- ◆ disposal considerations: dissolve or mix the material with a combustible solvent and burn in a chemical incinerator equipped with an afterburner and scrubber

SOURCES
- ◆ *Health Effects Notebook for Hazardous Air Pollutants*, U.S. Environmental Protection Agency, December 1994
- ◆ *Material Safety Data Sheets (MSDS)*
- ◆ *Merck Index*, 12th edition, 1996
- ◆ *Sax's Dangerous Properties of Industrial Materials*, 9th edition, 1996

ACETONITRILE (ec)

CAS #: 75-05-8
MOLECULAR FORMULA: C_2H_3N
FORMULA WEIGHT: 41.06
SYNONYMS: acetonitril (German, Dutch), cyanomethane, cyanure de methyl (French), ethanenitrile, ethyl nitrile, methanecarbonitrile, methane, cyano-, methyl cyanide, methylkyanid (Czech)

USES
- ◆ It is predominantly used as a solvent in the manufacture of pharmaceuticals, for spinning fibers, for casting and molding of plastic materials, in lithium batteries, for the extraction of fatty acids from animal and vegetable oils, and in chemical laboratories for the detection of materials such as pesticide residues.
- ◆ in organic synthesis, as the starting material for acetophenone, alphanaphthaleneacetic acid, thiamine, and acetamidine
- ◆ can be used to recrystallize steroids
- ◆ as a medium for promoting reactions involving ionization
- ◆ It is used in dyeing textiles, in coating compositions as a stabilizer for chlorinated solvents, and in perfume production as a chemical intermediate.
- ◆ as the starting material for many types of nitrogen-containing compounds

PHYSICAL PROPERTIES
- ◆ colorless, limpid liquid
- ◆ aromatic odor; ether-like odor
- ◆ burning, sweetish taste
- ◆ boiling point 81.6°C at 760 mm Hg
- ◆ melting point −5°C
- ◆ vapor pressure 87 mm Hg at 24°C
- ◆ specific gravity 0.786 at 15°C
- ◆ vapor density 1.41
- ◆ very soluble in ethyl alcohol
- ◆ freely soluble in water
- ◆ will slowly degrade to cyanides

6

- ♦ miscible with water, methanol, benzene, methyl acetate, acetone, ethyl acetate, ether, acetamide solutions, chloroform, carbon tetrachloride, ethylene chloride, and many unsaturated hydrocarbons
- ♦ immiscible with many saturated hydrocarbons
- ♦ heat of vaporization 313 Btu/lb = 174 cal/g = 7.29×10^5 J/kg
- ♦ surface tension 29.04 dynes/cm at 20°C
- ♦ Odor Threshold 70.0 mg/m^3, 875 mg/m^3, irritating
- ♦ Viscosity 0.43 cP at 0°C, 0.35 cP at 20°C, 0.30 cP at 40°C

CHEMICAL PROPERTIES

- ♦ Chemical action will slowly release cyanides.
- ♦ toxic vapors generated when heated
- ♦ Contact with strong oxidizers may cause explosions.
- ♦ dissolves somewhat in inorganic salts
- ♦ Heat contributes to instability.
- ♦ Liquid acetonitrile will attack some forms of plastics, rubber, and coatings.
- ♦ heat of combustion -13.360 Btu/lb = -420 cal/g = -10.7×10^5 J/kg
- ♦ flash point 5°C (42°F)
- ♦ autoignition temperature 522°C (973°F)
- ♦ explosion limits 4.4% lower and 16% upper

HEALTH RISK

- ♦ Toxic Exposure Guidelines
  - • OSHA 8-hr time-weighted average 40 ppm (70 mg/m^3)
  - • NIOSH 10-hr time-weighted average 20 ppm (34 mg/m^3)
  - • NIOSH TWA 40 ppm (34 mg/m^3)
  - • ACGIH STEL (105 mg/m^3)
- ♦ Toxicity Data
  - • oral-human TDLo 570 mg/kg
  - • inhalation-human TCLo 160 ppm/4H
  - • oral-rat LD$_{50}$ 2730 mg/kg
  - • inhalation-rat LC$_{50}$ 7551 ppm/8H
  - • inhalation-rat LC$_{50}$ 330 ppm/90d
  - • intraperitoneal-rat LD$_{50}$ 850 mg/kg
  - • inhalation-cat LC$_{50}$ 18,000 mg/m^3
  - • inhalation-guinea pig LC$_{50}$ 9497 mg/m^3
  - • inhalation-rabbit LC$_{50}$ 4749 mg/m^3
- ♦ Acute Risks
  - • Concentrations up to 500 ppm acetonitrile through inhalation exposure cause irritation of mucous membranes, and higher concentrations can produce weakness, nausea, convulsions, and death.
  - • Acetonitrile is considered toxic in humans when ingested or through skin contact.
  - • Tests involving acute exposure of animals have shown acetonitrile to have high to moderate acute toxicity from oral exposure, while the inhalation test has shown the chemical to have moderate acute toxicity.
  - • target organs: central nervous system, liver, kidneys

- Effects from severe exposures include irritability, skin eruptions, confusion, delirium, convulsions, paralysis, and death.
- ◆ Chronic Risks (Noncancer)
  - Chronic (long-term) inhalation exposure to acetonitrile results in cyanide poisoning from metabolic release of cyanide after absorption. The major effects consist of those on the central nervous system, such as headaches, numbness, and tremor.
  - Cyanide poisoning can also be produced through the ingestion of acetonitrile or from contact with the skin.
  - Application of acetonitrile to the skin may produce dermatitis.
  - Animal studies have shown that different species vary widely in susceptibility to acetonitrile by various routes.
  - EPA has calculated a provisional RfC of 0.05 mg/m³ for acetonitrile.

HAZARD RISK
- ◆ Acute Hazard Level
  - high; extreme in cases where heat or environmental conditions accelerate release of cyanide
- ◆ Chronic Hazard Level
  - No chronic problems have been noted; cyanides do pose chronic ingestion, inhalation, and irritation hazard.
- ◆ Degree of Hazard to Public Health
  - Both disaster potential and release of toxic vapors and solutions pose an extreme threat to public safety; major hazard is potential release of cyanide.
- ◆ flammable
- ◆ keep away from sources of ignition—no smoking

EXPOSURE ROUTES
- ◆ Sources of acetonitrile emissions into the air include manufacturing and industrial facilities, automobile exhaust, and volatilization from aquatic environments.
- ◆ Individuals may be exposed to acetonitrile through breathing contaminated air, from smoking tobacco or proximity to someone who is smoking, or through skin contact in the workplace.
- ◆ There are tests currently available to determine personal exposure to acetonitrile, such as the determination of blood cyanide or urinary thiocyanate.

REGULATORY STATUS/INFORMATION
- ◆ subject to SARA Section 313 reporting requirements
- ◆ RCRA Requirements
  - When acetonitrile, as a commercial chemical product or manufacturing chemical intermediate, becomes a waste, it must be managed according to federal or state hazardous waste regulations.
- ◆ DOT class: Class 3, flammable liquid and poison
- ◆ TSCA Requirements

- Section 8(d) requires manufacturers, importers, and processors of listed chemical substances and mixtures to submit to EPA copies and lists of unpublished health and safety studies; acetonitrile is included in this list.
- ◆ FIFRA Requirements
  - Acetonitrile is exempted from the requirement of a tolerance when used as a solvent for blended emulsifiers in all pesticides used before crop emerges from soil.

ADDITIONAL/OTHER COMMENTS
- ◆ handle and store under nitrogen
- ◆ First Aid
  - flush skin with water
  - if inhaled, remove to fresh air; if not breathing, give artificial respiration
  - remove and wash contaminated clothing promptly
  - if swallowed, wash out mouth with water provided person is conscious
- ◆ Fire Fighting
  - extinguishing media: carbon dioxide, dry chemical powder, or appropriate foam
  - special procedures: approach fire from upwind to avoid hazardous vapors and toxic decomposition products
- ◆ dangerous to aquatic life in high concentrations
- ◆ storage: outside or detached is preferred; inside storage should be in a standard flammable liquids storage warehouse, room, or cabinet

SOURCES
- ◆ *Environmental Contaminant Reference Databook*, volume 1, 1995
- ◆ *Fire Protection Guide to Hazardous Materials*, 11th edition, 1994
- ◆ *Health Effects Notebook for Hazardous Air Pollutants*, U.S. Environmental Protection Agency, December 1994
- ◆ *Material Safety Data Sheets (MSDS)*
- ◆ *NIOSH Pocket Guide to Chemical Hazards*, June 1994
- ◆ *Sax's Dangerous Properties of Industrial Materials*, 9th edition, 1996

ACETOPHENONE (ec)

CAS #: 98-96-2
MOLECULAR FORMULA: C_2H_8O
FORMULA WEIGHT: 120.16
SYNONYMS: 1-phenylethanone, acetophenon, acetylbenzene, benzene acetyl, benzoyl methide, ethanone 1 phenyl, hypnon, hypnone, ketone methyl phenyl, methyl phenyl ketone, phenyl methyl ketone

USES
- ◆ in perfumery to impart an orange-blossom–like odor
- ◆ catalyst for polymerization of olefins
- ◆ in organic synthesis, especially as a photosensitizer

- specialty solvent for plastics and resins
- chemical intermediate for the odorant ethyl methyl phenylglycidate, a riot control agent, 2-chloroacetophenone, 2-bromoacetophenone for dyes, 3-nitroacetophenone
- flavoring agent in nonalcoholic beverages, ice cream, candy, baked goods, gelatins, puddings, chewing gum
- fragrance ingredient in soaps, detergents, creams, lotions, perfumes
- flavorant in tobacco

PHYSICAL PROPERTIES
- monoclinic prisms or plates; slightly oily liquid; colorless liquid; liquid forms laminar crystals at low temperatures
- sweet pungent odor of acacia
- bitter, aromatic flavor
- boiling point 202°C
- melting point 20.5°C
- vapor pressure 0.44 mm Hg at 25°C
- specific gravity 1.030
- vapor density 4.14
- freely soluble in alcohol, chloroform, ether, fatty oils, glycerol
- soluble in acetone and benzene
- heat of vaporization 11,731.5 gcal/gmole
- surface tension 39.8 dynes/cm at 20°C
- viscosity 2.28 cP at 11.9°C, 1.617 cP at 25°C, 1.246 cP at 50°C
- odor threshold 0.01–0.025 mg/m^3
- water solubility 6130 mg/L at 25°C

CHEMICAL PROPERTIES
- stable under normal laboratory storage conditions
- heat of combustion 991.6 kcal/g at 25°C
- flash point, closed cup 105°C (221°F)
- autoignition temperature 1058°C
- Henry's Law constant 1.07×10^{-5} atm-cu m/mole

HEALTH RISK
- Toxicity Data
  - oral-rat LD$_{50}$ 815 mg/kg
  - intraperitoneal-mouse LD$_{50}$ 200 mg/kg
  - oral-mouse LD$_{50}$ 740 mg/kg
  - skin-rabbit LD$_{50}$ 15,900 µL/kg
  - RfD 0.1 mg/kg/d
  - NOAEL rats 423 mg/kg/d
- Acute Risks
  - Acute (short-term) exposure to acetophenone vapor may produce skin irritation and transient corneal injury. One study noted a decrease in light sensitivity in exposed humans.
  - Acute oral exposure has been observed to cause hypnotic or sedative effects, hematological effects, and a weakened pulse in humans.

- Congestion of the lungs, kidneys, and liver was reported in rats acutely exposed to high levels of acetophenone via inhalation.
- Tests involving acute exposure of animals have demonstrated acetophenone to have moderate acute toxicity from oral or dermal exposure.
- narcotic in high concentrations
◆ Chronic Risks (Noncancer)
  - No information is available on the chronic (long-term) effects of acetophenone in humans.
  - Degeneration of olfactory bulb cells was reported in rats chronically exposed via inhalation. In another study, chronic inhalation exposure of rats produced hematological effects and, at high doses, congestion of cardiac vessels and pronounced dystrophy of the liver.
  - In two studies, no effects were observed in rats chronically exposed to acetophenone in their diet.
  - EPA's Office of Air Quality Planning and Standards considers acetophenone to be a high concern pollutant based on severe chronic toxicity.

HAZARD RISK

◆ Acute Hazard Level
  - low-potential BOD problem; may smother benthic life
◆ Chronic Hazard Level
  - none expected except perhaps for local irritation; no effects noted from 0.102 g/kg in diet of rats for 30 days
◆ Degree of Hazard to Public Health
  - slight
◆ Action Level
  - avoid areas where narcotic vapors may be concentrated

EXPOSURE ROUTES

◆ poison by intraperitoneal and subcutaneous routes
◆ Occupational exposure to acetophenone may occur during its manufacture and use.
◆ Acetophenone has been detected in ambient air and drinking water; exposure of the general public may occur through the inhalation of contaminated air or the consumption of contaminated water.
◆ Hippuric acid may be monitored in the urine to determine whether or not exposure to acetophenone has occurred.

REGULATORY STATUS/INFORMATION

◆ EPA TSCA Sections 8(B) and 8(D)
◆ RCRA Requirements
  - As stipulated in 40CFR261.33, when acetophenone, as a commercial chemical product or manufacturing chemical intermediate, becomes a waste, it must be managed according to federal or state hazardous waste regulations.
◆ DOT class: Class 3, Label: flammable liquid
◆ FIFRA Requirements

- Acetophenone is exempted from the requirement of a tolerance when used as an attractant in accordance with good agricultural practice as inert ingredients in pesticide formulations applied to growing crops only.

ADDITIONAL/OTHER COMMENTS
- ◆ Cleanup
  - absorb on paper; evaporate on a glass or iron dish in hood; burn the paper
- ◆ First Aid
  - immediately flush eyes or skin with water
  - if inhaled, remove to fresh air; if not breathing, give artificial respiration
  - remove and wash contaminated clothing promptly
  - if swallowed, wash out mouth with water provided person is conscious
- ◆ fire fighting: water, foam, carbon dioxide, dry chemical
- ◆ major water use threatened: fisheries from BOD
- ◆ disposal method: spray into incinerator
- ◆ EPA has classified acetophenone as a Group D, not classifiable as to human carcinogenicity.

SOURCES
- ◆ *Environmental Contaminant Reference Databook*, volume 1, 1995
- ◆ *Health Effects Notebook for Hazardous Air Pollutants*, U.S. Environmental Protection Agency, December 1994
- ◆ *Material Safety Data Sheets (MSDS)*
- ◆ *Merck Index*, 12th edition, 1996
- ◆ *Sax's Dangerous Properties of Industrial Materials*, 9th edition, 1996

2-ACETYLAMINOFLUORENE

CAS #: 53-96-3
MOLECULAR FORMULA: $C_{15}H_{13}NO$
FORMULA WEIGHT: 223.29
SYNONYMS: AAF, 2-AAF, 2-acetamidofluorene, 2-acetaminofluorene, acetoaminofluorene, 2-Acetylamino-fluoren (German), N-acetyl-2-aminofluorene, Azetylaminofluoren (German), FAA, 2-FAA, 2-fluorenylacetamide, N-fluoren-2-yl acetamide, N-2-fluorenylacetamide

USES
- ◆ Used in the Laboratory by Biochemists and Technicians
  - study of liver enzymes
  - carcinogenesis of aromatic amines
  - mutagenicity of aromatic amines

PHYSICAL PROPERTIES
- ◆ tan, crystalline solid
- ◆ melting point 192–196°C
- ◆ insoluble in water
- ◆ K_{ow} 3.22

CHEMICAL PROPERTIES
- ◆ stable under normal temperatures and pressures
- ◆ incompatible with acids, acid anhydrides, and oxidizing agents

HEALTH RISK
- ◆ Toxic Exposure Guidelines
  - • OHSA PEL cancer suspect agent
  - • NIOSH REL TWA use 29 CFR 1910.1014
- ◆ Toxicity Data
  - • mmo-sat 100 µg/plate
  - • dnr-esc 5 µmol/L
  - • dnd-human:fbr 1 mmol/L
  - • sce-human:lym 4400 µg/L
  - • otr-rat:lvr 100 µg/L
  - • sce-mouse-intraperitoneal 20 mg/kg
  - • intraperitoneal-mouse TDLo 100 mg/kg
  - • oral-rat TDLo 672 mg/kg/8W-C:CAR
  - • skin-rat TDLo 260 mg/kg/71W-I:CAR
  - • parenteral-rat TDLo 1700 mg/kg/17W-I:ETA
  - • intraperitoneal-mouse TDLo 60 mg/kg/2W-I:NEO
  - • subcutaneous-mouse TDLo 400 mg/kg
  - • oral-rabbit TD 14 g/kg/56W-I:CAR,TER
  - • oral-mouse LD_{50} 810 mg/kg
  - • intraperitoneal-mouse LD_{50} 2200 mg/kg
- ◆ Acute Risks
  - • no information on effects on humans
  - • moderate acute toxicity on mice with oral exposure
- ◆ Chronic Risks
  - • no information on effects on humans
  - • Offspring of mice had skeletal defects, cleft lips and palates, and cerebral hernias.
  - • liver tumors in rats and mice
  - • bladder tumors in rats and mice
  - • renal pelvis tumors in rats
  - • zymbal gland tumors in rats
  - • colon and lung tumors in rats
  - • pancreas tumors in rats
  - • testis tumors in rats
  - • kidney tumors in mice
  - • EPA Group 2B: probable human carcinogen

HAZARD RISK
- ◆ incompatible with acids, acid anhydrides, and oxidizing agents
- ◆ hazardous decomposition products: nitrogen oxides, carbon monoxide, carbon dioxide, hydrogen fluoride, nitrogen
- ◆ decomposition: emits toxic fumes of NOx

EXPOSURE ROUTES
- ◆ inhalation
- ◆ dermal contact in laboratories
- ◆ Organic chemists, chemical stockroom workers, and biomedical researchers have greatest risk of exposure.

◆ Release to environment is insignificant (less than 20 lb/year are consumed in U.S.).

REGULATORY STATUS/INFORMATION
◆ Clean Air Act Amendment, Title III: hazardous air pollutants
◆ Resource Conservation and Recovery Act
  • list of hazardous constituents. 40CFR258, 40CFR261, 40CFR261.11
  • groundwater monitoring list. 40CFR264
  • U waste #: U005
◆ CERCLA and Superfund
  • designation of hazardous substances. 40CFR302.4
  • list of chemicals. 40CFR372.65
◆ OSHA 29 CFR1910.1000 Table Z1 and specially regulated substances
◆ chemical regulated by the State of California

ADDITIONAL/OTHER COMMENTS
◆ First Aid
  • wash eyes immediately with large amounts of water
  • flush skin with plenty of soap and water
  • remove to fresh air immediately
◆ fire fighting: use dry chemical, water spray or mist, chemical foam, or alcohol-resistant foam
◆ personal protection: wear self-contained breathing apparatus and chemical goggles

SOURCES
◆ *Health Effects Notebook for Hazardous Air Pollutants*, U.S. Environmental Protection Agency, December 1994
◆ *Material Safety Data Sheets (MSDS)*
◆ *Sax's Dangerous Properties of Industrial Materials, 9th edition*, 1996
◆ *Aldrich Chemical Company Catalog*, 1996–1997
◆ *Environmental Law Index to Chemicals*, 1994 edition
◆ *Merck Index*, 12th edition, 1996

ACROLEIN (rc)

CAS #: 107-02-8
MOLECULAR FORMULA: C_3H_4O
FORMULA WEIGHT: 56.064
SYNONYMS: acquinite, acraldehyde, acroleine, acrylaldehyde, acrylic aldehyde, allyl aldehyde, aqualin, aqualine, aquinite, biocide, crolean, EPA Pesticide Chemical Code 000701, ethylene aldehyde, magnacide, magnacide H, propenal, slimicide, transacrolein, many others

USES
◆ Manufacturing of Other Chemicals
  • pesticides
  • colloidal forms of metals
  • plastics
  • perfumes
  • organic synthesis

- glycerin
- acrylic acid
- esters
♦ warning agent in methyl chloride refrigerant
♦ military poison mixtures
♦ aquatic herbicide

PHYSICAL PROPERTIES
♦ colorless liquid
♦ melting point −87°C
♦ boiling point 53°C
♦ density 0.8389 g/mL at 20°C
♦ specific gravity 0.839 (water = 1)
♦ vapor density 1.94 (air = 1)
♦ surface tension 24 dynes/cm at 20°C
♦ viscosity 0.393 mP at 20°C
♦ heat of vaporization 120 cal/g (5.02×10^5 J/kg)
♦ piercing, disagreeable odor
♦ odor threshold 0.2 ppm
♦ vapor pressure 214 mm Hg at 20°C
♦ soluble in water, ether, ethanol
♦ miscible in benzene, acetone

CHEMICAL PROPERTIES
♦ flash point −6°C (−2°F)
♦ flammability limits 2.8% lower and 31% upper
♦ autoignition temperature 219°C (428°F)
♦ heat of combustion −6950 cal/g (-290×10^5 J/kg)
♦ incompatible with bases, oxidizers, reducing agents, and oxygen
♦ incompatible with amines, SO_2, metal salts, thiourea, or dimethylamine
♦ may polymerize on exposure to light

HEALTH RISK
♦ Toxic Exposure Guidelines
  - NIOSH REL TWA 0.1 ppm
  - NIOSH REL STEL 0.3 ppm
  - ACGIH TLW TWA 0.1 ppm
  - ACGIH TLW STEL 0.3 ppm
  - OSHA PEL TWA 0.1 ppm
  - OSHA PEL STEL 0.3 ppm
  - DFG MAK 0.1 ppm
♦ Toxicity Data
  - inhalation-human LCLo 5500 ppb
  - inhalation-human LCLo 153 ppm/10 months
  - oral-rat LD_{50} 26 mg/kg
  - inhalation-rat LC_{50} 18 mg/m³
  - intraperitoneal-rat LD_{50} 4 mg/kg
  - subcutaneous-rat LD_{50} 50 mg/kg
  - oral-mouse LD_{50} 28 mg/kg
  - intraperitoneal-mouse LD_{50} 9008 µg/kg
  - subcutaneous-mouse LD_{50} 30 mg/kg

- oral-rabbit LD_{50} 7 mg/kg
- skin-rabbit LD_{50} 200 mg/kg
- unreported-mammal LD_{50} 45 mg/kg
◆ Other Toxicity Indicators
  - EPA RfC 0.00002 mg/m^3
  - EPA RfD (provisional) 0.02 mg/kg/d
◆ Acute Risks
  - may be fatal if inhaled, swallowed, or absorbed through skin
  - extremely destructive to tissue of the mucous membranes and upper respiratory tract, eyes, and skin
  - Inhalation may be fatal as a result of spasm or inflammation of larynx and bronchi.
  - Inhalation may be fatal as a result of chemical pneumonitis or pulmonary edema.
  - burning sensation
  - coughing
  - wheezing
  - laryngitis
  - shortness of breath
  - headache
  - nausea and vomiting
  - allergic skin reaction
◆ Chronic Risks
  - upper respiratory tract irritation
  - eye, nose, and throat irritation
  - carcinogen
  - damage to cardiovascular system, liver, eyes, kidneys

HAZARD RISK
◆ flammable
◆ corrosive
◆ incompatible with bases, oxidizers, reducing agents, and oxygen
◆ incompatible with amines, SO_2, metal salts, thiourea, or dimethylamine
◆ sensitive to heat
◆ explosive hazard
◆ may polymerize on exposure to light
◆ combustion or decomposition products of carbon monoxide and/or carbon dioxide

EXPOSURE ROUTES
◆ inhalation
◆ ingestion
◆ absorption through skin
◆ intradermal absorption

REGULATORY STATUS/INFORMATION
◆ Clean Water Act Section 307: Priority Pollutant
◆ Clean Water Act Section 311: Hazardous Substance
◆ Clean Air Act Section 112r: Accidental Release Prevention Substances

- ◆ National Institute for Occupational Safety and Health (NIOSH) Recommendation Substances
- ◆ Resource Conservation and Recovery List (RCRA) Hazardous Substances (40 CFR Ch. 1)
- ◆ RCRA Groundwater Monitoring List (40 CFR Ch. 1)
- ◆ RCRA Land Disposal Restrictions: Second Third Wastes
- ◆ Comprehensive Environmental Response, Compensation, and Liability Act (Superfund): EPA Office of Solid Waste and Emergency Response
- ◆ Resource Conservation and Recovery Act/EPCRA (Section 302) Extremely Hazardous Substances
- ◆ Resource Conservation and Recovery Act/EPCRA (Section 313) Toxic Chemicals
- ◆ Department of Transportation Hazardous Materials Table (Source List 13A)

ADDITIONAL/OTHER COMMENTS
- ◆ First Aid
  - wash eyes and/or skin immediately with large amounts of water for 15 minutes
  - remove to fresh air if inhaled
  - wash out mouth with water if swallowed
- ◆ Fire Fighting
  - extinguishing media: carbon dioxide, dry chemical powder, or appropriate foam
  - special procedures: wear self-contained breathing apparatus and protective clothing
  - unusual fire and explosion hazards: extremely flammable; vapor may travel long distance to a source of ignition and flash back
- ◆ Personal Protection
  - use only in a chemical fume hood
  - wear chemical-resistant gloves, safety goggles, and other protective clothing
  - safety shower and eye bath
- ◆ storage: keep tightly closed; keep away from heat, sparks, and open flame

SOURCES
- ◆ *Agency for Toxic Substances and Disease Registry Public Health Statement* (http://atsdr1.atsdr.cdc.gov:8080), December 1990
- ◆ *MSDS (Acrolein)*, Aldrich Chemical Company, Inc. 1996
- ◆ *Environmental Contaminant Reference Databook*, volume 1, 1995
- ◆ *EPA Notebook Hap Index* (http://www.epa.gov/oar/oaqps/airtox/ hapindex.html)
- ◆ *Merck Index*, 12th edition, 1996
- ◆ *Sax's Dangerous Properties of Industrial Materials*, 9th edition, 1996
- ◆ *Suspect Chemicals Sourcebook*, 1996

ACRYLAMIDE (rc)

CAS #: 79-06-1
MOLECULAR FORMULA: C_3H_5NO
FORMULA WEIGHT: 71.08
SYNONYMS: acrylamide, acrylic amide, ethylenecarboxamide, propenamide, 2-propenamide, propenoic acid amide, RCRA waste number U007, UN2074 (DOT), vinyl amide

USES
- ◆ Industrial Applications
  - • production of polyacrylamide polymers
  - • used as a chemical intermediate in the production of N-methylol acrylamide and N-butoxyacrylamide
- ◆ absorbent in disposable diapers, medical products, and agricultural products
- ◆ sugar beet juice clarification
- ◆ adhesives
- ◆ printing ink emulsion stabilizers
- ◆ thickening agents for agricultural sprays
- ◆ latex dispersions
- ◆ textile printing paste
- ◆ water retention aids

PHYSICAL PROPERTIES
- ◆ white, crystalline chunks
- ◆ melting point 84.5°C
- ◆ boiling point 125°C at 25 mm Hg
- ◆ density 1.122 g/mL at 30°C
- ◆ specific gravity 1.122 at 30°C
- ◆ vapor density 2.45 (air = 1)
- ◆ odorless
- ◆ vapor pressure 1.6 mm Hg at 20°C
- ◆ soluble in ether, water, methanol, acetone, ethyl acetate, benzene, ethanol

CHEMICAL PROPERTIES
- ◆ flash point 138°C
- ◆ not flammable
- ◆ autoignition temperature 424°C
- ◆ Henry's Law constant 302×10^{-10} atm-m³/mol
- ◆ incompatible with bases, oxidizers, reducing agents, iron and iron salts, copper, aluminum, brass, free radical initiators
- ◆ air sensitive
- ◆ may polymerize on exposure to light

HEALTH RISK
- ◆ Toxic Exposure Guidelines
  - • NIOSH REL TWA 0.03 mg/m³ (skin)
  - • ACGIH TLW TWA 0.03 mg/m³ (skin)
  - • OSHA PEL TWA 0.03 mg/m³ (skin)
  - • MSHA TWA 0.3 mg/m³
- ◆ Toxicity Data
  - • oral-rat LD_{50} 124 mg/kg

- skin-rat LD_{50} 400 mg/kg
- intraperitoneal-rat LD_{50} 90 mg/kg
- unreported-rat LD_{50} 208 mg/kg
- oral-mouse LD_{50} 107 mg/kg
- intraperitoneal-mouse LD_{50} 170 ppm/6h
- unreported-mouse LD_{50} 156 mg/kg
- oral-rabbit LD_{50} 150 mg/kg
- skin-rabbit LD_{50} 1680 µL/kg
- unreported-rabbit LD_{50} 280 mg/kg
- oral-guinea pig LD_{50} 150 mg/kg
- subcutaneous-guinea pig LD_{50} 173 mg/kg
- unreported-guinea pig LD_{50} 173 mg/kg
- oral-quail LD_{50} 186 mg/kg
- oral-mammal LD_{50} 100 mg/kg

◆ Other Toxicity Indicators
- EPA RfD 0.0002 mg/kg/d
- EPA Group B2 probable human carcinogen of medium carcinogenic hazard $1/ED_{10}$ 16 per (mg/kg)/d
- EPA Cancer Risk Level (1 in a million excess lifetime risk) 8E-7 mg/m^3

◆ Acute Risks
- harmful if inhaled, swallowed, or absorbed through skin
- eye and skin irritation
- possible nervous system disturbances
- respiratory system
- drowsiness
- incoordination
- hallucinations
- confusion
- peripheral neuropathy

◆ Chronic Risks
- carcinogen
- nerve damage
- damage to cardiovascular system
- reddish rash
- adverse blood affects

HAZARD RISK
◆ incompatible with bases, oxidizers, reducing agents, iron and iron salts, copper, aluminum, brass, free radical initiators
◆ hazardous combustion products may include carbon monoxide, carbon dioxide, nitrogen oxides, and/or ammonia

EXPOSURE ROUTES
◆ adsorption through skin and mucous membranes
◆ ingestion

REGULATORY STATUS/INFORMATION
◆ Toxic Substances Control Act: EPA Office of Pollution Prevention and Toxics
◆ Emergency Planning and Community Right-to-Know Act (EPCRA) Toxics Release Inventory Data: EPA Office of Pollution Prevention and Toxics

- ◆ Clean Air Act
- ◆ Comprehensive Environmental Response, Compensation, and Liability Act (Superfund): EPA Office of Solid Waste and Emergency Response
- ◆ Resource Conservation and Recovery Act/EPCRA (Sec. 302/304/311/312/313): EPA Office of Solid Waste and Emergency Response
- ◆ Safe Drinking Water Act: Statutory Contaminants (Sec. 1412b)
- ◆ National Institute for Occupational Safety and Health (NIOSH) Recommendation Substances
- ◆ Resource Conservation and Recovery List (RCRA) Hazardous Substances (40 CFR Ch. 1)
- ◆ National Toxicology Program: Anticipated Carcinogen
- ◆ Department of Transportation Hazardous Materials Table (Source List 13A)

ADDITIONAL/OTHER COMMENTS

- ◆ First Aid
  - wash eyes and/or skin immediately with large amounts of water for 15 minutes
  - remove to fresh air if inhaled
  - wash out mouth with water if swallowed
- ◆ Fire Fighting
  - extinguishing media: carbon dioxide, dry chemical powder, or appropriate foam
  - special procedures: wear self-contained breathing apparatus and protective clothing
  - unusual fire and explosion hazards: extremely flammable; vapor may travel long distance to a source of ignition and flash back
- ◆ Personal Protection
  - use only in a chemical fume hood
  - wear chemical-resistant gloves, safety goggles, and other protective clothing
  - safety shower and eye bath
- ◆ storage: keep tightly closed and in a dry place

SOURCES

- ◆ *MSDS (Acrylamide)*, Aldrich Chemical Company, Inc., 1996
- ◆ *Environmental Contaminant Reference Databook*, volume 1, 1995
- ◆ *EPA Notebook Hap Index* (http://www.epa.gov/oar/oaqps/airtox/ hapindex.html)
- ◆ *Merck Index*, 12th edition, 1996
- ◆ *Chemicals in the Environment: Acrylamide*, Office of Pollution Prevention and Toxics, U.S. Environmental Protection Agency, September 1994
- ◆ *Chemical Summary for Acrylamide*, Office of Pollution Prevention and Toxics, U.S. Environmental Protection Agency, September 1994

◆ *Sax's Dangerous Properties of Industrial Materials*, 9th edition, 1996
◆ *Suspect Chemicals Sourcebook*, 1996

ACRYLIC ACID (rc)

CAS #: 79-10-7
MOLECULAR FORMULA: $C_3H_4O_2$
FORMULA WEIGHT: 72.06
SYNONYMS: acroleic acid, acrylic acid, acrylic acid, glacial, ethylene-carboxylic acid, glacial acrylic acid, propene acid, propenoic acid, 2-propenoic acid, RCRA waste number U008, vinylformic acid

USES
◆ Manufacture of other chemicals
  • acrylic esters
  • acrylic resins
  • superabsorbant polymers
  • detergents
◆ Treatment Processes
  • oil treatment chemicals
  • water treatment chemicals
  • water absorbent polyacrylic acid polymers
◆ Commercial
  • polishes
  • paints
  • coatings
  • rug backings
  • adhesives
  • plastics
  • textiles
  • paper finishes
  • leather finishing
◆ Natural
  • produced and used by several species of marine algae and in the stomach of sheep

PHYSICAL PROPERTIES
◆ colorless liquid
◆ melting point 14°C (at 1 atm)
◆ boiling point 141.6°C (at 1 atm)
◆ density 1.0511 at 20°C
◆ surface tension 28.1 dynes/cm at 30°C
◆ specific gravity 1.051 (water = 1)
◆ vapor density 2.5 (air = 1)
◆ aromatic odor, acrid, irritating
◆ odor threshold 1.04 ppm (in air)
◆ vapor pressure 3.2 mm Hg at 20°C
◆ miscible in water, benzene, chloroform, ether, and acetone

CHEMICAL PROPERTIES
◆ flash point 54°C (130°F)
◆ flammability limits 2% lower and 13.7% upper

- autoignition temperature 438°C (820°F)
- Henry's Law constant 3.2 E-7 atm-m³/mol
- dissociation constant K at 25°C = 5.6 E-5
- heat of combustion 327.0 kcal/g at 25°C
- heat of vaporization 10,955.1 gcal/gmole
- contributes to the production of photochemical smog

HEALTH RISK

- Toxic Exposure Guidelines
  - ACGIH TLV TWA 2 ppm (skin)
  - NIOSH REL TWA 2 ppm (skin)
  - OSHA PEL TWA 10 ppm (skin)
- Toxicity Data
  - oral-rat TDLo 100 g/kg
  - oral-rat LD_{50} 33,500 µg/kg
  - intraperitoneal-rat LD_{50} 22 mg/kg
  - unreported-rat LD_{50} 1250 mg/kg
  - oral-mouse LD_{50} 2400 mg/kg
  - intraperitoneal-mouse LD_{50} 144 mg/kg
  - inhalation-mouse LC_{50} 5300 mg/m³/2H
  - subcutaneous-mouse LD_{50} 1590 mg/kg
  - unreported-mouse LD_{50} 1590 mg/kg
  - skin-rabbit LD_{50} 280 mg/kg
  - eye-rabbit 250 µg/24H
  - unreported-rabbit LD_{50} 250 mg/kg
  - inhalation-rat LCLo 4000 ppm/4H
- Other Toxicity Indicators
  - EPA RfC 0.0003 mg/m³
  - EPA RfD 0.08 mg/kg/d
- Acute Risks
  - may be fatal if inhaled, swallowed, or absorbed through skin
  - extremely destructive to mucous membranes, upper respiratory tract, eyes, and skin
  - Inhalation may be fatal due to spasm, inflammation, and edema of the larynx and bronchi.
  - Inhalation may be fatal as a result of chemical pneumonitis and pulmonary edema.
  - burning sensation
  - coughing
  - wheezing
  - laryngitis
  - shortness of breath
  - headache
  - nausea and vomiting
- Chronic Risks
  - man: not known
  - animal: reduced birth weight in dogs

HAZARD RISK

- corrosive

- ◆ incompatible with or reacts strongly with strong oxidizers, strong bases
- ◆ decomposes to produce toxic fumes of carbon monoxide and carbon dioxide
- ◆ flammable
- ◆ combustible
- ◆ may undergo exothermic polymerization
- ◆ may become explosive if confined
- ◆ fire hazard when exposed to heat or flame

EXPOSURE ROUTES
- ◆ inhalation
- ◆ ingestion
- ◆ dermal contact
- ◆ absorption through mucous membranes

REGULATORY STATUS/INFORMATION
- ◆ Clean Air Act Amendments (1990) Hazardous Air Pollutant
- ◆ National Institute for Occupational Safety and Health (NIOSH)
- ◆ RCRA Land Disposal Restrictions: Second Third Wastes Recommendation Substances
- ◆ Comprehensive Environmental Response, Compensation, and Liability Act (Superfund): EPA Office of Solid Waste and Emergency Response
- ◆ Resource Conservation and Recovery Act/EPCRA (Section 302) Toxic Chemicals
- ◆ Department of Transportation Hazardous Materials Table (Source List 13A)

ADDITIONAL/OTHER COMMENTS
- ◆ First Aid
  - wash eyes and/or skin immediately with large amounts of water for 15 minutes
  - remove to fresh air if inhaled
  - wash out mouth with water if swallowed
- ◆ Fire Fighting
  - extinguishing media: carbon dioxide, dry chemical powder, or appropriate foam
  - special procedures: wear self-contained breathing apparatus and protective clothing
  - unusual fire and explosion hazards: emits toxic fumes under fire conditions; vapor may travel long distance to a source of ignition and flash back
- ◆ Personal Protection
  - use only in a chemical fume hood
  - wear chemical-resistant gloves, safety goggles, and other protective clothing
  - safety shower and eye bath
  - do not breathe vapor
- ◆ storage: keep tightly closed; keep away from heat or open flame; store in a cool, dry place. Do not refrigerate or freeze.

SOURCES
- *MSDS (Acrylic Acid)*, Aldrich Chemical Company, Inc., 1996
- *Environmental Contaminant Reference Databook*, volume 1, 1995
- *EPA Notebook Hap Index* (http://www.epa.gov/oar/oaqps/airtox/hapindex.html)
- *Merck Index*, 12th edition, 1996
- *Chemical Summary for Acrylic Acid*, Office of Pollution Prevention and Toxics, U.S. Environmental Protection Agency, September 1994
- *Chemical in the Environment: Acrylic Acid*, Office of Pollution Prevention and Toxics, U.S. Environmental Protection Agency, September 1994
- *Sax's Dangerous Properties of Industrial Materials*, 9th edition, 1996
- *Suspect Chemicals Sourcebook*, 1996

ACRYLONITRILE (rc)

CAS #: 107-13-1
MOLECULAR FORMULA: C_3H_3N
FORMULA WEIGHT: 53.06
SYNONYMS: acritet, acrylon, acrylonitrile monomer, carbacryl, cyanoethylene, fumigrain, miller's fumigrain, propenenitrile, 2-propenenitrile, RCRA waste number U009, TL 314, VCN, ventox, vinyl cyanide

USES
- Manufacture of Other Chemicals
  - acrylic fibers
  - plastics
  - rubber elastomers
  - plasticizers
  - solvents
  - polymeric materials
  - dyes
  - pharmaceuticals
  - insecticides
  - nylon
- Formation of High-Impact Resins
  - styrene-acrylonitrile
  - acrylonitrile-butadiene-styrene

PHYSICAL PROPERTIES
- colorless liquid
- melting point −83°C
- boiling point 77°C at 1 atm
- density 0.8004 g/mL at 25°C
- specific gravity 0.8004 (water = 1)
- vapor density 1.83 (air = 1)
- surface tension 27.3 dynes/cm at 24°C
- viscosity 0.34 cP at 25°C
- heat of vaporization 147 cal/g

- practically odorless, but can have a sweet odor
- odor threshold 40.4 mg/m³
- vapor pressure 100 mm Hg at 23°C
- soluble in isopropyl alcohol, alcohol, ether, acetone, benzene
- miscible with ethanol, carbon tetrachloride, ethyl acetate, liquid carbon dioxide, toluene, petroleum ether, and xylene

CHEMICAL PROPERTIES

- flash point 32°F
- flammability limits 3% lower and 17% upper
- autoignition temperature 480°C (897°F)
- heat of combustion −7930 cal/g
- Henry's Law constant 8.8E-5 atm-m³/mol
- incompatible with oxidizers, acids, bases, copper, copper alloys, and heat
- may polymerize due to exposure to light

HEALTH RISK

- Toxic Exposure Guidelines
  - ACGIH TLV TWA 2 ppm (skin)
  - NIOSH REL TWA 2 ppm (skin)
  - OSHA PEL TWA 2 ppm (skin)
- Toxicity Data
  - inhalation-man LCLo 1 g/m³/1H
  - skin-child LDLo 2015 mg/kg
  - oral-rat LD_{50} 78 mg/kg
  - inhalation-rat LC_{50} 428 ppm/4H
  - skin-rat LD_{50} 148 mg/kg
  - intraperitoneal-rat LD_{50} 65 mg/kg
  - unreported-rat LD_{50} 1250 mg/kg
  - subcutaneous-rat LD_{50} 75 mg/kg
  - oral-mouse LD_{50} 27 mg/kg
  - intraperitoneal-mouse LD_{50} 46 mg/kg
  - subcutaneous-mouse LD_{50} 25 mg/kg
  - skin-rabbit LD_{50} 250 µL/kg
  - oral-guinea pig LD_{50} 50 mg/kg
  - skin-guinea pig LD_{50} 202 mg/kg
  - subcutaneous-guinea pig LD_{50} 130 mg/kg
- Other Toxicity Indicators
  - EPA RfC 0.002 mg/m³
- Acute Risks
  - may be fatal if inhaled, swallowed, or absorbed through skin
  - OSHA regulated carcinogen
  - extremely destructive to mucous membranes, upper respiratory tract, eyes, and skin
  - Inhalation may be fatal due to spasm, inflammation, and edema of the larynx and bronchi.
  - Inhalation may be fatal as a result of chemical pneumonitis and pulmonary edema.
  - burning sensation
  - coughing

- wheezing
- laryngitis
- shortness of breath
- headache
- nausea and vomiting

HEALTH RISK

- ◆ highly flammable
- ◆ corrosive
- ◆ explosive
- ◆ incompatible with oxidizers, acids, bases, copper, copper alloys, and heat
- ◆ hazardous decomposition or combustion by-products include carbon monoxide, carbon dioxide, hydrogen cyanide, and/or nitrogen oxides

EXPOSURE ROUTES

- ◆ readily absorbed through the skin
- ◆ inhalation
  - cigarette smoke
  - automobile exhaust

REGULATORY STATUS/INFORMATION

- ◆ U.S. EPA B1 Probable Human Carcinogen
- ◆ Clean Water Act Section 304: Water Quality Criteria Substance
- ◆ Clean Water Act Section 307: Priority Pollutant
- ◆ Clean Water Act Section 311: Hazardous Substances
- ◆ Clean Air Act Section 112(r): Accidental Release Prevention Substances
- ◆ National Institute for Occupational Safety and Health (NIOSH) Recommendation Substances: Occupational Safety and Health Act (29 U.S.C. 1900)
- ◆ Resource Conservation and Recovery Act (RCRA) Hazardous Substance (40 CFR Part 261)
- ◆ Resource Conservation and Recovery Act (RCRA) Hazardous Constituents for Groundwater Monitoring
- ◆ Comprehensive Environmental Response, Compensation, and Liability Act (CERCLA) Hazardous Substances (40CFR302)
- ◆ Superfund Amendments and Reauthorization Act of 1986 (SARA)/Emergency Planning and Community Right-to-Know Amendments (EPCRA) Title III Section 302 Extremely Hazardous Substance (40 CFR Ch.1 Part 355.50)
- ◆ Superfund Amendments and Reauthorization Act of 1986 (SARA)/Emergency Planning and Community Right-to-Know Amendments (EPCRA) Title III Section 313 Toxic Chemicals (40 CFR Ch.1 Part 372)
- ◆ National Toxicity Program: Anticipated Carcinogen
- ◆ ACGIH Threshold Limit Value Chemicals
- ◆ OSHA Air Contaminants
- ◆ Department of Transportation Hazardous Materials Table

ADDITIONAL/OTHER COMMENTS
- ◆ First Aid
  - • wash eyes and/or skin immediately with large amounts of water for 15 minutes
  - • remove to fresh air if inhaled
  - • wash out mouth with water if swallowed
- ◆ Fire Fighting
  - • extinguishing media: carbon dioxide, dry chemical powder, or appropriate foam
  - • special procedures: wear self-contained breathing apparatus and protective clothing
  - • unusual fire and explosion hazards: flammable liquid; vapor may travel long distance to a source of ignition and flash back
- ◆ Personal Protection
  - • use only in a chemical fume hood
  - • wear chemical-resistant gloves, safety goggles, and other protective clothing
  - • safety shower and eye bath
- ◆ storage: keep tightly closed; keep away from heat, sparks, or open flame; store in a cool, dry place; light sensitive

SOURCES
- ◆ Agency for Toxic Substances and Disease Registry Public Health Statement (http://atsdr1.atsdr.cdc.gov:8080), December 1990
- ◆ *MSDS (Acrylonitrile)*, Aldrich Chemical Company, Inc., 1996
- ◆ *Environmental Contaminant Reference Databook*, volume 1, 1995
- ◆ *EPA Notebook Hap Index* (http://www.epa.gov/oar/oaqps/airtox/ hapindex.html)
- ◆ *Merck Index*, 12th edition, 1996
- ◆ *Chemicals in the Environment: Acrylonitrile*, Office of Pollution Prevention and Toxics, U.S. Environmental Protection Agency, September 1994
- ◆ *Sax's Dangerous Properties of Industrial Materials*, 9th edition, 1996
- ◆ *Suspect Chemicals Sourcebook*, 1996

ALLYL CHLORIDE

CAS #: 107-05-1
MOLECULAR FORMULA: $H_2C=CHCH_2Cl$
FORMULA WEIGHT: 76.53
SYNONYMS: allile (cloruro di) (Italian), Allylchlorid (German), allyle (chlorure d') (French), chlorallylene, chloroallylene, 3-chloroprene, 3-chloro-1-propene, 3-chloropropene, 1-chloro propene-2, 1-chloro-2-propene, α-chloropropylene, 3-chloro-1-propylene, 3-Chloropropylene, 3-Chlorpropen (German), 2-propenyl chloride

USES
- ◆ Used in the Production Process

- epichlorohydrin
- glycerin
- varnish
- plastics
- adhesives
- perfumes
- pharmaceuticals
- insecticides
◆ Used in the Synthesis of Allyl Compounds
  - allyl alcohol
  - allyl amines
  - allyl esters
  - polyesters

PHYSICAL PROPERTIES
◆ colorless to pale-yellow liquid
◆ garlic-onion odor
◆ odor threshold 1.2 ppm
◆ melting point −136.4°C
◆ boiling point 44.6°C
◆ density 0.938 at 20°/4°

◆ vapor pressure 362 mm Hg at 25°C
◆ vapor density 2.64
◆ solubility <0.1 in water
◆ log K_{ow} 1.45
◆ evaporation rate 0.14
◆ specific gravity 0.94

CHEMICAL PROPERTIES
◆ stable
◆ flash point −36°C
◆ autoignition temperature 905°F
◆ flammability limits 2.9% lower and 11.2% upper

◆ incompatibility with strong oxidizers, amines, metals, and acids
◆ Polymerization may occur upon heating or when in contact with acids or galvanized metals.

HEALTH RISK
◆ Toxic Exposure Guidelines
  - ACGIH TLV TWA 1 ppm
  - OSHA PEL TWA 1 ppm
  - DFG MAK 1 ppm
  - NIOSH REL TWA 1 ppm
◆ Toxicity Data
  - mmo-esc 20 µL/plate
  - mmo-omi 10 µL/plate
  - skin-rabbit 10 mg/24H
  - eye-rabbit 469 mg
  - eye-guinea pig 290 ppm/6H
  - oral-mouse TDLo 4 g/kg
  - intraperitoneal-rat TDLo 1200 mg/kg:REP
  - oral-mouse TDLo 50 g/kg/78W-I:ETA
  - oral-rat LD_{50} 700 mg/kg
  - inhalation-rat LC_{50} 11 g/m³/2H
  - oral-mouse LD_{50} 425 mg/kg
  - inhalation-mouse LC_{50} 11,500 mg/m³/2H
  - intraperitoneal-mouse LD_{50} 155 mg/kg
  - intravenous-dog LD_{50} 7150 µg/kg
  - skin-rabbit LD_{50} 2066 mg/kg
◆ Acute Risks
  - irritation of eyes
  - irritation of respiratory passages
  - unconsciousness
  - death
  - conjunctivitis
  - reddening of eyelids
  - corneal burn

- local vasoconstriction
- numbness
- headache
- dizziness
- inflammation of lungs in animals
- swelling of kidneys in animals

◆ Chronic Risks
  - liver and kidney damage
  - affects the central nervous system
  - motor and sensory neurotoxic damage
  - pulmonary edema
  - decreased maternal weight gain in rabbits
  - increased maternal heart, liver, spleen, and kidney weights in rats
  - EPA Group C: possible human carcinogen agent (cancer causing agent)

HAZARD RISK

◆ dangerous fire and explosion hazard when exposed to heat, flame, or oxidizers

◆ extremely flammable

◆ vigorous or explosive reaction above −70°C with alkyl aluminum chlorides and aromatic hydrocarbons

◆ violently exothermic polymerization reaction with Lewis acids and metals

◆ Vapors may cause flash fire.

◆ incompatible with HNO_3, ethylene imine, ethylenediamine, chlorosulfonic acid, oleum, and NaOH

◆ Combustion will produce carbon dioxide, carbon monoxide, and hydrogen chloride.

EXPOSURE ROUTES

◆ production and processing facilities

◆ production of allyl chloride and epichlorohydrin

◆ production of glycerin

◆ breathing contaminated air

◆ skin contact

REGULATORY STATUS/INFORMATION

◆ Clean Air Act Amendment, Title III: hazardous air pollutants

◆ Resource Conservation and Recovery Act
  - list of hazardous constituents. 40CFR258
  - groundwater monitoring list. 40CFR264

◆ CERCLA and Superfund
  - designation of hazardous substances. 40CFR302.4
  - list of chemicals. 40CFR372.65

◆ Clean Water Act
  - designation of hazardous substances. 40CFR116.4
  - determination of reportable quantities. 40CFR117.3

◆ Occupational Health and Safety Act
  - 29CFR1910.1000 Table Z1
  - 29CFR1910.119 list of highly hazardous chemicals, toxics, and reactives

◆ chemical regulated by the State of California

ADDITIONAL/OTHER COMMENTS
◆ Symptoms
  • inhalation: headache, dizziness, unconsciousness, irritation to eyes, nose, and throat
  • contact: rapid absorption, local vasoconstriction, numbness
  • ingestion: burns, severe irritation of gastrointestinal tract
◆ First Aid
  • wash eyes and skin immediately with large amounts of water
  • remove to fresh air immediately or provide respiratory support
◆ fire fighting: use carbon dioxide, alcohol foam, or dry chemical
◆ personal protection: wear self-contained breathing apparatus, protective clothing, and chemical goggles or gas mask if necessary
◆ storage: keep cool and away from heat and use adequate ventilation
SOURCES
◆ *Health Effects Notebook for Hazardous Air Pollutants*, U.S. Environmental Protection Agency, December 1994
◆ *Material Safety Data Sheets (MSDS)*
◆ *Sax's Dangerous Properties of Industrial Materials*, 9th edition, 1996
◆ *Hawley's Condensed Chemical Dictionary*, 12th edition, 1993
◆ *Aldrich Chemical Company Catalog*, 1996–1997
◆ *Environmental Law Index to Chemicals*, 1996
◆ *Merck Index*, 12th edition, 1996

4-AMINOBIPHENYL (rs)

CAS #: 92-67-1
MOLECULAR FORMULA: $C_{12}H_{11}N$ or $C_6H_5C_6H_4NH_2$
FORMULA WEIGHT: 169.24
SYNONYMS: p-aminobiphenyl, 4-aminodiphenyl, p-aminodiphenyl, anilinobenzene, biphenylamine, (1,1'-biphenyl)-4-amine, 4-biphenylamine, p-biphenylamine, paraaminodiphenyl, 4-phenylaniline, xenylamine

USES
◆ Used as a Component or in the Manufacturing Process
  • detection of sulfates
  • rubber antioxidant
  • carcinogen in cancer research
  • organic research
  • dye intermediate
PHYSICAL PROPERTIES
◆ colorless crystals
◆ melting point 52–54°C (125.6–129.2°F)
◆ boiling point 302.2°C (576) at 760 mm Hg, 191°C (375.8°F) at 15 mm Hg

- soluble in hot water, alcohol, chloroform; slightly soluble in water
- specific gravity 1.16 at 68° (ref to water at 4°C)
- density 1.16 g/cm^3
- vapor pressure 1 mm HG at 68°F
- floral odor
- volatile with steam

CHEMICAL PROPERTIES

- lower explosive limit (LEL) not determined
- upper explosive limit (UEL) not determined
- autoignition temperature 450°C (842°F)
- combustible solid, but must be preheated before ignition
- oxidized by air; incompatible with acids, acid anhydrides, oxidizing agents

HEALTH RISK

- Toxic Exposure Guidelines
  - OSHA cancer suspect agent
  - NIOSH REL carcinogen
  - IDLH carcinogen, no level detected
  - ACGIH confirmed human carcinogen, TWA
- Toxicity Data
  - microsomal mutagenicity assay, salmonella typhemurium 2 μg/plate
  - DNA damage, e coli 30 μmol/L
  - mutation in mammalian somatic cells, human fibroblast 60 mg/L
  - DNA damage, rat liver 30 μmol/L
  - unscheduled DNA synthesis, mouse-oral 200 mg/kg
  - oral-rat TDLo 4524 mg/kg, 48W continuous, equivocal tumorigenic agent
  - subcutaneous-mouse TDLo 216 mg/kg/3d intermittently, carcinogenic
  - oral-mouse TD 5460 μg/kg, carcinogenic
  - oral-rat LD$_{50}$ 500 mg/kg
  - oral-mouse LD$_{50}$ 205 mg/kg
  - intraperitoneal-mouse LDLo 250 mg/kg
  - oral-dog LDLo 25 mg/kg
  - oral-rabbit LD$_{50}$ 690 mg/kg
  - target organs bladder, skin
- Acute Risks
  - toxic (may be fatal) by ingestion, inhalation, and skin contact (absorption)
  - causes skin and eye irritation
  - irritating to mucous membrane and upper respiratory tract
  - Absorption into body may cause cyanosis.
- Chronic Risks
  - highly toxic suspected human carcinogen
  - may alter genetic material

HAZARD RISK
- ◆ slight to moderate fire hazard when exposed to heat, flames (sparks), or powerful oxidizers
- ◆ at decomposition, emits toxic fumes of NOx

EXPOSURE ROUTES
- ◆ dermal
- ◆ inhalation

REGULATORY STATUS/INFORMATION
- ◆ Clean Air Act
  - • Title III: regulated hazardous air pollutant
- ◆ Resource Conservation and Recovery Act
  - • 40CFR258.40, Appendixes 1 and 2, MSW landfills
  - • 40CFR264, Appendix 9, groundwater monitoring chemicals list
  - • 40CFR261, Appendix 8, hazardous constituents, see also 40CFR261.11
- ◆ CERCLA and Superfund
  - • 40CFR372.65, list of substances subject to Community-Right-to-Know, designation as a specific toxic chemical
- ◆ Occupational Health and Safety Act
  - • specially regulated substances (carcinogen). 29CFR1910.1001–1910.1048
  - • limits for air contaminants. 29CFR1910.1000 Table Z1
  - • regulated carcinogen. 29CFR1910.1011
- ◆ chemical regulated by the State of California

ADDITIONAL/OTHER COMMENTS
- ◆ Symptoms
  - • dizziness
  - • headache
  - • lethargy
  - • dyspnea
  - • ataxia, coma, seizure
  - • weakness
  - • methemoglobinemia
  - • urinary burning
  - • acute hemorrhagic cystitis
- ◆ First Aid
  - • eyes: irrigate immediately
  - • inhalation: artificial respiration
  - • skin: remove all contaminated clothing; wash with soap and water immediately
  - • ingestion: seek immediate medical attention
- ◆ Fire Fighting
  - • extinguishing media: use dry chemical, foam, or carbon dioxide
  - • unusual fire and explosion hazards: emits toxic fumes
- ◆ Personal protection: full protective clothing and self-contained breathing apparatus
- ◆ Spill or Leak

- cover spill with dry lime or soda ash; keep in closed container and dispose of properly
- ventilate area and wash spill site after complete material pickup

SOURCES
- *Aldrich Chemical Company Catalog*, 1996–1997
- *Environmental Law Index to Chemicals*, 1994 edition
- *MSDS*, Aldrich Chemical Company, Inc., 1996
- *Merck Index*, 12th edition, 1996
- *NIOSH Pocket Guide to Chemical Hazards*, June 1994
- *Sax's Dangerous Properties of Industrial Materials*, 9th edition, volumes 1 and 2, 1996
- *Hawley's Condensed Chemical Dictionary*, 12th edition, 1993
- *Health Effects Notebook for Hazardous Air Pollutants*, U.S. Environmental Protection Agency, December 1994

ANILINE

CAS #: 62-53-3
MOLECULAR FORMULA: $C_6H_5NH_2$
FORMULA WEIGHT: 93.14
SYNONYMS: aminobenzene, aminophen, anilin (Czech), anilina (Italian, Polish), aniline-oil, anyvim, benzamine, benzenamine, benzene, amino-, blue-oil, C.I. 76000, huile d'aniline (French), phenyl-amine

USES
- Used in Manufacturing Process
  - resins
  - varnishes
  - perfumes
  - shoe backs
  - printing inks
  - cloth marking inks
  - paint removers
  - photographic chemicals
  - explosives
  - herbicides
  - fungicides
  - rigid polyurethanes
  - isocyanate synthesis
  - urethane foams
  - petroleum refining
  - antiknock compound for gasolines
  - teryl and optical whitening agents
- Used as a Chemical Intermediate
  - methylenediisocyanate
  - rubber processing chemicals
  - accelerators
  - dyes and pigments
  - hydroquinone
  - pesticides
  - pharmaceuticals
  - 4-anilinophenol
  - specialty resins
  - cyclohexylamine
  - corrosion inhibitors
  - aniline salts
  - artificial sweeteners
- Used as a Component
  - lacquers and wood stains
  - skin stains
  - vulcanization accelerator
  - catalysts and stabilizers in the synthesis for hydrogen peroxide and cellulose
- used as a solvent

PHYSICAL PROPERTIES
- colorless, oily liquid
- aromatic amine-like odor
- odor threshold 0.5 ppm
- burning taste
- melting point $-6.3°C$
- boiling point $184–186°C$
- density 1.022 at 20°C
- surface tension 44.1 dynes/cm at 10°C in contact with air
- vapor pressure 0.67 mm Hg at 25°C
- vapor density 3.22
- specific gravity 1.0216
- log K_{ow} 0.90
- heat of vaporization 198 Btu/lb
- pK_b 9.30
- soluble in water, alcohol, ether, and benzene
- viscosity 4.423–4.435 cP at 20°C

CHEMICAL PROPERTIES
- stable but combustible
- flash point 158°F
- autoignition temperature 615°C
- flammability limits 1.3% lower, upper unknown
- heat of combustion $-14,980$ Btu/lb
- explosive limits 1.3% lower and 20–25% upper
- polymerizes to a resinous mass
- volatile with steam
- reacts vigorously with oxidizers

HEALTH RISK
- Toxic Exposure Guidelines
  - ACGIH TLV TWA 2 ppm
  - ACGIH STEL 20 mg/m³
  - OSHA PEL TWA 2 ppm
  - DFG MAK 2 ppm
  - BAT 1 mg/L
  - MSHA standard 19 mg/m³
- Toxicity Data
  - skin-rabbit 500 mg/24H MOD
  - eye-rabbit 102 mg SEV
  - mma-sat 100 µg/plate
  - dnr-esc 25 µL/well/16H
  - bfa-rat/sat 300 mg/kg
  - oral-mouse TDLo 4480 mg/kg
  - oral-rat TDLo 11 g/kg
  - unknown-human LDLo 357 mg/kg
  - unknown-man LDLo 150 mg/kg
  - oral-rat LD₅₀ 250 mg/kg
  - inhalation-rat LCLo 250 ppm/4H
  - skin-rat LD₅₀ 1400 mg/kg
  - intraperitoneal-rat LD₅₀ 420 mg/kg
  - inhalation-mouse LC₅₀ 175 ppm/7H
  - intraperitoneal-mouse LD₅₀ 492 mg/kg
  - subcutaneous-mouse LD₅₀ 200 mg/kg
  - oral-dog LD₅₀ 195 mg/kg
  - skin-dog LDLo 1540 mg/kg
  - inhalation-cat LCLo 180 ppm/8H

- ◆ Acute Risks
  - irritation of skin
  - severe irritation of the eye
  - mild sensitizer
  - headache
  - weakness
  - dizziness
  - irritability
  - cyanosis
  - drowsiness
  - unconsciousness
  - blue discoloration of fingertips, lips, and nose
  - nausea
  - vomiting
  - delirium
  - coma
  - shock
  - affects the lung
  - upper respiratory tract irritation
  - congestion
  - convulsions
  - tachycardia
  - dyspnea
  - hemolysis of the red blood cells
  - affects the liver
- ◆ Chronic Risks
  - formation of methemoglobin
  - cyanosis
  - irritation to mucous membranes
  - decrease in red blood cell count, hemoglobin levels, and hematocrit
  - malignant bladder growths
  - headache
  - dizziness
  - insomnia
  - loss of appetite
  - loss of weight
  - visual disturbances
  - skin lesions
  - EPA Group B2: probable human carcinogen

HAZARD RISK
- ◆ combustible
- ◆ reacts vigorously with oxidizing materials
- ◆ Spontaneously Explosive Reactions Occur with
  - benzenediazonium-2-carboxylate
  - dibenzol peroxide
  - fluorine nitrate
  - nitrosal perchlorate
  - red fuming nitric acid
  - peroxodisulfuric acid
  - tetranitromethane
- ◆ Violent Reactions with
  - boron chloride
  - peroxyformic acid
  - diisopropyl peroxydicarbonate
  - fluorine
  - trichloronitromethane
  - acetic anhydride
  - chlorosulfonic acid
  - hexachloromelamine
  - oleum
  - perchromates
  - K_2O_2
  - β-propriolactone
  - $AgClO_4$
  - Na_2O_2
  - H_2SO_4
  - trichloromelamine
  - acids
  - peroxydisulfuric acid
  - FO_3Cl
  - diisopropyl peroxydicarbonate
  - n-halomides
  - trichloronitromethane
- ◆ ignites on contact with sodium peroxide and water
- ◆ Forms Heat- or Shock-Sensitive Explosive Mixtures with
  - anilinium chloride
  - nitromethane

- hydrogen peroxide
- 1-chloro-2,3-epoxypro-
  pane
- peroxomonosulfuric acid

◆ Reactions with perchloryl fluoride, perchloric acid, and ozone
  form explosive products.
◆ Decomposition emits highly toxic fumes of NO_x.

EXPOSURE ROUTES
- ◆ breakdown of pollutants in
    air
- ◆ burning of plastics
- ◆ burning tobacco
- ◆ breathing contaminated air
- ◆ smoking tobacco
- ◆ proximity to someone smok-
    ing
- ◆ near industrial sources using
    large amounts of aniline
- ◆ corn
- ◆ grains
- ◆ rhubarb
- ◆ apples
- ◆ beans
- ◆ rapeseed cake (animal feed)
- ◆ volatile component of black
    tea
- ◆ drinking water
- ◆ surface water

REGULATORY STATUS/INFORMATION
◆ Clean Air Act Amendment, Title III: hazardous air pollutants
◆ Resource Conservation and Recovery Act
  - list of hazardous constituents. 40CFR261, 40CFR261.11
  - groundwater monitoring list. 40CFR264
  - risk specific doses. 40CFR266
  - health based limits for exclusion of waste-derived residues.
    40CFR266
  - U waste #: U012
◆ CERCLA and Superfund
  - designation of hazardous substances. 40CFR302.4
  - list of extremely hazardous substances. 40CFR355-AB
  - list of chemicals. 40CFR372.65
◆ Clean Water Act
  - designation of hazardous substances. 40CFR116.4
  - determination of reportable quantities. 40CFR117.3
◆ Toxic Substance Control Act
  - chemicals subject to test rules or consent orders for which
    the testing reimbursement period has passed.
    40CFR799.18
◆ OSHA 29CFR1910.1000 Table Z1
◆ chemical regulated by the State of California

ADDITIONAL/OTHER COMMENTS
◆ Symptoms
  - ingestion: irritation of digestive tract
  - inhalation: irritation of respiratory tract, blue discoloration
    of fingertips, cheeks, nose, and lips, nausea, vomiting,
    headache, drowsiness, delirium, coma, shock
◆ First Aid
  - wash eyes immediately with large amounts of water
  - flush skin with plenty of soap and water
  - remove to fresh air immediately or provide respiratory sup-
    port
  - give 2–4 cupfuls of milk or water

- fire fighting: use dry chemical, water spray, chemical foam, or carbon dioxide
- personal protection: wear self-contained breathing apparatus, protective clothing, a face shield, and chemical goggles
- storage: keep from sources of ignition and contact with oxidizing materials; tightly closed container; cool, dry, well-ventilated area

SOURCES
- *Health Effects Notebook for Hazardous Air Pollutants*, U.S. Environmental Protection Agency, December 1994
- *Material Safety Data Sheets (MSDS)*
- *Sax's Dangerous Properties of Industrial Materials*, 9th edition, 1996
- *Hawley's Condensed Chemical Dictionary*, 12th edition, 1993
- *Aldrich Chemical Company Catalog*, 1996–1997
- *Environmental Law Index to Chemicals*, 1996
- *Merck Index*, 12th edition, 1996
- *Environmental Contaminant Reference Databook*, volume 1, 1995

o-ANISIDINE (ch)

CAS #: 90-04-0
MOLECULAR FORMULA: C_7H_8NO
FORMULA WEIGHT: 123.17
SYNONYMS: o-aminoanisole, 2-aminoanisole, 1-amino-2-methoxybenzene, 2-anisidine, o-anisylamine, benzenamine, 2-methoxy-(9CI), 2-methoxy-1-aminobenzene, o-methoxyaniline, 2-methoxyaniline, 2-methoxybenzenamine, o-methoxyphenylamine

USES
- Used as a Component or in the Manufacturing Process
  - azo dyes
  - guaiacol

PHYSICAL PROPERTIES
- red or yellow, oily liquid
- melting point 5–6°C (40–42°F) at 760 mm Hg
- boiling point 225°C (437°F) at 760 mm Hg
- density 1.092 g/cm³ at 20°C
- specific gravity 1.10 (water = 1.0)
- amine-like odor
- vapor pressure 0.212 mm Hg at 20°C
- insoluble in water
- soluble in dilute mineral acid, alcohol, and ether

CHEMICAL PROPERTIES
- volatile with steam
- incompatible with strong oxidizers
- flammability limits not available

HEALTH RISK
- Toxic Exposure Guidelines

- ACGIH TLV TWA 0.5 mg/m³
- OSHA PEL TWA 0.5 mg/m³
- NIOSH 0.5 mg/m³
- IDLH 50 mg/m³
- ◆ Toxicity Data
  - microsomal mutagenicity assay salmonella typhimurium 333 µg/plate
  - oral-rat LD_{50} 2000 mg/kg
  - oral-mouse LD_{50} 1400 mg/kg
  - oral-rabbit LD_{50} 870 mg/kg
  - oral-bird (wild species) LD_{50} 422 mg/kg
- ◆ Acute Risks
  - harmful if swallowed, inhaled, or absorbed through skin
  - eye and skin irritation
  - irritating to mucous membranes and upper respiratory tract
  - cyanosis
- ◆ Chronic Risks
  - allergic respiratory and skin reactions
  - carcinogen

HAZARD RISK

- ◆ NFPA rating Health 2, Flammability 1, Reactivity 0
- ◆ combustible liquid
- ◆ Thermal decomposition may produce carbon monoxide, carbon dioxide, and nitrogen oxides.
- ◆ incompatible with oxidizing agents

EXPOSURE ROUTES

- ◆ may occur for those workers in the dye industry
- ◆ detected in tobacco smoke

REGULATORY STATUS/INFORMATION

- ◆ Clean Air Act hazardous air pollutants. Title III. (The original list was published in 42USC7412.)
- ◆ CERCLA and Superfund
  - chemicals and chemical categories to which this part applies (CAS # listing). 40CFR372.65
- ◆ Toxic Substance Control Act
  - OSHA chemicals in need of dermal absorption testing. 40CFR712.30.e10
- ◆ chemical regulated by the State of California

ADDITIONAL/OTHER COMMENTS

- ◆ Symptoms
  - headache
  - dizziness
  - cyanosis
- ◆ First Aid
  - wash eyes and skin immediately with large amount of water
  - remove contaminated clothing
  - if swallowed, wash out mouth (if person is conscious) and call physician
- ◆ fire fighting: use dry chemical, foam, carbon dioxide, or water spray

◆ personal protection: wear full protective clothing and positive self-contained breathing apparatus
◆ storage: store in cool, dry location; keep away from heat and open flame.

SOURCES
◆ *Environmental Contaminant Reference Databook*, volume 1, 1995
◆ *Environmental Law Index to Chemicals*, 1996 edition
◆ *Fire Protection Guide to Hazardous Materials*, 11th edition, 1994
◆ *Health Effects Notebook for Hazardous Air Pollutants*, U.S. Environmental Protection Agency, December 1994
◆ *Material Safety Data Sheets (MSDS)*
◆ *Merck Index*, 12th edition, 1996
◆ *NIOSH Pocket Guide to Chemical Hazards*, June 1994
◆ *Sax's Dangerous Properties of Industrial Materials*, 9th edition, volume 2, 1996

ANTIMONY COMPOUNDS

CAS #: Antimony compounds have variable CAS #s. The CAS # for antimony is 7440-36-0.

MOLECULAR FORMULA: Antimony compounds have variable molecular formulas. The molecular formula for antimony is Sb. The molecular formula for antimony oxide is O_3Sb_2.

Note: For all hazardous air pollutants (HAPs) that contain the word "compounds" in the name, and for glycol ethers, the following applies: Unless otherwise specified, these HAPs are defined by the 1990 Amendments to the Clean Air Act as including any unique chemical substance that contains the named chemical (i.e., antimony, arsenic, etc.) as part of that chemical's infrastructure.

FORMULA WEIGHT: Antimony compounds have variable formula weights. The formula weight for antimony is 121.75. The formula weight for antimony oxide is 135.76.

SYNONYMS: antimony: antimony black, antimony regulus, antymon (Polish), C.I. 77050, stibium
antimony oxide: antimonious oxide, antimony peroxide, antimony sesquioxide, antimony trioxide, antimony white, C.I. pigment white 11, dechlorane-A-O, diantimony trioxide, flowers of antimony

USES
◆ Antimony
  • white metal
  • bullets
  • bearing metal
  • thermoelectric piles
  • blackening iron
  • semiconductor devices
  • medicine
◆ Antimony Alloys
  • lead storage batteries
  • solder
  • sheet and pipe metal
  • bearings
  • castings
  • pewter
  • antimony lead
  • infrared devices

- diodes
- ◆ Antimony Oxides
  - fire retardant
  - plastics
  - paints
  - ceramics
  - fireworks
  - enamels for plastics, metals, and glass

- Hall effect devices

- textiles
- rubber
- adhesives
- pigments
- paper

PHYSICAL PROPERTIES (Antimony)
- ◆ silvery-white metal
- ◆ very brittle
- ◆ moderately hard metal
- ◆ melting point 630°C
- ◆ boiling point 1635°C
- ◆ density 6.684 at 25°C

- ◆ heat of vaporization 195,100 J/mol
- ◆ vapor pressure 1 mm Hg at 886°C
- ◆ insoluble in water
- ◆ soluble in hot, concentrated H_2SO_4

PHYSICAL PROPERTIES (Antimony Oxide)
- ◆ white cubes
- ◆ melting point 650°C
- ◆ boiling point 1550°C

- ◆ insoluble in water
- ◆ soluble in KOH and HCl

CHEMICAL PROPERTIES (Antimony)
- ◆ slightly oxidized in air
- ◆ avoid heat and light
- ◆ incompatible with strong oxidizing agents, strong acids, halogen acids, chlorine, and fluorene

- ◆ Polymerization will not occur.

HEALTH RISK
- ◆ Toxic Exposure Guidelines
  - OSHA PEL TWA 0.5 mg(Sb)/m³
  - ACGIH TLV TWA 0.5 mg(Sb)/m³
  - DFG MAK 0.5 mg(Sb)/m³
  - NIOSH REL TWA 0.5 mg(Sb)/m³
- ◆ Toxicity Data (Antimony)
  - inhalation-rat TCLo 50 mg/m³/7H/52W-I:CAR
  - oral-rat LD_{50} 7 g/kg
  - intraperitoneal-rat LD_{50} 100 mg/kg
  - intraperitoneal-mouse LD_{50} 90 mg/kg
  - intraperitoneal-guinea pig LD_{50} 150 mg/kg
- ◆ Toxicity Data (Antimony Oxide)
  - inhalation-rat TCLo 270 µg/m³
  - inhalation-rat TCLo 4200 µg/m³/52W-I:CAR
  - inhalation-rat TC 4 mg/m³/1Y-I:ETA
  - inhalation-rat TD 1600 µg/m³/52W-I:NEO
  - inhalation-rat TC 50 mg/m³/7H/52W-I:CAR
  - intraperitoneal-mouse LD_{50} 172 mg/kg
  - intravenous-dog LDLo 3 mg/kg
  - subcutaneous-rabbit LDLo 2500 µg/kg

◆ Acute Risks
  • irritation of eyes and lungs
  • heart and lung problems
  • stomach pain
  • diarrhea
  • vomiting
  • stomach ulcers
  • liver and kidney damage
  • hair loss
  • affects the skin and eyes
  • antimony spots (rash)
  • ocular conjunctivitis
  • gastrointestinal effects
  • effects on cardiovascular system
  • inflammation of membranes
  • fatigue

◆ Chronic Risks
  • respiratory effects
  • antimony pneumoconiosis
  • alterations in pulmonary function
  • chronic bronchitis
  • chronic emphysema
  • inactive tuberculosis
  • pleural adhesions
  • irritation
  • increased blood pressure
  • altered EKG readings
  • heart muscle damage
  • gastrointestinal effects
  • lung cancer in rats
  • dermatitis
  • keratitis
  • nasal septum ulceration
  • increase in spontaneous abortions
  • disturbances in menstrual cycle
  • decrease in offspring of rats
  • metaplasia in uterus in rats
  • disturbances in ovum-maturing process in rats
  • EPA: high concern pollutant

HAZARD RISK
◆ moderate fire and explosion hazard in forms of dust and vapor
◆ Antimony reacts violently with NH_4NO_3, halogens, BrN_3, BrF_3, $HClO_3$, ClO, ClF_3, HNO_3, KNO_3, $KMnO_4$, K_2O_2, $NaNO_3$, oxidants.
◆ Electrolysis of acid sulfides and antimony halide yields explosive antimony.
◆ when heated or on contact with acid, emits toxic fumes of SbH_3

EXPOSURE ROUTES
◆ factories that convert antimony ores into metal or make antimony oxide
◆ soil
◆ hazardous waste sites
◆ antimony-processing sites: i.e., smelters, coal fired plants, refuse incinerators
◆ food
◆ drinking water
◆ breathing dust or skin contact in industrial sites
◆ attached to small particles in air
◆ rivers, lakes, streams

REGULATORY STATUS/INFORMATION
◆ CERCLA and Superfund
  • list of chemicals. 40CFR372.65

ADDITIONAL/OTHER COMMENTS
◆ Symptoms

- inhalation: irritation of eyes and lungs, stomach pain, diarrhea, vomiting, stomach ulcer
  - ingestion: vomiting, gastrointestinal upset
- First Aid
  - wash eyes and skin immediately with large amounts of water
  - remove to fresh air immediately or provide respiratory apparatus if necessary
- fire fighting: use extinguishing media appropriate for surrounding fire
- personal protection: filter mask, protective clothing, and respiratory support if necessary
- storage: keep away from foodstuffs

SOURCES

- *Health Effects Notebook for Hazardous Air Pollutants*, U.S. Environmental Protection Agency, December 1994
- *Material Safety Data Sheets (MSDS)*
- *Sax's Dangerous Properties of Industrial Materials*, 9th edition, 1996
- *Hawley's Condensed Chemical Dictionary*, 12th edition, 1993
- *Environmental Law Index to Chemicals*, 1996 edition
- *Environmental Contaminant Reference Databook*, volume 1, 1995
- *Merck Index*, 12th edition, 1996
- Agency for Toxicological Substances and Disease Registry

ARSENIC COMPOUNDS (inorganic including Arsine)

CAS #: Arsenic compounds have variable CAS #s. The CAS # for arsenic is 7440-38-2. The CAS # for arsine is 7784-42-1.

MOLECULAR FORMULA: Arsenic compounds have variable molecular formulas. The molecular formula for arsenic is As. The molecular formula for arsine is AsH_3.

Note: For all hazardous air pollutants (HAPs) that contain the word "compounds" in the name, and for glycol ethers, the following applies: Unless otherwise specified, these HAPs are defined by the 1990 Amendments to the Clean Air Act as including any unique chemical substance that contains the named chemical (i.e., antimony, arsenic, etc.) as part of that chemical's infrastructure.

FORMULA WEIGHT: Arsenic compounds have variable formula weights. The formula weight for arsenic is 74.92. The formula weight for arsine is 77.93.

SYNONYMS: arsenic: arsen (German, Polish), arsenicals, arsenic-75, arsenic black, colloidal arsenic, grey arsenic, metallic arsenic
arsine: arsenic hydride, arsenic trihydride, arseniuretted hydrogen, arsenous hydride, arsenowodor (Polish), Arsenwasserstoff (German), hydrogen arsenide

USES

- Inorganic Arsenic
  - wood preservation
  - weed killers

- treatment of syphilis
- treatment of psoriasis
- antiparasitic agent in veterinary medicine
- homeopathic and folk remedies
- metallurgy for hardening copper, lead, alloys
- manufacture of glass
- artificial isotope as radioactive tracer in toxicology
- battery grids
- cable sheaths
- boiler tubes
- gallium arsenide for dipoles
- doping agent in germanium and silicon solid-state products
- medicine
- insecticides
- herbicides
- silvicides
- defoliants
- desiccants
- rodenticides

◆ Arsine
  - microelectronics industry
  - semiconductor manufacture

PHYSICAL PROPERTIES (Arsenic)
◆ silvery to black, crystalline amorphous metalloid
◆ brittle
◆ when heated gives off fumes of garlic
◆ melting point 814°C
◆ boiling point sublimes at 612°C
◆ vapor pressure 1 mm Hg at 372°C
◆ heat of vaporization 11.2 kcal/g-atom
◆ insoluble in water
◆ soluble in HNO_3

PHYSICAL PROPERTIES (Arsine)
◆ colorless gas
◆ mild garlic odor
◆ density 2.695 g/L
◆ melting point −116°C
◆ boiling point −62.5°C
◆ vapor pressure 217 mm Hg at 70°F
◆ vapor density 2.66
◆ soluble in water, benzene, and chloroform

CHEMICAL PROPERTIES (Arsenic)
◆ heat of sublimation 30.5 kcal/g-atom
◆ heat of fusion 22.4 kcal/g-atom
◆ darkens in moist air
◆ attacked by hydrochloric acid in presence of oxidant
◆ reacts with nitric oxide
◆ low thermal conductivity: semiconductor

CHEMICAL PROPERTIES (Arsine)
◆ stable
◆ lower explosive limit 5.8%, upper explosive limit 64%
◆ Polymerization will not occur.

HEALTH RISK
◆ Toxic Exposure Guidelines
  - OSHA PEL TWA inorganic 0.01 mg(As)/m^3
  - OSHA PEL TWA organic 0.5 mg(As)/m^3
  - ACGIH TLV 0.2 mg(As)/m^3
  - NIOSH REL Cl 2 μg(As)/m^3/15M

- ◆ Toxicity Data (Arsenic)
  - oral-rat TDLo 605 µg/kg
  - oral-rat TDLo 580 µg/kg
  - oral-man TDLo 76 mg/kg/12Y-I:CAR
  - implant-rabbit TDLo 75 mg/kg:ETA
  - oral-man TDLo 7857 mg/kg/55Y
  - oral-rat LD_{50} 763 mg/kg
  - oral-mouse LD_{50} 145 mg/kg
  - intraperitoneal-mouse LD_{50} 46,200 µg/kg
  - subcutaneous-rabbit LDLo 300 mg/kg
  - intraperitoneal-guinea pig LDLo 10 mg/kg
  - subcutaneous-guinea pig LDLo 300 mg/kg
- ◆ Toxicity Data (Arsine)
  - inhalation-human TCLo 3 ppm:RBC
  - inhalation-human LCLo 225 ppm/30M
  - inhalation-man TDLo 338 ppt
  - inhalation-rat LCLo 300 mg/m³/15M
  - inhalation-mouse LCLo 70 mg/m³/3H
  - inhalation-dog LCLo 400 mg/m³/15M
  - inhalation-monkey LCLo 70 mg/m³/15M
  - inhalation-cat LCLo 150 mg/m³/20M
  - inhalation-rabbit LCLo 500 mg/m³/15M
  - inhalation-frog LCLo 4500 mg/m³/3H
- ◆ Other Toxicity Indicators
  - EPA Cancer Risk Level (1 in a million excess lifetime risk) 2×10^{-7} mg/m³
- ◆ Acute Risks (Inorganic Arsenic)
  - gastrointestinal effects
  - nausea
  - diarrhea
  - abdominal pain
  - hemolysis
  - central and peripheral nervous system disorders
  - death
  - vomiting
  - headaches
  - weakness
  - delirium
  - cardiovascular system effects
  - hypotension
  - shock
  - liver, kidney, and blood effects
  - anemia
  - leukopenia
  - decreased production of red and white blood cells
  - abnormal heart rhythm
  - blood vessel damage
- ◆ Acute Risks (Arsine)
  - headache
  - nausea
  - vomiting
  - tightness of chest
  - pain in abdomen and loins
  - death
  - hemolytic anemia
  - hemoglobinuria
  - jaundice
  - kidney failure
  - extreme acute toxicity
- ◆ Chronic Risks (Inorganic Arsenic)

- irritation of skin and mucous membranes
- dermatitis
- conjunctivitis
- pharyngitis
- rhinitis
- gastrointestinal effects
- anemia
- peripheral neuropathy
- skin lesions
- hyperpigmentation
- gangrene of the extremities
- vascular lesions
- liver or kidney damage
- exfoliation
- herpes
- polyneuritis
- ◆ Chronic Risks (Arsine)
  - hemolysis
  - abnormal blood cell morphology
  - increased spleen weight in animals
  - anorexia
- altered hematopoiesis
- appearance of small corns or warts
- increase in spontaneous abortion rates
- lower than normal birth weights
- lung cancer
- increased risk of nonmelanoma skin cancer
- bladder and liver cancer
- tumors of mouth, esophagus, larynx, bladder, para nasal sinus
- EPA Group A: human carcinogen

- paretesia
- hematemesis
- renal failure
- pulmonary edema
- increase in spontaneous abortion rates

HAZARD RISK
- ◆ Dangerous
  - when heated to decomposition
  - when metallic arsenic contacts acids or acid fumes
  - when water solutions of arsenicals are in contact with active metals such as Fe, Zn, Al
- ◆ emits highly toxic fumes of arsenic
- ◆ Arsenic is flammable in the form of dust when exposed to heat or flame.
- ◆ Arsenic is flammable by chemical reaction with
  - bromates
  - chlorates
  - iodates
  - peroxides
  - lithium
  - NCl_3
  - HNO_3
  - $KMnO_4$
  - Rb_2C_2
  - $AgNO_4$
  - NOCl
  - IF_5
  - CrO_3
  - ClF_3
  - ClO
  - BrF_3
  - BrF_5
  - BrN_3
  - RbC_3BCH
  - CsC_3BCH
- ◆ Arsenic reacts vigorously with oxidizing materials.

EXPOSURE ROUTES (Inorganic Arsenic)
- ◆ released into air by volcanoes
- ◆ weathering of arsenic-containing minerals and ores

- ◆ commercial or industrial processes
- ◆ Food is largest source.
- ◆ drinking water
- ◆ soil
- ◆ natural mineral deposits
- ◆ contamination from human activities
- ◆ dermal or ingestion exposure
- ◆ workers in metal smelters and nearby residents
- ◆ burning plywood
- ◆ fish and shellfish

REGULATORY STATUS/INFORMATION
- ◆ CERCLA and Superfund
  - designation of hazardous substances. 40 CFR 302.4
- ◆ chemical regulated by the State of California

ADDITIONAL/OTHER COMMENTS
- ◆ Symptoms
  - inhalation: sore throat and irritated lungs
  - skin contact: redness and swelling
  - ingestion: gastrointestinal effects, vomiting, diarrhea, shock
- ◆ First Aid
  - move victim to uncontaminated atmosphere
  - keep warm
  - provide respiration if necessary
- ◆ fire fighting: allow arsine to burn itself out
- ◆ personal protection: safety goggles, protective clothing, respiration
- ◆ storage: cool, dry, ventilated area

SOURCES
- ◆ *Health Effects Notebook for Hazardous Air Pollutants*, U.S. Environmental Protection Agency, December 1994
- ◆ *Material Safety Data Sheets for Arsine (MSDS)*
- ◆ *Sax's Dangerous Properties of Industrial Materials*, 9th edition, 1996
- ◆ *Hawley's Condensed Chemical Dictionary*, 12th edition, 1993
- ◆ *Environmental Law Index to Chemicals*, 1996
- ◆ *Merck Index*, 12th edition, 1996
- ◆ Agency for Toxicological Substances and Disease Registry

ASBESTOS

CAS #: 1332-21-4
MOLECULAR FORMULA: most common form is chrysotile [$Mg_6(Si_4O_{10})(OH)_8$]
FORMULA WEIGHT: not applicable for asbestos
SYNONYMS: 4TO4, 7NO5, 7RF10, asbestos (asbestiform minerals), asbestos, chrysotile ($3MgO$, $2SiO_2$, $2H_2O$), asbestos, fibers, asbestos, friable, asbestos, synthetic fibers, AT 7-1, calidrea-hop, calidria-R-G 244, chloro-bestos 25, fapm-410-120, ferodo C3C, K-6-20, M-3-60, M-4-5, M-5-60, mountain-cork, mountain-leather, mountain-wood, P-5-50, synthetic fibers, asbestos, white-asbestos

USES
- ◆ Used in Products
  - building materials
  - paper products

- asbestos-cement products
- friction products
- textiles
- packings
- gaskets
- pipes
- ducts
- valves, flanges
- clutch/transmission components
- electronic motor components
- molten glass handling equipment
- hoods
- vents for corrosive materials
- chemical tanks
- vessel manufacturing
- electrical switchboards
- laboratory furniture
- cooling tower components
- flat and corrugated sheets
- asbestos-reinforced plastics and rubber
- mechanical material
- floor tiles
- reinforcement material in vinyl and asphalt flooring products
- reinforcing pigment in surface coatings and pigments
- insulator
- fireproof fabrics
- brake lining
- roofing compositions
- paint filler
- chemical filter
- inert filler medium
- component of paper dryer felts
- diaphragm cells

◆ ascarite used to absorb CO_2 in combustion analysis

PHYSICAL PROPERTIES

◆ a group of six different minerals occurring naturally in the environment
◆ Most common type is white.
◆ Others are blue, gray, or brown.
◆ long, thin fibers similar to fiberglass
◆ not volatile
◆ not soluble
◆ Small fibers may occur in suspension in both air and water.

◆ serpentine asbestos: chrysotile mineral, strong, flexible fibers for spinning
◆ amphibole asbestos: brittle fibers, resistant to chemicals and heat

CHEMICAL PROPERTIES

◆ noncombustible
◆ fire-resistant fibers

HEALTH RISK

◆ Toxic Exposure Guidelines
  - ACGIH TLV 2 fibers > 5 µm/cm³ (chrysotile and other forms of asbestos)
  - ACGIH TLV 0.5 fibers > 5 µm/cm³
  - ACGIH TLV 0.1 fiber < 5 µm/cm³ (crocidolite)
  - OSHA PEL 0.2 fibers > 5 µm/cm³
  - NIOSH REL 0.1 fiber < 5 µm/cm³
◆ Toxicity Data
  - intraperitoneal-rat TDLo 50 mg/kg:ETA
  - intrapleural-rat TD 100 mg/kg:NEO
  - inhalation-human TCLo 1.2 fibers/cc/19Y-C:PUL
◆ Other Toxicity Indicators

- EPA Cancer Risk Level (1 in a million excess lifetime risk) 4 fibers/cm^3
◆ Acute Risks
  - Usually at least 4–7 years of exposure are required before fibrosis results.
  - irritation
◆ Chronic Risks
  - lung disease: asbestosis
  - diffuse fibrous scarring of the lungs
  - shortness of breath
  - difficulty breathing
  - coughing
  - pulmonary hypertension
  - immunological effects
  - lung cancer
  - mesothelioma
  - gastrointestinal cancer
  - Long- and intermediate-range fibers are more carcinogenic than short fibers.
  - cancer of the esophagus
  - cancer of the stomach
  - cancer of the intestines
  - cancer of the pleura and peritoneum
  - cancer of the bronchus and or pharynx.
  - clubbing of the fingers
  - "asbestos bodies" found in sputum and alveolar walls
  - death
  - EPA Group A: human carcinogen (cancer causing agent)

HAZARD RISK
◆ Material does not burn or burns with difficulty.
◆ common air contaminant

EXPOSURE ROUTES
◆ erosion of natural deposits in asbestos-bearing rocks
◆ asbestos-related industries
◆ clutches and brakes on cars and trucks
◆ outdoor air
◆ insulation
◆ ceiling and floor tiles
◆ corrosion from asbestos-cement pipes
◆ disintegration of asbestos roofing materials
◆ water

REGULATORY STATUS/INFORMATION
◆ Clean Air Act Amendment, Title III: hazardous air pollutants
◆ Safe Drinking Water Act
  - 83 contaminants required to be regulated under SDWA of 1986. 47FR9352
  - inorganic chemical sampling and analytical requirements. 40CFR141.23
  - public notification. 40CFR141.32
  - maximum contaminant level goal for inorganic constituents. 40CFR141.51
  - maximum contaminant level for inorganic constituents. 40CFR141.62
◆ CERCLA and Superfund
  - designation of hazardous substances. 40CFR302.4
  - list of chemicals. 40CFR372.65
◆ Clean Water Act
  - toxic pollutants. 40CFR401.15
  - 126 priority pollutants. 40CFR423

- ◆ Toxic Substance Control Act
  - • 40CFR716.120.c6 related to asbestos and varieties of asbestos
- ◆ Occupational Safety and Health Act
  - • 29CFR1910.1000 Table Z1 and specially regulated substances
- ◆ chemical regulated by the State of California

ADDITIONAL/OTHER COMMENTS
- ◆ Symptoms
  - • inhalation: irritation
- ◆ fire fighting: use agent suitable for surrounding fire (material itself does not burn or burns with difficulty)
- ◆ personal protection: respiratory apparatus

SOURCES
- ◆ *Health Effects Notebook for Hazardous Air Pollutants*, U.S. Environmental Protection Agency, December 1994
- ◆ *Material Safety Data Sheets (MSDS)*
- ◆ *Sax's Dangerous Properties of Industrial Materials*, 9th edition, 1996
- ◆ *Hawley's Condensed Chemical Dictionary*, 12 edition, 1993
- ◆ *Environmental Law Index to Chemicals*, 1996
- ◆ *Merck Index*, 12th edition, 1996
- ◆ *Environmental Contaminant Reference Databook*, volume 1, 1995

BENZENE

CAS #: 71-43-2
MOLECULAR FORMULA: C_6H_6
FORMULA WEIGHT: 78.12
SYNONYMS: annulene, benzeen, benzine (obsolete), benzol, benzole, coal naphtha, cyclohexatriene, phene, phenyl-hydride, polystream, pyrobenzol, pyrobenzole, many others

USES
- ◆ Manufacture of Other Chemicals
  - • dyes
  - • organic compounds
  - • linoleum
  - • varnishes
  - • lacquers
  - • resins
  - • medicines
  - • cumene
  - • cyclohexane
  - • nitrobenzene
  - • chlorobenzenes
  - • sulfonic acid
- ◆ Used as a Component or in the Manufacturing Process
  - • printing
  - • paint
  - • dry cleaning
  - • coatings
  - • adhesives
  - • rubber
  - • detergents
  - • inks
  - • paint thinners
  - • degreasing agent
  - • tires
  - • shoes

PHYSICAL PROPERTIES

- clear, colorless liquid
- melting point 5.5°C (41.9°F)
- boiling point 80°C (176°F)
- density 0.8787 g/mL at 15°C
- specific gravity 0.88 (water = 1)
- vapor density 2.77 (air = 1)
- surface tension 28.9 dynes/cm
- viscosity 0.6468 mP at 20°C
- heat capacity 32.4 cal/g mol (liquid at 25°C)
- heat of vaporization 94.1 cal/g or 3.94×10^5 J/kg
- aromatic odor
- odor threshold 1.5 ppm (mg/m^3)
- vapor pressure: 1 mm at -36.7°C, 40 mm at 7.6°C, 400 mm at 60.6°C
- negligible solubility in water
- miscible with alcohol, ether, acetone, carbon tetrachloride

CHEMICAL PROPERTIES

- generally very stable
- will not polymerize
- will react vigorously with strong oxidizing agents, sulfuric acid, nitric acid, chlorine, oxygen, ozone, permanganates, peroxides, perchlorates
- flash point -11°C (12°F)
- flammability limits 1.3% lower and 7.1% upper
- autoignition temperature 562°C
- heat of combustion -9.698 cal/g or -406.0×10^5 J/kg
- enthalpy (heat) of formation at 25°C (liquid) 49.1 kJ/mol

HEALTH RISK

- Toxic Exposure Guidelines
  - ACGIH TLV TWA 10 ppm (32 mg/m^3)
  - OSHA PEL TWA 1 ppm (3.26 mg/m^3)
  - NIOSH REL TWA 0.32 mg/m^3
  - IDLH 500 ppm
- Toxicity Data
  - oral-human TDLo 130 mg/kg. Toxic effect: central nervous system
  - inhalation-human LCLo 20,000 ppm/5 months
  - inhalation-human TCLo 210 ppm. Toxic effect: blood
  - inhalation-human TCLo 100 ppm/10Y-intermittent. Toxic effect: carcinogenic
  - oral-rat LD_{50} 3800 mg/kg
  - inhalation-rat LC_{50} 31,951 mg/m^3
  - skin-mouse TDLo 1200 g/kg/49W-intermittent. Toxic effect: neoplastic
  - intraperitoneal-mouse LD_{50} 468 mg/kg
- Other Toxicity Indicators
  - EPA Cancer Risk Level (1 in a million excess lifetime risk) 1.0×10^{-4} mg/m^3
- Acute Risks
  - irritation of skin, eyes, and upper respiratory tract
  - drowsiness
  - dizziness
  - headaches
  - unconsciousness
  - death in high exposures
- Chronic Risks

- blood disorders
- bone marrow disease
- aplastic anemia
- excessive bleeding
- damage to immune system (changes in blood levels of antibodies and loss of white blood cells)
- chromosomal aberrations
- menstrual disorders
- leukemia (cancer of tissues that form white blood cells)
- EPA Group A: known human carcinogen

HAZARD RISK

- ◆ NFPA rating Health 2, Flammability 3, Reactivity 0, Special Hazards none
- ◆ dangerous fire hazard
- ◆ explodes on contact with diborane, BrF_5, permanganic acid
- ◆ forms sensitive explosive mixtures with IF_4, $AgClO_4$, nitric acid, liquid O_2
- ◆ ignites on contact with sodium peroxide and water
- ◆ incompatible or reacts strongly with strong oxidizers, many fluorides and perchlorates

EXPOSURE ROUTES

- ◆ primarily by inhalation
- ◆ Absorption through skin and eyes is possible.
- ◆ ingestion
- ◆ Cigarette smoke is a major route of exposure.
- ◆ emissions from burning coal and oil
- ◆ motor vehicle exhaust
- ◆ evaporation of gasoline at service stations
- ◆ occupational exposure in industries that use benzene in the manufacturing process
- ◆ drinking of contaminated water

REGULATORY STATUS/INFORMATION

- ◆ Clean Air Act Amendments of 1990 Section 112 Hazardous Air Pollutant
- ◆ Safe Drinking Water Act
  - one of 83 contaminants required to be regulated under the 1986 act
  - maximum contaminant level goal (MCLG) for organic contaminants. 40CFR141.50(a)
  - variances and exemptions from maximum contaminant levels (MCLs). 40CFR141.62
  - waste specific prohibitions. 40CFR148.10
- ◆ Resource Conservation and Recovery Act
  - constituents for detection monitoring for municipal solid waste landfills. 40CFR258
  - list of hazardous constituents. 40CFR258, 40CFR261, and 40CFR261.11
  - groundwater monitoring list. 40CFR264
  - risk specific doses. 40CFR266
  - health based limits for exclusion of waste-derived residues. 40CFR266
  - D waste #: D018

- reportable quantity (RQ): 10 lbs (4.54 kg)
- U waste #: U019
◆ CERCLA and Superfund
  - designation of hazardous substances. 40CFR302.4
  - list of chemicals. 40CFR372.65
◆ Clean Water Act
  - designation of hazardous substances. 40CFR116.4
  - reportable quantity list. 40CFR117.3
  - toxic pollutants list. 40CFR401.15
  - total toxic organics (TTOs). 40CFR413.02
  - 126 priority pollutants. 40CFR423
◆ OSHA 29CFR1910.1000 Table Z1, Table Z2, and specially regulated substances
◆ Department of Transportation
  - hazard class/division: 3
  - identification number: UN1114
  - labels: flammable liquid
  - emergency response guide #:130
◆ Community Right-to-Know Threshold Planning Quantity: not listed

ADDITIONAL/OTHER COMMENTS
◆ Symptoms
  - inhalation: irritates eyes, skin, and nose
  - skin absorption: respiratory system, giddiness, headache, nausea
◆ First Aid
  - wash eyes immediately with large amounts of water
  - wash skin immediately with soap and water
  - provide respiratory support
◆ fire fighting: use dry chemical, foam, or carbon dioxide
◆ personal protection: in high vapor concentrations, wear self-contained breathing apparatus
◆ storage: outside preferred; inside should be in a standard flammable liquids storage room or cabinet

SOURCES
◆ *ACGIH Threshold Limit Values (TLVs) for 1996*
◆ *CRC Handbook of Chemistry and Physics*, 77th edition, 1996–1997
◆ *Environmental Contaminant Reference Databook*, volume 1, 1995
◆ *Environmental Law Index to Chemicals*, 1996 edition
◆ *Fire Protection Guide to Hazardous Materials*, 11th edition, 1994
◆ *Health Effects Notebook for Hazardous Air Pollutants*, U.S. Environmental Protection Agency, December 1994
◆ *Material Safety Data Sheets (MSDSs)*
◆ *NIOSH Pocket Guide to Chemical Hazards*, June 1994
◆ *Sax's Dangerous Properties of Industrial Materials*, 9th edition, 1996
◆ *Suspect Chemicals Sourcebook*, 1996 edition

BENZIDINE (rtc)

CAS #: 92-87-5
MOLECULAR FORMULA: $C_{12}H_{12}N_2$
FORMULA WEIGHT: 184.26 g/mol
SYNONYMS: p-benzine, 4,4'-bianiline, biphenyl-4,4'-diamine, (1,1'-biphenyl)-4,4'-diamine, 4,4'-biphenyl diamine, biphenyl,4,4'-diamino, CI-azoic-diazo-component-112, 4,4'-diaminobiphenyl, 4,4'-diamino-1-1'-biphenyl, 4,4'-diaminodiphenyl, 4,4'-diphenylinediamine, fast corinth base B, and others

USES
◆ Manufacture of Other Chemicals
  • AZO dyes
  • intermediate for synthesis of various colors
◆ Used as a Component or in the Manufacturing Process
  • plastic films
  • determination of nicotine (quantitative)
  • hardener in rubber compounds
  • reagent
  • spray reagent for sugars
  • used for detection of hydrogen cyanide and sulfate
  • detection of blood stains
  • lignification measurement

PHYSICAL PROPERTIES
◆ grayish, yellow, white, or slightly reddish crystal powder
◆ melting point 117°C
◆ boiling point 400°C
◆ vapor density 6.36
◆ specific gravity 1.25 at 68°F
◆ density 1.250 g/cm³ at 20°C
◆ insoluble in water
◆ vapor pressure 5×10^{-4} mm Hg at 25°C

CHEMICAL PROPERTIES
◆ Chemical will darken when exposed to air or light.
◆ octanal/water partition coefficient log K_{ow} 1.34
◆ may be sublimed
◆ found as a free base in natural waters
◆ forms insoluble salts with sulfuric acid
◆ can be diazotized and oxidized
◆ Amino groups can be acetylated and alkylated.

HEALTH RISK
◆ Toxic Exposure Guidelines
  • ACGIH TWA 0 ppm, confirmed human carcinogen
  • NIOSH TWA 0 ppm
  • OSHA TWA 0 ppm
◆ Toxicity Data
  • inhalation-man TCLo 17,600 µg/m³/14Y
  • inhalation-rat TCLo 10 mg/m³/56W
  • oral-rat TDLo 108 mg/kg/27d
  • subcutaneous-rat TDLo 2025 mg/kg/27W
  • intratracheal-rat TDLo 315 mg/kg/34W
  • oral-rat LD_{50} 309 mg/kg
◆ Other Toxicity Indicators

- EPA Cancer Risk Level (1 in a million excess lifetime risk) 2.0×10^{-8} mg/m^3
- DOT ID number 1885
- RTECS number DC9625000
- EPA hazardous waste number U021

◆ Acute Risks
  - cyanosis
  - mental confusion
  - vertigo
  - skin rashes
  - fatigue
  - possible heart attack
  - headaches
  - nausea
  - skin irritation
  - central nervous system depression

◆ Chronic Risks
  - bladder damage
  - possibly cancerous
  - affects on blood, liver, kidney, and central nervous system

HAZARD RISK
◆ flammable
◆ emits highly toxic fumes under fire conditions
◆ violent reaction with strong oxidants
◆ forms explosive mixtures in the air from small, dispersed particles
◆ can ignite by electric sparks
◆ sinks in water
◆ decomposes on heating
◆ explosion risk above flash point temperature

EXPOSURE ROUTES
◆ inhalation of fine powders
◆ direct contact with skin or eyes
◆ ingestion
◆ absorption through the skin
◆ occupational exposure in the dye and chemical manufacturing industries
◆ drinking of contaminated water
◆ inhalation of dust or mist of azo dyes

REGULATORY STATUS/INFORMATION
◆ Clean Air Act
  - Title 3: hazardous air pollutants 42USC7412
◆ Resource Conservation Recovery Act
  - discarded commercial chemical products, off-specification species, container residues, and spill residues thereof. 40CFR261.33
  - Appendix 8: hazardous constituents. 40CFR261
  - Appendix 5: risk specific doses 40 CFR266
  - Appendix 7: health based limits for exclusion of waste-derived residues 40CFR266
◆ CERCLA and Superfund
  - designation of hazardous substances. 40CFR302.4
  - Appendix A: sequential CAS registry number list of CERCLA hazardous substances

- Appendix B: radionuclides
- chemicals and chemical categories to which this part applies. 40 CFR372.65
◆ Clean Water Act
  - toxic pollutants. 40CFR401.15
  - total toxic organics. 40CFR413.02
  - Appendix A: 126 priority pollutants
◆ Occupational Safety and Health Act
  - Table Z1: limits for air contaminants under the OSHA 29CFR1910.1000
  - OSHA specially regulated substances. 29CFR1910.1001-1048
◆ chemicals regulated by the State of California

ADDITIONAL/OTHER COMMENTS
◆ Symptoms
  - caustic effects
  - coughing
  - redness or tingling sensation of the skin
  - bad sight (eye contact)
  - sore throat
  - labored breath
  - nausea
  - abdominal pain (if ingested)
◆ First Aid
  - flush eyes or skin with large amounts of water; use soap on skin
  - remove contaminated clothes
  - remove to fresh air
  - give oxygen
  - if ingested, wash out mouth with water providing person is conscious and have person drink water
◆ fire fighting: extinguish with dry chemical powder, carbon dioxide, appropriate foam, or a water spray
◆ personal protection: use a gastight, fireproof suit and self-contained breathing equipment
◆ storage: keep away from light and strong oxidants

SOURCES
◆ *ACGIH Threshold Limit Values (TLVs) for 1996*
◆ *Environmental Contaminant Reference Databook*, volume 1, 1995
◆ *Environmental Law Index to Chemicals*, 1996 edition
◆ *Health Effects Notebook for Hazardous Air Pollutants*, U.S. Environmental Protection Agency, December 1994
◆ *MSDS*, NIOSH
◆ *1997 MSDS*, University of Kentucky, 1996–1997
◆ *Sax's Dangerous Properties of Industrial Materials*, 9th edition, 1996

BENZOTRICHLORIDE (rtc)

CAS #: 98-07-7
MOLECULAR FORMULA: $C_7H_5Cl_3$
FORMULA WEIGHT: 195.47 g/mol
SYNONYMS: alpha-alpha-alpha-trichlorotoluene, benzene (trichloro-methyl), benzenyl chloride, benzenyl trichloride, benzoic trichloride, benzyl trichloride, benzylidyne chloride, omega-omega-omega-trichlo-rotoluene, phenylchloroform, phenyltrichloromethane, many others

USES
- ◆ Manufacture of Other Chemicals
  - dyes
  - benzoyl chloride
  - chlorinated toluenes
  - benzotrifluoride

PHYSICAL PROPERTIES
- ◆ clear, colorless liquid
- ◆ boiling point 219°C
- ◆ specific gravity 1.380
- ◆ vapor density 6.77
- ◆ viscosity (dynamic) 2.4 mPa at 20°C
- ◆ penetrating odor
- ◆ melting point −7°C
- ◆ density 1.38 g/cm³ at 20°C
- ◆ surface tension 38.03 mN/m at 20°C
- ◆ vapor pressure 0.4137 mm Hg at 25°C
- ◆ index of refraction 1.558 at 20°C
- ◆ heat of vaporization 12,169 kcal/kmol

CHEMICAL PROPERTIES
- ◆ hydrolyzes in the presence of moisture
- ◆ autoignition temperature 210°C
- ◆ reacts with strong oxidizing agents
- ◆ flash point 97°C
- ◆ heat of combustion 3684 kJ/mole at constant pressure.
- ◆ water partition coefficient log K_{ow} 2.92

HEALTH RISK
- ◆ Toxic Exposure Guidelines
  - data not available
- ◆ Toxicity Data
  - oral-rat LD_{50} 6 g/kg
  - inhalation-rat LC_{50} 19 ppm/2H
  - oral-mouse LD_{50} 702 mg/kg
  - inhalation-mouse LC_{50} 8 ppm/2H
  - oral-mammal LD_{50} 1300 mg/kg
  - inhalation-mammal LC_{50} 60 mg/m³
- ◆ Other Toxicity Indicators
  - EPA hazardous waste number U023
  - RTECS number XT9275000
- ◆ Acute Risks
  - irritation of eyes, mucous membranes, and respiratory tract
  - pulmonary edema
  - headaches
  - Ingestion can burn mouth, throat, esophagus, and gastrointestinal tract.
  - severe eye injury or lacrimation
  - abdominal cramping

- nausea
- diarrhea
◆ Chronic Risks
  - weakness
  - irritability
  - mild leukopenia
  - anorexia
  - insomnia
  - tremors in the digits
  - liver function disturbances

HAZARD RISK
◆ flammable
◆ explosion risk above flash point
◆ can ignite by electric sparks
◆ reacts with water
◆ decomposes on heating
◆ emits toxic fumes under fire conditions

EXPOSURE ROUTES
◆ primarily by inhalation
◆ ingestion
◆ absorption through the skin

REGULATORY STATUS/INFORMATION
◆ Clean Air Act
  - Title 3: hazardous air pollutants
◆ Resource Conservation Recovery Act
  - discarded commercial chemical products, off-specification species, container residues, and spill residues thereof. 261.33.e. (P waste)
  - Appendix 8: hazardous constituents. 40CFR261
◆ CERCLA and Superfund
  - designation of hazardous substances. 40CFR302.4
  - Appendix A: sequential CAS registry number list of CERCLA hazardous substances
  - Appendix B: radionuclides
  - list of extremely hazardous substances and their threshold planning quantities. 40CFR355-AB

ADDITIONAL/OTHER COMMENTS
◆ Symptoms
  - labored breath
  - coughing
  - sore throat
  - abdominal pains
  - vomiting
  - redness of the skin
  - etching wounds
◆ First Aid
  - flush eyes or skin with large amounts of water
  - remove contaminated clothes
  - remove to fresh air
  - if not breathing, give artificial respiration; if breathing is difficult, give oxygen
  - if ingested, wash out mouth with water provided person is conscious
◆ fire fighting: extinguish with dry chemical powder, carbon dioxide, or appropriate foam
◆ Personal Protection

- wear approved respirator, chemical-resistant gloves, safety goggles, and other protective clothing
- use only in a chemical fume hood
◆ storage: keep containers tightly closed and store in a cool, dry area

SOURCES
◆ *Environmental Contaminant Reference Databook*, volume 1, 1995
◆ *Environmental Law Index to Chemicals*, 1996 edition
◆ *Health Effects Notebook for Hazardous Air Pollutants*, U.S. Environmental Protection Agency, December 1994
◆ *MSDS*, Aldrich Chemical Company, Inc., 1996
◆ *MSDS*, NIOSH
◆ *1997 MSDS*, University of Kentucky, 1996–1997
◆ *Sax's Dangerous Properties of Industrial Materials*, 9th edition, 1996

BENZYL CHLORIDE (rtc)

CAS #: 100-44-7
MOLECULAR FORMULA: C_7H_7Cl
FORMULA WEIGHT: 126.59 g/mol
SYNONYMS: alpha chlorotoluene, alpha-tolyl chloride, benzene (chloromethyl), chloromethylbenzene, chlorophenylmethane, omega-chlorotoluene, toluene (alpha-chloro), tolylchloride, and others

USES
◆ Manufacture of Other Chemicals
- benzyl compounds
- synthetic tannins
- dyes
- perfumes
- artificial resins
- pharmaceutical products
- benzyl alcohol, benzyl esters, and benzyl amines
- fungicides, pesticides, and bactericides
- irritant gas in chemical warfare
◆ Component or Used in the Manufacturing Process
- photographic developer
- penicillin precursors
- rubber accelerators
- gasoline gum inhibitor
- pickling inhibitors
- lubricants
- plastics
- odorants
- plasticizers

PHYSICAL PROPERTIES
◆ colorless to slightly yellow liquid
◆ boiling point 179°C
◆ specific gravity 1.100
◆ surface tension 0.0375 N/m at 20°C
◆ vapor density 4.36
◆ specific heat 135.4 kPa
◆ odor threshold low 0.235 mg/m³, high 1.55 mg/m³
◆ solubility 10% in ethanol, 10% in ethyl ether, 10% in chloroform, 493 ppm in water at 20°C
◆ melting point −47°C
◆ density 1.100 at 20°C
◆ unpleasant, irritating odor

- vapor pressure 1 mm Hg at 22°C, 10 mm Hg at 60.8°C, 100 mm Hg at 114.2°C, 400 mm Hg at 155.8°C
- index of refraction 1.558 at 20°C

- heat of vaporization 70 cal/g
- critical temperature 411°C
- critical pressure 567 psia
- index of refraction 1.5415 at 15°C
- corrosive to metal

CHEMICAL PROPERTIES

- explosion limits upper 14%, lower 1.1%
- flammability limits lower 1.1%
- flashpoint 74°C
- autoignition temperature 585°C

- fairly stable, stable during transport
- water partition coefficient log K_{ow} 2
- heat of combustion 6700 cal/g
- reacts with steam and oxidizing agents

HEALTH RISK

- Toxic Exposure Guidelines
  - ACGIH TLV-TWA 1 ppm
  - MSHA standard-air TWA 1 ppm
  - DFG-MAK confirmed carcinogen
  - NIOSH REL-air TWA 1 ppm/15 min
  - OSHA TWA 1 ppm
- Toxicity Data
  - oral-rat LD_{50} 1231 mg/kg
  - inhalation-rat LD_{50} 150 ppm/2H
  - subcutaneous-rat LD_{50} 1 g/kg
  - oral-mouse LD_{50} 150 mg/kg
  - inhalation-mouse LC_{50} 80 ppm/2H
  - adsorption-skin-mouse TDLo 9200 mg/kg
  - intraperitoneal-mouse TDLo 250 mg/kg
  - oral-mammal LD_{50} 1500 mg/kg
  - inhalation-mammal LC_{50} 390 mg/m^3
- Other Toxicity Indicators
  - DOT number 1738
  - RTECS number *S8925000
- Acute Risks
  - irritation of eyes, mucous membranes, skin (contact), and upper respiratory tract (inhaled)
  - cause severe skin burns
  - depression of the central nervous system
  - slight conjunctivitis
  - weeping and twitching of eyelids
- Chronic Risks
  - disturbances of liver functions
  - may cause cancer
  - may cause genetic damage
  - decreased number of leukocytes in blood
  - pulmonary edema
  - susceptible to illnesses similar to colds and allergic rhinitis

HAZARD RISK

- moderate fire hazard

- reacts with water or steam to produce toxic and corrosive hydrochloric acid fumes
- reacts violently with oxidizing materials
- releases heat and toxic hydrochloric acid vapors when it comes into contact with many metals

EXPOSURE ROUTES
- inhalation
- ingestion
- absorption through the skin
- eyes exposed to the vapor
- air contaminated from floor tile manufacturers using butyl benzyl phthalate
- occupations producing or using dyes

REGULATORY STATUS/INFORMATION
- Clean Air Act
  - Title 3: hazardous air pollutants. 42USC7412
- Resource Conservation Recovery Act
  - discarded commercial chemical products, off-specification species, container residues, and spill residues thereof. 40CFR261.33
- Clean Water Act
  - designation of hazardous substances. 40CFR116.4
  - Table 116.4a: list of hazardous substances
  - Table 116.4b: list of hazardous substances by CAS #
  - determination of reportable quantities. 40CFR117.3
  - Table 117.3: reportable quantities of hazardous substances designated pursuant to sec. 311 of CWA
- Toxic Substance Control Act
  - OSHA chemicals in need of dermal absorption testing. 40CFR712.30.e10
  - OSHA chemicals in need of dermal absorption testing. 40CFR716.120.d10
- Occupational Safety and Health Act
  - limits for air contaminants under the OSHA act. 29CFR1910.1000
- chemical regulated by the State of California

ADDITIONAL/OTHER COMMENTS
- Symptoms
  - headaches
  - sleeplessness
  - diarrhea
  - blurred vision (eye contact)
  - nausea
  - irritation of eyes, skin, and nose
  - irritability
  - redness (contact)
  - weakness
- First Aid
  - flush eyes or skin with large amounts of water
  - remove contaminated clothes
  - remove to fresh air

- give oxygen if breathing is difficult or person is not breathing
- if ingested, wash out mouth with water provided person is conscious
◆ fire fighting: do not use water; extinguish with dry chemical powder, carbon dioxide, or appropriate foam
◆ personal protection: wear approved respirator, chemical-resistant gloves, safety goggles, and other protective clothing
◆ storage: keep containers tightly closed and store in a cool, dry area; keep away from heat and open flame

SOURCES
◆ *ACGIH Threshold Limit Values (TLVs) for 1996*
◆ *Environmental Contaminant Reference Databook*, volume 1, 1995
◆ *Environmental Law Index to Chemicals*, 1996 edition
◆ *Health Effects Notebook for Hazardous Air Pollutants*, U.S. Environmental Protection Agency, December 1994
◆ *MSDS*, Aldrich Chemical Company, Inc., 1996
◆ *MSDS*, NIOSH
◆ *1997 MSDS*, University of Kentucky, 1996–1997
◆ *Sax's Dangerous Properties of Industrial Materials*, 9th edition, 1996

BERYLLIUM COMPOUNDS

CAS #: Beryllium compounds have variable CAS #s. The CAS # for beryllium is 7440-41-7.

MOLECULAR FORMULA: Beryllium compounds have variable molecular formulas. The molecular formula for beryllium is Be.

Note: For all hazardous air pollutants (HAPs) that contain the word "compounds" in the name, and for glycol ethers, the following applies: Unless otherwise specified, these HAPs are defined by the 1990 Amendments to the Clean Air Act as including any unique chemical substance that contains the named chemical (i.e., antimony, arsenic, etc.) as part of that chemical's infrastructure.

FORMULA WEIGHT: Beryllium compounds have variable formula weights. The formula weight for beryllium is 9.0121.

SYNONYMS: beryllium-9, beryllium, metal powder (DOT), glucinum

USES
◆ Beryllium and Its Metal Alloys Used as Components
  - electrical components
  - tools
  - aircraft
  - missiles
  - satellites
  - metal-fabricating uses
  - molds for plastics
◆ Beryllium Used in Consumer Products Including
  - televisions
  - calculators
  - personal computers
  - X-ray machines
  - mirrors
◆ Beryllium oxide is used to make specialty ceramics for electrical and high technology applications.

PHYSICAL PROPERTIES
◆ gray metal
◆ melting point 1278°C

- boiling point 2970°C
- density 1.8477
- high permeability to X-rays
- hard, brittle
- soluble in acids (except nitric) and alkalies
- insoluble in water
- Some compounds of beryllium are soluble in water.
- vapor pressure 1 mm Hg at 1520°C
- Brinell hardness 60–125

CHEMICAL PROPERTIES
- heat capacity 0.437 cal/g/°C at 30°C
- latent heat of fusion 3.5 kcal/mole
- similar chemical properties to aluminum
- metal resistant to attack by acid due to the formation of a thin oxide film
- finely divided or amalgamated metal reacts with HCl, dil H_2SO_4, and dil HNO_3
- attacked by strong bases with evolution of H_2
- resistant to oxidation at ordinary temperatures
- high thermal conductivity

HEALTH RISK
- Toxic Exposure Guidelines
  - OSHA PEL (transitional: TWA 0.002 mg(Be)/m³; CL 0.005; Pk 0.025/30M/8H) TWA 0.002 mg(Be)/m³
  - OSHA STEL 0.005 mg(Be)/m³/30M; CL 0.025 mg(Be)/m³
  - ACGIH TLV TWA 0.002 mg/m³, suspected human carcinogen
  - DFG MAK grinding of beryllium metal and alloys 0.005 mg/m³
  - DFG MAK other beryllium compounds 0.002 mg/m³
  - NIOSH REL 0.005 mg/m³
- Toxicity Data (Beryllium Metal)
  - dnd-esc 30 µmol/L
  - dni-nml:intravenous 30 µmol/L
  - dnd-human:hla 30 µmol/L
  - dnd-mouse:ast 30 µmol/L
  - intratracheal-rat TDLo 13 mg/kg:NEO
  - intravenous-rabbit TDLo 20 mg/kg:ETA
  - inhalation-human TCLo 300 mg/m³:PUL
  - intravenous-rat LD_{50} 496 µg/kg
- Other Toxicity Indicators
  - EPA Cancer Risk Level (1 in a million excess lifetime risk) 4×10^{-7} mg/m³
- Acute Risks
  - inflammation of the lungs
  - acute pneumonitis
  - reddening and swelling of the lungs: acute beryllium disease
  - Beryllium compounds vary in acute toxicity, ranging from high to extreme acute toxicity.
  - dermatitis
  - coughing
  - shortness of breath
  - loss of appetite
  - loss of weight
  - fatigue

◆ Chronic Risks
  • chronic beryllium disease (berylliosis)
  • pulmonary granulomatous disease
  • irritation of mucous membranes
  • reduced lung capacity
  • shortness of breath
  • fatigue
  • anorexia
  • dyspnea
  • malaise
  • weight loss
  • death
  • chronic pneumonitis
  • effects on adrenal gland
  • skin allergy
  • acute conjunctivitis
  • lung cancer
  • dermatitis
  • chronic skin ulcers
  • rhinitis
  • nasopharyngitis
  • epistaxis
  • bronchitis
  • scanty sputum
  • low-grade fever
  • rales
  • substernal pain
  • diffuse haziness throughout both lungs
  • appearance of soft, ill-defined opacities
  • lung fibrosis
  • pulmonary edema
  • hemorrhagic extravasation
  • large numbers of plasma cells
  • increase in pulse and respiratory rates
  • EPA Group B2: probable human carcinogen

HAZARD RISK
  ◆ moderate fire hazard in form of dust or powder
  ◆ moderate fire hazard when exposed to flame or by spontaneous chemical reaction
  ◆ slight explosion hazard in the form of powder or dust
  ◆ incompatible with halocarbons
  ◆ reacts incandescently with fluorine or chlorine
  ◆ Mixtures of the powder with CCl_4 or trichloroethylene will flash or spark on impact.
  ◆ Decomposition emits very toxic fumes of BeO.

EXPOSURE ROUTES
  ◆ where it is mined
  ◆ where it is processed
  ◆ where it is converted into alloys and chemicals
  ◆ inhalation of beryllium dust
  ◆ inhalation of fumes from the burning of coal or fuel oil
  ◆ tobacco smoke
  ◆ ingestion of fruits and vegetables and water
  ◆ natural occurrence in soils
  ◆ fly ash through chimney stacks
  ◆ electronic devices

REGULATORY STATUS/INFORMATION
  ◆ CERCLA and Superfund
    • list of chemicals. 40CFR372.65

ADDITIONAL/OTHER COMMENTS
  ◆ Symptoms
    • can start during exposure or be delayed up to 5 years after leaving the workplace
    • coughing, shortness of breath, loss of appetite, loss of weight, fatigue

♦ First Aid
  • wash eyes immediately with large amounts of water
  • flush skin with plenty of soap and water
  • remove to fresh air immediately
♦ fire fighting: use sand or other methods for metal powder fires
♦ personal protection: use safety glasses and gloves and avoid contact with skin
♦ storage: package in impervious containers

SOURCES
♦ *Health Effects Notebook for Hazardous Air Pollutants*, U.S. Environmental Protection Agency, December 1994
♦ *Material Safety Data Sheets (MSDS)*
♦ *Sax's Dangerous Properties of Industrial Materials*, 9th edition, 1996
♦ *Hawley's Condensed Chemical Dictionary*, 12th edition, 1993
♦ *Environmental Law Index to Chemicals*, 1996
♦ *Merck Index*, 12th edition, 1996
♦ *Toxicological Profile for Beryllium*
♦ Agency for Toxicological Substances and Disease Registry

BIPHENYL (jl)

CAS #: 92-52-4
MOLECULAR FORMULA: $C_{12}H_{10}$
FORMULA WEIGHT: 154.22
SYNONYMS: bibenzene, 1,1-biphenyl, diphenyl (OSHA), lemonese, phenador-x, phenylbenzene, PHPH, xenene

USES
♦ intermediate for polychlorinated biphenyls
♦ organic synthesis
♦ heat transfer agent
♦ fungistat in packaging of citrus fruit
♦ plant disease control
♦ manufacturing of benzidine
♦ dyeing assistant for polyesters

PHYSICAL PROPERTIES
♦ white scales; leaflets from dilute alcohol; colorless leaflets
♦ melting point 69–71°C
♦ boiling point 254–255°C at 1 atm
♦ density 991.9 kg/m³
♦ specific gravity 0.992 (water = 1)
♦ vapor density 5.31 (air = 1)
♦ surface tension not available
♦ viscosity not available
♦ heat capacity not available
♦ heat of vaporization not available
♦ aromatic, pleasant odor
♦ vapor pressure 9.46 mm Hg at 115°C

- negligible solubility in water
- soluble in alcohol, ether, and most other organic solvents; very soluble in benzene and methanol

CHEMICAL PROPERTIES
- flash point 113°C
- flammability limits 0.6% lower and 5.8% upper
- autoignition temperature 1004°C
- heat of combustion not available
- enthalpy (heat) of formation not available
- will not polymerize
- incompatible with strong oxidizing agents

HEALTH RISK
- Toxic Exposure Guidelines
  - ACGIH TLV TWA 0.2 ppm
  - OSHA PEL TWA 0.2 ppm
  - DFG MAK 0.2 ppm
- Toxicity Data
  - LC_{50} (oral rat) 2400 mg/kg
  - TCL_o 4400 μg/m$_3$
  - LD_{50} (rats and rabbits) 2400 mg/kg
  - LD_{50} (mice) 1900 mg/kg
  - LOAEL (250 mg/kg/d)
  - NOAEL (50 mg/kg/d)
  - RfD (0.05 mg/kg/d)
- Acute Risks
  - harmful if swallowed
  - causes eye and skin irritation
  - Material is irritating to mucous membranes and upper respiratory tract.
  - may cause nervous system disturbances
  - damage to the liver
  - Exposure can cause nausea, headache, vomiting, and gastrointestinal disturbances.
- Chronic Risks
  - Laboratory experiments have shown mutagenic effects.
  - target organs: liver, central nervous system, peripheral nervous system

HAZARD RISK
- Biphenyl is a combustible solid.
- Poisonous gases are produced in fire.
- If employees are expected to fight fires, they must be trained and equipped as stated in OSHA 1910.156.

EXPOSURE ROUTES
- Individuals may be exposed to biphenyl in the workplace.
- Paper impregnated with biphenyl is used in citrus packing to reduce fruit damage by fungus during shipment and storage. Biphenyl residue on citrus fruits has been detected, and individuals may be exposed by ingestion of contaminated fruit.
- Biphenyl has been detected in drinking water. Individuals may be exposed by the consumption of contaminated water.

◆ Biphenyl has been reported in diesel exhaust.

REGULATORY STATUS/INFORMATION

◆ TSCA Section 4(a) Final Test Rule and Consequent Agreement Substances
◆ TSCA Section 4(e) Interagency Testing Committee (ITC) Priority List
◆ TSCA Section 12(b) One-Time Export Notification Substances (Section 4)
◆ Clean Air Act Section 111 Volatile Organic Compounds (40 CFR 60)
◆ Clean Air Act Section 112 Statutory Air Pollutants (42 USC 7412)
◆ CERCLA Hazardous Substances (40 CFR 302)
◆ SARA Title III Section 313 Toxic Chemicals (40 CFR 372)
◆ SARA Section 110 Priority List of CERCLA Hazardous Substances
◆ OSHA Air Contaminants (29 CFR 1910)
◆ ACGIH Threshold Limit Value (TLV) Chemicals
◆ Massachusetts Substance List

ADDITIONAL/OTHER COMMENTS

◆ First Aid
  • Eye Contact
    • immediately flush with large amounts of water for at least 15 minutes, occasionally lifting upper and lower lids
    • seek medical attention
  • Skin Contact
    • quickly remove contaminated clothing
    • immediately wash area with large amounts of soap and water
    • seek medical attention immediately
  • Inhalation
    • remove the person from exposure
    • begin rescue breathing if breathing has stopped and CPR if heart action has stopped
    • transfer promptly to a medical facility
◆ Handling and Storage
  • Prior to working with biphenyl, workers should be trained in its proper handling and storage.
  • Biphenyl must be stored to avoid contact with oxidizers, such as perchlorates, peroxides, chlorates, nitrates, and permanganates, since violent reactions occur.
◆ Spills or Leaks
  • if biphenyl is spilled, take the following steps.
  • restrict persons not wearing protective equipment from area of spill until cleanup is complete
  • remove all ignition sources
  • ventilate area of spill
  • collect powdered material in the most convenient and safe manner and deposit in sealed containers

66

- It may be necessary to contain and dispose of biphenyl as a hazardous waste. Contact your state environmental program for specific recommendations.

SOURCES
- *Aldrich Chemical Company Catalog*, 1996–1997
- *Suspect Chemicals Sourcebook*, 1996 edition
- *Hawley's Condensed Chemical Dictionary*, 9th edition, 1993
- *Sax's Dangerous Properties of Industrial Materials*, 9th edition 1996
- *Merck Index*, 12th edition 1996

BIS(2-ETHYLHEXYL)PHTHALATE

CAS #: 117-81-7
MOLECULAR FORMULA: C_6H_4-1,2-$[CO_2CH_2CH(C_2H_5)(CH_2)_3CH_3]_2$
FORMULA WEIGHT: 390.62
SYNONYMS: BEHP, bis(2-ethylhexyl)-1,2-benzenedicarboxylate, bisoflex 81, bisoflex DOP, compound 889, DAF 68, DEHP, di(2-ethylhexyl)orthophthalate, di(2-ethylhexyl)phthalate, dioctyl phthalate, di-sec-octyl phthalate, DOP, ergoplast FDO, ethylhexyl phthalate, 2-ethylhexyl phthalate, eviplast 80, eviplast 81, fleximel, flexol DOP, flexol plasticizer DOP, good-rite GP 264, hatcol DOP, hercoflex 260, kodaflex DOP, mollan O, nuoplaz DOP, octoil, octyl phthalate, palatinol AH, phthalic acid dioctyl ester, Pittsburg PX-138, platinol AH, platinol DOP, RC plasticizer DOP, reomol DOP, reomol D 79P, sicol 150, staflex DOP, truflex DOP, vestinol AH, vinicizer 80, witcizer 312

USES
- Used in the Production Process
  - polyvinyl chloride
  - added to plastics to make them flexible
  - liquid used in vacuum pumps

PHYSICAL PROPERTIES
- colorless liquid with almost no odor
- melting point −50°C
- boiling point 384°C
- density 0.9810
- vapor pressure 6.2×10^{-8} mm Hg at 25°C
- solubility 0.1 g/L at 20°C
- viscosity 80 cP at 20°C
- log K_{ow} 4.2−5.11

CHEMICAL PROPERTIES
- stable under normal temperature and pressure
- flash point 199°C
- autoignition temperature 390°C
- explosion limits lower 10%, upper not available
- incompatible with strong oxidants

HEALTH RISK
- Toxic Exposure Guidelines
  - ACGIH TLV STEL 10 mg/m³
  - OSHA PEL 5 mg/m³
  - ACGIH TLV 5 mg/m³
  - NIOSH REL 5 mg/m³

- ◆ Toxicity Data
  - skin-rabbit 500 mg/24H MLD
  - eye-rabbit 500 mg
  - eye-rabbit 500 mg/24H MLD
  - dns-rat:lvr 500 μmol/L
  - sln-hamster:lvr 50 mg/L
  - oral-mouse TDLo 1 g/kg
  - intraperitoneal-mouse TDLo 24 g/kg
  - oral-rat TDLo 216 g/kg/2Y-C:CAR
  - oral-mouse TDLo 260 g/kg/2Y-C:CAR
  - oral-mouse TD 120 g/kg/24W-C:ETA
  - oral-man TDLo 143 mg/kg:GIT
  - oral-rat LD_{50} 30,600 mg/kg
  - intraperitoneal-rat LD_{50} 30,700 mg/kg
  - intravenous-rat LD_{50} 250 mg/kg
  - oral-mouse LD_{50} 30 g/kg
  - intraperitoneal-mouse LD_{50} 14 g/kg
  - intravenous-mouse LD_{50} 1060 mg/kg
  - oral-rabbit LD_{50} 34 g/kg
  - skin-rabbit LD_{50} 25 g/kg
  - skin-guinea pig LD_{50} 10 g/kg
- ◆ Acute Risks
  - gastrointestinal distress
  - mild skin and eye irritation
  - effects on liver and kidney
  - adverse effects on weight gain and food consumption
  - low acute toxicity from oral exposure
- ◆ Chronic Risks
  - no information available on chronic effects in humans
  - increased lung weights in animals
  - increased liver weights in animals
  - effects on the liver
  - birth defects in rats and mice
  - decrease in testicular weights
  - tubular atrophy
  - liver tumors in rats and mice
  - EPA Group B2: probable human carcinogen

HAZARD RISK
- ◆ irritating and toxic gases emitted during a fire
- ◆ incompatible with strong oxidants
- ◆ hazardous decomposition products: carbon dioxide, toxic gases, and carbon monoxide
- ◆ Decomposition emits acrid smoke.

EXPOSURE ROUTES
- ◆ most probable through food
- ◆ from plastics during processing and storage
- ◆ blood transfusions
- ◆ kidney dialysis
- ◆ use of respirators
- ◆ drinking water
- ◆ ambient air
- ◆ air in a newly painted room
- ◆ air in room with recently installed flooring
- ◆ factories that manufacture or use DEHP

REGULATORY STATUS/INFORMATION
- ◆ Clean Air Act Amendment, Title III: hazardous air pollutants
- ◆ Resource Conservation and Recovery Act
  - list of hazardous constituents. 40CFR258
  - groundwater monitoring list. 40CFR264
  - risk specific doses. 40CFR266
  - health based limits for exclusion of waste-derived residues. 40CFR266
  - PICs found in stack effluents. 40CFR266
- ◆ CERCLA and Superfund
  - designation of hazardous substances. 40CFR302.4
- ◆ Clean Water Act
  - total toxic organics (TTOs). 40CFR413.02
  - 126 priority pollutants. 40CFR423
- ◆ chemical regulated by the State of California

ADDITIONAL/OTHER COMMENTS
- ◆ Symptoms
  - ingestion: gastrointestinal effects
  - inhalation: mild skin and eye irritation
- ◆ First Aid
  - wash eyes immediately with large amounts of water
  - flush skin with plenty of soap and water
  - remove to fresh air immediately
- ◆ fire fighting: use dry chemical, water spray or mist, chemical foam, or alcohol-resistant foam
- ◆ personal protection: wear self-contained breathing apparatus and full protective gear

SOURCES
- ◆ *Health Effects Notebook for Hazardous Air Pollutants*, U.S. Environmental Protection Agency, December 1994
- ◆ *Material Safety Data Sheets (MSDS)*
- ◆ *Sax's Dangerous Properties of Industrial Materials*, 9th edition, 1996
- ◆ *Hawley's Condensed Chemical Dictionary*, 12th edition, 1993
- ◆ *Aldrich Chemical Company Catalog*, 1996–1997
- ◆ *Environmental Law Index to Chemicals*, 1996
- ◆ *Merck Index*, 12th edition, 1996

BIS(CHLOROMETHYL)ETHER

CAS #: 542-88-1
MOLECULAR FORMULA: $(CH_2Cl)O(CH_2Cl)$
FORMULA WEIGHT: 114.96
SYNONYMS: BCME, bis(chloromethyl)(sym-), chloro(chloromethoxy)-methane, chloro-methyl-ether, dichloro-dimethyl-ether, dichloro-di-methyl-ether(sym-), dichloro-dimethyl-ether(1,1'-), dichloro-dimethyl-ether(alpha, alpha'-), dichloro-methyl-ether, dichloro-methyl-ether-

(sym-), dimethyl-1,1'-dichloroether, methane, oxybis(chloro-, mono-chloromethyl ether, oxybis(chloromethane)

USES
◆ Variety of Uses
  • lab reagent
  • chloromethylation (no longer used for this)
  • monitoring indicator for chloromethyl ether
  • alkylating agent in manufacture of polymers
  • research chemical
  • intermediate in synthesis of anionic exchange strong-base resins of the quaternary ammonium type

PHYSICAL PROPERTIES
◆ colorless liquid
◆ suffocating odor
◆ melting point −41.5°C
◆ boiling point 106°C
◆ density 1.323 at 15°C

◆ vapor pressure 30 mm Hg at 22°C
◆ vapor density 4.0
◆ miscible with alcohol, ether, organic solvents
◆ log K_{ow} −0.38

CHEMICAL PROPERTIES
◆ volatile liquid
◆ unstable in moist air
◆ flash point <19°C
◆ flammable
◆ forms spontaneously from formaldehyde and chloride ions in moist air

◆ reacts with acids
◆ reacts with water
◆ attacks plastics
◆ reacts with rubber

HEALTH RISK
◆ Toxic Exposure Guidelines
  • ACGIH TLV TWA 0.001 ppm
  • OSHA cancer suspect agent
  • DFG MAK human carcinogen
◆ Toxicity Data
  • otr-hamster:kdy 80 µg/L
  • mma-sat 20 µg/plate
  • dns-human:fbr 160 µg/L
  • dns-mouse-skin 360 µmol/kgL
  • inhalation-rat TCLo 100 ppb/6H/4W-I:CAR
  • subcutaneous-rat TDLo 375 mg/kg/43W-I:CAR
  • inhalation-mouse TCLo 100 ppb/6H/26W-I:NEO
  • skin-mouse TDLo 5520 mg/kg/23W-T:ETA
  • inhalation-man TCLo 3 ppm:EYE
  • inhalation-man LCLo 100 ppm/3M:PUL
  • oral-rat LD_{50} 210 mg/kg
  • inhalation-rat LC_{50} 7 ppm/7H
  • inhalation-mouse LC_{50} 25 mg/m³/6H
  • skin-rabbit LD_{50} 280 mg/kg
  • inhalation-hamster LC_{50} 7 ppm/7H
◆ Acute Risks
  • skin irritation
  • mucous membrane irritation

- respiratory tract irritation
- lung irritation
- congestion
- edema
- dizziness
- headache
◆ Chronic Risks
  - chronic bronchitis
  - chronic cough
  - impaired respiratory function
  - respiratory distress
  - lung edema
  - lung cancer

- nausea
- vomiting
- sore throat
- abdominal cramps
- hemorrhage
- corneal opacity in rabbits

- respiratory tract tumors in rats
- pulmonary adenomas in mice
- skin papillomas and carcinomas in mice
- EPA Group A: known human carcinogen

HAZARD RISK
◆ flammable
◆ dangerous fire hazard
◆ explosion from air mixtures of ether vapors
◆ can form peroxides which can detonate when heated
◆ decomposes on heating
◆ may generate electrostatic charges
◆ can ignite by electric sparks
◆ formation of corrosive vapors and toxic fumes of Cl^-

EXPOSURE ROUTES
◆ probability of human exposure is low; not used in the United States
◆ rapidly degrades in the environment
◆ inhalation in workplace during production and use

REGULATORY STATUS/INFORMATION
◆ Clean Air Act Amendment, Title III: hazardous air pollutants
◆ Resource Conservation and Recovery Act
  - risk specific doses. 40CFR266
  - health based limits for exclusion of waste-derived residues. 40CFR266
◆ CERCLA and Superfund
  - designation of hazardous substances. 40CFR302.4
  - list of chemicals. 40CFR372.65
◆ OSHA 29CFR1910.119 list of highly hazardous chemicals, toxics, and reactives
◆ chemical regulated by the State of California

ADDITIONAL/OTHER COMMENTS
◆ Symptoms
  - contact: redness, prickling, burning
  - ingestion: labored breath, vomiting, sore throat, abdominal pain, stupefying
◆ First Aid
  - wash eyes immediately with large amounts of water
  - flush skin with plenty of soap and water
  - remove to fresh air immediately and drink water if ingested

- fire fighting: use carbon dioxide, foam (alcohol proof), dry chemical, and halogenated extinguishing agents
- personal protection: compressed air/oxygen apparatus, gas-tight suit, fireproof suit

SOURCES
- *Health Effects Notebook for Hazardous Air Pollutants*, U.S. Environmental Protection Agency, December 1994
- *Material Safety Data Sheets (MSDS)*
- *Sax's Dangerous Properties of Industrial Materials*, 9th edition, 1996
- *Hawley's Condensed Chemical Dictionary*, 12th edition, 1993
- *Environmental Law Index to Chemicals*, 1996
- *Merck Index*, 12th edition, 1996
- *Environmental Contaminant Reference Databook*, volume 1, 1995

BIS(2-ethylhexyl)phthalate. *See* **p. 67 following biphenyl.**

BROMOFORM

CAS #: 75-25-2
MOLECULAR FORMULA: $CHBr_3$
FORMULA WEIGHT: 252.75
SYNONYMS: bromoforme (French), bromoformio (Italian), methyl Tribromide, NCI-C55130, tribrommethaan (Dutch), Tribrommethan (German), tribromometan (Italian), tribromomethane

USES
- Used as a Solvent
  - waxes
  - greases
  - oils
  - liquid-solvent extractions
  - nuclear magnetic resonance studies
- Used in Chemical Synthesis
  - pharmaceuticals
  - fire-resistant chemicals
  - gauge fluid
  - reagent for graphite
  - ore faction
  - heavy liquid floatation agent in mineral separation
  - cough suppression agent
- Used in Industry
  - shipbuilding
  - aircraft
  - aerospace industries
  - medication
  - sedative
  - antitussive medication
  - antiseptic
  - vulcanization of rubber
- Used in Polymer Reactions
  - catalyst
  - sensitizer
  - indicator

PHYSICAL PROPERTIES
- colorless to pale-yellow liquid
- sweetish odor
- odor threshold 1.3 ppm
- sweetish taste
- melting point 8.3°C

- boiling point 149°C at 15 mm Hg
- density 2.8899 at 20°C
- surface tension 41.53 dynes/cm at 20°C
- heat of vaporization 9673.3 gcal/gmol
- vapor pressure 5 mm Hg at 20°C
- vapor density 8.7
- soluble in benzene, chloroform, alcohol, ether, solvent naphtha, fixed and volatile oils
- slightly soluble in water
- log K_{ow} 2.38

CHEMICAL PROPERTIES
- gradually decomposes, acquiring yellow color
- attacks some forms of plastics, rubber, and coatings
- solidifies at 7.5°C
- incompatible with metals, strong oxidants, bases, lithium, and sodium-potassium alloy
- nonflammable

HEALTH RISK
- Toxic Exposure Guidelines
  - ACGIH TLV TWA 0.5 ppm
  - OSHA PEL TWA 0.5 ppm
  - MSHA TWA 5 mg/m³
- Toxicity Data
  - sln-dmg-oral 3000 ppm
  - sce-human:lym 80 μmol/L
  - sce-hamster:ovr 290 μg/L
  - intraperitoneal-mouse TDLo 1100 mg/kg/8W-I:NEO
  - oral-human LDLo 143 mg/kg
  - oral-rat LD_{50} 1147 mg/kg
  - inhalation-rat LCLo 45 g/m³/4H
  - intraperitoneal-rat LD_{50} 414 mg/kg
  - oral-mouse LD_{50} 1400 mg/kg
  - subcutaneous-mouse LD_{50} 1820 mg/kg
  - subcutaneous-rabbit LDLo 410 mg/kg
- Other Toxicity Indicators
  - EPA Cancer Risk Level (1 in a million excess lifetime risk) 9×10^{-4} mg/m³
- Acute Risks
  - central nervous system depression
  - slowing of normal brain activities
  - liver and kidney injury
  - unconsciousness
  - death
  - irritation, provoking flow of tears and saliva
  - reddening of face
  - areflexia
  - convulsions
  - dizziness
  - headache
  - pulmonary edema
  - shock
  - amnesia
- Chronic Risks
  - effects on liver, kidney, and central nervous system
  - increase in liver and intestinal tumors in animals
  - EPA Group B2: probable human carcinogen

HAZARD RISK

- toxic gases and vapors (hydrogen bromide, bromine, and carbon monoxide) released in fire
- explosive reaction with crown ethers or potassium hydroxide
- violent reaction with Li or NaK alloy
- violent reaction with acetone or bases
- incompatible with metals, strong oxidants, and strong bases
- decomposition emits highly toxic fumes of Br^-

EXPOSURE ROUTES

- drinking water that has been disinfected with bromine or bromine compounds
- swimming pools disinfected with bromine or bromine compounds
- inhalation of bromoform in evaporated air near pools
- through skin from bromoform in water
- inhalation of ambient air near factories or laboratories
- near a chemical waste site where bromoform leaked into soil or water

REGULATORY STATUS/INFORMATION

- Clean Air Act Amendment, Title III: hazardous air pollutants
- Safe Drinking Water Act
  - priority list of drinking water contaminants. 55FR1470
  - community water systems and nontransient, noncommunity water systems. 40CFR141.40.e
- Resource Conservation and Recovery Act
  - constituents for detection monitoring (for municipal solid waste landfill). 40CFR258
  - list of hazardous constituents. 40CFR258, 40CFR261, 40CFR261.11
  - groundwater monitoring list. 40CFR264
  - health based limits for exclusion of waste-derived residues. 40CFR266
  - PICs found in stack effluents. 40CFR266
  - U waste #: U225
- CERCLA and Superfund
  - designation of hazardous substances. 40CFR302.4
  - list of chemicals. 40CFR372.65
- Clean Water Act
  - total toxic organics (TTOs). 40CFR413.02
  - 126 priority pollutants. 40CFR423
- Toxic Substance Control Act
  - OSHA chemicals in need of dermal absorption testing. 40CFR712.30.e10
  - hazardous waste constituents subject to testing. 40CFR799.5055
- OSHA 29CFR1910.1000 Table Z1
- chemical regulated by the State of California

ADDITIONAL/OTHER COMMENTS

- Symptoms

- inhalation: irritation of nose and throat, provokes flow of tears and saliva, reddening of face
- ingestion: dizziness, disorientation, slurred speech, unconsciousness, death

◆ First Aid
  - wash eyes immediately with large amounts of water
  - flush skin with plenty of soap and water
  - remove to fresh air immediately or provide respiratory apparatus if necessary

◆ fire fighting: bromoform is noncombustible; use agent appropriate to extinguish surrounding fire.

◆ personal protection: wear self-contained breathing apparatus, chemical goggles, and protective clothing

◆ storage: tightly closed container; cool, dry, well-ventilated area away from incompatible substances

SOURCES

◆ *Health Effects Notebook for Hazardous Air Pollutants*, U.S. Environmental Protection Agency, December 1994

◆ *Material Safety Data Sheets (MSDS)*

◆ *Sax's Dangerous Properties of Industrial Materials*, 9th edition, 1996

◆ *Hawley's Condensed Chemical Dictionary*, 12th edition, 1993

◆ *Aldrich Chemical Company Catalog*, 1996–1997

◆ *Environmental Law Index to Chemicals*, 1996

◆ *Environmental Contaminant Reference Databook*, volume 1, 1995

◆ *Merck Index*, 12th edition, 1996

1,3-BUTADIENE (bh)

CAS #: 106-99-0
MOLECULAR FORMULA: C_4H_6
FORMULA WEIGHT: 54.10
SYNONYMS: biethylene, bivinyl, butadieen (Dutch), buta-1,3-dieen (Dutch), butadien (Polish), Buta-1,3-dien (German), butadiene, buta-1,3-diene, 1,3-butadiene (ACGIH:OSHA), alpha, gamma-butadiene, butadiene (OSHA), divinyl, erthrene, NCI-560602, pyrrolyene, vinylethylene

USES

◆ Used as Component or in the Manufacturing Process
  - styrene-butadiene rubbers
  - plastics
  - acrylics
  - latex paints
  - resins
  - organic intermediate in adiponitrile production

PHYSICAL PROPERTIES

◆ colorless gas

◆ mild aromatic odor

◆ melting point −109°C (−164°F)

- ◆ boiling point −4.5°C (−24°F) at 1 atm
- ◆ specific gravity 0.621
- ◆ vapor density 1.81 kg/m³ at 15°C
- ◆ vapor pressure 2100 mm Hg at 25°C
- ◆ sparingly soluble in water
- ◆ slightly soluble in methanol, ethanol
- ◆ soluble in organic solvents
- ◆ odor threshold 1.6 ppm
- ◆ octanol/water partition coefficient (log K_{ow}) 1.99

CHEMICAL PROPERTIES

- ◆ flammable gas
- ◆ highly reactive
- ◆ explosion limits 2.0% lower and 11.5% upper by volume in air
- ◆ autoignition temperature 419°C (788°F)
- ◆ flash point −76°C (−105°F)
- ◆ polymerizes and copolymerizes easily
- ◆ stabilization with o-dihydroxybenzene

HEALTH RISK

- ◆ Toxic Exposure Guidelines
  - ACGIH TLV TWA 2–10 ppm (4.4–22 mg/m³)
  - OSHA PEL TWA 1000 ppm (2210 mg/m³)
  - NIOSH REL TWA reduce to lowest possible level
  - DFG MAK animal carcinogen, suspected human carcinogen
  - EPA Cancer Risk Level (1 in a million excess lifetime risk) 4.0E-06 mg/m³
- ◆ Toxicity Data
  - muscle-mouse lym 20 pph
  - inhalation-rat TCLo 8000 ppm/6H (6–15 D preg): TER
  - inhalation-rat TCLo 625 ppm/6H/61W:CAR
  - inhalation-mouse TCLo 1250 ppm/6H/60W-I:CAR
  - inhalation-rat TC 1000 ppm/6H/2Y-I:CAR
  - inhalation-rat TC 8000 ppm/6H/2Y-I:NEO
  - inhalation-rat TC 8000 ppm/6H/15W-I:CAR
  - inhalation-human TCLo 2000 ppm/7H:EYE
  - inhalation-human TCLo 8000 ppm/EYE, PUL
  - oral-rat LD_{50} 5480 mg/kg
  - inhalation-rat LC_{50} 270 g/m³/4H
  - inhalation-mouse LC_{50} 270 g/m³/2H
  - inhalation-rabbit LCLo 25 pph/23M
- ◆ Acute Risks
  - irritation of eyes, nasal passages, throat, and lungs
  - blurred vision
  - fatigue
  - headaches
  - vertigo
  - Dermal exposure causes a sensation of cold, followed by a burning sensation, which may lead to frostbite.
  - nausea
  - unconsciousness

- ◆ Chronic Risks
  - • cardiovascular diseases (e.g. rheumatic and arteriosclerotic heart disease)
  - • blood disorders
  - • respiratory paralysis
  - • reproductive and developmental effects in animals
  - • EPA Group B2: probable human carcinogen

HAZARD RISK
- ◆ dangerous fire hazard
- ◆ incompatible with strong oxidizers, halogens, oxygens, and copper alloys
- ◆ autopolymerizes in the presence of sodium
- ◆ forms carbon monoxide and carbon dioxide upon combustion
- ◆ may form explosive mixture with air
- ◆ Vapor may travel considerable distance to source of ignition and flash back.

EXPOSURE ROUTES
- ◆ primarily by inhalation
- ◆ ingestion: found in drinking water and in plastic and rubber food containers
- ◆ Motor vehicle exhaust is a minor route of exposure.
- ◆ highly industrialized cities or near oil refineries and chemical factories

REGULATORY STATUS/INFORMATION
- ◆ Clean Air Act
  - • Title III: hazardous air pollutant
- ◆ Resource Conservation Recovery Act
  - • risk specific sources. 40CFR266
- ◆ CERCLA and Superfund
  - • chemicals and chemical categories to which this part applies (CAS # listing). 40CFR372.65
- ◆ Occupational Safety and Health Act of 1970
  - • limits for air contaminants applicable for the transitional period and to the extent set forth in paragraph 1910.1000.f under OSHA. 29CFR1910.1000
- ◆ chemical regulated by the State of California

ADDITIONAL/OTHER COMMENTS
- ◆ First Aid
  - • flush eyes or skin with copious amounts of water for at least 15 minutes
  - • remove contaminated clothing
  - • if inhaled, remove to fresh air
  - • if not breathing, give artificial respiration
  - • if breathing is difficult, give oxygen
  - • if ingested, wash out mouth with water and call physician
- ◆ Spill or Leak
  - • evacuate area and keep personnel upwind
  - • shut off all sources of ignition
  - • wear self-contained breathing apparatus, rubber boots, and heavy rubber gloves

- shut off leak if there is no risk
- ventilate area and wash spill site after material pickup is complete

SOURCES

- *Encyclopedia of Chemical Technology*, 4th edition, 1991
- *Environmental Law Index to Chemicals*, 1996 edition
- *Health Effects Notebook for Hazardous Air Pollutants*, U.S. Environmental Protection Agency, December 1994
- *Handbook of Environmental Data on Organic Chemicals*, 2nd edition, 1983
- *Material Safety Data Sheets (MSDS)*
- *Merck Index*, 12th edition, 1996
- *Sax's Dangerous Properties of Industrial Materials*, 9th edition, 1996

CADMIUM COMPOUNDS

CAS #: Cadmium compounds have variable CAS #s. The CAS # for cadmium is 7440-43-9.

MOLECULAR FORMULA: Cadmium compounds have variable molecular formulas. The molecular formula for cadmium is Cd.

Note: For all hazardous air pollutants (HAPs) that contain the word "compounds" in the name, and for glycol ethers, the following applies: Unless otherwise specified, these HAPs are defined by the 1990 Amendments to the Clean Air Act as including any unique chemical substance that contains the named chemical (i.e., antimony, arsenic, etc.) as part of that chemical's infrastructure.

FORMULA WEIGHT: Cadmium compounds have variable formula weights. The formula weight for cadmium is 112.40.

SYNONYMS: cadmium: C.I. 77180, colloidal cadmium, Kadmium (German)

USES

- Used in industry
  - electrodeposited and dipped coatings on metals
  - bearing and low metal alloys
  - brazing alloys
  - fire protection systems
  - nickel-cadmium storage batteries
  - power transmission wire
  - TV phosphors
  - basis of pigments in ceramic glazes
  - machinery enamels
  - baking enamels
  - Weston-standard-cell control of atomic fission in nuclear reactors
  - fungicide
  - photography and lithography
  - selenium rectifiers
  - electrodes for cadmium-vapor lamps and photoelectric cells
  - soft solder and solder for aluminum
  - deoxidizer in Ni plating
  - process engraving
  - powder used as an amalgam in dentistry
  - to charge Jones reductors

PHYSICAL PROPERTIES

- soft, silver-white metal
- usually found in combination with other elements such as oxygen (cadmium oxide), chlorine (cadmium chloride), or sulfur (cadmium sulfide)
- malleable or powder
- becomes brittle at 80°C
- density 8.642
- melting point 320.9°C
- boiling point 767°C
- vapor pressure 1 mm Hg at 394°C
- Mohs hardness 2.0
- soluble in acids, especially nitric acid, and in ammonium nitrate solution
- Cadmium compounds range in solubility in water from quite soluble to practically insoluble.

PHYSICAL PROPERTIES (Cadmium Chloride)

- white, solid
- melting point 568°C
- boiling point 960°C
- density 4.05 g/cm³
- odorless
- soluble in acetone
- vapor pressure 10 mm Hg at 656°C

PHYSICAL PROPERTIES (Cadmium Oxide)

- red or brown, solid
- melting point <1426°C
- decomposes at 900°C
- density 8.15 g/cm³
- vapor pressure 1 mm Hg at 1000°C

CHEMICAL PROPERTIES (Cadmium)

- tarnishes in moist air
- corrosion resistance poor in industrial atmospheres
- resistant to alkalies
- lowers melting point of certain alloys when used in low percentages
- combustible

HEALTH RISK

- Toxic Exposure Guidelines
  - ACGIH TLV (cadmium dust and cadmium oxide) 0.05 mg/m³
  - OSHA PEL (cadmium dust) 0.2 mg/m³
  - OSHA PEL (cadmium fumes) 0.1 mg/m³
  - NIOSH IDLH (cadmium dusts or fumes) 40 mg/m³
  - NIOSH REL (cadmium) reduce to lowest feasible level
- Toxicity Data
  - inhalation-human TCLo 1500 μg/m³/14Y-I:CAR,PUL
  - rat LC$_{50}$ 500 mg/m³
- Acute Risks
  - effects on the lung
  - bronchial and pulmonary irritation
  - long-lasting impairment of lung function
  - irritation of stomach
  - vomiting
  - diarrhea
  - increases salivation
  - choking
  - anemia
  - renal dysfunction
  - tenesmus
- Chronic Risks
  - kidney effects
  - proteinuria
  - a decrease in glomerular filtration rate

- increased frequency of kidney stone formation
- bronchiotitis
- emphysema
- effects on the skeleton
- decreased birth weights
- skeletal malformations
- impaired neurological development
- decreased reproduction in animals
- testicular damage in animals
- excess risk of lung cancer
- high blood pressure
- fragile bones
- tissues injured in animals: liver, testes, immune system, nervous system, and blood
- EPA Group B1: probable human carcinogen

HAZARD RISK

- ◆ Cadmium is flammable in powder form.
- ◆ Cadmium compounds emit toxic fumes of Cd when heated to decomposition.
- ◆ Cadmium and its salts are highly toxic.
- ◆ Soluble compounds of cadmium are highly toxic.
- ◆ incompatible with ignition sources, dust generation, moisture, excess heat

EXPOSURE ROUTES

- ◆ Primary source is burning of fossil fuels such as coal or oil.
- ◆ incineration of municipal waste materials
- ◆ from zinc, lead, or copper smelters
- ◆ food
- ◆ application of phosphate fertilizers
- ◆ sewage sludge
- ◆ soil
- ◆ smoking; smokers have twice as much exposure as non-smokers
- ◆ spills or leaks at hazardous waste sites
- ◆ fish, plants, animals
- ◆ breathing contaminated workplace air
- ◆ drinking contaminated water

REGULATORY STATUS/INFORMATION

- ◆ CERCLA and Superfund
  - list of chemicals. 40CFR372.65

ADDITIONAL/OTHER COMMENTS

- ◆ Symptoms
  - ingestion: increased salivation, choking, vomiting, abdominal pain, anemia, renal dysfunction, diarrhea, tenesmus
  - inhalation: throat dryness, cough, headache, vomiting, chest pain, extreme restlessness and irritability, pneumonitis, bronchopneumonia, metal-fume fever
- ◆ First Aid
  - wash eyes immediately with large amounts of water
  - flush skin with plenty of soap and water
  - remove to fresh air immediately or provide respiratory apparatus if necessary
- ◆ fire fighting: use dry chemical and carbon dioxide
- ◆ personal protection: wear self-contained breathing apparatus, chemical goggles, and protective clothing

♦ storage: keep away from heat, flame, and sources of ignition; tightly closed container; cool, dry, well-ventilated area

SOURCES

♦ *Health Effects Notebook for Hazardous Air Pollutants*, U.S. Environmental Protection Agency, December 1994
♦ *Material Safety Data Sheets (MSDS)*
♦ *Sax's Dangerous Properties of Industrial Materials*, 9th edition, 1996
♦ *Hawley's Condensed Chemical Dictionary*, 12th edition, 1993
♦ *Environmental Law Index to Chemicals*, 1996
♦ *Merck Index*, 12th edition, 1996
♦ *Toxicological Profile for Cadmium*
♦ Agency for Toxicological Substances and Disease Registry

CALCIUM CYANAMIDE

CAS #: 156-62-7
MOLECULAR FORMULA: CaNCN
FORMULA WEIGHT: 80.11
SYNONYMS: alzodef, calcium-carbimide, calcium salt (1:1), CCC, Cy-L 500, Cyanamide, Cyanaml, lime-nitrogen, nitrogen-lime, Nitrolim, Nitrolime

USES
♦ Used in Products
  • fertilizer
  • defoliant
  • herbicide
  • fungicide
  • pesticide
  • larvicide
  • nitrogen products
  • antiethylic agent
  • desulfurizing agent
♦ Used in Manufacturing Process
  • refining iron
  • calcium cyanide
  • melamine
  • dicyandiamide
  • hardening steel

PHYSICAL PROPERTIES
♦ pure: glistening hexagonal crystals, colorless
♦ commercial grades: grayish-black lumps of powder
♦ melting point 1300°C
♦ sublimes >1500°C
♦ density 1.083
♦ insoluble in water

CHEMICAL PROPERTIES
♦ heat of formation −69 kcal/mole at 25°C
♦ heat of fusion 1.29 cal/g
♦ pure: nonvolatile
♦ pure: noncombustible
♦ decomposes in water, liberating ammonia and acetylene

HEALTH RISK
♦ Toxic Exposure Guidelines
  • ACGIH TLV TWA 0.5 mg/m^3
  • OSHA PEL TWA 0.5 mg/m^3
  • DFG MAK 1 mg/m^3
♦ Toxicity Data

- mmo-sat 1 mg/plate
- mma-sat 100 μg/plate
- oral-mouse TDLo 170 g/kg/2Y-C:ETA
- oral-human LDLo 571 mg/kg
- oral-rat LD_{50} 158 mg/kg
- inhalation-rat LCLo 86 mg/m³/4H
- skin-rat LD_{50} 84 mg/kg
- intravenous-rat LD_{50} 125 mg/kg
- unr-rat LD_{50} 1000 mg/kg
- oral-mouse LD_{50} 334 mg/kg
- intraperitoneal-mouse LD_{50} 100 mg/kg
- intravenous-mouse LD_{50} 282 mg/kg
- oral-cat LD_{50} 100 mg/kg
- oral-rabbit LD_{50} 1400 mg/kg
- skin-rabbit LD_{50} 590 mg/kg
- ◆ Acute Risks
  - eye irritant
  - skin irritant
  - respiratory tract irritation
  - gastris
  - rhinitis
  - pharyngitis
  - laryngitis
  - tracheobronchitis
  - vasomotor reaction
  - intense localized erythe-matous flushing of the face, upper body, and arms
  - headache
  - dizziness
  - fatigue
  - vertigo
  - congestion of the mucosa
  - nausea
  - vomiting
  - tachycardia
  - hypotension
  - peripheral neuropathy
  - effects on liver enzymes in rats
- ◆ Chronic Risks
  - chronic rhinitis
  - perforation of the nasal septum
  - slow-healing dermal ulceration

HAZARD RISK
- ◆ flammable
- ◆ Reaction with water forms explosive acetylene gas.
- ◆ fire risk with moisture or combined with calcium carbide
- ◆ Decomposition emits toxic fumes of NO_x and CN^-.

EXPOSURE ROUTES
- ◆ manufacture and use of calcium cyanimide

REGULATORY STATUS/INFORMATION
- ◆ Clean Air Act Amendment, Title III: hazardous air pollutants
- ◆ CERCLA and Superfund
  - list of chemicals. 40CFR372.65
- ◆ chemical regulated by the State of California

ADDITIONAL/OTHER COMMENTS
- ◆ Symptoms
  - skin contact: skin burns, irritation, redness
  - ingestion: headache, nausea, drowsiness
  - inhalation: burns eyes, sore throat
- ◆ First Aid

- wash eyes immediately with large amounts of water
- flush skin with plenty of soap and water
- remove to fresh air immediately or provide respiratory support if necessary
◆ fire fighting: use dry chemical, dry sand. No water and no foam.
◆ personal protection: compressed air/oxygen apparatus, gas-tight suit
◆ storage: keep away from flames or sparks

SOURCES
◆ *Health Effects Notebook for Hazardous Air Pollutants*, U.S. Environmental Protection Agency, December 1994
◆ *Material Safety Data Sheets (MSDS)*
◆ *Sax's Dangerous Properties of Industrial Materials*, 9th edition, 1996
◆ *Hawley's Condensed Chemical Dictionary*, 12th edition, 1993
◆ *Aldrich Chemical Company Catalog*, 1996–1997
◆ *Environmental Law Index to Chemicals*, 1996
◆ *Merck Index*, 12th edition, 1996

CAPROLACTAM

CAS #: 105-60-2
MOLECULAR FORMULA: $C_6H_{11}NO$
FORMULA WEIGHT: 113.18
SYNONYMS: aminocaproic lactam, 6-aminohexanoic acid cyclic lactam, 2-azacycloheptanone, 6-caprolactam, omega-caprolactam (MAK), caprolattame (French), cyclohexanone iso-oxime, epsylon kaprolaktam (Polish), hexahydro-2-azepinone, hexahydro-2H-azepin-2-one, 6-hexanelactam, hexanone isoxime, Hexanonisoxim (German), 1,6-hexolactam, e-kaprolaktam (Czech), 2-ketohexamethylenimine, NCI-C50646, 2-oxohexamethylenimine, 2-perhydroazepinone

USES
◆ Used in the Manufacturing process
- synthetic fibers
- nylon 6
- brush bristles
- textile stiffeners
- film coatings
- synthetic leather
- plastics
- plasticizers
- paint vehicles
- cross-linking for polyure-thanes
- synthesis of lysine
- blocking agent for coating polymers

PHYSICAL PROPERTIES
◆ white, hygroscopic, crystal-line solid or leaflets
◆ unique, unpleasant odor
◆ odor threshold 59.7 ppm
◆ melting point 69°C
◆ viscosity 9 cP at 78°C
◆ freely soluble in water, methanol, ethanol, ether, tetrahydrofurfuryl alcohol, dimethylformamide
◆ soluble in chlorinated hydrocarbons, cyclohexene, petroleum fractions

◆ vapor pressure 1.9×10^{-3} mm Hg at 25°C   ◆ log K_{ow} −0.19

CHEMICAL PROPERTIES
◆ flash point 110°C

HEALTH RISK
◆ Toxic Exposure Guidelines
  • OSHA PEL (dust) 1 mg/m^3
  • OSHA PEL STEL (dust) 3 mg/m^3
  • OSHA PEL (vapor) 5 ppm
  • OSHA PEL STEL (vapor) 10 ppm
  • ACGIH TLV (dust) 1 mg/m^3
  • ACGIH TLV STEL (dust) 3 mg/m^3
  • ACGIH TLV (vapor) 4.3 ppm
  • ACGIH TLV STEL (vapor) 8.6 ppm
  • DFG MAK 25 mg/m^3
  • NIOSH REL (dust and vapor) 1 mg/m^3
◆ Toxicity Data
  • skin-rabbit 500 mg/24H MLD
  • eye-rabbit 20 mg/24H MOD
  • slt-dmg-oral 5 mmol/L
  • mmo-smc 100 mg/L
  • cyt-human:lym 270 mg/L
  • oral-rabbit TDLo 3450 mg/kg
  • oral-rat TDLo 10 g/kg
  • inhalation-human TCLo 100 ppm:PUL
  • oral-rat LD$_{50}$ 1210 mg/kg
  • inhalation-rat LC$_{50}$ 300 mg/m^3/2H
  • oral-mouse LD$_{50}$ 930 mg/kg
  • inhalation-mouse LC$_{50}$ 450 mg/m^3
  • intraperitoneal-mouse LD$_{50}$ 650 mg/kg
  • subcutaneous-mouse LDLo 750 mg/kg
  • skin-rabbit LDLo 1438 mg/kg
  • subcutaneous-frog LDLo 2800 mg/kg
◆ Acute Risks
  • irritation of eyes, nose, throat, skin
  • burning of eyes, nose, throat, skin
  • headaches
  • malaise
  • confusion
  • nervous irritation
  • dermatitis
  • fever
  • grand mal seizures
◆ Chronic Risks
  • peeling of hands
  • eye, nose, and throat irritation
  • neurological effects
  • gastrointestinal effects
  • cardiovascular effects
  • dermatological changes
  • immunological changes
  • weight gain depression in rats
  • increased liver and kidney weights in rats
  • gynecological effects
  • dysmenorrhea
  • menorrhagia
  • oligomenorrhea

- postpartum hemorrhage
- toxemia of pregnancy
- premature birth
- inadequate uterine contractions during labor
- depressed fetal body weights in rats and mice
- increased fetal resorptions in rats
- adverse effects on spermatogenesis in rats
- evidence of noncarcinogenicity for humans

HAZARD RISK
- ◆ potentially explosive reaction with acetic acid + dinitrogen trioxide
- ◆ Decomposition emits toxic fumes of NO_x.

EXPOSURE ROUTES
- ◆ dermal contact through manufacture of compound
- ◆ inhalation through manufacture of compound
- ◆ emissions and effluents from manufacture of compound
- ◆ emissions and effluents from use of compound
- ◆ water
- ◆ ingestion of contaminated drinking water

REGULATORY STATUS/INFORMATION
- ◆ Clean Air Act Amendment, Title III: hazardous air pollutants
- ◆ chemical regulated by the State of California

ADDITIONAL/OTHER COMMENTS
- ◆ Symptoms
  - oral: cramps, frequent urination, nausea, vomiting, diarrhea, headache, prickling
- ◆ First Aid
  - wash eyes and skin immediately with large amounts of water
  - remove to fresh air immediately or provide respiratory apparatus if necessary
  - drink water and induce vomiting
- ◆ fire fighting: use water spray, carbon dioxide, foam, or dry chemical
- ◆ personal protection: compressed air/oxygen apparatus, gastight suit, fireproof suit

SOURCES
- ◆ *Health Effects Notebook for Hazardous Air Pollutants*, U.S. Environmental Protection Agency, December 1994
- ◆ *Material Safety Data Sheets (MSDS)*
- ◆ *Sax's Dangerous Properties of Industrial Materials*, 9th edition, 1996
- ◆ *Environmental Law Index to Chemicals*, 1996
- ◆ *Merck Index*, 12th edition, 1996

CAPTAN

CAS #: 133-06-2
MOLECULAR FORMULA: $C_9H_8Cl_3NO_2S$
FORMULA WEIGHT: 300.59
SYNONYMS: aacaptan, amercide, bangton, captaf, captan-50W, captex, cyclohexene-1,2-dicarboximide, n-(trichloromethyl)-mercapto(4-), ENT-26538, esso-fungicide 406, flit-406, fungus-ban type II, glyodex-37-22, hexacap, isoindole-1,3(2h)-dione(1h-),3a,4,7,7a-tetrahydro-2-((trichloromethyl)-thio)-, kaptan, malipur, merpan, micro-check 12, neracid, orthocide, orthocide-406, orthocide-50, orthocide-7.5, orthocide-75, orthocide-83, osocide, SR-406, stauffer-captan, tetra-hydro-2-((trichloromethyl)-thio)-1h-isoindole-1,3-(2h)-dione(3a,4,7,7a-), tetrahydro-n-(trichloromethanesulphnyl)-phthalimide (3a,4,7,7a-), trichloro-methyl-mercapto-4-cyclohexene-1,2-dicarboximide(n-), tetrahydro-n-(trichloromethylthio)-phthalimide(1,2,3,6-), trichloro-methyl-mercapto-delta (sup 4)-tetrahydro-phthalimide(n-), trichloro-methyl-thiocyclohex-4-ene-1,2-dicarboximide(n-), trichloro-methyl-thio-4-cyclohexene-1,2-dicarboximide(n-), trichloro-methyl-thio-3a,4,7,7a-tetrahydrophthalimide(n-), trichloro-methyl-thio-delta-4-tetrahydro-drophthalimide(n-), trichloro-methyl-thio-tetrahydrophthalimide(n-), trichloro-methyl-thio-3a,4,y,7a-tetrahydrophthalimide(n-), tri chloro-methyl-thio-1,2,5,6-tetrahydrophthalimide, trimegol, vancide-89, vancide-89RE, vanguard-K

USES

- Used in Agriculture as a Fungicide
  - on fruits
  - on vegetables
  - ornamentals on plant seeds
  - food crop packaging boxes
  - cotton seeds
  - leaves
  - paints
  - plastics
  - leather
  - fabrics
- Used in Products
  - cosmetics
  - pharmaceuticals
  - oil-based paints
  - lacquers
  - wallpaper paste
  - plasticizers
  - polyethylene
  - vinyl
  - rubber stabilizers
  - textiles
  - bacteriostat
  - dermatological anti-infective
  - bacteriostatic agent in soaps

PHYSICAL PROPERTIES

- pure: white, crystalline substance
- technical grade: yellow to buff
- pure: odorless
- technical grade: pungent smell
- melting point 178°C
- density 1.5
- practically insoluble in water
- soluble in acetone, benzene, and toluene
- slightly soluble in ethylene dichloride, and chloroform

- ◆ vapor pressure $<1\times10^{-5}$ mm Hg at 25°C
- ◆ log K_{ow} 2.35–2.54

CHEMICAL PROPERTIES
- ◆ combustible
- ◆ attacks metals
- ◆ reacts with bases
- ◆ reacts with acids

HEALTH RISK
- ◆ Toxic Exposure Guidelines
  - ACGIH TLV TWA 5 mg/m^3
  - OSHA PEL TWA 5 mg/m^3
- ◆ Toxicity Data
  - mmo-sat 310 ng/plate
  - cyt-human:lng 10 mg/L
  - sce-human:lym 30 μmol/L
  - oms-ctl:lvr 1 mmol/L
  - oral-mouse TDLo 250 mg/kg
  - oral-hamster TDLo 200 mg/kg
  - oral-mouse TDLo 1075 g/kg/80W-C:NEO
  - oral-mouse TD 540 g/kg/80W-C:ETA
  - oral-human TDLo 1071 mg/kg
  - oral-rat LD$_{50}$ 9 g/kg
  - inhalation-mouse LC$_{50}$ 5000 mg/m^3/2H
  - intraperitoneal-mouse LD$_{50}$ 30 mg/kg
- ◆ Acute Risks
  - dermatitis
  - conjunctivitis
  - vomiting
  - diarrhea
  - irritation of eyes, skin, nose, and mouth
  - irritation of oral and esophageal mucous membranes
  - pyrosis
  - abdominal pains
  - blurred vision
  - low to moderate acute toxicity
- ◆ Chronic Risks
  - systemic effects on central nervous system
  - gonadotropic effects
  - mutagenic and teratogenic properties
  - decreased body weight in rodents
  - excitability in mice
  - fetotoxic in rabbits
  - slight reduction in fetal weight in rats
  - tumors of the duodenum in rats
  - EPA Group B2: probable human carcinogen

HAZARD RISK
- ◆ combustible
- ◆ fine dust explosive with air
- ◆ very toxic
- ◆ decomposes on heating and burning
- ◆ produces irritating and toxic gases: sulfur oxides, nitrogen oxides
- ◆ formation of corrosive vapors
- ◆ can ignite by electric sparks

EXPOSURE ROUTES

- manufacture and formulation of captan
- application of captan
- fruits and vegetables
- air near treated plants
- field workers

REGULATORY STATUS/INFORMATION

- Clean Air Act Amendment, Title III: hazardous air pollutants
- Federal Insecticide, Fungicide, and Rodenticide Act
  - tolerances and exemptions from tolerances for pesticide chemicals in or on raw agricultural commodities. 40CFR180.102-1147
  - tolerances for pesticides in food. 40CFR185
- CERCLA and Superfund
  - designation of hazardous substances. 40CFR302.4
  - list of chemicals. 40CFR372.65
- Clean Water Act
  - designation of hazardous substances. 40CFR116.4
  - determination of reportable quantities. 40CFR117.3
- chemical regulated by the State of California

ADDITIONAL/OTHER COMMENTS

- Symptoms
  - inhalation: irritation of eyes, skin, nose, and mouth, blurred vision
  - ingestion: irritation of oral and esophageal mucous membranes, pyrosis, diarrhea, vomiting, nausea
  - skin contact: redness, dermatitis, prickling
- First Aid
  - wash eyes immediately with large amounts of water
  - flush skin with plenty of soap and water
  - remove to fresh air immediately
  - induce vomiting
- fire fighting: use water spray and powder
- personal protection: protective clothing and gloves, safety goggles, respiratory apparatus
- storage: floor-level ventilation, separated from strong bases

SOURCES

- *Health Effects Notebook for Hazardous Air Pollutants*, U.S. Environmental Protection Agency, December 1994
- *Material Safety Data Sheets (MSDS)*
- *Sax's Dangerous Properties of Industrial Materials*, 9th edition, 1996
- *Hawley's Condensed Chemical Dictionary*, 12th edition, 1993
- *Environmental Law Index to Chemicals*, 1996
- *Merck Index*, 12th edition, 1996

CARBARYL

CAS #: 63-25-2
MOLECULAR FORMULA: $C_{12}H_{11}NO_2$
FORMULA WEIGHT: 201.24
SYNONYMS: arilat, arilate, arylam, atoxan, bercema nmc50, caprolin, carbamic-acid methyl-1-naphthyl ester, carbaryl-1-naphthalenol, methylcarbamate, carbatox, carbatox 60, carbatox-75, carbavur, carbomate, carpolin, compound-7744, denapon, dicarbam, dyna-carbyl, ENT-23969, germain-S, karbatox, karbatox-75, karbosep, menaphtam, methyl-carbamate-1-naphthalenol, methyl-carbamate-1-naphthol, methyl-carbamic-acid, methyl-1-naphthyl-carbamate(n-), methyl-alpha-naphthylcarbamate(n-), monsur, mugan, murvi-N, NAC, naphthalenol-methyl-carbamate(1-), naphthalenyl-methylcarbamate(alpha-), naphthalenyl-methylcarbamate(1-), naphthol-n-methylcarbamate(1-), naphthyl-n-methylcarbamate(alpha-), naphthyl-n-methylcarbamate (1-), NMC-50, oltitox, oms 629, oms-×9, panam, pomex, prosevor-85, ravyon, seffein, sevimol, sevin, sevin-4, sewin, vioxin

USES
◆ Used as an Insecticide
- corn
- soybean
- cotton
- fruit
- nuts
- alfalfa
- livestock
- poultry
- gardens
- lawns
- commercial/industrial use

◆ Used in Agriculture
- acaricide
- molluscicide
- to control lepidoptera
- to control coleoptera of apples
- control of earthworms in turf
- growth regulator for fruit thinning
- animal ectoparasiticide

◆ Used to Control Pests on Animals, Poultry, and Premises
- sarcoptic mange on buffalo
- to control fleas, lice, ticks, and mites
- medication

◆ Used in medical facilities and sewage treatment plants

PHYSICAL PROPERTIES
◆ white, crystalline solid
◆ essentially odorless
◆ melting point 142°C
◆ density 1.232 at 20°C
◆ vapor pressure 0.000041 mm Hg at 25°C

◆ slightly soluble in water
◆ soluble in DMF, acetone, isophorone, and cyclohexanone
◆ log K_{ow} 2.36

CHEMICAL PROPERTIES
◆ stable to heat, light, acids
◆ hydrolyzed in alkalies

◆ decomposes at its boiling point
◆ noncorrosive

HEALTH RISK
◆ Toxic Exposure Guidelines

- ACGIH TLV TWA 5 mg/m^3
- OSHA PEL TWA 5 mg/m^3
- DFG MAK 5 mg/m^3
- NIOSH REL TWA 5 mg/m^3
- MSHA standard 5 mg/m^3

◆ Toxicity Data
- skin-rabbit 12 mg/24H SEV
- eye-rabbit 500 mg/24H MOD
- mmo-sat 250 µg/plate
- mma-human:fbr 1 µmol/L
- dns-human:fbr 1 µmol/L
- cyt-human:emb 40 µg/kg
- oral-rat TDLo 27,500 µg/kg
- oral-hamster TDLo 250 mg/kg
- oral-rat TDLo 5640 mg/kg/94W-I:ETA
- implant-rat TDLo 80 mg/kg:CAR
- oral-man TDLo 500 mg/kg:PNS
- oral-rat LD$_{50}$ 230 mg/kg
- skin-rat LD$_{50}$ 4000 mg/kg
- intraperitoneal-rat LD$_{50}$ 64 mg/kg
- subcutaneous-rat LD$_{50}$ 1400 mg/kg
- intravenous-rat LD$_{50}$ 41,900 µg/kg
- oral-mouse LD$_{50}$ 128 mg/kg
- intraperitoneal-mouse LD$_{50}$ 25 mg/kg
- subcutaneous-mouse LD$_{50}$ 6717 mg/kg
- skin-rabbit LD$_{50}$ 2000 mg/kg

◆ Acute Risks
- cholinesterase inhibition
- affects central nervous system
- nausea
- vomiting
- bronchoconstriction
- blurred vision
- headache
- stomachache
- excessive salivation
- cyanosis
- diarrhea
- slow pulse
- ataxia
- convulsions
- coma
- respiratory failure
- eye and skin irritation
- sensory change
- muscle weakness

◆ Chronic Risks
- cholinesterase inhibition
- headaches
- memory loss
- muscle weakness
- cramps
- anorexia
- kidney and liver effects
- respiratory and heart failure
- dyspnea
- dermatitis
- reduced fertility and litter size in rats
- increased mortality in offspring of rats
- EPA Group D: not classifiable as to human carcinogenicity

HAZARD RISK
◆ decomposes on heating
◆ fine dust explosive with air

- ◆ formation of toxic vapors
- ◆ may generate electrostatic charges
- ◆ decomposition emits toxic fumes of NO_x

EXPOSURE ROUTES
- ◆ manufacture and formulation of the pesticide
- ◆ application of the pesticide
- ◆ spray drift in regions surrounding agricultural areas
- ◆ surface water
- ◆ food

REGULATORY STATUS/INFORMATION
- ◆ Clean Air Act Amendment, Title III: hazardous air pollutants
- ◆ Federal Insecticide, Fungicide, and Rodenticide Act
  - • tolerances and exemptions from tolerances for pesticide chemicals in or on raw agricultural commodities. 40CFR180.102-1147
  - • class of cholinesterase-inhibiting pesticides. 180.3(5)
  - • tolerances for pesticides in animal feeds. 40CFR186
- ◆ Resource Conservation and Recovery Act
  - • list of hazardous constituents. 40CFR261, 40CFR261.11
  - • U waste #: U279
- ◆ CERCLA and Superfund
  - • designation of hazardous substances. 40CFR302.4
  - • list of chemicals. 40CFR372.65
- ◆ Clean Water Act
  - • designation of hazardous substances. 40CFR116.4
  - • determination of reportable quantities. 40CFR117.3
- ◆ OSHA 29CFR1910.1000 Table Z1
- ◆ chemical regulated by the State of California

ADDITIONAL/OTHER COMMENTS
- ◆ Symptoms
  - • ingestion: vomiting, diarrhea, headache, cramps, nausea, narrowing of pupils, abdominal pain
  - • contact: prickling
- ◆ First Aid
  - • wash eyes immediately with large amounts of water
  - • flush skin with plenty of soap and water
  - • remove to fresh air immediately
  - • induce vomiting
- ◆ fire fighting: use water spray, foam, dry chemical, or carbon dioxide
- ◆ personal protection: compressed air/oxygen apparatus, gas-tight suit
- ◆ storage: keep from flames or sparks; keep container closed when not in use

SOURCES
- ◆ *Health Effects Notebook for Hazardous Air Pollutants*, U.S. Environmental Protection Agency, December 1994
- ◆ *Material Safety Data Sheets (MSDS)* for arsine
- ◆ *Sax's Dangerous Properties of Industrial Materials*, 9th edition, 1996
- ◆ *Hawley's Condensed Chemical Dictionary*, 12th edition, 1993

♦ *Environmental Law Index to Chemicals*, 1996
♦ *Merck Index*, 12th edition, 1996
♦ *Environmental Contaminant Reference Databook*, volume 1, 1995

CARBON DISULFIDE

CAS #: 75-15-0
MOLECULAR FORMULA: CS_2
FORMULA WEIGHT: 76.13
SYNONYMS: carbon bisulfide (DOT), carbon bisulphide, carbon disulphide, carbone (sulfure de) (French), carbonio (solfuro di) (Italian), carbon sulfide, carbon sulphide (DOT), dithiocarbonic anhydride, Kohlendisulfid (schwefelkohlenstoff) (German), koolstofdisulfide (zwavelkoolstof) (Dutch), NCI-C04591, Schwefelkohlenstoff (German), solfuro di carbonio (Italian), sulphocarbonic anhydride, weevltox, wegla dwusiarczek (Polish)

USES
♦ Used in the Manufacturing Process
  • rayon
  • carbon tetrachloride
  • xanthogenates
  • floating agents
  • soil disinfectants
  • electronic vacuum tubes
  • agent in metal treatment and plating
  • corrosion inhibitor
  • polymerization inhibitor for vinyl chloride
  • agent in removal of metals in wastewater
  • regenerator for transition metal sulfide catalysts
  • development restrainer for instant color photography
  • flame lubricant in cutting glass
  • generating petroleum catalysts
  • optical glass
  • paint removers
  • tallow
  • explosives
  • varnishes
  • paints
  • enamels
  • rocket fuel
  • putty preservatives
  • rubber cement
♦ Used as a Solvent
  • phosphorus
  • sulfur
  • selenium
  • bromine
  • iodine
  • fats
  • resins
  • rubbers
  • for cleaning and extractions
  • waxes
  • lacquers
  • camphor
  • in extraction of growth inhibitors
♦ Used as a Chemical Intermediate
  • cellophane
  • rubber compounds
  • fumigant
  • sulfur and carbonyl sulfide
  • xanthates
  • vegadex herbicide via a dithiocarbamate
  • adhesives used in food packaging
  • rare earth sulfides

PHYSICAL PROPERTIES

- pure: colorless liquid
- impure: yellowish
- pure: sweet, pleasant, chloro-form-like odor
- commercial: foul odor, like rotten eggs
- odor threshold 0.1–0.2 ppm
- melting point −111.5°C
- boiling point 46.5°C at 760 mm Hg

CHEMICAL PROPERTIES

- flash point −5°C
- autoignition temperature 99°C
- heat of combustion −5814 Btu/lb
- heat of vaporization 84.1 cal/g at boiling point
- explosion limits in air 1.3% lower and 50% upper

- density 1.2632 at 20°C
- surface tension 32 dynes/cm
- coefficient of viscosity 0.363 at 20°C
- vapor pressure 352.6 mm Hg
- vapor density 2.67
- soluble in alcohol, benzene, and ether
- slightly soluble in water
- log K_{ow} 1.84–2.16

- heat of fusion 1.049 kcal/mole
- heat capacity 18.17 cal/mol/degree at 24.3°C
- ebullioscopic constant 2.35°
- dielectric constant 2.641 at low frequencies
- critical temperature 280°C
- critical pressure 72.9 atm at 20°C

HEALTH RISK

- Toxic Exposure Guidelines
  - ACGIH TLV 31 mg/m³
  - OSHA PEL 12 mg/m³
  - NIOSH REL 3 mg/m³
  - DFG MAK 30 mg/m³
- Toxicity Data
  - mmo-sat 100 µL/plate
  - sce-human:lym 10,200 µg/L
  - inhalation-man TCLo 40 mg/m³
  - inhalation-rat TCLo 100 mg/m³/8H
  - inhalation-human LCLo 4000 ppm/30M
  - inhalation-human LCLo 2000 ppm/5M
  - unr-man LDLo 186 mg/kg
  - oral-rat LD_{50} 3188 mg/kg
  - inhalation-rat LC_{50} 25 g/m³/2H
  - oral-mouse LD_{50} 2780 mg/kg
  - inhalation-mouse LC_{50} 10 g/m³/2H
  - oral-guinea pig LD_{50} 2125 mg/kg
  - intraperitoneal-guinea pig LDLo 400 mg/kg
- Acute Risks
  - changes in breathing
  - chest pains
  - nausea
  - vomiting
  - dizziness
  - fatigue
  - headache
  - mood changes
  - lethargy
  - blurred vision
  - delirium
  - euphoria
  - convulsions
  - respiratory failure

- irritation to eyes, mucous membranes, and upper respiratory tract
- thyroid hypofunction
- anesthetic

- brain chemistry changes in rats
- sensory and motor nerve conduction alterations in rats

◆ Chronic Risks
  - neurotoxic effects
  - behavioral changes
  - neurophysiological changes
  - reduced nerve conduction velocity
  - polyneuropathy
  - psychic abnormalities
  - coronary heart disease
  - increase incidence of angina
  - muscle pain
  - headaches
  - general fatigue
  - ocular effects
  - blisters

  - eczematous lesions
  - blood dyscrasias
  - affects liver enzymes of animals
  - decreased sperm count and libido in men
  - menstrual disturbances in women
  - congenital malformations in animals
  - embryotoxicity in animals
  - increased incidence in lymphatic leukemia
  - EPA Group D: not classifiable as to human carcinogenicity

HAZARD RISK
◆ flammable liquid
◆ dangerous fire hazard when exposed to heat, flame, sparks, friction, oxidizing materials
◆ severe explosive hazard when exposed to heat or flame
◆ ignition and potentially explosive reaction when heated in contact with rust or iron
◆ Mixtures with sodium or potassium-sodium alloys are powerful, shock-sensitive explosives.
◆ explodes on contact with permanganic acid
◆ potentially explosive reaction with nitrogen oxide
◆ Mixtures with dinitrogen tetraoxide are heat-, spark-, and shock-sensitive explosives.
◆ reacts with metal azides to produce shock- and heat-sensitive, explosive metal azidodithioformates
◆ Aluminum powder ignites in CS_2 vapor.
◆ Vapor ignites on contact with fluorine.
◆ reacts violently with azides
◆ incompatible with air, metals, oxidants
◆ Decomposition emits highly toxic fumes of SO_x.

EXPOSURE ROUTES
◆ Main route is in the workplace.
◆ use of carbon disulfide in manufacturing process
◆ release to air from industrial processes
◆ drinking water
◆ low amounts emitted from volcanoes and marshes

REGULATORY STATUS/INFORMATION
◆ Clean Air Act Amendment, Title III: hazardous air pollutants

- Safe Drinking Water Act: waste specific prohibitions—solvent wastes. 40CFR148.10
- Federal Insecticide, Fungicide, and Rodenticide Act
  - tolerances and exemptions from tolerances for pesticide chemicals in or on raw agricultural commodities. 40CFR180.102-1147
- Resource Conservation and Recovery Act
  - constituent for detection monitoring (for municipal solid waste landfill). 40CFR258
  - list of hazardous constituents. 40CFR258, 40CFR261, 40CFR261.11
  - groundwater monitoring list. 40CFR264
  - reference air concentrations. 40CFR266
  - health based limits for exclusion of waste-derived residues. 40CFR266
  - P waste #: 022
- CERCLA and Superfund
  - designation of hazardous substances. 40CFR302.4
  - list of extremely hazardous substances and their threshold planning quantities. 40CFR355-AB
  - list of chemicals. 40CFR372.65
- Clean Water Act
  - designation of hazardous substances. 40CFR116.4
  - determination of reportable quantities. 40CFR117.3
- Toxic Substance Control Act
  - OSHA chemicals in need of dermal absorption testing. 40CFR712.30.e10, 40CFR716.120.d10
- OSHA 29CFR1910.1000 Table Z1 and Table Z2
- chemical regulated by the State of California

ADDITIONAL/OTHER COMMENTS

- Symptoms
  - inhalation of vapors: irritation of skin, eyes, and mucous membranes, headache, garlicky breath, nausea, vomiting, diarrhea, abdominal pain, weak pulse, palpitations, fatigue, vertigo, hallucinations
- First Aid
  - wash eyes and skin immediately with large amounts of water
  - remove to fresh air immediately and provide respiratory support
  - wash mouth out with water
- fire fighting: use water spray, carbon dioxide, dry chemical powder, or foam
- personal protection: wear self-contained breathing apparatus, chemical goggles, and protective clothing
- Storage
  - keep tightly closed, away from sparks, heat, and open flame
  - can be stored in iron, aluminum, glass, porcelain, and teflon

SOURCES
◆ *Health Effects Notebook for Hazardous Air Pollutants*, U.S. Environmental Protection Agency, December 1994
◆ *Material Safety Data Sheets (MSDS)*
◆ *Sax's Dangerous Properties of Industrial Materials*, 9th edition 1996
◆ *Hawley's Condensed Chemical Dictionary*, 12th edition, 1993
◆ *Aldrich Chemical Company Catalog*, 1996–1997
◆ *Environmental Law Index to Chemicals*, 1996
◆ *Environmental Contaminant Reference Databook*, volume 1, 1995
◆ *Merck Index*, 12th edition, 1996

CARBON TETRACHLORIDE

CAS #: 56-23-5
MOLECULAR FORMULA: CCl_4
FORMULA WEIGHT: 153.81
SYNONYMS: benzinoform, carbona, carbon-chloride, carbon-tet, flukoids, freon-10, halon-104, methane-tetrachloride, methane, tetrachloro-, necatorina, perchloro-methane, R-10 (refrigerant), R-10, teraform, tetra-chloro-methane, tetra-finol, tetra-form, tetra-sol, univerm, vermoestricid
USES
◆ Used in Industry
  • recovery of tin in tin plating waste
  • formulation of gasoline additives
  • refrigerants and propellants for aerosol cans
  • metal degreasing
  • production of semiconductors
  • to reduce fire hazard in combinations with carbon disulfide or ethylene dichloride intended as grain fumigants
  • cleaning agent for machinery and electrical equipment
  • formerly a dry cleaning agent
  • formerly a fire extinguisher
  • formerly a pesticide
◆ Used as a Solvent
  • rubber cement
  • cable and semiconductor manufacture
  • oils
  • fats
  • lacquers
  • varnishes
  • rubber waxes
  • resins
◆ Used in Synthesis Processes
  • nylon-7
  • organic chlorination processes
  • polymer technology as reaction medium
  • catalyst
  • chlorination of organic compounds
  • soap perfumery
  • insecticides

96

- starting material in manufacture of organic compounds
◆ Used as an Intermediate
  - fluorocarbons
  - pesticides
  - hexachloroethane
  - tetrabromomethane
  - pyrosulfuryl chloride

PHYSICAL PROPERTIES
◆ colorless, clear, heavy liquid
◆ sweetish, aromatic smell
◆ odor threshold >10 ppm
◆ melting point −23°C
◆ boiling point 76.54°C
◆ density 1.5940 at 20°C
◆ surface tension (liquid-water) 45 dynes/cm
◆ surface tension (liquid) 270 dynes/cm
◆ viscosity 1.329 cP at 0°C
◆ vapor pressure 91.3 mm Hg at 20°C
◆ vapor density 5.32
◆ specific gravity 1.594
◆ insoluble in water
◆ miscible with alcohol, ether, chloroform, benzene, solvent naphtha, and most fixed and nonvolatile oils
◆ log K_{ow} 2.64

CHEMICAL PROPERTIES
◆ noncombustible
◆ generally inert
◆ heat of vaporization 8271.5 gcal/gmol
◆ heat of combustion 37.3 kcal/gmol at 20°C
◆ nonexplosive
◆ incompatible with alkali metals, finely powdered metals, and oxidizing agents

HEALTH RISK
◆ Toxic Exposure Guidelines
  - OSHA PEL (transitional TWA 10 ppm; CL 25 ppm; PK 200 ppm/5 min) TWA 2 ppm
  - ACGIH TLV TWA 5 ppm
  - ACGIH TLV STEL 30 ppm (skin)
  - DFG MAK 10 ppm
  - NIOSH REL 12.6 mg/m³
  - NIOSH REL CL 2 ppm/60M
  - MSHA standard 63 mg/m³
◆ Toxicity Data
  - skin-rabbit 4 mg MLD
  - skin-rabbit 500 mg/24H MLD
  - eye-rabbit 2200 µg/30S MLD
  - eye-rabbit 500 mg/24H MLD
  - mmo-sat 20 µL/L
  - mmo-asn 5000 ppm
  - inhalation-rat TCLo 250 ppm/8H
  - oral-rat TDLo 3 g/kg
  - subcutaneous-rat TDLo 15,600 mg/kg/12W-I:ETA
  - oral-mouse TDLo 4400 mg/kg/19W-I:NEO
  - subcutaneous-rat TD 182 g/kg/70W-I:CAR
  - inhalation-human TCLo 20 ppm:GIT
  - oral-woman TDLo 1800 mg/kg:EYE,CNS

- oral-man TDLo 1700 mg/kg:CNS, PUL, GIT
- inhalation-human LCLo 1000 ppm
- inhalation-human TCLo 45 ppm/3D:CNS,GIT
- inhalation-human TCLo 317 ppm/30M:GIT
- inhalation-human LCLo 5 pph/5M
- unknown-man LDLo 93 mg/kg
- oral-rat LD_{50} 2350 mg/kg
- inhalation-rat LC_{50} 8000 ppm/4H
- skin-rat LD_{50} 5070 mg/kg
- intraperitoneal-rat LD_{50} 1500 mg/kg
- oral-mouse LD_{50} 8263 mg/kg
- inhalation-mouse LC_{50} 9526 ppm/8H
- intraperitoneal-mouse LD_{50} 572 mg/kg
- oral-dog LDLo 1000 mg/kg
- inhalation-dog LCLo 14,620 ppm/8H
- intraperitoneal-dog LD_{50} 1500 mg/kg
- intravenous-dog LDLo 125 mg/kg
- inhalation-cat LCLo 38,110 ppm/2H
- subcutaneous-cat 300 mg/kg

◆ Other Toxicity Indicators
  - EPA Cancer Risk Level (1 in a million excess lifetime risk) 7×10^{-5} mg/m^3

◆ Acute Risks
  - primarily damages liver and kidneys
  - depression of central nervous system
  - headache
  - weakness
  - dizziness
  - lethargy
  - nausea
  - vomiting
  - delayed pulmonary edema
  - stomach pains
  - dermatitis
  - irritation of eyes with lacrimation and burning
  - dark urine
  - jaundice and liver enlargement

◆ Chronic Risks
  - liver and kidney damage
  - pupillary constriction
  - coma
  - antipsychotic effects
  - tremors
  - somnolence
  - anorexia
  - gastrointestinal effects
  - kidney and hepatic damage
  - acute nephrosis
  - production of polyneuritis
  - narrowing of visual fields
  - neurological changes
  - cirrhosis of the liver
  - decreased fertility in rats
  - decreased sperm production in rats
  - possible liver cancer
  - EPA Group B2: probable human carcinogen

HAZARD RISK

- forms impact-sensitive explosive mixtures with particulates of many metals
- violent or explosive reactions on contact with fluorine
- explosive mixtures with ethylene
- potentially explosive reaction on contact with boranes
- potentially dangerous reaction with dimethyl formamide
- caused explosions when used as a fire extinguisher on wax and uranium fires
- incompatible with aluminum trichloride, dibenzol peroxide, and potassium-tert-butoxide
- vigorous exothermic reaction with allyl alcohol
- Decomposition emits toxic fumes of Cl⁻ and phosgene.

EXPOSURE ROUTES

- air from accidental release from production and uses
- air from disposal in landfills
- building materials
- cleaning agents
- manufacture of carbon tetrachloride
- drinking contaminated water
- ingestion of bread or other products made with carbon-tetra-chloride-fumigated grain (in past)

REGULATORY STATUS/INFORMATION

- Clean Air Act Amendment, Title III: hazardous air pollutants
- Safe Drinking Water Act
  - 83 contaminants required to be regulated under the Safe Drinking Water Act of 1986. 40FR9352
  - public notification. 40CFR141.32
  - maximum contaminant level goal for organic contaminants. 40CFR141.50(a)
  - maximum contaminant level goal for organic chemicals. 40CFR141.61
  - variances and exemptions from the maximum contaminant levels for chemicals. 40CFR142.62
  - waste specific prohibitions—solvent wastes. 40CFR148.10
- Federal Insecticide, Fungicide, and Rodenticide Act
  - tolerances and exemptions from tolerances for pesticide chemicals in or on raw agricultural commodities. 40CFR180.102-1147
- Resource Conservation and Recovery Act
  - design criteria (for municipal solid waste landfill). 40CFR258.40
  - constituent for detection monitoring (for municipal solid waste landfill). 40CFR258
  - list of hazardous constituents. 40CFR258, 40CFR261, 40CFR261.11
  - groundwater monitoring list. 40CFR264
  - risk specific doses. 40CFR266
  - health based limits for exclusion of waste-derived residues. 40CFR266

- PICs found in stack effluents. 40CFR266
- D waste #: 019
- U waste #: 211
◆ CERCLA and Superfund
  - designation of hazardous substances. 40CFR302.4
  - list of chemicals. 40CFR372.65
◆ Clean Water Act
  - designation of hazardous substances. 40CFR116.4
  - determination of reportable quantities. 40CFR117.3
  - toxic pollutants. 40CFR401.15
  - total toxic organics (TTOs). 40CFR413.02
  - 126 priority pollutants. 40CFR423
◆ OSHA 29CFR1910.1000 Table Z1 and Table Z2
◆ chemical regulated by the State of California

ADDITIONAL/OTHER COMMENTS
◆ Symptoms
  - abdominal pain, anemia, blurred vision, coma, cyanosis, dermatitis, diarrhea, dizziness, dyspnea, headache, hematemesis, hematuria, hepatomegaly, jaundice, mental confusion, nausea, vomiting, optic atrophy, optic neuritis, proteinuria, pulmonary edema, weight loss
◆ First Aid
  - wash eyes and skin immediately with large amounts of water
  - remove to fresh air immediately and provide respiratory support
  - wash mouth out with water
◆ fire fighting: noncombustible; use extinguishing media appropriate for surrounding fire
◆ personal protection: wear self-contained breathing apparatus, chemical goggles, and protective clothing
◆ storage: keep tightly closed in a cool, dry place

SOURCES
◆ *Health Effects Notebook for Hazardous Air Pollutants*, U.S. Environmental Protection Agency, December 1994
◆ *Material Safety Data Sheets (MSDS)*
◆ *Sax's Dangerous Properties of Industrial Materials*, 9th edition, 1996
◆ *Hawley's Condensed Chemical Dictionary*, 12th edition, 1993
◆ *Aldrich Chemical Company Catalog*, 1996–1997
◆ *Environmental Law Index to Chemicals*, 1996
◆ *Environmental Contaminant Reference Databook*, volume 1, 1995

CARBONYL SULFIDE (gd)

CAS #: 463-58-1
MOLECULAR FORMULA: COS
FORMULA WEIGHT: 60.08 g/mol
SYNONYMS: carbon monoxide monosulfide, carbon oxide sulfide, carbon oxysulfide, carbonyl sulfide, oxycarbon sulfide, many others

USES
+ intermediate for herbicides
+ intermediate in the synthesis of organic sulfur compounds and alkyl carbonates
+ reaction of diluted sulfuric acid with ammonium thiocynate

PHYSICAL PROPERTIES
+ colorless gas
+ typical sulfur odor when pure
+ liquefies under pressure
+ hydrolyzed by water
+ melting point −138.2°C (−58.36°F)
+ boiling point −50.2°C (−216.76°F)
+ vapor pressure 9412 mm Hg at 25°C
+ vapor density 2.1 g/cm³ at 20°C
+ density 1.24 g/cm³ at −87°C
+ critical temperature 102°C (215.6°F)
+ critical pressure 58 atm
+ gas/vapor heavier than air
+ log K_{ow} .08009
+ very slightly soluble in water, alcohols, and toluene

CHEMICAL PROPERTIES
+ explosion limits 11.9% lower and 29% upper (by volume in air)
+ strong oxidizing agent
+ moderately water soluble
+ reacts with oxidizing agents, acids, water
+ Hazardous Combustion or Decomposition Products
  • carbon monoxide
  • carbon dioxide
  • sulfur oxides
  • hydrogen sulfide gas

HEALTH RISK
+ Toxicity Data
  • inhalation-rat LC_{50} 1110 ppm/4H
  • intraperitoneal-rat LD_{50} 23 mg/kg
  • inhalation-mouse LC_{50} 2770 mg/m³
  • inhalation-mouse LCLo 1200 ppm/35M
+ Acute Risks
  • Inhalation may be fatal as a result of spasm, inflammation and edema of the larynx and bronchi, chemical pneumonitis, and pulmonary edema.
  • harmful if swallowed or absorbed through skin

- extremely destructive to tissues of the mucous membranes and upper respiratory tract, eyes, and skin
  - Symptoms include burning sensation, coughing, wheezing, laryngitis, headache, shortness of breath, nausea and vomiting.
◆ Chronic Risks
  - serious nervous system impairment with narcotic effects
  - EPA has not established an RfC or an RfD.

HAZARD RISK
◆ highly flammable
◆ highly toxic
◆ poisonous
◆ corrosive
◆ freezing of tissues
◆ pricking
◆ gas/vapor explosive with air
◆ stupefying/asphyxiating

EXPOSURE ROUTES
◆ primarily by inhalation during its production and use
◆ released naturally into the atmosphere by volcanoes
◆ possible absorption through skin
◆ ingestion
◆ poison by intraperitoneal route

REGULATORY STATUS/INFORMATION
  - Clean Air Act, Title III
  - hazardous air pollutant. 42USC7412
◆ CERCLA and Superfund
  - chemicals and chemical categories for which a CAS # applies. 40CFR372.65
  - chemical regulated by the State of California
◆ Department of Transportation
  - classification: 2.3
  - label: poisonous gas; flammable gas
◆ Community Right-to-Know List. Reported in EPA TSCA Inventory.
◆ subject to SARA Section 313 reporting requirements

ADDITIONAL/OTHER COMMENTS
◆ Symptoms
  - irritating
  - skin burns
  - freezing of tissues
◆ First Aid
  - in case of accident seek medical advice immediately
  - wash eyes immediately with large amounts of water
  - discard contaminated clothing and shoes
  - if inhaled, remove to fresh air
  - if breathing is difficult, give oxygen
  - if swallowed, wash out mouth with water
◆ Fire Fighting

- extinguishing methods: stop flow of gas or use carbon dioxide, dry chemical, or water spray
- special procedures: wear self-contained breathing apparatus and protective clothing
◆ Personal Protection
- in case of insufficient ventilation, wear suitable respiratory equipment
- wear chemical safety goggles and rubber gloves

SOURCES

◆ *Sax's Dangerous Properties of Industrial Materials*, 9th edition, 1996
◆ *MSDS*, Aldrich Chemical Company Inc.
◆ http://ulisse.ei.jrc.it/cgibin_edc/sub_query_name
◆ *Environmental Law Index to Chemicals*, 1996 edition
◆ *Health Effects Notebook for Hazardous Air Pollutants*, U.S. Environmental Protection Agency, December 1994

CATECHOL

CAS #: 120-80-9
MOLECULAR FORMULA: $C_6H_4(OH)_2$
FORMULA WEIGHT: 110.12
SYNONYMS: o-benzenediol, 1,2-benzenediol, catechin, C.I. 76500, C.I. oxidation base 26, o-dihydroxybenzene, 1,2-dihydroxybenzene, o-dioxybenzene, o-diphenol, durafur developer C, fouramine PCH, fourrine 68, o-hydroquinone, o-hydroxyphenol, 2-hydroxyphenol, NCI-C55856, oxyphenic acid, pelagol grey C, o-phenyl-lenediol, pyrocatechin, pyrocatechinic acid, pyrocatechol, pyrocatechuic acid

USES

◆ Used in Industry
- photographic developer
- developer for fur dyes
- antiseptic
- electroplating
- specialty inks
- antioxidants
- light stabilizers
- organic synthesis
- polymerization inhibitors
- lubricating oils

PHYSICAL PROPERTIES

◆ colorless tablets or monoclinic crystals
◆ discolor in air
◆ faint, phenolic odor
◆ melting point 105°C
◆ boiling point 240°C
◆ density 1.341 at 15°C

◆ vapor pressure 0.03 mm Hg at 20°C
◆ vapor density 3.79
◆ soluble in water, chloroform, and benzene
◆ very soluble in alcohol and ether
◆ log K_{ow} 0.88

CHEMICAL PROPERTIES

◆ combustible
◆ flash point 261°F

◆ sublimes
◆ volatile with steam

HEALTH RISK

◆ Toxic Exposure Guidelines
- OSHA PEL TWA 5 ppm (skin)
- ACGIH TLV TWA 5 ppm (skin)

- ◆ Toxicity Data
  - mrc-smc 300 mg/L
  - dni-human:hla 200 μmol/L
  - dns-rat-oral 1 g/kg
  - subcutaneous-rat TDLo 5 mg/kg
  - oral-rat TDLo 437 g/kg/2Y-C:CAR
  - oral-mouse TDLo 645 g/kg/96W-C:CAR
  - oral-rat LD_{50} 260 mg/kg
  - subcutaneous-rat LDLo 110 mg/kg
  - oral-mouse LD_{50} 260 mg/kg
  - intraperitoneal-mouse LD_{50} 68 mg/kg
  - subcutaneous-mouse LD_{50} 247 mg/kg
  - intravenous-dog LDLo 40 mg/kg
  - skin-rabbit LD_{50} 800 mg/kg
  - intraperitoneal-guinea pig LDLo 150 mg/kg
  - parenteral-frog LDLo 160 mg/kg
- ◆ Acute Risks
  - eczematous dermatitis from skin contact
  - convulsions
  - illness similar to that from phenol
  - central nervous system depression
  - prolonged rise of blood pressure
  - peripheral vasoconstriction
  - respiratory failure
  - death
- ◆ Chronic Risks
  - No information is available on the chronic effects in humans.
  - increases effects of benzo[a]pyrene on the skin in mice
  - IARC Group 3: not classifiable as to carcinogenicity to humans

HAZARD RISK
- ◆ combustible when exposed to heat or flame
- ◆ can react vigorously with oxidizing materials
- ◆ hypergolic reaction with concentrated nitric acid
- ◆ Decomposition emits acrid smoke and irritating fumes.

EXPOSURE ROUTES
- ◆ manufacture and use
- ◆ consumption of contaminated drinking water
- ◆ ingestion of contaminated food

REGULATORY STATUS/INFORMATION
- ◆ Clean Air Act Amendment, Title III: hazardous air pollutants
- ◆ CERCLA and Superfund
  - designation of hazardous substances. 40CFR302.4
- ◆ chemical regulated by the State of California

ADDITIONAL/OTHER COMMENTS
- ◆ Symptoms
  - ingestion: vomiting, diarrhea, abdominal pain, sore throat, coughing
- ◆ First Aid

- wash eyes and skin immediately with large amounts of water
- remove to fresh air immediately or provide respiratory apparatus if necessary
- drink water or milk
- ◆ fire fighting: use water, carbon dioxide, and dry chemical
- ◆ personal protection: compressed air/oxygen apparatus, gas-tight suit

SOURCES
- ◆ *Health Effects Notebook for Hazardous Air Pollutants*, U.S. Environmental Protection Agency, December 1994
- ◆ *Material Safety Data Sheets (MSDS)*
- ◆ *Sax's Dangerous Properties of Industrial Materials*, 9th edition, 1996
- ◆ *Environmental Law Index to Chemicals*, 1996
- ◆ *Merck Index*, 12th edition, 1996
- ◆ *Hawley's Condensed Chemical Dictionary*, 12th edition, 1993
- ◆ *Aldrich Chemical Company Catalog*, 1996–1997

CHLORAMBEN (gd)

CAS #: 133-90-4
MOLECULAR FORMULA: $C_7H_5Cl_2NO_2$
FORMULA WEIGHT: 206.02 g/mol
SYNONYMS: 3-amino-2,5-dichloroacide, benzoic acid, chlorambeen, chlorambene, chlorambeno, clorambene, clorambeno, many others

USES
- ◆ pesticides
- ◆ herbicides
- ◆ selective weed control in crops such as soybeans
- ◆ plant growth regulation

PHYSICAL PROPERTIES
- ◆ colorless crystalline solid
- ◆ odorless
- ◆ melting point 200–201°C (392–393.8°F)
- ◆ vapor pressure 0.007 mm Hg at 100°C
- ◆ very soluble in ethyl alcohol
- ◆ very soluble in water
- ◆ Alkali salts are water soluble.

CHEMICAL PROPERTIES
- ◆ not available

HEALTH RISK
- ◆ Toxic Exposure Guidelines
  - LOAEL 15 mg/kg/d
  - Rfd 0.015 mg/kg/d
- ◆ Toxicity Data
  - oral-rat LD_{50} 3500 mg/kg
  - dermal-rat LD_{50} 2200 mg/kg
  - skin-rabbit LD_{50} 3136 mg/kg
- ◆ Acute Risks

- unlikely to present acute hazard in normal use
- Exposure to high levels results in mild to moderate dermal irritation.
◆ Chronic Risks
  - liver tumors reported in animals

HAZARD RISK
- very toxic
- decomposes on heating

EXPOSURE ROUTES
◆ inhalation
◆ contaminated drinking water
◆ absorption through skin during its use as a herbicide

REGULATORY STATUS/INFORMATION
◆ Clean Air Act Title III
  - hazardous air pollutant. 42USC7412
◆ Federal Insecticide, Fungicide and Rodenticide Act
  - pesticide classified for restriction. 40CFR152.175
◆ CERCLA and Superfund
  - chemicals and chemical categories for which a CAS # applies. 40CFR372.65
◆ chemical regulated by the State of California

ADDITIONAL/OTHER COMMENTS
◆ Personal Protection
  - compressed air/oxygen apparatus
  - gastight suit
  - keep upwind
◆ Spill or Leak
  - consider evacuation
  - collect rest carefully

SOURCES
◆ *Environmental Law Index to Chemicals*, 1996 edition
◆ *Health Effects Notebook for Hazardous Air Pollutants*, U.S. Environmental Protection Agency, December 1994
◆ http://ulisse.ei.jrc.it/cgibin_edc/sub_query_name
◆ *Material Safety Data Sheets (MSDS)*

CHLORDANE (gd)

CAS #: 57-74-9
MOLECULAR FORMULA: $C_{10}H_6Cl_8$
FORMULA WEIGHT: 409.78
SYNONYMS: aspon-chlordane, chloordaan, chlordan, chlorindan, chlor kil, chlortox, clordano, 1,2,4,5,6,7,8,8-octachlor-3a,4,7,7a-tetra-hydro-4,7-methanoindin chlordan, many others

USES
◆ pesticide
◆ insecticide
◆ for household and veterinary uses
◆ chlorination of chlordene
◆ fumigant

PHYSICAL PROPERTIES
- colorless to amber-colored viscous liquid
- pungent chlorine-like odor
- melting point $-116°C$ $(-176.8°F)$
- boiling point $175°C$ $(347°F)$ at 2 mm Hg
- vapor pressure 0.00001 mm Hg at $25°C$
- vapor density 3.9 g/cm^3 at $20°C$
- density 1.6 g/cm^3 at $25°C$
- specific gravity 0.69 g/mL
- not water soluble
- viscosity SSU 100 seconds at $38°C$
- infinitely soluble in kerosene and many organic solvents
- decomposes in weak alkalies
- refractive index 5.75 at $25°C$

CHEMICAL PROPERTIES
- explosion limits 1.1% lower and 6.7% upper (by volume in air)
- flash point $10°F$
- stable chemical
- incompatible with reducing agents
- produces HCl upon decomposition
- will not polymerize
- attacks plastics, rubber, and coatings
- strong oxidizer and alkaline reagent

HEALTH RISK
- Toxicity Data
  - inhalation-rat LC_{50} 100 mg/m
  - intraperitoneal-rat LD_{50} 343 mg/kg
  - oral-mouse LD_{50} 145 mg/kg
  - intraperitoneal-mouse LDLo 240 mg/kg
  - intravenous-mouse LD_{50} 100 mg/kg
  - inhalation-cat LC_{50} 100 mg/kg
  - skin-rabbit LD_{50} 780 mg/kg
- Acute Risks
  - irritates eyes, skin, nose, and throat
  - dizziness
  - nausea
  - abdominal pain
  - vomiting
  - facial congestion
  - limb convulsions
  - fever
  - unconsciousness
  - labored breathing
  - tachycardia
  - epileptiform seizures with frothing at the mouth
  - Rapid skin absorption results in blurred vision, cough, confusion, ataxia, and delirium.
- Chronic Risks
  - known to cause cancer in the State of California
  - liver and kidney damage

- anorexia
- loss of weight
- hepatocellular carcinomas

HAZARD RISK
- ◆ dangerous fire hazard

EXPOSURE ROUTES
- ◆ digging in soils in areas where it has been applied
- ◆ occupational exposure in the manufacture and formulation
- ◆ workers who apply the chemical, such as farmers, lawn-care workers, and pest-control workers
- ◆ from eating contaminated food

REGULATORY STATUS/INFORMATION
- ◆ Clean Air Act Title III
  - hazardous air pollutant. 42USC7412
- ◆ Safe Drinking Water Act
  - 83 contaminants required to be regulated under the SDWA of 1986. 47FR9352
  - organic chemicals other than total trihalomethanes, sampling and analytical requirements. 40CFR141.24
  - public notification. 40CFR141.32
  - maximum contaminant level goal (MCLG) for organic contaminants. 40CFR141.50(a)
  - maximum contaminant level (MCL) for organic chemicals. 40CFR141.61
  - variances and exemptions from the maximum contaminant levels for organic and inorganic chemicals. 40CFR142.62
- ◆ Resource Conservation and Recovery Act
  - Appendix 8: hazardous constituents. 40CFR261; see also 40CFR261.11
  - Appendix 9: groundwater modeling chemicals list. 40CFR264
  - Appendix 5: risk specific doses. 40CFR266
  - Appendix 7: health based limits for exclusion of waste-derived residues. 40CFR266
  - Table 1: maximum concentration of contaminants for the toxicity characteristic D020. 40CFR261.24
  - Table 302.4: list of hazardous substances and reportable quantities for designation. 40CFR302.4
  - discarded commercial chemical products, off-specification species, container residues, and spill residues thereof. characteristic P036 waste. 40CFR261.33
- ◆ CERCLA and Superfund
  - Appendix A: sequential CAS registry # list of extremely hazardous substances. 40CFR302.4
  - Appendix B: radionuclides. 40CFR302.4
  - designation of hazardous substances. 40CFR302.4
  - Appendix B: list of extremely hazardous substances and their threshold planning quantities. 40CFR355-AB
  - chemicals and chemical categories for which a CAS # applies. 40CFR372.65

◆ Clean Water Act
  • Table 116.4A: list of hazardous substances. 40CFR116.4A
  • Table 116.4B: list of hazardous substances by CAS #. 40CFR116.4B
  • Table 117.3: reportable quantities of hazardous substances designated pursuant to section 311 of CWA. 40CFR117.3
  • Appendix A: 126 priority pollutants. 40CFR423
◆ Federal Insecticide, Fungicide, and Rodenticide Act
  • class of chlorinated organic pesticides. 180.3(4)
◆ Occupational Safety and Health Act
  • Table Z1 limits for air contaminants under OSHA. 29CFR119.1000
◆ chemical regulated by the State of California
◆ Subject to SARA Title III, Section 313 reporting requirements
  • Community Right-to-Know List. EPA Extremely Hazardous Substances List.

ADDITIONAL/OTHER COMMENTS
◆ Spill or Leak
  • take up with absorbent material
  • ventilate area
  • eliminate all ignition sources
◆ First Aid
  • flush skin with large amounts of water
  • wash eyes immediately with large amounts of water
  • discard contaminated clothing and shoes
  • if inhaled, remove to fresh air
  • if breathing is difficult, give oxygen
  • if swallowed, seek medical attention
◆ Fire Fighting
  • extinguishing media: carbon dioxide, foam, and dry chemicals
  • special procedures: wear self-contained breathing apparatus when fighting chemical fire
◆ Personal Protection
  • in case of insufficient ventilation, wear face mask with organic vapor canister
  • wear chemical safety face shield and rubber gloves
  • wear impervious clothing

SOURCES
◆ *Sax's Dangerous Properties of Industrial Materials*, 9th edition, 1996
◆ *MSDS*, Supelco, Inc.
◆ http://ulisse.ei.jrc.it/cgibin_edc/sub_query_name
◆ *Environmental Law Index to Chemicals*, 1996 edition
◆ *Hawley's Condensed Chemical Dictionary*, 12th edition, 1993
◆ *CRC Handbook of Chemistry and Physics*, 77th edition, 1996–1997
◆ *NIOSH Pocket Guide to Chemical Hazards*, June 1994
◆ *Health Effects Notebook for Hazardous Air Pollutants*, U.S. Environmental Protection Agency, December 1994

CHLORINE (gd)

CAS #: 7782-50-5
MOLECULAR FORMULA: Cl_2
FORMULA WEIGHT: 70.90
SYNONYMS: bertholite, chloor, chlor, chlore, chlorine mol., chloro, molecular chlorine, a few others

USES
- ◆ Manufacture of Other Chemicals
  - carbon tetrachloride
  - trichloroethylene
  - chlorinated hydrocarbons
  - polychloroprene (neoprene)
  - polyvinylchloride
  - hydrogen chloride
  - ethylene dichloride
  - hypochlorous acid
  - metallic chlorides
  - chloroacetic acid
  - chlorobenzene
  - chlorinated lime
  - chlorates
  - chloroform
  - the extraction of bromine
- ◆ Used as a Component or in the Manufacturing Process
  - water purification
  - shrinkproofing wool
  - flame retardant compounds
  - special batteries (lithium or zinc)
  - paper products
  - dyestuffs
  - textiles
  - petroleum products
  - medicines
  - antiseptics
  - insecticides
  - solvents
  - paints
  - plastics
  - pulp bleaching
  - disinfectants
  - synthetic rubber
  - catalyst
  - processing meat, fish, vegetables, and fruits

PHYSICAL PROPERTIES
- ◆ dense, greenish-yellow, diatomic gas
- ◆ clear, amber liquid
- ◆ pungent, very irritating odor
- ◆ nonmetallic halogen element
- ◆ critical temperature 144.0°C

- ◆ critical pressure 78.525 atm absolute
- ◆ critical volume 1.763 L/kg
- ◆ melting point −101°C (−149.8°F)
- ◆ boiling point −34.9°C (30.8°F)
- ◆ liquefaction pressure 7.86 atm at 25°C
- ◆ liquid density 1.47 g/L at 0°C and 3.65 atm
- ◆ vapor density 3.21 g/L at 0°C
- ◆ vapor pressure 4800 mm Hg (6.8 atm) at 20°C
- ◆ water solubility 0.64 g Cl per 100 g water
- ◆ very low electrical conductivity
- ◆ slightly soluble in cold water
- ◆ soluble in chlorides and alcohols

CHEMICAL PROPERTIES
- ◆ oxidizing gas under pressure
- ◆ noncombustible but supports combustion
- ◆ flash point 0°C
- ◆ extremely strong oxidizing agent
- ◆ strongly electronegative
- ◆ Dangerous on Contact with
  - • turpentine
  - • ether
  - • ammonia
  - • hydrocarbons
  - • hydrogen
  - • powdered metals
  - • and other reducing materials
- ◆ combines with moisture to form HCl

HEALTH RISK
- ◆ Toxic Exposure Guidelines
  - • ACGIH TLV TWA 0.5 ppm; STEL 1 ppm
  - • OSHA PEL TWA 0.5 ppm; STEL 1 ppm
  - • NIOSH REL (chlorine) .5 ppm for 15 minutes
  - • DFG MAK 0.5 ppm (1.5 mg/m^3)
- ◆ Toxicity Data
  - • inhalation-rat LC$_{50}$ 293 ppm for 1 hour
  - • inhalation-human LCLo 500 ppm for 5 minutes
  - • inhalation-mouse LC$_{50}$ 137 ppm for 1 hour
  - • inhalation-dog LCLo 800 ppm for 30 minutes
  - • inhalation-cat LCLo 660 ppm for 4 hours
  - • inhalation-rabbit LDLo 660 ppm for 4 hours
- ◆ Acute Risks
  - • poisonous and corrosive
  - • moderately toxic to humans by inhalation
  - • very irritating by inhalation
  - • human respiratory effects by inhalation
    - • changes in the trachea or bronchi
    - • emphysema
    - • chronic pulmonary edema or congestion
  - • strong irritant to eyes and mucous membranes
  - • Liquid burns the skin.

- ◆ Chronic Risks
- ◆ effects limited to minor changes in pulmonary function

HAZARD RISK
- ◆ NFPA rating: Health 2, Flammability 0, Reactivity 1, No special hazards
- ◆ Explodes on Contact with
  - • molten aluminum
  - • ammonia
  - • amidosulfuric acid
  - • biuret
  - • tert-butanol
  - • 3-chloropropyne
  - • diborine
  - • diethyl ether
  - • and many, many others
- ◆ ignition or explosion reaction with metals
- ◆ violently reacts with many alcohols
- ◆ a military poison

EXPOSURE ROUTES
- ◆ inhalation
- ◆ occupational health risk in industries where produced
- ◆ drinking water and swimming water
- ◆ accidental releases
- ◆ rapid absorption through skin
- ◆ absorption by ingestion

REGULATORY STATUS/INFORMATION
- ◆ Clean Air Act Title III
  - • hazardous air pollutant. 42USC7412
- ◆ Safe Drinking Water Act
  - • priority list of drinking water contaminants. 55FR1470
- ◆ Resource Conservation and Recovery Act
  - • Appendix 4: reference air concentrations. 40CFR266
- ◆ CERCLA and Superfund
  - • Appendix A: sequential CAS registry # list of extremely hazardous substances. 40CFR302.4
  - • Appendix B: radionuclides. 40CFR302.4
  - • designation of hazardous substances. 40CFR302.4
  - • Appendix B: list of extremely hazardous substances and their threshold planning quantities. 40CFR355-AB
  - • chemicals and chemical categories for which a CAS # applies. 40CFR372.65
- ◆ Clean Water Act
  - • Table 117.3: reportable quantities of hazardous substances designated pursuant to section 311 of CWA. 40CFR117.3
- ◆ Occupational Safety and Health Act
  - • Table Z1 limits for air contaminants under OSHA. 29CFR119.1000
  - • Appendix A: list of highly hazardous chemicals, toxics, reactives. 57FR6407
- ◆ Department of Transportation

- classification: 2.3
- label: poisonous gas
◆ chemical regulated by the State of California
◆ Community Right-to-Know List. EPA Extremely Hazardous Substances List.

ADDITIONAL/OTHER COMMENTS
◆ Spill or Leak
  - take up with absorbent material and wash site afterward
  - ventilate area
  - eliminate all ignition sources
◆ First Aid
  - flush skin with large amounts of water
  - wash eyes immediately with large amounts of water
  - discard contaminated clothing and shoes
  - if inhaled, remove to fresh air
  - if breathing is difficult, give oxygen
  - if swallowed, seek medical attention
◆ Fire Fighting
  - extinguishing media: water spray or fog nozzle
  - special procedures: wear self-contained breathing apparatus and protective clothing
◆ Personal Protection
  - in case of insufficient ventilation, wear face mask with canister
  - wear chemical safety face shield and rubber gloves
  - wear impervious clothing

SOURCES
◆ *Sax's Dangerous Properties of Industrial Materials*, 9th edition, 1996
◆ *MSDS*, Aldrich Chemical Company, Inc.
◆ http://ulisse.ei.jrc.it/cgibin_edc/sub_query_name
◆ *Environmental Law Index to Chemicals*, 1996 edition
◆ *Hawley's Condensed Chemical Dictionary*, 12th edition, 1993
◆ *CRC Handbook of Chemistry and Physics*, 77th edition, 1996–1997
◆ *NIOSH Pocket Guide to Chemical Hazards*, June 1994
◆ *Health Effects Notebook for Hazardous Air Pollutants*, U.S. Environmental Protection Agency, December 1994

CHLOROACETIC ACID (bd)

CAS #: 79-11-8
MOLECULAR FORMULA: $ClCH_2COOH$
FORMULA WEIGHT: 94.50
SYNONYMS: acide chloracetique, acide monochloracetique, acidomonocloroacetico, alpha chloroacetic acid, chloroacetic acid, chloroacetic acid, liquid (UN1750) (DOT), chloroacetic acid, solid (UN1751) (DOT), chloroethanoic acid, mca, monochloracetic acid, monochloroacetic acid, monochloroethanoic acid, NCI-C60231

USES

◆ Used as a Chemical Intermediate in the Production of Other Chemicals
  • pharmaceuticals (e.g., vitamin A)
  • indigoid dyes
  • carboxymethylcellulose
  • ethyl chloroacetate
  • glycine
  • synthetic caffeine
  • sarcosine
  • thioglycolic acid
  • ethylenediaminetetraacetic acid (EDTA)
◆ Other Common Uses
  • bacteriostat (analytical chemistry)
  • preservative
  • herbicide
  • a defoliant

PHYSICAL PROPERTIES

◆ colorless or white crystals; colorless to light-brownish crystals
◆ melting point: alpha 63°C (145°F), beta 55–56°C (131–133°F), gamma 50°C (122°F) at 760 mm Hg
◆ boiling point: alpha, beta, gamma 189°C (372°F) at 760 mm Hg
◆ density 1.580 g/mL at 20°C, 1.4043 g/mL at 40°C (solid)
◆ specific gravity 1.40 (water = 1.0)
◆ vapor density 3.26 (air = 1.0)
◆ surface tension 33 dynes/cm (0.033 N/m) at 80°C (176°F)
◆ heat of vaporization 139 cal/g (5.82×10^5 J/kg) (250 Btu/lb)
◆ vinegar odor
◆ vapor pressure 1 mm Hg at 43°C, 3 mm Hg at 55°C, less than 1 mm Hg at 20°C
◆ octanol/water partition coefficient log k_{ow} 0.22
◆ very soluble in water
◆ slight solubility in chloroform; soluble in acetone, carbon disulfide, benzene, ethanol, diethyl ether, carbon tetrachloride

CHEMICAL PROPERTIES

◆ stable
◆ Polymerization will not occur.
◆ flash point 126°C (259°F)
◆ autoignition temperature >500°C (>932°F)
◆ flammability limits 8% lower and upper not determined
◆ heat of combustion −1.008 cal/g (-342.17×10^5 J/kg) (−1.814 Btu/lb) (solid)
◆ enthalpy (heat) of formation at 25°C (crystals) −510.5 kJ/mol
◆ enthalpy of fusion at 334.4 K (61.3°C) 12.3 kJ/mol
◆ dissociation constants pk_a 2.86
◆ Compound absorbs water from the air and forms a syrup.
◆ will react on contact with strong bases and most common metals
◆ corrosive to metals

HEALTH RISK
- Toxic Exposure Guidelines
  - Threshold Limit Value (TLV/TWA) not established
  - Short-Term Exposure Limit (STEL) not established
  - Permissible Exposure Limit (PEL) not established
- Toxicity Data
  - oral-rat LD_{50} 76 mg/kg
  - inhalation-rat LC_{50} 180 mg/m^3
  - subcutaneous-rat LD_{50} 5 mg/kg
  - oral-mouse LD_{50} 255 mg/kg
  - subcutaneous-mouse LD_{50} 250 mg/kg
  - aquatic toxicity rating TLm 96: 100–1000 ppm
  - intraperitoneal-mouse LDLo 500 mg/kg
- Other Toxicity Indicators
  - probable oral lethal dose (human) 50–500 mg/kg, between 1 teaspoon and 1 oz for 150 lb person
- Acute Risks
  - irritation of skin, eyes, and upper respiratory tract
  - destructive to tissue of the mucous membranes
  - headaches
  - coughing
  - wheezing
  - difficult breathing
  - chest pains
  - severe lung irritation
  - pulmonary edema
  - chemical pneumonitis
  - inflammation of the larynx
  - convulsions
  - nausea
  - vomiting
  - gastrointestinal irritation
- Chronic Risks
  - blood disorders
  - liver disorders
  - kidney damage
  - damage to central nervous system
  - heart damage
  - damage to skeletal muscles
  - chromosomal mutation data reported
  - questionable carcinogen with experimental tumorigenic data

HAZARD RISK
- NFPA rating: Health 3, Flammability 1, Reactivity 0
- slight fire hazard
- combustible liquid when exposed to heat or flame
- corrosive and combustible solid
- When heated to decomposition it emits toxic gases such as hydrogen chloride, phosgene, carbon monoxide, and carbon dioxide.

- ◆ incompatible or reacts violently with most common metals, strong bases, alkalies, amines and ammonia, strong oxidizing agents and strong reducing agents
- ◆ Closed containers exposed to heat may explode.

EXPOSURE ROUTES

- ◆ inhalation
- ◆ contact with skin and eyes
- ◆ ingestion
- ◆ readily absorbed through skin
- ◆ drinking of contaminated water
- ◆ job related exposure in industries that use chloroacetic acid in the manufacturing process
- ◆ subcutaneous and intravenous routes

REGULATORY STATUS/INFORMATION

- ◆ Clean Air Act Amendments of 1990, Title III: hazardous air pollutant
- ◆ CERCLA and Superfund
  - Reportable Quantity (RQ): 1 lb
  - list of extremely hazardous substances and their threshold planning quantities. 40CFR355-AB: 40CFR 355 Appendix B
  - list of chemicals. 40CFR372.65
  - threshold planning quantity (TPQ): 100/10,000 lb
- ◆ chemical regulated by the State of California
- ◆ Department of Transportation
  - hazard class/division: 6.1
  - identification number: UN1751
  - labels: poison, corrosive
  - regulatory references: 49CFR 172.101
- ◆ International
  - hazard class: 6.1, 8
  - identification number: UN1751
  - labels: 6 toxic, corrosive
  - regulatory references: 49CFR Part 176; IMDG code
- ◆ Air
  - hazard class: 6.1, 8
  - identification number: UN1751
  - labels: 6 poison, corrosive
  - regulatory references: 49CFR Part 175; ICAO

ADDITIONAL/OTHER COMMENTS

- ◆ Symptoms
  - inhalation: headache, coughing, shortness of breath, chest pains, mucous membrane irritation
  - skin contact: severe irritation or burns
  - eye contact: severe irritation or burns
  - skin absorption: dermatitis
  - ingestion: burns to mouth, throat, and stomach; headache, nausea, vomiting, convulsions
- ◆ First Aid
  - flush eyes immediately with running water for 15 minutes

- flush skin with large amounts of water
- if inhaled, remove to fresh air and give artificial respiration
- if swallowed, do not induce vomiting; give large amounts of water; may also give milk or milk of magnesia
◆ fire fighting: use water spray, dry chemical, alcohol-resistant foam, or carbon dioxide
◆ personal protection: if intense heat or flame is involved, wear special protective clothing and positive pressure self-contained breathing apparatus
◆ storage: store in a cool, dry, well-ventilated, corrosion-proof area; keep container tightly closed and separate from alkalies, alcohols, oxidizing materials, reducing agents, and metals

SOURCES
◆ *ACGIH Threshold Limit Values (TLVs) for 1996*
◆ *Aldrich Chemical Company Catalog*, 1996–1997
◆ *CRC Handbook of Chemistry and Physics*, 77th edition, 1996–1997
◆ *CRC Handbook of Thermophysical and Thermochemical Data*, 1994
◆ *Environmental Contaminant Reference Databook*, Volume 1, 1995
◆ *Environmental Law Index to Chemicals*, 1996 edition
◆ *Fire Protection Guide to Hazardous Materials*, 11th edition, 1994
◆ *Hawley's Condensed Chemical Dictionary*, 12th edition, 1993
◆ *Hazardous Substances Data Bank (HSDB) Fact Sheet*
◆ *Material Safety Data Sheets (MSDS)*
◆ *Merck Index*, 12th edition, 1996
◆ *NIOSH Pocket Guide to Chemical Hazards*, June 1994
◆ *Sax's Dangerous Properties of Industrial Materials*, 9th edition, volumes 1 and 2, 1996
◆ *Suspect Chemicals Sourcebook*, 1996 edition

2-CHLOROACETOPHENONE

CAS #: 532-27-4
MOLECULAR FORMULA: $C_6H_5COCH_2Cl_2$
FORMULA WEIGHT: 154.60
SYNONYMS: CAF, CAP, chloroacetophenone, chloro-acetophenone, liquid, chloroacetophenone(1-), chloro-acetophenone(alpha-), chloroacetophenone(omega-), chloro-acetophenone(a-), chloro-methyl-phenyl-ketone, 2-chloro-1-phenylethanone, chloro-1-phenylethanone(2-), CN, mace, mace (lacrimator), phenacyl-chloride, phenyl-chloromethyl ketone

USES
◆ Used in Manufacturing Process
  - riot-control agent (tear gas)
  - chemical mace
  - pharmaceutical intermediate
  - alcohol denaturant

PHYSICAL PROPERTIES
- colorless to gray crystals
- odor resembling apple blossoms
- odor threshold 0.035 ppm
- melting point 56.5°C
- boiling point 244–245°C
- vapor pressure 5.4×10^{-3} mm Hg at 20°C
- log K_{ow} 2.09
- practically insoluble in water
- soluble in alcohol, ether, and benzene
- specific gravity 1.32

CHEMICAL PROPERTIES
- stable
- heat of combustion −9340 Btu/lb
- incompatible with strong oxidizers, water, and alkali
- will not polymerize

HEALTH RISK
- Toxic Exposure Guidelines
  - OSHA PEL TWA 0.05 ppm
  - ACGIH TLV TWA 0.05 ppm
  - NIOSH REL 0.05 ppm
- Toxicity Data
  - skin-rat 12%/6H open MOD
  - skin-rabbit 5 mg/24H MLD
  - skin-rabbit 12%/6H open MOD
  - eye-rabbit 1 mg MLD
  - eye-rabbit 3 mg SEV
  - skin-guinea pig 12%/6H open MOD
  - skin-mouse TDLo 2400 mg/kg/27W-I:NEO
  - inhalation-human LCLo 159 mg/m³/20M
  - inhalation-human TCLo 93 mg/m³/3M:EYE
  - inhalation-human TCLo 20 mg/m³:EYE
  - oral-rat LD$_{50}$ 127 mg/kg
  - inhalation-rat LCLo 417 mg/m³/15M
  - intraperitoneal-rat LD$_{50}$ 36 mg/kg
  - intravenous-rat LD$_{50}$ 41 mg/kg
  - oral-mouse LD$_{50}$ 139 mg/kg
  - inhalation-mouse LCLo 600 mg/m³/15M
  - intraperitoneal-mouse LD$_{50}$ 60 mg/kg
  - intravenous-mouse LD$_{50}$ 81 mg/kg
  - oral-rabbit LD$_{50}$ 118 mg/kg
  - inhalation-rabbit LCLo 465 mg/m³/20M
  - intravenous-rabbit LD$_{50}$ 30 mg/kg
  - oral-guinea pig LD$_{50}$ 158 mg/kg
  - inhalation-guinea pig LCLo 490 mg/m³/30M
  - intraperitoneal-guinea pig LD$_{50}$ 17 mg/kg
- Acute Risks
  - potent eye, throat, and skin irritant
  - burning of the eyes with lacrimation
  - blurred vision
  - possible corneal damage
  - irritation of nose, throat, and skin
  - burning of nose, throat, and skin
  - burning in chest with dyspnea
  - laryngotracheobronchitis

- first, second, and third degree burns
- conjunctiva irritation
- difficulty breathing
◆ Chronic Risks
  - no information available for humans
  - squamous hyperplasia of the nasal respiratory epithelium in rats
  - nasal lesions in rats

- agitation
- coma
- contraction of pupils
- loss of reflexes

- increase in fibroadenomas of the mammary gland in rats
- lung injury
- acute pulmonary edema

HAZARD RISK
◆ Combustion will produce carbon dioxide, carbon monoxide, and hydrogen chloride.
◆ capable of creating dust explosions
◆ incompatible with strong oxidizers, water, and alkali
◆ Decomposition emits toxic fumes of Cl^-.

EXPOSURE ROUTES
◆ use as a tear gas
◆ use as a chemical mace to control riots
◆ manufacture and use by inhalation and dermal contact

REGULATORY STATUS/INFORMATION
◆ chemical regulated by the State of California

ADDITIONAL/OTHER COMMENTS
◆ Symptoms
  - inhalation: tearing, burning of eyes, difficulty in breathing
  - contact: irritation of skin and eyes
  - ingestion: agitation, coma, contraction of pupils, loss of reflexes
◆ First Aid
  - wash eyes and skin immediately with large amounts of water
  - remove to fresh air or provide respiratory support
  - induce vomiting
◆ fire fighting: use dry chemical, water spray or mist, or carbon dioxide
◆ personal protection: wear self-contained breathing apparatus and protective clothing and goggles
◆ Storage
  - keep from contact with oxidizing materials
  - keep container tightly closed and away from water or alkali

SOURCES
◆ *Health Effects Notebook for Hazardous Air Pollutants*, U.S. Environmental Protection Agency, December 1994
◆ *Material Safety Data Sheets (MSDS)*
◆ *Sax's Dangerous Properties of Industrial Materials*, 9th edition, 1996
◆ *Hawley's Condensed Chemical Dictionary*, 12th edition 1993
◆ *Aldrich Chemical Company Catalog*, 1996-1997
◆ *Environmental Law Index to Chemicals*, 1996

◆ *Merck Index*, 12th edition, 1996
◆ *Environmental Contaminant Reference Databook*, volume 1, 1995

CHLOROBENZENE (bd)

CAS #: 108-90-7
MOLECULAR FORMULA: C_6H_5Cl
FORMULA WEIGHT: 112.56
SYNONYMS: benzene chloride, chloorbenzeen, chlorbenzene, chlorbenzol, chlorobenzen, chlorobenzol, clorobenzene, mcb, monochloorbenzeen, monochlorbenzene, monochlorbenzol, monochlorobenzene, monoclorobenzene, NCI-C54886, phenyl chloride, RCRA waste number UO37

USES
◆ Manufacture of Other Chemicals
  • dyestuffs
  • phenol
  • aniline
  • cumene
  • DDT
  • o- and p-dichlorobenzene isomers
  • chloronitrobenzenes
  • insecticides
◆ Used as a Solvent or in the Manufacturing Process
  • adhesives
  • paints
  • polishes
  • waxes
  • pharmaceuticals
  • diisocyanates
  • textiles
  • dry cleaning
  • surface coatings
  • tar and grease remover
  • surface coating removers
  • natural rubber
◆ used as a heat transfer medium

PHYSICAL PROPERTIES
◆ clear, colorless liquid
◆ almond-like odor
◆ odor threshold 0.210 ppm (mg/m³)
◆ melting point −45°C (−49°F) at 760 mm Hg
◆ boiling point 132°C (269°F) at 760 mm Hg
◆ density 1.1058 g/mL at 20°C
◆ specific gravity 1.11 (water = 1.0)
◆ vapor density 3.9 (air = 1.0)
◆ vapor pressure 12 mm Hg at 25°C
◆ surface tension 33 dynes/cm at 25°C
◆ heat of vaporization 75 cal/g (3.14×10^5 J/kg)
◆ viscosity 0.790 cP at 21.1°C (70°F)
◆ negligible solubility in water (448 ppm in water)
◆ soluble in alcohol, ether, benzene, chloroform, and carbon disulfide; miscible with nearly all organic solvents
◆ heat capacity 150.1 J/K-mol at 298.15K (25°C)
◆ % volatiles by volume 100 (21°C)
◆ octanol/water partition coefficient log K_{ow} 2.18–2.84

CHEMICAL PROPERTIES
◆ generally very stable

- Polymerization will not occur.
- will react vigorously with strong oxidizers
- Combustion by-products include phosgene and hydrogen chloride gases.
- flash point 28°C (84°F)
- flammability limits 1.3% lower and 7.1% upper
- autoignition temperature 592°C (1099°F)
- heat of combustion 6700 cal/g (280×10^5 J/kg)
- enthalpy (heat) of formation at 25°C (liquid) 11.0 kJ/mol
- enthalpy of fusion 9.61 kJ/mol at 227.9K (-45.3°C)
- enthalpy of vaporization 40.97 kJ/mol at 25°C
- thermal conductivity 0.1266 W/m-K at 298.15K (25°C)

HEALTH RISK

- Toxic Exposure Guidelines
  - ACGIH TLV TWA 10 ppm (46 mg/m^3)
  - OSHA PEL TWA 75 ppm (350 mg/m^3)
  - NIOSH REL TWA not established
  - IDLH + 1000 ppm
- Toxicity Data
  - oral-rat LD$_{50}$ 2.29 g/kg
  - inhalation-rat TCLo 210 ppm/6H
  - intraperitoneal-rat LD$_{50}$ 1655 mg/kg
  - oral-mouse LD$_{50}$ 1.44 g/kg
  - oral-guinea pig LD$_{50}$ 5.06 g/kg
  - oral-rabbit LD$_{50}$ 2250 mg/kg
- Other Toxicity Indicators
  - Clinical examination of workers exposed to chlorobenzene in the manufacture of polyvinyl chloride showed that some workers suffered from poisoning at concentrations in the atmosphere close to 50 mg/m^3.
- Acute Risks
  - irritation of skin, eyes, and upper respiratory tract
  - nausea
  - vomiting
  - dizziness
  - headaches
  - gastrointestinal effects
  - drowsiness
  - unconsciousness
  - rapid respiration
  - red urine
  - irregular pulse
  - twitching of extremities
- Chronic Effects
  - kidney damage
  - liver damage
  - central nervous system depression
  - blood disorders
  - bone marrow disease
  - lung damage

- glandular disorders
- may alter genetic material
- affects the formation of platelets which are necessary for blood clotting

HAZARD RISK
- ◆ NFPA rating: Health 2, Flammability 3, Reactivity 0
- ◆ dangerous fire hazard when exposed to heat or flame
- ◆ moderate explosion hazard when exposed to heat or flame
- ◆ Contact with strong oxidizers may cause fire.
- ◆ Vapors may flow to distant ignition sources and flash back.
- ◆ Closed containers exposed to heat may explode.
- ◆ forms explosive mixtures with powdered sodium or phosphorus trichloride and sodium
- ◆ violent reaction with silver perchlorate and dimethyl sulfoxide
- ◆ when heated to decomposition, emits toxic gases such as hydrogen chloride, phosgene, carbon monoxide, and carbon dioxide

EXPOSURE ROUTES
- ◆ primarily by inhalation
- ◆ ingestion
- ◆ eye contact
- ◆ skin contact
- ◆ occupational exposure in industries that use chlorobenzene to manufacture other chemicals
- ◆ contaminated drinking water
- ◆ by intraperitoneal route

REGULATORY STATUS/INFORMATION
- ◆ Clean Air Act Amendments of 1990, Title III: hazardous air pollutants
- ◆ Safe Drinking Water Act
  - one of 83 contaminants required to be regulated under the 1986 act
  - waste specific prohibitions—solvent wastes. 40CFR148.10
- ◆ Resource Conservation and Recovery Act
  - constituents for detection monitoring for municipal solid waste landfills. 40CFR258
  - list of hazardous constituents. 40CFR258, 40CFR261, and 40CFR261.11
  - groundwater monitoring list. 40CFR264
  - health based limits for exclusion of waste-derived residues. 40CFR266
  - PICs found in stack effluents. 40CFR266
  - D waste #: D021
  - U waste #: U037
- ◆ CERCLA and Superfund
  - Reportable Quantity (RQ): 100 lbs
  - designation of hazardous substances. 40CFR302.4
  - list of chemicals. 40CFR372.65
- ◆ Clean Water Act

- designation of hazardous substances. 40CFR116.4
- reportable quantity list. 40CFR117.3
- total toxic organics (TTOs). 40CFR413.02
- 126 priority pollutants. 40CFR423
◆ Toxic Substance Control Act
  - OSHA chemicals in need of dermal absorption testing. 40CFR712.30.e10, 40CFR716.120.d10
  - reporting on precursor chemical substances. 40CFR766.38
◆ OSHA, 29CFR1910.1000 Table Z1: limits for air contaminants.
◆ chemical regulated by the State of California
◆ Department of Transportation
  - hazard class/division: 3
  - identification number: UN1134
  - Reportable Quantity (RQ): 100 lbs.
  - labels: 3 flammable liquid
  - Regulatory References: 49CFR172.101

ADDITIONAL/OTHER COMMENTS
◆ Symptoms
  - inhalation: irritation of eyes, skin, and mucous membranes
  - ingestion: headache, nausea, vomiting, dizziness, gastrointestinal irritation
◆ First Aid
  - flush eyes immediately with plenty of water for at least 15 minutes
  - wash skin immediately with large amounts of water for at least 15 minutes
  - dilute by drinking water, if swallowed; do not induce vomiting
  - if inhaled, remove to fresh air; provide artificial respiration and administer oxygen as needed
◆ fire fighting: use alcohol foam, dry chemical, or carbon dioxide. (Water may be ineffective.)
◆ personal protection: wear self-contained breathing apparatus as well as safety goggles and impervious clothing for splash protection
◆ storage: Outside or detached storage is preferred; inside storage should be in a standard flammable liquids storage room.

SOURCES
◆ *ACGIH Threshold Limit Values (TLVs) for 1996*
◆ *Aldrich Chemical Company Catalog*, 1996–1997
◆ *CRC Handbook of Chemistry and Physics*, 77th edition, 1996–1997
◆ *CRC Handbook of Thermophysical and Thermochemical Data*, 1994
◆ *Environmental Contaminant Reference Databook*, volume 1, 1995
◆ *Environmental Law Index to Chemicals*, 1996 edition
◆ *Fire Protection Guide to Hazardous Materials*, 11th edition, 1994

- *Hawley's Condensed Chemical Dictionary*, 12th edition, 1993
- *Hazardous Substances Data Bank (HSDB) Fact Sheet*
- *Material Safety Data Sheets (MSDS)*
- *Merck Index*, 12th edition, 1996
- *NIOSH Pocket Guide to Chemical Hazards*, June 1994
- *Sax's Dangerous Properties of Industrial Materials*, 9th edition, volumes 1 and 2, 1996
- *Suspect Chemicals Sourcebook*, 1996 edition

CHLOROBENZILATE (bd)

CAS #: 510-15-6
MOLECULAR FORMULA: $C_{16}H_{14}Cl_2O_3$
FORMULA WEIGHT: 325.20
SYNONYMS: acar, acaraben, acaraben-4E, akar, akar-50, akar-338, benzilan, benz-o-chlor, caswell-no. 434, chlorbenzilate, chlorobenzylate, compound 338, 4,4'-dichlorobenzilate, 4,4'-dichlorobenzilic acid ethyl ester, ENT-18,596, EPA-pesticide-code-028801, ethyl-4-chloro-alpha-(4-chlorophenyl)-alpha-hydroxybenzeneacetate, ethyl-p,p'-dichlorobenzilate, ethyl-4,4'-dichlorobenzilate, ethyl-4,4'-dichlorodiphenyl glycollate, ethyl-4,4'-dichlorophenyl glycollate, ethyl ester of 4,4'-dichlorobenzilic acid, ethyl-2-hydroxy-2,2-bis-(4-chlorophenyl) acetate, folbex, folbex smoke strips, G-338, G-23992, geigy 338, kop mite, NCI-C00408, NCI-C60413, RCRA waste number U038

USES
- Used as a Component in Agriculture
  - acaricide in spider mite control (former use in U.S.A.)
  - acaricide for citrus crops, ornamentals, cotton, (noncitrus) fruits, and nuts (former use in U.S.A.)
  - on premise and for plant mite control (former use in U.S.A.)
  - synergist for DDT (former use in U.S.A.)
  - in hives against the bee mite *Acarapis woodi* (for use outside U.S.A.)
  - organochlorine pesticide (use has been restricted)

PHYSICAL PROPERTIES
- colorless solid (pure)
- thick, dark-yellow liquid (technical product)
- aromatic odor
- melting point 36–37.3°C
- boiling point 146–148°C at 0.04 mm Hg
- density 1.2816 g/cm³ at 20°C
- specific gravity 1.28 at 20°C
- vapor pressure 2.2×10^{-6} mm Hg (1.20×10^{-6} mbar) at 20°C
- slightly soluble in water 10.0 mg/L at 20°C
- soluble in most organic solvents, such as acetone (1 kg/kg acetone), dichloromethane, methanol, and toluene; 600 g/kg hexane; 700 g/kg 1-octanol; slightly soluble in benzene
- index of refraction 1.5727 at 20°C

CHEMICAL PROPERTIES
- very stable

- ◆ Hazardous polymerization will not occur.
- ◆ flash point >200°F
- ◆ autoignition temperature 470°C
- ◆ susceptible to gas phase reaction with hydroxyl radicals, if released to the atmosphere; rate constant for the vapor phase reaction with photochemically produced hydroxyl radicals 4.92×10^{-12} cm³/mol-sec at 25°C
- ◆ hydrolyzed by alkali and by strong acids to the inactive p,p'-dichlorobenzilic acid and ethanol
- ◆ incompatible with lime
- ◆ high affinity for activated charcoal
- ◆ dehalogenated by sodium in isopropyl alcohol

HEALTH RISK
- ◆ Toxic Exposure Guidelines
  - no occupational exposure standards or recommendations listed in the United States or any foreign countries
- ◆ Toxicity Data
  - oral-rat LD_{50} 2784–3880 mg/kg
  - oral-female rat LD_{50} 1220 mg/kg. Toxic effect malignant neoplasms
  - oral-male rat LD_{50} 1040 mg/kg. Toxic effect malignant neoplasms
  - inhalation-rat LC_{50} >2.28 mg/L air/4H
  - oral-mouse LD_{50} 729 mg/kg. Toxic effect malignant neoplasms
  - oral-mouse TDLo 71 g/kg/82W
  - percutaneous-rabbit LD_{50} >10,000 mg/kg
  - eye-rabbit 25 mg MOD
  - skin-rabbit 125 mg open MLD
  - oral-dog LD_{50} 500 mg/kg. Toxic effect arteriosclerosis
- ◆ Other Toxicity Indicators
  - rainbow trout LC_{50} 0.7 mg/L/96H at 13°C
  - bobwhite quail LC_{50} 3375 ppm/7d
  - mallard duck LC_{50} >8000 ppm/5d
  - sheepshead minnow LC_{50} 1.0 mg/L/48H
- ◆ Acute Risks
  - headaches
  - tiredness
  - nausea
  - diarrhea
  - vomiting
  - convulsions
  - dizziness
  - confusion
  - tremors
  - fatal in high exposures
  - respiratory failure
  - tingling of the tongue, lips, or parts of face
  - a skin and eye irritant
  - apprehension
  - excitability
  - breathing difficulty
  - mild delirium and fever
  - muscle pains
- ◆ Chronic Risks
  - adverse testicular effects
  - affects liver functions
  - kidney damage

- affects protein synthesis, lipid synthesis, detoxification, excretion
- observed microsomal enzyme induction
- increased alkaline phosphatase and aldolase activity
- damage to central nervous system
- blood disorders

HAZARD RISK
- ◆ NFPA rating: Health 2, Flammability 0, Reactivity 0
- ◆ When heated to decomposition it emits toxic fumes of Cl^-.
- ◆ Runoff from fire control or dilution water may give off poisonous gases and cause water pollution.
- ◆ incompatible with lime
- ◆ easily hydrolyzed in strong alkali or acid; the dichlorobenzilic acid is unstable and readily decarboxylates

EXPOSURE ROUTES
- ◆ inhalation
- ◆ eye contact
- ◆ dermal absorption
- ◆ ingestion
- ◆ contaminated drinking water
- ◆ occupational exposure among persons associated with the production, formulation, and agricultural application of chlorobenzilate

REGULATORY STATUS/INFORMATION
- ◆ Clean Air Act Amendments of 1990, Title III: hazardous air pollutant
- ◆ Federal Insecticide, Fungicide, and Rodenticide Act
  - class of chlorinated organic pesticides, 180-3(4)
- ◆ Resource Conservation and Recovery Act
  - list of hazardous organic constituents. 40CFR258, 40CFR261, 40CFR261.11
  - groundwater monitoring list. 40CFR264
  - U waste #: U038
- ◆ CERCLA and Superfund
  - designation of hazardous substances. 40CFR302.4
  - list of chemicals 40CFR372.65
  - Reportable Quantity (RQ): 10 lbs. (4.54 kg)
- ◆ chemical regulated by the State of California
- ◆ Department of Transportation
  - hazard class/division: Poison B
  - identification number: UN 2761
  - labels required: none

ADDITIONAL/OTHER COMMENTS
- ◆ Symptoms
  - inhalation: irritates skin and eyes, respiratory failure
  - ingestion: convulsions, vomiting, diarrhea
- ◆ First Aid
  - move victim from contaminated air to fresh air; apply artificial respiration if necessary; if breathing is difficult, give oxygen

- flush eyes immediately with large amounts of running water
- wash skin immediately with plenty of soap and water, including hair and under fingernails
- if ingested, immediately give 1 or 2 glasses of water to drink and induce vomiting
  - Valium may be useful in controlling convulsions.
◆ fire fighting: For small fires, use dry chemical, carbon dioxide, halon, water spray, or standard foam extinguishing media; for large fires, water spray, fog, or standard foam is recommended.
◆ personal protection: Wear full chemical protective clothing and a NIOSH-approved pesticide respirator or supplied-air respirator.
◆ storage: Chlorobenzilate should be stored in a dry, well-ventilated, secure area, away from alkali or strong acids.

SOURCES
◆ *CRC Handbook of Chemistry and Physics*, 77th edition, 1996–1997
◆ *Environmental Law Index to Chemicals*, 1996 edition
◆ *Hazardous Substances Data Bank (HSDB) Fact Sheet*
◆ *Material Safety Data Sheet (MSDS)*
◆ *Merck Index*, 12th edition, 1996
◆ *Sax's Dangerous Properties of Industrial Materials*, 9th edition, volumes 1 and 2, 1996
◆ *Suspect Chemicals Sourcebook*, 1996 edition

CHLOROFORM (bd)

CAS #: 67-66-3
MOLECULAR FORMULA: $CHCl_3$
FORMULA WEIGHT: 119.38
SYNONYMS: chloroforme, cloroformio, formyl trichloride, freon-20, methane trichloride, methane, trichloro-, methenyl chloride, methenyl trichloride, methyl trichloride, NCI-CO2686, R-20, RCRA waste number U044, tcm, trichloormethaan, trichloroform, trichloromethane

USES
◆ Used as a Pharmaceutical
  - general anesthetic (former use)
  - component of cough syrups
  - component of toothpastes (former use)
  - component of toothache compound (former use)
  - component of liniments
◆ Used as a General Solvent for Other Compounds

- adhesives
- pesticides
- fats
- oils
- rubbers
- alkaloids
- waxes
- gutta-percha (latex)
- resins
- penicillins

◆ Other uses
- cleansing agent
- rubber industry
- in fire extinguishers to lower the freezing temperature of carbon tetrachloride
- insecticidal fumigant
- chemical intermediate for fluorocarbon
- chemical intermediate for dyes and pesticides
- chemical intermediate for tribromomethane
- dry cleaning agent
- polymer chain transfer agent

PHYSICAL PROPERTIES
◆ clear, colorless liquid
◆ melting point −63°C (−82°F) at 760 mm Hg
◆ boiling point 62°C (143°F) at 760 mm Hg
◆ density 1.4832 g/cm^3 at 20°C
◆ specific gravity 1.48 (water = 1.0)
◆ surface tension 27.1 dynes/cm at 20°C
◆ viscosity 5.63 mP at 20°C, 5.10 mP at 30°C
◆ octanol/water partition coefficient log K_{ow} 1.97
◆ heat capacity 114.2 J/mol-K (liquid at 298.15K)
◆ vapor density 4.13 (air = 1.0)
◆ heat of vaporization 59.3 cal/g (2.483×10^5 J/kg)
◆ vapor pressure 100 mm Hg at 10.4°C, 159 mm Hg at 20°C, 245 mm Hg at 30°C
◆ pleasant, sweet odor
◆ odor threshold 3.30 mg/L
◆ slight solubility in water
◆ % volatiles by volume 100 (21°C)
◆ miscible with alcohol, ether, benzene, carbon disulfide, petroleum ether, carbon tetrachloride, fixed and volatile oils
◆ very refractive liquid

CHEMICAL PROPERTIES
◆ generally stable
◆ Hazardous polymerization will not occur.
◆ develops acidity from prolonged exposure to air and light
◆ Addition to acetone in the presence of a base will result in a highly exothermic reaction.
◆ reacts vigorously with disilane (Si_2H_6) when exposed to sunlight
◆ Addition to methanol in the presence of sodium hydroxide will result in an exothermic reaction.
◆ enthalpy (heat) of formation at 25°C (liquid) 134.5 kJ/mol
◆ Gibbs energy of formation at 25°C (liquid) 73.7 kJ/mol
  - standard molar entropy at 25°C (liquid) 201.7 J/K-mol
  - enthalpy of vaporization at 25°C 31.28 kJ/mol
  - thermal conductivity at 25°C 0.1172 W/m-K
  - enthalpy of fusion at 209.5K 8.8 kJ/mol
  - practically nonflammable

HEALTH RISK
◆ Toxic Exposure Guidelines

- ACGIH TLV TWA 10 ppm (49 mg/m^3) suspected human carcinogen
- OSHA PEL (transitional: ceiling level 50 ppm) TWA 2 ppm (9.78 mg/m^3)
- NIOSH REL (waste anesthetic gases and vapors) ceiling level 2 ppm/1 hour; (chloroform) ceiling level 2 ppm/60 min
- IDLH 500 ppm

◆ Toxicity Data
- oral-human LDLo 140 mg/kg
- inhalation-human LCLo 25,000 ppm/5 months
- inhalation-human TCLo 5000 mg/m^3/7 months
- oral-rat LD$_{50}$ 908 mg/kg
- inhalation-rat LC$_{50}$ 47,702 mg/m^3/4H
- subcutaneous-mouse LD$_{50}$ 704 mg/kg
- intraperitoneal-mouse LD$_{50}$ 1 g/kg
- inhalation-mouse LC$_{50}$ 28 g/m^3
- oral-dog LD$_{50}$ 2250 mg/kg
- oral-rabbit LD$_{50}$ 9827 mg/kg

◆ Other Toxicity Indicators
- rainbow trout LC$_{50}$ 43,800 µg/L/96H
- largemouth bass LC$_{50}$ 51 ppm/96H
- channel catfish LC$_{50}$ 75 ppm/96H
- pink shrimp LC$_{50}$ 81,500 µg/L/96H
- Acute responses from exposure at various concentrations of chloroform in man have been reported to be fainting sensation and vomiting from 4096 ppm; dizziness and salivation after a few minutes at 1475 ppm; increased intracranial pressure and nausea in 7 minutes; aftereffects, fatigue and headache for several hours at 1024 ppm.
- In six acute fatalities due to the intentional or forced inhalation of chloroform, blood levels of 10–48 mg/L and urine levels of 0–60 mg/L were observed.

◆ Acute Risks
- irritation of skin, eyes, and mucous membranes
- dryness of mouth and throat
- headaches
- fatigue
- dizziness
- mental dullness
- nausea
- unconsciousness
- will defat tissue
- vomiting
- hallucinations
- nervous aberrations
- death in high exposures

◆ Chronic Risks
- kidney and liver damage
- cardiac-respiratory failure
- paralysis

- hypotension
- toxemia
- may alter genetic material
- EPA: suspected carcinogen for man

HAZARD RISK
- NFPA rating Health 2, Flammability 0, Reactivity 0
- slight fire hazard when exposed to high heat; otherwise practically nonflammable
- will react violently with (acetone and a base), aluminum disilane, lithium, magnesium, nitrogen tetroxide, potassium, (perchloric acid and phosphorus pentoxide), (potassium hydroxide and methanol), potassium-tert-butoxide, sodium, (sodium hydroxide and methanol), sodium methylate
- incompatible with acetone, strong alkalies, chemically active metals, dinitrogen tetraoxide, fluorine, potassium-tert-butoxide, sodium, sodium hydroxide, methanol, sodium methoxide, triisopropyl-phosphine
- When heated to decomposition, it emits toxic fumes of hydrogen chloride gas, phosgene gas, carbon monoxide, carbon dioxide.
- forms explosive mixtures with chemically active metals; weak explosion with lithium; fairly strong with sodium; strong with potassium; and violent with sodium-potassium alloys
- explodes when in contact with aluminum powder or magnesium powder
- can be explosive when confined with water; poses explosive hazard if present in boiler feed or cooling water

EXPOSURE ROUTES
- inhalation
- ingestion
- eye contact
- skin contact and absorption
- drinking of water contaminated with chloroform due to its indirect production in the chlorination of drinking water; spills and releases on land will also leach into the groundwater where it will reside for long periods of time
- occupational exposure in industries that use chloroform as an industrial solvent, extractant, and chemical intermediate
- automobile exhaust
- atmospheric decomposition of trichloroethylene

REGULATORY STATUS / INFORMATION
- Clean Air Act, Title III: hazardous air pollutants. The original list was published in 42USC7412
- Safe Drinking Water Act
  - priority list of drinking water contaminants. 55FR1470
  - community water systems and nontransient, noncommunity water systems. 40CFR141.40.e
- Resource Conservation and Recovery Act
  - constituents for detection monitoring for municipal solid waste landfills. 40CFR258

- list of hazardous constituents. 40CFR258, 40CFR261, 40CFR261.11
- groundwater monitoring chemicals list. 40CFR264
- risk specific doses. 40CFR266
- health based limits for exclusion of waste-derived residues. 40CFR266
- PICs found in stack effluents 40CFR266
- D waste #: D022
- U waste #: U044; discarded commercial chemical products, off-specification species, container residues, and spill residues thereof. 40CFR261.33

◆ CERCLA and Superfund
  - Reportable Quantity (RQ): 10 lbs (4.54 kg)
  - Threshold Planning Quantity (TPQ): 10,000 lbs.
  - designation of hazardous substances. 40CFR302.4
  - list of extremely hazardous substances and their threshold planning quantities. 40CFR355
  - list of chemicals. 40CFR372.65

◆ Clean Water Act
  - designation of hazardous substances. 40CFR116.4
  - reportable quantity list. 40CFR117.3
  - list of toxic pollutants. 40CFR401.15
  - total organic toxics (TTOs). 40CFR413.02
  - 126 priority pollutants. 40CFR423

◆ OSHA, 29CFR1910.1000 Table Z1: limits for air contaminants

◆ Department of Transportation
  - hazard class/division: 6.1
  - identification number: UN1888
  - labels: 6 keep away from food (domestic)

ADDITIONAL/OTHER COMMENTS

◆ Symptoms
  - inhalation: headache, dizziness, drowsiness, and irritation of upper respiratory tract
  - skin contact: irritation; prolonged contact may cause dermatitis
  - eye contact: irritation; may cause temporary corneal damage
  - ingestion: nausea, vomiting, gastrointestinal irritation, burns to mouth and throat

◆ First Aid
  - flush eyes immediately with plenty of water for at least 15 minutes
  - in case of contact, flush skin with soap and water
  - provide artificial respiration; give oxygen if breathing is difficult
  - if swallowed, drink water or milk and induce vomiting
  - keep victim warm if unconscious and having convulsions

- fire fighting: use extinguishing agent suitable for type of surrounding fire; use water to keep fire exposed containers cool
- personal protection: in high concentrations of vapor, wear goggles and self-contained breathing apparatus
- storage: store in cool, dry, well-ventilated location; keep container tightly closed and isolate from strong alkalies and strong mineral acids

SOURCES
- *ACGIH Threshold Limit Values (TLVs) for 1996*
- *Aldrich Chemical Company Catalog*, 1996–1997
- *CRC Handbook of Chemistry and Physics*, 77th edition, 1996–1997
- *CRC Handbook of Thermophysical and Thermochemical Data*, 1994
- *Environmental Contaminant Reference Databook*, volume 1, volumes 1 and 2, 1996
- *Environmental Law Index to Chemicals*, 1996 edition
- *Fire Protection Guide to Hazardous Materials*, 11th edition, 1994
- *Hawley's Condensed Chemical Dictionary*, 12th edition, 1993
- *Hazardous Substances Data Bank (HSDB) Fact Sheet*
- *Material Safety Data Sheets (MSDS)*
- *Merck Index*, 12th edition, 1996
- *NIOSH Pocket Guide to Chemical Hazards*, June 1994
- *Sax's Dangerous Properties of Industrial Materials*, 9th edition, volumes 1 and 2, 1996
- *Suspect Chemicals Sourcebook*, 1996 edition

CHLOROMETHYL METHYL ETHER (yf)

CAS #: 107-30-2
MOLECULAR FORMULA: C_2H_5ClO
FORMULA WEIGHT: 80.51
SYNONYMS: chlorodimethyl ether, chloromethoxymethane, CMME, dimethylchloroether, methyl chloromethyl ether, monochlorodimethyl ether, monochloromethyl ether

USES
- chloroalkylating agent in the preparation of anionic exchange resins
- lachrimatory agent (military poison)
- intermediate for chloromethylated compounds, especially ion-exchange resins
- used in the process of hydrochlorination of methanol and formaldehyde
- used in the production of chloromethylated compounds

PHYSICAL PROPERTIES
- colorless liquid
- boiling point 59.15°C at 760 mm Hg
- decomposes easily in water
- vapor pressure 3.55 psi at 20°C

♦ specific gravity 1.060 (water = 1.0)

♦ melting point −103.5°C at 760 mm Hg
♦ volatile liquid

CHEMICAL PROPERTIES
♦ flash point 15°C
♦ highly flammable

HEALTH RISK
♦ Toxicity Data
  • inhalation-rat 55 ppm/7H
  • inhalation-mouse 1030 mg/m³/2H
  • inhalation-hamster 65 ppm/7H
  • oral-rat 817 mg/kg
  • target organs: lungs, eyes, kidneys
  • mutagenetic/chromosomal aberrations
  • LC_{50} mice 1030 mg/m³
  • LC_{50} hamsters 214 mg/m³
  • LC_{50} rats 181 mg/m³
♦ Acute Risks
  • OSHA-regulated carcinogen
  • dermatitis
  • causes burns
  • symptoms include burning sensation, coughing, wheezing, laryngitis, shortness of breath, headache, nausea and vomiting
  • may be fatal if inhaled, swallowed, or absorbed through skin
  • destructive to tissues of mucous membranes, upper respiratory tract, eyes, and skin
  • Inhalation may be fatal as a result of spasm, inflammation and edema of larynx and bronchi, chemical pneumonitis, and pulmonary edema.
♦ Chronic Risks
  • carcinogen (cancer causing agent)
  • chronic bronchitis
  • tumors

HAZARD RISK
♦ extremely flammable: on combustion, forms toxic gases and vapors such as phosgene, carbon monoxide, and hydrogen chloride
♦ Vapor/air mixtures are explosive.
♦ decomposes in water producing hydrogen chloride and formaldehyde
♦ caustic to metals present in water
♦ decomposes upon heating; forms peroxide
♦ reacts with rubber
♦ readily hydrolyzed

EXPOSURE ROUTES
♦ inhalation
♦ ingestion
♦ absorption through skin and eyes

REGULATORY STATUS/INFORMATION
♦ Occupational Safety and Health Act

- regulated carcinogen under 29CFR1910, subpart z
- Highly Hazardous Chemicals 29CFR1910
- OSHA TQ 500 lb
◆ EPA's genetic toxics program, 1988 includes CMME
◆ EPA's Toxic Substance Control Act Sections 8(b) and (d)
◆ NTP 7th Annual Report on Carcinogens, 1992, lists CMME as known to be carcinogenic.
◆ Clean Air Act
  - section 112 Statutory Air Pollutants (42USC7401, 7412, 7416, 7601)
  - section 112 High Risk Pollutant 40CFR63; section 112(r) Accidental Release Prevention Substances (40CFR part 68)
◆ CERCLA and Superfund
  - Hazardous Substances 40CFR302
  - CERCLA RQ 10 lbs.; TPQ 100 lbs.
  - Superfund Amendments Reclamation Act Title III Section 302 Extremely Hazardous Substances (40CFR Chapter 1, Part 355.50)
  - Superfund Amendments Title III Section 313 Toxic Chemicals (40CFR Chapter 1 Part 372)
◆ American Conference of GIH Threshold Limit Values Chemicals: suspected human carcinogen
◆ Department of Traffic Appendix A 49CFR172.101

ADDITIONAL/OTHER COMMENTS
◆ fire fighting: use carbon dioxide, dry chemical powder, or foam
◆ disposal: mix with a combustible solvent and burn in chemical incinerator equipped with afterburner and scrubber
◆ Personal Protection
  - when working with chloromethyl methyl ether, wear NIOSH/MSHA approved respirator, chemical-resistant gloves, safety goggles, and protective clothing
  - use only in a chemical fume hood
◆ Storage: store in a cool, dry place
◆ confirmed carcinogen with experimental carcinogenic, tumorigenic, and neoplastigenic data
◆ When heated to decomposition it emits toxic fumes of Cl^-.

SOURCES
◆ *Material Safety Data Sheets (MSDS)*
◆ University of Kentucky web page
  http://www.chem.uky.edu/resources/msds.html
◆ *Environmental Law Index to Chemicals*, 1996 edition
◆ *Health Effects Notebook for Hazardous Air Pollutants*, U.S. Environmental Protection Agency, December 1994
◆ *Sax's Dangerous Properties of Industrial Materials*, 9th edition, 1996

CHLOROPRENE (ch)

CAS #: 126-99-8
MOLECULAR FORMULA: C_4H_5Cl
FORMULA WEIGHT: 88.54
SYNONYMS: 2-chloor-1,3-butadieen (Dutch), 2-Chlor-1,3-butadien (German), chlorobutadiene, 2-chlorobuta-1,3-diene, 2-chloro-1,3-butadiene, chloropreen (Dutch), Chloropren (German, Polish), chloroprene, beta-chloroprene (OSHA, MAK), chloroprene inhibited (DOT), chloroprene uninhibited (DOT), neoprene

USES
- ◆ Manufacture of Other Chemicals
  - • neoprene
- ◆ Used as a Component or in the Manufacturing Process
  - • artificial rubber
  - • adhesives for food packaging
  - • wire and cable jackets
  - • gaskets
  - • roof coatings
  - • binder for fibers

PHYSICAL PROPERTIES
- ◆ colorless liquid
- ◆ melting point −130°C (−202.6°F) at 760 mm Hg
- ◆ boiling point 59.4°C (139°F) at 760 mm Hg
- ◆ density 0.958 g/cm³ at 20°C
- ◆ specific gravity 1.0 (water = 1.00)
- ◆ vapor density 3.0 (air = 1.00)
- ◆ Vapor has pungent, ethereal odor.
- ◆ odor threshold 15 ppm (55 mg/m³)
- ◆ vapor pressure 174 mm Hg at 20°C
- ◆ very slightly soluble in water
- ◆ soluble with ether, acetone, benzene; soluble in alcohol and diethyl ether

CHEMICAL PROPERTIES
- ◆ polymerizes on standing
- ◆ will react with peroxides and other oxidizers, liquid or gaseous fluorine
- ◆ flammability limits 4.0% lower and 20.0% upper (percent by volume in air)
- ◆ autoignition temperature 39°F (4.0°C)

HEALTH RISK
- ◆ Toxic Exposure Guidelines
  - • ACGIH TLV TWA 10 ppm (skin)
  - • OSHA PEL TWA 10 ppm (skin)
  - • NIOSH REL 1 ppm/15M
  - • IDLH 300 ppm
- ◆ Toxicity Data
  - • microsomal mutagenicity assay-salmonella typhimurium 2 pph/4H
  - • cytogenetic analysis-human-dose 1 mg/m³

- inhalation-rat TCLo 4 mg/m^3/24H (female 3–4D post): TER
- inhalation-rat TCLo 10 ppm/4H (female 3–20D post): REP
- oral-rat LD$_{50}$ 450 mg/kg
- oral-rat TDLo 9100 µg/kg/26W-I
- inhalation-rat LC$_{50}$ 11,800 mg/m^3/4H
- subcutaneous-rat LDLo 500 mg/kg
- inhalation-cat LCLo 1290 mg/m^3/8H
- subcutaneous-cat LDLo 100 mg/kg
- intravenous-rabbit LDLo 96 mg/kg

◆ Acute Risks
- poison by ingestion, intravenous, and subcutaneous routes
- Vapor first causes irritation of the respiratory tract.
- depression of respiration
- asphyxia
- dermatitis
- conjunctivitis
- corneal necrosis
- anemia
- temporary loss of hair
- nervousness
- irritability

◆ Chronic Risks
- highly toxic with repeated ingestion or inhalation
- questionable carcinogen
- causes severe degenerative changes in the vital organs
- reproductive effects

HAZARD RISK

◆ NFPA rating Health 2, Flammability 3, Reactivity 0
◆ very dangerous fire hazard when exposed to heat or flame
◆ explosive in vapor form when exposed to heat or flame
◆ autooxidizes in air to form an unstable peroxide that catalyzes exothermic polymerization of the monomer
◆ when heated to decomposition emits toxic fumes of Cl$^-$
◆ reacts with oxidizing materials
◆ incompatible or reacts violently with chromic anhydride

EXPOSURE ROUTES

◆ Workers may be exposed by inhalation or dermal exposure.
◆ during its manufacture, transport, and storage and during the manufacture of polychloroprene elastomers and polychloroprene-containing products

REGULATORY STATUS/INFORMATION

◆ Clean Air Act hazardous air pollutants. Title III. (The original list was published in 42USC7412.)
◆ Resource Conservation and Recovery Act
- list of hazardous inorganic and organic constituents. 40CFR258 Appendix 2
- hazardous constituents. 40CFR261 Appendix 8. see also 40CFR261.11

- groundwater monitoring chemicals list. 40CFR264 Appendix 9
◆ CERCLA and Superfund
  - chemicals and chemical categories to which this part applies (CAS number listing). 40CFR372.65
◆ Toxic Substance Control Act
  - list of substances. 40CFR716.120.a
◆ chemical regulated by the State of California

ADDITIONAL/OTHER COMMENTS
◆ Symptoms
  - irritation of respiratory tract, depression of respiration
  - asphyxia
  - irritation of skin
◆ First Aid
  - call a doctor
  - if inhaled, prompt removal from exposure
  - flush eyes with water
  - clean skin with soap and water
  - if ingested, gastric lavage followed by saline catharsis
◆ fire fighting: carbon dioxide, dry chemicals, water spray, alcohol foam; shut off source of fuel
◆ personal protection: wear skin protection and self-contained breathing apparatus
◆ storage: *not stored;* must be destroyed in special waste incinerators and flue gases must be scrubbed

SOURCES
◆ *Environmental Contaminant Reference Databook*, volume 1, 1995
◆ *Environmental Law Index to Chemicals*, 1996 edition
◆ *Fire Protection Guide to Hazardous Materials*, 11th edition, 1994
◆ *Health Effects Notebook for Hazardous Air Pollutants*, U.S. Environmental Protection Agency, December 1994
◆ *Material Safety Data Sheets (MSDS)*
◆ *Merck Index*, 12th edition, 1996
◆ *NIOSH Pocket Guide to Chemical Hazards*, June 1994
◆ *Sax's Dangerous Properties of Industrial Materials*, 9th edition, volume 2, 1996

CHROMIUM COMPOUNDS

CAS #: Chromium compounds have variable CAS #s. The CAS # for chromium is 7440-47-3.
MOLECULAR FORMULA: Chromium compounds have variable molecular formulas. The molecular formula for chromium is Cr.

Note: For all hazardous air pollutants (HAPs) that contain the word "compounds" in the name, and for glycol ethers, the following applies: Unless otherwise specified, these HAPs are defined by the 1990 Amendments to the Clean Air Act as including any unique chemical substance that contains the named chemical (i.e., antimony, arsenic, etc.) as part of that chemical's infrastructure.

FORMULA WEIGHT: Chromium compounds have variable formula weights. The formula weight for chromium is 52.00.

SYNONYMS: chrome

USES
- ◆ Chromium
  - making steel
  - alloys
  - bricks in furnaces
  - increases resistance and durability of metals
  - protective coating for automotive and equipment accessories
  - nuclear and high temperature research
- ◆ Chromium Compounds (in Either Chromium(III) or Chromium(IV) Forms)
  - chrome plating
  - manufacture of dyes and pigments
  - leather and wood preservation
  - treatment of cooling tower water
  - drilling muds
  - textiles
  - toner for copying machines

PHYSICAL PROPERTIES
- ◆ steel-gray metallic pieces, powder, and flakes
- ◆ odorless
- ◆ Compounds have strong and varied colors.
- ◆ melting point 1890°C
- ◆ boiling point 2672°C
- ◆ density 7.14
- ◆ vapor pressure 1 mm Hg/70°F at 1616°C
- ◆ vapor density 7.1 g/cm^3
- ◆ Chromium(III) compounds are sparingly soluble in water.
- ◆ Chromium(IV) compounds are readily soluble in water.
- ◆ Cr ion forms many coordination compounds.
- ◆ exists in active and passive forms
- ◆ soluble in acids (except nitric) and strong alkalies

CHEMICAL PROPERTIES
- ◆ heat capacity 5.58 cal/g-atom deg at 25°C
- ◆ latent heat of fusion 3.5 kcal/g-atom
- ◆ latent heat of vaporization 81.7 kcal/g-atom
- ◆ reacts with dilute HCl and H_2SO_4
- ◆ attacked by caustic alkalies and alkali carbonates
- ◆ not oxidized by air, even in presence of much moisture
- ◆ Active form reacts readily with dilute acids to form chromous salts.

HEALTH RISK

◆ Toxic Exposure Guidelines
  - OSHA PEL (Cr metal and insoluble compounds) 1 mg/m³
  - OSHA PEL (Cr (II) and Cr (III) compounds) 0.5 mg/m³
  - ACGIH TLV (Cr metal and Cr (II) and Cr (III) compounds) 0.5 mg/m³
  - ACGIH TLV (water soluble Cr (VI) compounds) 0.05 mg/m³
  - NIOSH REL (ceiling) (chromic acid) 0.2 mg/m³
  - NIOSH REL (chromic acid) 0.05 mg/m³
  - NIOSH REL (other Cr (VI) compounds) 0.025 mg/m³
  - NIOSH REL (carc. Cr (VI) compounds) 0.001 mg/m³

◆ Toxicity Data
  - rat LC_{50} 87 mg/m³ (chromium trioxide)
  - rat LC_{50} 45 mg/m³ (Cr (VI))
  - intravenous-rat TDLo 2160 µg/kg/6W-I:ETA (chromium)
  - oral-human LDLo 71 mg/kg:GIT (chromium)

◆ Acute Risks (Principally Chromium (VI))
  - Chromium(VI) is much more toxic than chromium(III).
  - Chromium(III) is an essential element in humans; daily intake 50 to 200 µg/d.
  - respiratory tract effects from chromium(VI)
  - dypsnea
  - coughing
  - wheezing
  - gastrointestinal effects
  - neurological effects
  - skin burns
  - irritation of nose, lungs, stomach, and intestine
  - perforation of the nasal septum

◆ Chronic Risks (Principally Chromium (VI))
  - effects on respiratory tract
  - perforations and ulcerations of septum
  - bronchitis
  - decreased pulmonary function
  - pneumonia
  - asthma
  - nasal itching and soreness
  - effects on the liver and kidney
  - effects on gastrointestinal and immune systems
  - effects on the blood
  - contact dermatitis
  - sensitivity
  - ulceration of the skin
  - complications during pregnancy and childbirth
  - severe developmental effects in mice: gross abnormalities, decreased litter size, reduced sperm count, degeneration of the outer cellular layer of the seminiferous tubules
  - Inhaled chromium is a human carcinogen.
  - increased risk of lung cancer
  - lung tumors in animals
  - Chromate salts are suspected carcinogens.

- Chromate salts produce tumors of the lungs, nasal cavity, and paranasal sinus.
- Chromic acid and its salts have a corrosive action on the skin and mucous membranes.
- EPA Group A: human carcinogen (chromium (VI))
- EPA Group D: not classifiable as to carcinogenicity (chromium (III))

HAZARD RISK

- ◆ Powder form of chromium is combustible.
- ◆ When powdered form is exposed to heat or ignition sources, it is a moderate fire and explosion hazard.
- ◆ incompatible with strong oxidizers, mineral acids, ammonium nitrate, hydrogen peroxide, chlorates, and sulfur dioxide
- ◆ Reaction with mineral acids may liberate hydrogen gas.
- ◆ reacts readily with dilute acids to form chromous salts

EXPOSURE ROUTES

- ◆ Chromium occurs in the environment in two major states as Cr(III) and Cr(IV).
- ◆ wood treated with copper dichromate
- ◆ leather tanned with chromic sulfate
- ◆ chromate production
- ◆ stainless steel production
- ◆ chrome plating
- ◆ working in tanning industries
- ◆ chromium waste disposal sites
- ◆ chromium manufacturing and processing plants

EXPOSURE ROUTES (Chromium (III))

- ◆ eating food
- ◆ drinking water
- ◆ inhaling air
- ◆ essential dietary element
- ◆ chromium(IV) detoxification to chromium(III) in the body

REGULATORY STATUS/INFORMATION

- ◆ CERCLA and Superfund
  - list of chemicals. 40CFR372.65

ADDITIONAL/OTHER COMMENTS

- ◆ Symptoms
  - inhalation: eye and skin irritation
  - contact: corrosive on skin, dermatitis, burning
- ◆ First Aid
  - wash eyes immediately with large amounts of water
  - flush skin with plenty of soap and water
  - remove to fresh air immediately or provide respiratory support if necessary
  - give water to drink and induce vomiting
- ◆ fire fighting: use water spray or fog, dry chemical, carbon dioxide, or sand
- ◆ personal protection: wear self-contained breathing apparatus, chemical goggles, and protective clothing

◆ storage: cool, dry, well-ventilated area away from heat or ignition sources and oxidizing agents
SOURCES
◆ *Health Effects Notebook for Hazardous Air Pollutants*, U.S. Environmental Protection Agency, December 1994
◆ *Material Safety Data Sheets for Arsine (MSDS)*
◆ *Sax's Dangerous Properties of Industrial Materials*, 9th edition, 1996
◆ *Hawley's Condensed Chemical Dictionary*, 12th edition, 1993
◆ *Environmental Law Index to Chemicals*, 1996
◆ *Merck Index*, 12th edition, 1996
◆ Agency for Toxicological Substances and Disease Registry

COBALT COMPOUNDS

CAS #: Cobalt compounds have variable CAS #s. The CAS # for cobalt is 7440-48-4.
MOLECULAR FORMULA: Cobalt compounds have variable molecular formulas. The molecular formula for cobalt is Co.
Note: For all hazardous air pollutants (HAPs) that contain the word "compounds" in the name, and for glycol ethers, the following applies: Unless otherwise specified, these HAPs are defined by the 1990 Amendments to the Clean Air Act as including any unique chemical substance that contains the named chemical (i.e., antimony, arsenic, etc.) as part of that chemical's infrastructure.
FORMULA WEIGHT: Cobalt compounds have variable formula weights. The formula weight for cobalt is 58.93.
SYNONYMS: aquacat, C.I. 77320, cobalt-59, Kobalt (German, Polish), super cobalt

USES
◆ Used in Manufacturing
  • superalloys
  • pigments
  • cobalt salts
  • nuclear technology
  • experimental medicine
  • cancer research
  • cobalt bomb
  • electroplating ceramics
  • lamp filaments
  • catalyst
  • trace element in fertilizers
  • glass
  • drier in painting inks
  • paints
  • varnishes
  • colors
  • cements
  • cemented carbides
  • jet engines
  • coordination and complexing agent
  • treatment for anemia

PHYSICAL PROPERTIES
◆ steel-gray, shiny, hard metal
◆ Hydrated salts of cobalt are red.
◆ Soluble salts form red solutions which become blue on adding concentrated HCl.
◆ magnetic
◆ ductile
◆ somewhat malleable
◆ ferromagnetic
◆ permeability two-thirds that of iron
◆ exists in two allotropic forms
◆ melting point 1493°C

- ◆ boiling point 3100°C
- ◆ density 8.92
- ◆ Brinell hardness 125
- ◆ insoluble in water
- ◆ readily soluble in diluted HNO_3

CHEMICAL PROPERTIES
- ◆ Hexagonal form is more stable than the cubic form at room temperature.
- ◆ stable in air or toward water at ordinary temperature
- ◆ latent heat of fusion 62 cal/g
- ◆ latent heat of vaporization 1500 cal/g
- ◆ specific heat 0.1056 cal/g/°C at (15–100°C)
- ◆ slowly attacked by HCl or cold H_2SO_4
- ◆ corrodes readily in air
- ◆ noncombustible except in powder

HEALTH RISK
- ◆ Toxic Exposure Guidelines
  - OSHA PEL (cobalt carbonyl and cobalt hydrocarbonyl) 0.1 mg/m³
  - ACGIH TLV (cobalt carbonyl and cobalt hydrocarbonyl) 0.1 mg/m³
  - ACGIH STEL (cobalt metal, dust and fume) 0.1 mg/m³
  - OSHA PEL (cobalt metal, dust and fume) 0.05 mg/m³
  - ACGIH TLV (cobalt metal, dust and fume) 0.05 mg/m³
- ◆ Toxicity Data
  - implant-rabbit TDLo 75 mg/kg:ETA
  - oral-rat LDLo 1500 mg/kg
  - intraperitoneal-rat LDLo 250 mg/kg
  - intravenous-rat LDLo 100 mg/kg
  - intraperitoneal-mouse LDLo 100 mg/kg
  - oral-rabbit LDLo 750 mg/kg
  - intravenous-rabbit 100 mg/kg
  - rat LC_{50} 165 mg/m³
- ◆ Acute Risks
  - respiratory effects
  - significant decrease in ventilatory function
  - congestion
  - edema
  - hemorrhage of the lung
  - nausea and vomiting from ingestion of soluble salts
  - liver and kidney damage
- ◆ Chronic Risks
  - effects on respiratory system
  - respiratory irritation
  - wheezing
  - asthma
  - pneumonia
  - fibrosis
  - cardiac effects
  - functional effects on the ventricles
  - enlargement of the heart
  - congestion of the liver, kidneys, and conjunctiva
  - immunological effects that include cobalt sensitization
  - cardiomyopathy
  - cardiogenic shock
  - sinus tachycardia
  - left ventricle failure
  - gastrointestinal effects
  - nausea
  - vomiting

- diarrhea
- effects on the blood
- liver injury
- allergic dermatitis
- decreased body weight in animals
- necrosis of the thymus in animals
- increase in total red cell mass of the blood
- hematologic, digestive, and pulmonary changes
- testicular atrophy in animals
- decrease in sperm motility in animals
- significant increase in the length of the estrus cycle in animals
- stunted growth and decreased survival of newborn pups
- increased deaths due to lung cancer

HAZARD RISK

- ◆ Powdered cobalt ignites spontaneously in air.
- ◆ flammable when exposed to heat or flame
- ◆ explosive reaction with hydrazinium nitrate, ammonium nitrate + heat, and 1,3,4,7-tetramethylisoindole (at 390°C)
- ◆ ignites on contact with bromine pentafluoride
- ◆ incandescent reaction with acetylene or nitryl fluoride

EXPOSURE ROUTES

- ◆ natural element found throughout the environment
- ◆ air
- ◆ drinking water
- ◆ food
- ◆ workers in the hard metal industry
- ◆ burning coal and oil
- ◆ soil
- ◆ dust
- ◆ seawater

REGULATORY STATUS/INFORMATION

- ◆ CERCLA and Superfund
  - list of chemicals. 40CFR372.65

ADDITIONAL/OTHER COMMENTS

- ◆ Symptoms
  - inhalation of dust: pulmonary symptoms; powder may cause dermatitis
  - ingestion of soluble salts: nausea and vomiting by local irritation

SOURCES

- ◆ *Health Effects Notebook for Hazardous Air Pollutants*, U.S. Environmental Protection Agency, December 1994
- ◆ *Sax's Dangerous Properties of Industrial Materials*, 9th edition, 1996
- ◆ *Hawley's Condensed Chemical Dictionary*, 12th edition, 1993
- ◆ *Environmental Law Index to Chemicals*, 1996
- ◆ *Merck Index*, 12th edition, 1996
- ◆ Agency for Toxicological Substances and Disease Registry

COKE OVEN EMISSIONS

CAS #: A CAS # for coke oven emissions is not applicable.
MOLECULAR FORMULA: Coke oven emissions are made up of 53%
hydrogen, 26% methane, 11% nitrogen, 7% carbon monoxide, and 3%
heavier hydrocarbons.
FORMULA WEIGHT: The formula weight for coke oven emissions is
not applicable.
SYNONYMS: coke oven gas

USES
- ◆ Used as Raw Materials
  - plastics
  - solvents
  - dyes
  - drugs
  - waterproofing
  - paints
  - pipe coating
  - roads
  - roofing
  - insulation
  - pesticides
  - sealants
  - to produce hydrogen
  - to extract metals from
    their ores, especially iron
  - to synthesize calcium car-
    bide
  - to manufacture graphite
    and electrodes
  - coal tar used in treatment
    of skin disorders

PHYSICAL PROPERTIES
- ◆ mixture of coal tar, coal tar pitch, and creosote
- ◆ contain chemicals such as benzo(a)pyrene, benzanthracene,
  chrysene, and phenanthrene
- ◆ condensed coke oven emissions: brownish, thick liquid or
  semisolid
- ◆ condensed: naphthalene-like odor
- ◆ uncondensed: vapors that escape when the ovens are changed
  and emptied
- ◆ OSHA: benzene-soluble fraction of total particulate matter
  present during destructive distillation or carbonization
  of coal to produce coke

CHEMICAL PROPERTIES
- ◆ not available

HEALTH RISK
- ◆ Toxic Exposure Guidelines
  - OSHA PEL (coal tar pitch volatiles, benzene soluble) 0.2
    mg/m^3
  - ACGIH TLV (coal tar pitch volatiles, benzene soluble) 0.2
    mg/m^3
  - OSHA TLV (coke oven emissions) 0.150 mg/m^3
  - NIOSH REL (coal tar pitch volatiles, benzo(a)pyrene) 0.1
    mg/m^3
- ◆ Toxicity Data
  - not available
- ◆ Other Toxicity Indicators
  - EPA Cancer Risk Level (10^{-6} excess risk) 2×10^{-6} mg/m^3
- ◆ Acute Risks

- No information is available on the acute effects in humans.
- Acute Risks (Animals)
  - weakness
  - depression
  - dyspnea
  - general edema
  - effects on liver
- Chronic Risks
  - conjunctivitis
  - severe dermatitis
  - lesions of the respiratory and digestive systems
  - effects on liver in animals
  - increase in cancer of the lung
  - increase in cancer of the trachea
  - increase in cancer of the bronchus
  - increase in cancer of the kidney
  - increase in cancer of the prostate and other sites
  - tumors of the lung and skin in animals
  - induced skin carcinomas in mice
  - papillomas in mice
  - lung adenomas and squamous cell lung tumors in mice
  - EPA Group A: human carcinogen

HAZARD RISK
- not available

EXPOSURE ROUTES
- workers in aluminum, steel, graphite, electrical, and construction industries
- inhalation and dermal contact
- production of coke from coal
- use of coke to extract metals from ores
- use of coal tar to treat skin disorders
- workers at coking plants and coal tar production plants

REGULATORY STATUS/INFORMATION
- Clean Air Act Amendment, Title III: hazardous air pollutants
- CERCLA and Superfund
  - designation of hazardous substances. 40CFR302.4
- Occupational Safety and Health Act
  - 29CFR1910.1000 Table Z1
  - 29CFR1910.1001-1048 specially regulated substances
- chemical regulated by the State of California

ADDITIONAL/OTHER COMMENTS
- personal protection: protective clothing, respirators, protective equipment, and flame-resistant gloves

SOURCES
- *Health Effects Notebook for Hazardous Air Pollutants*, U.S. Environmental Protection Agency, December 1994
- *Hawley's Condensed Chemical Dictionary*, 12th edition, 1993
- *Environmental Law Index to Chemicals*, 1996
- *Technological Processes: Coke Oven Emissions*
- U.S. Department of Labor, Occupational Safety and Health Administration, OSHA Computerized Information System (OCIS)

◆ *Coke Oven Emissions: A Case Study of Technology-Based Regulation*, John D. Graham, David R. Holtgrave

CRESOLS (yf)

CAS #: 1319-77-3
MOLECULAR FORMULA: C_7H_8O
FORMULA WEIGHT: 108.14
SYNONYMS: cresylic acid, cresylsyre, crysylol, methylphenol, pure cresol, tricresol

USES
◆ laboratory reagent
◆ solvent, mainly for wire enamels
◆ ore flotation agent
◆ drug, disinfectant
◆ preservative
◆ insecticidal coating agent
◆ intermediate for
  • cresol phosphates
  • tricresylphosphate
  • phenolic resins
  • resin additives
  • cleaning compounds
◆ chemical processes
  • By-product of Catalytic Cracking of Petroleum, Obtained by Fractional Distillation
    • main products: phenol, mixed isomers of cresol, xylenols, other phenolic compounds
  • Catalytic Oxidation of Cymene
    • main products: cresol, acetone
◆ lube oil refining and additives
◆ carbon removal from engines
◆ wire enamel solvent

PHYSICAL PROPERTIES
◆ colorless, yellowish, or pinkish liquid
◆ specific gravity 1.04 (water = 1)
◆ melting point 11–35°C
◆ vapor density 3.7 (water = 1)
◆ critical temperature 321.4°C
◆ density 0.7275 g/cm³ at 20°C
◆ odor threshold 0.00028 ppm
◆ phenolic odor
◆ boiling point 201°C at 760 mm Hg
◆ vapor pressure 0.1 mm Hg at 20°C
◆ normal physical state liquid
◆ critical pressure 21.4 atm
◆ very soluble in ethyl alcohol

CHEMICAL PROPERTIES
◆ flash point 43°C, open cup; 86°C, closed cup
◆ flammability limits 1.1% lower, 1.4% upper (by volume in air)
◆ latent heat of vaporization 10,801 cal/gmol at 760 mm Hg
◆ ionization potential 8.93 eV
◆ latent heat of fusion 3095 cal/gmol
◆ autoignition temperature 598°C
◆ specific heat 1.05 cal/g at 0°C

HEALTH RISK

- ◆ Toxic Exposure Guidelines
  - OSHA threshold limit value 5 ppm
  - OSHA permissible exposure limit 5 ppm
  - ACGIH threshold limit value 5 ppm (skin)
- ◆ Toxicity Data
  - oral-rat 1454 mg/kg
  - skin-rabbit 2000 mg/kg
  - target organs: nasal septum, respiratory system, liver, eyes, kidneys, skin
  - noncarcinogenic
  - o-cresol, m-cresol reference dose 0.05 (mg/kg)/d
- ◆ Acute Risks
  - headache
  - vomiting
  - drowsiness
  - unconsciousness
  - vascular collapse
  - necrosis of liver and kidneys
  - gastrointestinal irritation
  - lesions of the mouth and esophagus
  - pancreatic complications
  - nausea
  - dizziness
  - irritation of upper respiratory tract
  - pulmonary edema
  - shock
  - peripheral neuritis
  - burns to mouth and throat
  - digestive tract bleeding
  - vesiculation ulceration
- ◆ Chronic Risks
  - damage to liver
  - damage to lungs
  - damage to central nervous system
  - damage to kidneys
  - damage to blood

HAZARD RISK

- ◆ flammable
- ◆ incompatible with strong oxidizing agents
- ◆ Decomposition Products
  - carbon monoxide
  - carbon dioxide
- ◆ avoid heat, flame, other sources of ignition and light
- ◆ causes severe burns
- ◆ combustible

EXPOSURE ROUTES

- ◆ inhalation
- ◆ skin contact/absorption
- ◆ car exhaust
- ◆ electrical power plant emissions
- ◆ ingestion
- ◆ eye contact
- ◆ cigarette smoke
- ◆ municipal solid waste incinerators
- ◆ Found in Foods Such as
  - tomatoes
  - butter
  - raw and roasted coffee
  - asparagus
  - bacon
  - black tea
  - cheeses
  - red wine

REGULATORY STATUS/INFORMATION
- ◆ regulated under EPA's TSCA inventory
- ◆ CERCLA and Superfund
  - • hazardous substance. 40CFR302
  - • CERCLA RQ 1000 lbs.
- ◆ Clean Water Act
  - • section 304 Water Quality Criteria Substances. 33USC1251 et seq
  - • section 307 Priority Pollutants. 40CFR423.17
  - • section 311 Hazardous Substances. 40CFR116.4
  - • section 111 Volatile Organic Compounds. 40CFR60
  - • subject to SARA Section 133 reporting requirements
- ◆ NIOSH Recommended Substance 29USC1900, 30USC80
- ◆ Resource Conservation and Recovery Act
  - • Hazardous Substance 40CFR261
  - • Land Disposal Restrictions 40CFR268.12
  - • OSHA Air Contaminants 29CFR1910 Subpart z

ADDITIONAL/OTHER COMMENTS
- ◆ fire fighting: water spray, alcohol foam, carbon dioxide, or dry chemical powder
- ◆ Storage
  - • store in a well-ventilated room in a tightly closed container
  - • Storage color code is red.
- ◆ Personal Protection
  - • When working with chloromethyl methyl ether, wear chemical-resistant gloves, safety goggles and shield, and protective clothing
  - • use only in a chemical fume hood near a fire extinguisher
- ◆ reacts violently with HNO_3, oleum, or chlorosulfonic acid
- ◆ When heated to decomposition, it emits highly toxic and irritating fumes.

SOURCES
- ◆ *Material Safety Data Sheets (MSDS)*
- ◆ University of Kentucky web page http://www.chem.uky.edu/resources/msds/html
- ◆ *Sax's Dangerous Properties of Industrial Materials*, 9th edition, 1996
- ◆ *Health Effects Notebook for Hazardous Air Pollutants*, U.S. Environmental Protection Agency, December 1994
- ◆ Industrial Chemicals, 2nd edition, 1957

m-CRESOL. *See* **p. 152 following o-cresol.**

o-CRESOL (rs)

CAS #: 95-48-7
MOLECULAR FORMULA: C_7H_8O or $CH_3C_6H_4OH$
FORMULA WEIGHT: 108.15
SYNONYMS: 2-cresol, o-cresylic acid, 1-hydroxy-2-methyl-benzene, o-hydroxytoluene, o-methylphenol, 2- methylphenol, orthocresol, o-oxytoluene, RCRA waste number u052, o-toluol, others

USES
- ◆ Manufacture of Other Chemicals
  - • phosphate
  - • coumarin
  - • salicylaldehyde
- ◆ Used as a Component or in the Manufacturing Process of
  - • disinfectants
  - • solvents
  - • resins
  - • scouring agents
  - • extraction of ore
  - • herbicides

PHYSICAL PROPERTIES
- ◆ normal state colorless crystalline, white crystals (liquid above 88°F)
- ◆ melting point 30.9°C (87.6°F)
- ◆ boiling point 191°C (375.8°F) at 760 mm Hg
- ◆ flash point 81.1° (178°F)
- ◆ specific gravity 1.05
- ◆ ionization potential 8.93 eV
- ◆ solubility 2% by weight (in water at 68°F, g/100 mL)
- ◆ vapor pressure 1 mm Hg at 38.2–53°C (.357 mm Hg at 20°C)
- ◆ density 1.047 g/cm³ at 20°C
- ◆ vapor density 3.72 g/cm³
- ◆ viscosity 4.49 cP at 40°C
- ◆ heat of vaporization 52.25 kJ/mole (12,487.6 cal/mole)
- ◆ concentration in saturated air 0.0323%; density of saturated air 1.00089 (air = 1.0) at 25°C
- ◆ soluble in alcohol, chloroform, ether, hot water, fixed alkali hydroxides, benzene, and carbon
- ◆ tetrachloride, acetone, organic solvents, vegetable oil at 30°F
- ◆ phenolic odor (sweet, tarry odor)
- ◆ odor threshold in air 0.26 ppm

CHEMICAL PROPERTIES
- ◆ lower explosive limit (LEL) 1.4% (by volume in air) at 149°C (300°F)
- ◆ upper explosive limit (UEL) not available
- ◆ autoignition temperature 599°C (1110°F)
- ◆ liquid heat of combustion 3692.8 kJ/g (882.6 kcal/g)
- ◆ solid heat of combustion 3714.2 kJ/g (882.7 kcal/g)

- ◆ darkens when exposed to air or light or with age
- ◆ highly corrosive
- ◆ combustible solid
- ◆ class IIIA combustible liquid
- ◆ reacts with oxidizing materials, acids, bases
- ◆ highly toxic

HEALTH RISK

- ◆ Toxic Exposure Guidelines
  - OSHA PEL TWA 5 ppm (22 mg/m³) (skin)
  - NIOSH REL TWA 2.3 ppm (10 mg/m³)
  - ACGIH THV TWA 5 ppm (22 mg/m³)
  - IDLH 250 ppm
- ◆ Irritation Data
  - skin-rabbit 524 mg/24H SEV
  - eye-rabbit 105 mg SEV
- ◆ Toxicity Data
  - skin-mouse TDLo 4800 mg/kg/12W-Intermittent
  - oral-rat LD_{50} 121 mg/kg
  - skin-rat LDLo 620 mg/kg
  - subcutaneous-rat LDLo 65 mg/kg
  - oral-mouse LD_{50} 344 mg/kg
  - inhalation-mouse LC_{50} 179 mg/m³/2H
  - skin-mouse LD_{50} 620 mg/kg
  - injection, peritoneal cavity-mouse LDLo 200 mg/kg
  - subcutaneous-mouse LDLo 410 mg/kg
  - intravenous-dog LDLo 80 mg/kg
  - subcutaneous-cat LDLo 55 mg/kg
  - EPA RfD (reference dose) 0.05 mg/kg/d
- ◆ Target Organs
  - central nervous system
  - lungs
  - liver
  - kidneys
  - pancreas
  - spleen
  - cardiovascular system
  - skin
- ◆ Acute Risks
  - extremely corrosive allergen
  - toxic via all routes of exposure
  - corrosive to body tissues
  - severe eye and skin irritant
  - possible human mutation
  - may cause skin eruptions
  - 8 grams can be fatal to man.
- ◆ Chronic Risks
  - kidney and liver damage
  - dermatitis
  - suspected carcinogen

HAZARD RISK
- ◆ flammable when exposed to heat, flame, or oxidants

EXPOSURE ROUTES
- ◆ inhalation
- ◆ oral
- ◆ dermal

REGULATORY STATUS/INFORMATION
- ◆ Clean Air Act
  - Title III: regulated under hazardous pollutant and accident prevention provisions
- ◆ Resource Conservation and Recovery Act
  - 40CFR258.40, Appendixes 1 and 2, MSW landfills
  - 40CFR264, requirement for groundwater monitoring, Appendix 9
  - 40CFR261.24, maximum concentration for toxicity
  - 40CFR302.4, list of reportable quantities
- ◆ CERCLA and Superfund
  - 40CFR302.4, designation as hazardous substance
  - Table 302.4, list of hazardous substances and reportable quantities
  - 40CFR355-AA, Appendix A, list of extremely hazardous substances and threshold planning quantities
  - 40CFR372.65, list of substances subject to Community-Right-to-Know
- ◆ Toxic Substance Control Act
  - 40CFR716.120.a, list of substances regulated under TSCA
  - TSCA 8(A) preliminary assessment information
- ◆ DOT Class 6.1, Poisonous material

ADDITIONAL/OTHER COMMENTS
- ◆ Symptoms
  - central nervous system depression
  - muscular weakness
  - gastroenteric disturbances
  - convulsions
  - death
  - burns to skin and eyes
- ◆ First Aid
  - inhalation: fresh air (artificial respiration if needed)
  - respiratory distress: oxygen inhalation
  - eyes: irrigate with running water for 15 minutes, keep eyelids open; get to physician; if physician is not available irrigate for an additional 15 minutes
  - skin: remove all contaminated clothing, wash with soap and water until odor is gone, proceed to wash with alcohol or glycerin, then rinse with additional water
  - ingestion: drink large quantities of saltwater, weak sodium bicarbonate solution, milk, or gruel, followed by raw egg or cornstarch paste; induce vomiting
- ◆ Preventative Measures
  - Contact lenses should not be worn.

◆ Fire Fighting
  • use water spray, dry chemical, foam, or carbon dioxide; use water spray on exposed containers to control temperature
  • flammable toxic vapor emitted from fire (see PPE)
◆ Personal Protection
  • special protective clothing, i.e., rubber boots, chemical safety glasses, face shield, rubber coveralls/apron, rubber shoes or boots, and positive, self-contained breathing apparatus
  • Protective creams are not adequate.
◆ Spill or Leak
  • approach from upwind; contain spill and dispose properly
◆ NFPA Diamond
  • Health 3, Flammability 2, Reactivity 0
◆ DOT: UN 2076

SOURCES
◆ *Aldrich Chemical Company Catalog*, 1996–1997
◆ *Environmental Contaminant Reference Databook*, volume 1, 1995
◆ *Environmental Law Index to Chemicals*, 1994 edition
◆ *Fire Protection Guide to Hazardous Materials*, 11th edition, 1994
◆ *MSDS*, Aldrich Chemical Company, Inc., 1996
◆ *Merck Index*, 12th edition, 1996
◆ *NIOSH Pocket Guide to Chemical Hazards*, June 1994
◆ *Sax's Dangerous Properties of Industrial Materials*, 9th edition, volumes 1 and 2, 1996
◆ *Hawley's Condensed Chemical Dictionary*, 12th edition, 1993
◆ *Health Effects Notebook for Hazardous Air Pollutants*, U.S. Environmental Protection Agency, December 1994

m-CRESOL

CAS #: 108-39-4
MOLECULAR FORMULA: $CH_3C_6H_4OH$
FORMULA WEIGHT: 108.15
SYNONYMS: 3-cresol, m-cresylic acid, 1-hydroxy-3-methylbenzene, m-hydroxytoluene, m-kresol, m-methylphenol, 3-methylphenol, m-oxytoluene, m-toluol

USES
◆ Used in the Production Process
  • herbicides
  • precursor to the pyrethroid insecticides
  • antioxidants
  • explosive 2,4,6-nitro-m-cresol
  • disinfectants
  • fumigants
  • photographic developers
  • explosives

PHYSICAL PROPERTIES
◆ colorless to yellow liquid
◆ phenolic odor
◆ odor threshold 0.00028 ppm
◆ melting point 10.9°C
◆ boiling point 202.8°C
◆ specific gravity 1.034
◆ vapor pressure 0.138 mm Hg at 25°C

♦ vapor density 3.72
♦ soluble in 40 parts water, in solutions of fixed alkali hydroxides
♦ miscible with alcohol, chloroform, and ether
♦ log K_{ow} 1.96

CHEMICAL PROPERTIES
♦ flash point 86°C
♦ autoignition temperature 557°C
♦ explosion limits in air 1.06% lower and 1.35% upper at 150°C
♦ incompatible with oxidizing agents and bases

HEALTH RISK
♦ Toxic Exposure Guidelines
  • OSHA PEL TWA 5 ppm (skin)
  • ACGIH TLV TWA 5 ppm
  • NIOSH REL (cresol) TWA 10 mg/m³
♦ Toxicity Data
  • skin-rabbit 517 mg/24H SEV
  • eye-rabbit 103 mg SEV
  • dni-human:hla 10 μmol/L/4H
  • subcutaneous-rabbit TDLo 134 g/kg
  • skin-mouse TDLo 2280 mg/kg/20W-I:NEO
  • oral-rat LD$_{50}$ 242 mg/kg
  • skin-rat LD$_{50}$ 1100 mg/kg
  • subcutaneous-rat LD$_{50}$ 900 mg/kg
  • oral-mouse LD$_{50}$ 828 mg/kg
  • intraperitoneal-mouse LD$_{50}$ 168 mg/kg
  • subcutaneous-mouse LDLo 450 mg/kg
  • intravenous-dog LDLo 150 mg/kg
  • subcutaneous-cat LDLo 180 mg/kg
  • oral-rabbit LDLo 1400 mg/kg
  • skin-rabbit LD$_{50}$ 2050 mg/kg
  • intravenous-rabbit LDLo 280 mg/kg
  • intraperitoneal-guinea pig LDLo 100 mg/kg
♦ Acute Risks (m-Cresol)
  • extremely destructive to tissue of the mucous membranes
  • destructive to upper respiratory tract
  • irritation to eyes and skin
  • damage to the kidney
  • target organs: central nervous system, lungs, liver, and kidneys
♦ Acute Risks (Mixed Cresols)
  • respiratory tract irritation
  • dryness
  • nasal constriction
  • throat irritation
  • dermal irritants
  • death
  • effects on gastrointestinal system
  • effects on blood
  • effects on liver, kidney, and central nervous system
♦ Chronic Risk
  • tumor promoter
  • effects on blood, liver, kidney, and central nervous system
  • reduced body weight in animals

- EPA Group C: possible human carcinogen

HAZARD RISK
- flammable when exposed to heat or flame
- moderately explosive in the form of vapor when exposed to heat or flame
- incompatible with oxidizing agents and bases
- hazardous decomposition products: carbon monoxide and carbon dioxide

EXPOSURE ROUTES (Mixed Cresols)
- ambient air
- car exhaust
- electrical power plants
- municipal solid waste incinerators
- oil refineries
- cigarettes
- homes heated with coal, oil, or wood
- tomatoes
- ketchup
- asparagus
- cheeses
- butter
- bacon
- smoked foods
- red wine
- raw and roasted coffee
- black tea
- production and use of mixed cresols and/or cresol

REGULATORY STATUS/INFORMATION
- Clean Air Act Amendment, Title III: hazardous air pollutants
- Resource Conservation and Recovery Act
  - list of hazardous constituents. 40CFR258
  - groundwater monitoring list. 40CFR264
  - D waste #: DO24
- CERCLA and Superfund
  - designation of hazardous substances. 40CFR302.4
  - list of chemicals. 40CFR372.65
- Toxic Substance Control Act
  - list of substances. 40CFR716.120.a
  - chemicals subject to test rules or consent orders for which the testing reimbursement period has passed. 40CFR799.18
- chemical regulated by the State of California

ADDITIONAL/OTHER COMMENTS
- Symptoms
  - exposure: burning sensation, coughing, wheezing, laryngitis, shortness of breath, headache, nausea, vomiting
- First Aid
  - wash eyes and skin immediately with large amounts of water
  - remove to fresh air immediately or provide respiratory apparatus if necessary
  - wash out mouth with water
- fire fighting: use dry chemical, water spray, chemical powder, or appropriate foam
- personal protection: self-contained breathing apparatus, protective clothing, and goggles

◆ storage: keep tightly closed away from heat and open flame in a cool, dry place

SOURCES

◆ *Health Effects Notebook for Hazardous Air Pollutants*, U.S. Environmental Protection Agency, December 1994
◆ *Material Safety Data Sheets (MSDS)*
◆ *Sax's Dangerous Properties of Industrial Materials*, 9th edition, 1996
◆ *Hawley's Condensed Chemical Dictionary*, 12th edition, 1993
◆ *Aldrich Chemical Company Catalog*, 1996–1997
◆ *Environmental Law Index to Chemicals*, 1996
◆ *Merck Index*, 12th edition, 1996

o-CRESOL. *See* p. 149 following cresols.

p-CRESOL

CAS #: 106-44-5
MOLECULAR FORMULA: $CH_3C_6H_4OH$
FORMULA WEIGHT: 108.15
SYNONYMS: 4-cresol, p-cresylic acid, -hydroxy-4-methylbenzene, p-hydroxytoluene, 4-hydroxytoluene, p-kresol, 1-methyl-4-hydroxybenzene, p-methylphenol, 4-methylphenol, p-oxytoluene, paramethyl phenol, p-toluol, p-tolyl alcohol

USES
◆ Used in the Production Process
  • disinfectants
  • explosives
  • synthetic perfumery materials
  • metal cleaning agent
  • agent in ore flotation
  • monomer for phenolic resins
  • synthetic flavor
  • 2,6-di-tert-butyl-para-cresol
  • synthetic resin
  • petroleum
  • photography
  • paint
  • agricultural industries
◆ Used as a Chemical Intermediate
  • tricresyl phosphate
  • cresyl diphenyl phosphate
◆ Used as a Solvent
  • wire enamels

PHYSICAL PROPERTIES
◆ white or colorless crystals
◆ phenolic odor
◆ odor threshold 0.2 ppm
◆ melting point 34.8°C
◆ boiling point 201.9°C
◆ density 1.0178 at 20°C
◆ surface tension 41.8 dynes/cm
◆ viscosity 7.0 cP at 40°C
◆ soluble in water, organic solvents, alcohol, ether, acetone, and benzene
◆ vapor pressure 0.11 mm Hg at 25°C
◆ vapor density 3.72

CHEMICAL PROPERTIES
◆ flash point 86°C

- ◆ autoignition temperature 558°C
- ◆ heat of combustion 882.5 kg/g at 20°C
- ◆ heat of vaporization 13,611.7 cal/mol
- ◆ flammability limits 1.1% by volume at 302°C
- ◆ volatile in steam

HEALTH RISK

- ◆ Toxic Exposure Guidelines
  - OSHA PEL TWA 5 ppm (skin)
  - ACGIH TLV TWA 5 ppm
  - NIOSH REL (cresol) TWA 10 mg/m^3
- ◆ Toxicity Data
  - skin-rabbit 517 mg/24H SEV
  - eye-rabbit 103 mg SEV
  - skin-mouse TDLo 2280 mg/kg/20W-I:NEO
  - oral-rat LD$_{50}$ 207 mg/kg
  - skin-rat LD$_{50}$ 750 mg/kg
  - subcutaneous-rat LD$_{50}$ 500 mg/kg
  - unr-rat LD$_{50}$ 1440 mg/kg
  - oral-mouse LD$_{50}$ 344 mg/kg
  - intraperitoneal-mouse LD$_{50}$ 25 mg/kg
  - subcutaneous-mouse LDLo 150 mg/kg
  - unknown-mouse LD$_{50}$ 160 mg/kg
  - subcutaneous-cat LDLo 80 mg/kg
  - oral-rabbit LDLo 620 mg/kg
  - skin-rabbit LD$_{50}$ 301 mg/kg
  - subcutaneous-rabbit LDLo 300 mg/kg
  - intravenous-rabbit LDLo 180 mg/kg
  - subcutaneous-guinea pig LDLo 200 mg/kg
  - subcutaneous-frog LDLo 150 mg/kg
- ◆ Acute Risks (p-Cresol)
  - eye burns
  - skin burns
  - digestive and respiratory tract burns
  - pulmonary edema
  - chemical pneumonitis
  - kidney and liver damage
  - abdominal pain
  - vomiting
  - muscular weakness
  - gastroenteric disturbances
  - severe depression
  - collapse
  - injury of spleen
  - injury of pancreas
- ◆ Acute Risks (Mixed Cresols)
  - respiratory tract irritation
  - dryness
  - nasal constriction
  - throat irritation
  - dermal irritants
  - death
  - effects on gastrointestinal system
  - effects on blood
  - effects on liver, kidney, and central nervous system
- ◆ Chronic Risks
  - promoter for tumors of the forestomach in animals
  - promoter for tumors
  - abdominal pain
  - vomiting
  - chemical pneumonitis

- severe digestive tract burns
- liver and kidney damage
- irritation of upper respiratory tract

- effects on blood, liver, kidney, and central nervous system
- reduced body weight in animals
- EPA Group C: possible human carcinogen

HAZARD RISK
- combustible when exposed to heat or flame
- moderately explosive in the form of vapor when exposed to heat or flame
- incompatible with bases, chlorosulfonic acid, nitric acid, oleum, and strong oxidizers
- hazardous decomposition products: carbon monoxide and carbon dioxide

EXPOSURE ROUTES (Mixed Cresols)
- ambient air
- car exhaust
- electrical power plants
- municipal solid waste incinerators
- oil refineries
- cigarettes
- homes heated with coal, oil, or wood
- tomatoes
- ketchup
- asparagus
- cheeses
- butter
- bacon
- smoked foods
- red wine
- raw and roasted coffee
- black tea
- production and use of mixed cresols and/or cresol

REGULATORY STATUS/INFORMATION
- Clean Air Act Amendment, Title III: hazardous air pollutants
- Resource Conservation and Recovery Act
  - list of hazardous constituents. 40CFR258
  - groundwater monitoring list. 40CFR264
  - D waste #: DO25
- CERCLA and Superfund
  - designation of hazardous substances. 40CFR302.4
  - list of chemicals. 40CFR372.65
- Toxic Substance Control Act
  - list of substances. 40CFR716.120.a
  - chemicals subject to test rules or consent orders for which the testing reimbursement period has passed. 40CFR799.18
- chemical regulated by the State of California

ADDITIONAL/OTHER COMMENTS
- Symptoms
  - inhalation: irritation of nose, throat, and eyes, swelling of conjunctiva, corneal damage
  - skin: intense burning, loss of feeling, white discoloration, softening, gangrene
  - ingestion: burning sensation in mouth and esophagus, vomiting

- ◆ First Aid
  - wash eyes immediately with large amounts of water
  - wash skin immediately with soap and water
  - remove to fresh air immediately or provide respiratory apparatus if necessary
  - wash out mouth with water or milk
- ◆ fire fighting: use dry chemical, water spray, carbon dioxide, or alcohol foam
- ◆ personal protection: self-contained breathing apparatus, protective clothing, and goggles
- ◆ storage: keep in a tightly closed container in a cool, dry, well-ventilated area

SOURCES
- ◆ *Health Effects Notebook for Hazardous Air Pollutants*, U.S. Environmental Protection Agency, December 1994
- ◆ *Material Safety Data Sheets (MSDS)*
- ◆ *Sax's Dangerous Properties of Industrial Materials*, 9th edition, 1996
- ◆ *Hawley's Condensed Chemical Dictionary*, 12th edition, 1993
- ◆ *Aldrich Chemical Company Catalog*, 1996–1997
- ◆ *Environmental Law Index to Chemicals*, 1996
- ◆ *Environmental Contaminant Reference Databook*, volume 1, 1995
- ◆ *Merck Index*, 12th edition, 1996

CUMENE (yf)

CAS #: 98-82-8
MOLECULAR FORMULA: C_9H_{12}
FORMULA WEIGHT: 120.21
SYNONYMS: cumol, 2-fenilpropano, isopropyl benzene, isopropylbenzol, (1-methylethyl)-(9CI) benzene, 2-phenylpropane, propane

USES
- ◆ thinner for paints, lacquers, and enamels
- ◆ Constituent of
  - petroleum based solvents
  - high octane aviation fuel
  - naphtha
- ◆ used in production of styrene
- ◆ used as a solvent
- ◆ Manufacture of Other Chemicals
  - phenol
  - acetophenone
  - polymerization catalysts
  - acetone
  - alphamethylstyrene
  - diisopropylbenzene
- ◆ catalyst for acrylic and polyester type resins
- ◆ used in gasoline blending
- ◆ raw material for peroxides and oxidation catalysts

- ◆ chemical intermediate for dicumyl peroxide

PHYSICAL PROPERTIES
- ◆ colorless liquid
- ◆ sharp, gasoline-like odor
- ◆ specific gravity 0.862 at 20°C (water = 1)
- ◆ surface tension 28.2 dynes/cm at 20°C
- ◆ boiling point 152.4°C at 760 mm Hg
- ◆ melting point −96°C
- ◆ heat of vaporization 10,335.3 gcal/gmol
- ◆ vapor pressure 10 mm Hg at 38.3°C
- ◆ vapor density 4.1 (air = 1)
- ◆ maximum solubility in water 50 mg/L
- ◆ soluble in ethyl alcohol at 20°C
- ◆ volatile
- ◆ odor threshold 0.012 ppm
- ◆ critical pressure 31.7 atm
- ◆ critical temperature 357.9°C
- ◆ refractive index 1.49
- ◆ viscosity at 20°C 0.8 cP

CHEMICAL PROPERTIES
- ◆ flash point 46°C
- ◆ autoignition temperature 795°F
- ◆ flammability limits 0.9% lower and 6.5% upper (by volume in air)
- ◆ explosive limits 1.1% lower and 8.0% upper (by volume in air)
- ◆ heat of combustion 1247.3 kcal/mol at 20°C
- ◆ heat of formation at 25°C −9.9 kcal/mol
- ◆ ionization potential 8175 eV
- ◆ specific heat at 25°C 0.46 cal/g° C

HEALTH RISK
- ◆ Toxic Exposure Guidelines
  - • American Conference of GIH TWA 50 ppm-skin
- ◆ Toxicity Data
  - • minimum lethal concentration mice 2000 ppm in air
  - • human adverse reflex response 0.028 mg/m^3
  - • oral-rat 1400 mg/kg
  - • oral-mouse 12,750 mg/kg
  - • inhalation-mouse 10 g/m^3
  - • skin-rabbit 12,300 μL/kg
  - • bioconcentration factor 3.5
- ◆ Other Toxic Indicators
  - • reference concentration 0.009 mg/m^3
  - • reference dose 0.04 (mg/kg)/d
- ◆ Acute Risks
  - • irritation to the eyes and mucous membranes
  - • central nervous system depression
  - • skin and upper respiratory tract
  - • dermatitis
  - • narcotic effects
- ◆ Chronic Risks

- damage to lungs
- damage to liver
- damage to kidneys
- target organ: central nervous system

HAZARD RISK
- USCG-grade D combustible liquid
- incompatibility with oxidizing agent
- hazardous combustion/decombustion by-products
  - carbon monoxide
  - carbon dioxide

EXPOSURE ROUTES
- inhalation
- swallowing
- absorption through skin and eyes

REGULATORY STATUS/INFORMATION
- Toxic Substances Control Act Section 8(d), manufacturers, importers and processors are required to submit to EPA copies of unpublished health and safety studies
- Resource Conservation and Recovery Act under 40CFR261.33, cumene must be disposed of in the proper manner, as set forth in the RCRA
- Cumene is exempted from the tolerance requirements under FIFRA when used as an inert ingredient in agricultural applications such as pesticides.
- Cumene is subject to Superfund Amendments Reauthorization Act Section 113 reporting requirements TSCA Section 4(a) Final Test Rule and Consent Agreement Substances 40CFR799 Subpart B.
- Toxic Substances Control Act Section 12(b) One Time Export Notification Substances 40CFR207
- Safe Drinking Water Act Special Monitoring Substances 40CFR141.40
- Clean Air Act Section 111 Volatile Organic Compounds 40CFR60
- Clean Air Act Section 112 Statutory Air Pollutants 42USC7401, 7412, 7416, 7601
- Resource Conservation and Recovery Act Hazardous Substances 40CFR261
- CERCLA Hazardous Substance 40CFR302
- Superfund Amendments Reauthorization Act Toxic Chemicals 40CFR Chapter 1 part 372
- Occupational Safety and Health Act Air Contaminants 29CFR1910. Subpart z
- Dpepartment of Traffic Appendix A 49CFR172.101
- Department of Traffic Marine Pollutants 49CFR172.101 Appendix B
- Occupational Safety and Health Act 8 hour threshold limit 50 ppm skin absorption

ADDITIONAL/OTHER COMMENTS
◆ fire fighting: use carbon dioxide, dry chemical powder, or foam
◆ disposal: mix with a combustible solvent and burn in chemical incinerator equipped with afterburner and scrubber
◆ personal protection: when working with cumene, wear NIOSH/MSHA approved respirator, chemical-resistant gloves, safety goggles, and protective clothing
◆ storage: store in a cool, dry place

SOURCES
◆ *Material Safety Data Sheets (MSDS)*
◆ University of Kentucky web page http://www.chem.uky.edu/resources/msds.html
◆ *Environmental Contaminants Reference Databook*, volume 1, 1995
◆ *Sax's Dangerous Properties of Industrial Materials*, 9th edition, 1996
◆ *Encyclopedia of Chemical Technology*, 4th edition, volume 7
◆ Encyclopedia of Chemical Processing and Design

CYANIDE COMPOUNDS

CAS #: Cyanide compounds have variable CAS #s. The CAS # for cyanide is 57-12-5.

MOLECULAR FORMULA: Cyanide compounds have variable molecular formulas. The formula for cyanide is $(CN)^{-1}$. The molecular formula for hydrogen cyanide is HCN.

Note: The 1990 Amendments to the Clean Air Act define the hazardous air pollutant "Cyanide Compounds" as follows: "X'CN where X = H' or any other group where a formal dissociation may occur. For example KCN or $Ca(CN)_2$."

FORMULA WEIGHT: Cyanide compounds have variable formula weights. The formula weight for cyanide is 26.02. The formula weight for hydrogen cyanide is 27.03.

SYNONYMS: cyanide: carbon nitride ion (CN^{1-}), cyanide anion, cyanure (French), isocyanide
hydrogen cyanide: formonitrile, hydrocyanic acid, prussic acid
sodium cyanide: cyanide of sodium, hydrocyanic acid, sodium salt
potassium cyanide: cyanide of potassium, hydrocyanic acid, potassium salt

USES
◆ Hydrogen Cyanide
  • production of organic chemicals
  • insecticide for fumigating enclosed spaces
  • gas chamber executions
◆ Cyanide Salts
  • electroplating
  • metal treatment

PHYSICAL PROPERTIES (Hydrogen Cyanide)
◆ colorless gas   ◆ faint bitter almond odor

- odor threshold (air) 0.58 ppm, (water) 0.17 ppm
- melting point $-13.24°C$
- boiling point $25.70°C$
- density $0.6884 g/cm^3$ at $20°C$
- vapor pressure 264.3 mm Hg at $0°C$
- miscible in water
- soluble in ethanol
- log K_{ow} 0.66

PHYSICAL PROPERTIES (Sodium Cyanide)
- white solid
- odorless when dry
- melting point $563.7°C$
- boiling point $1500°C$
- density $1.60 g/cm^3$
- solubility in water 48 g/100 mL at $10°C$
- slightly soluble in ethanol and formamide
- log K_{ow} -0.44
- vapor pressure 0.76 at $800°C$

PHYSICAL PROPERTIES (Potassium Cyanide)
- white solid
- faint bitter almond odor
- melting point $634.5°C$
- solubility in water 71.6 g/100 mL at $25°C$
- slightly soluble in ethanol

CHEMICAL PROPERTIES (Hydrogen Cyanide)
- autoignition temperature $538°C$
- flash point $-17.8°C$ (closed cup)
- flammability limits 6–41 vol% in air at $20°C$
- explosive limits 40% upper, 5.6% lower

HEALTH RISK
- Toxic Exposure Guidelines
  - ACGIH TLV 11 mg/m³
  - NIOSH REL 5 mg/m³
  - OSHA PEL 5 mg/m³
  - DFG MAK 5 mg/m³
- Toxicity Data
  - intraperitoneal-mouse LD$_{50}$ 3 mg/kg
  - rat LC$_{50}$ 500 mg/m³
  - human death 100 mg/m³
- Acute Risks
  - Cyanide is extremely toxic to humans.
  - Inhalation can be rapidly lethal.
  - effects on the brain, lungs, and heart
  - coma
  - harms central nervous system
  - harms respiratory system
  - harms cardiovascular system
  - rapid, deep breathing
  - shortness of breath
  - convulsions
  - loss of consciousness
  - salivation
  - nausea without vomiting
  - anxiety
  - confusion
  - vertigo
  - giddiness
  - lower jaw stiffness
  - convulsions
  - opisthotonos
  - paralysis
  - cardiac arrhythmias
  - transient respiratory stimulation
  - bradycardia
  - hydrogen cyanide: weakness, nausea, increased rate of respiration, eye and skin irritation

◆ Chronic Risks
  • effects on the central nervous system
  • headaches
  • numbness
  • tremor
  • weakness
  • dizziness
  • loss of appetite
  • deafness
  • loss of muscle coordination
  • loss of visual acuity
  • cardiovascular effects
  • respiratory effects
  • enlarged thyroid gland
  • cretinism
  • irritation to eyes and skin
  • congenital hypothyroidism in newborns
  • malformations in fetus and low fetal body weights in animals

HAZARD RISK
◆ flammable by chemical reaction with heat, moisture, and acid
◆ Many cyanides evolve hydrogen cyanide easily.
◆ Hydrogen cyanide is flammable and highly toxic.
◆ Carbon dioxide from air liberates HCN from cyanide solutions.
◆ Reaction with hypochlorite solutions may be violent at pH 10.0–10.3.
◆ explodes if melted with nitrates or chlorates at about 450°C
◆ violent reaction with F_2, Mg, nitrates, HNO_3, nitrites
◆ Metal cyanides are easily oxidized and may be thermally unstable.
◆ N-cyano derivatives may be reactive or unstable.
◆ Many organic nitriles may be very reactive.
◆ Upon decomposition or contact with acid, acid fumes, water, or steam, it emits toxic and flammable vapors of CN^-.

EXPOSURE ROUTES
◆ car exhaust
◆ emissions from chemical processing
◆ emissions from municipal waste incinerators
◆ smoking
◆ water from discharges from organic chemical industries, ironworks, and steelworks
◆ wastewater facilities
◆ electroplating work
◆ metallurgical work
◆ fire fighting
◆ steel manufacturing
◆ metal cleaning industries
◆ petroleum refineries
◆ pharmaceuticals
◆ tannery work
◆ blacksmithing
◆ photoengraving
◆ photography
◆ use of cyanide-containing pesticides
◆ groundwater from landfills
◆ cyanide-containing road salts

REGULATORY STATUS/INFORMATION
◆ CERCLA and Superfund
  • list of chemicals. 40CFR372.65

ADDITIONAL/OTHER COMMENTS
◆ Symptoms
  • ingestion: salivation, nausea, deep breathing, shortness of breath, anxiety, confusion, vertigo, giddiness, lower jaw stiffness, convulsions, opisthotonos, paralysis, coma, cardiac arrhythmias, transient respiratory stimulation

◆ First Aid
  • wash eyes and skin immediately with large amounts of water
  • get medical attention immediately
◆ personal protection: wear self-contained breathing apparatus, chemical goggles, and protective clothing
◆ storage: keep away from heat, flame, and sources of ignition; tightly closed container; cool, dry, well-ventilated area

SOURCES
◆ *Health Effects Notebook for Hazardous Air Pollutants*, U.S. Environmental Protection Agency, December 1994
◆ *Material Safety Data Sheets (MSDS)*
◆ *Sax's Dangerous Properties of Industrial Materials*, 9th edition, 1996
◆ *Environmental Law Index to Chemicals*, 1996
◆ *Toxicological Profile for Cyanide*
◆ Agency for Toxicological Substances and Disease Registry
◆ *Environmental Writer, Cyanide Compounds Chemical Backgrounder*, Environmental Health Centers

2,4-D, SALTS AND ESTERS

CAS #: 94-75-7
MOLECULAR FORMULA: $C_8H_6Cl_2O_3$
FORMULA WEIGHT: 221.04
SYNONYMS: acide-2,4-dichlorophenoxyacetique (French), acido (2,4-dicloro-fenossi)-acetico (Italian), agrotect, amidox, amoxone, aquakleen, BH 2,4-D, chipco turf herbicide "D," chloroxone, crop rider, crotilin, D 50, dacamine, 2,4-D acid, debroussaill-ant 600, decamine, ded-weed, ded-weed LV-69, desormone, (2,4-dichloor-fenoxy)-azijnz-uur (Dutch), dichlorophenoxyacetic acid, 2,4-dichlorophenoxyacetic acid (DOT), 2,4-dichlorphenoxyacetic acid, (2,4-Dichlor-phenoxy)-essigsaeure (German), dicopur, dicotox, dinoxol, DMA-4, dormone, 2,4-dwuchlorofenoksyoctosy kwas (Polish), emulsamine bk, emulsamine E-3, ENT 8,538, envert, envert DT, esteron, farmco, fernesta, fernimine, fernoxone, foredex 75, formola 40, hedonal, herbidal, ipaner, krotiline, lawn-keep, macrondray, miracle, monosan, moxone, netagrone 600, pennamine, phenox, pielik, planotox, plantgard, rhodia, salvo, spritz-hormin/2,4-D, super D weedone, supermone concentrate, transamine, tributon, trinoxol, vergemaster, verton D, weed tox, weedtrol

USES
◆ Used for Herbicide Control
  • broadleaf plants
  • grasses
  • wheat
  • barley
  • oats
  • sorghum
  • corn
  • sugarcane
  • non–crop areas pasture
  • range land
  • lawns and turf
  • jungle defoliation
  • weed control in cereals
  • control of Canada thistle

- used on rice
- used as a plant-growth regulator

◆ Used in Formulations
  - pine release
  - water hyacinth control
  - prevention of seed formation
◆ Used on Vegetables
  - increases red color in potatoes
  - causes all plants to ripen at the same time when used on tomatoes
◆ Used in Forest Management
  - brush control
  - conifer release
  - tree injection
  - increases latex output of old rubber trees
  - fruit drop control
◆ Agricultural Use
  - wheat
  - corn
  - grain sorghum
  - rice and other grains
◆ basic material from which the soluble esters and salts are produced

PHYSICAL PROPERTIES
  - white to yellow crystalline powder
  - slight phenolic odor, odorless when pure
  - melting point 138°C
  - boiling point 160°C at 0.4 mm Hg
  - density 1.416 at 25°C
  - odor threshold 3.13 mg/kg
  - surface tension 66.5 dynes/cm at 25°C and pH 14
  - vapor pressure 53 Pa at 160°C
  - pK_a 2.73
  - K_a 2.3×10^{-3}
  - vapor density 7.63
  - log K_{ow} 2.81
  - 540 ppm solubility in water at 20°C

CHEMICAL PROPERTIES
  ◆ stable under normal conditions
  ◆ heat of combustion −7700 Btu/lb
  ◆ will not polymerize

HEALTH RISK
  ◆ Toxic Exposure Guidelines
    - ACGIH TLV TWA 10 mg/m^3
    - OSHA PEL TWA 10 mg/m^3
    - DFG MAK 10 mg/m^3
    - NIOSH REL TWA 10 mg/m^3
  ◆ Toxicity Data
    - skin-rabbit 500 mg/24H MLD
    - eye-rabbit 750 µg/24H SEV
    - sce-human:lym 10 mg/L
    - dni-ham:ovr 1 mmol/L
    - oral-rat TDLo 500 mg/kg
    - oral-mouse TDLo 707 mg/kg
    - oral-human TDLo 80 mg/kg

- oral-man LDLo 93 mg/kg
- oral-rat LD_{50} 370 mg/kg
- skin-rat LDLo 1500 mg/kg
- intraperitoneal-rat LDLo 666 mg/kg
- oral-mouse LD_{50} 368 mg/kg
- intraperitoneal-mouse LDLo 125 mg/kg
- oral-dog LD_{50} 100 mg/kg
- oral-rabbit LDLo 800 mg/kg
- skin-rabbit LD_{50} 1400 mg/kg
- oral-guinea pig LD_{50} 469 mg/kg
- oral-hamster LD_{50} 500 mg/kg

◆ Acute Risks
  - severe irritation of eyes and skin
  - gastroenteric distress
  - nausea
  - vomiting
  - diarrhea
  - mild central nervous system depression
  - dysphagia
  - ventricular fibrillation
  - transient liver and kidney injury

◆ Chronic Risks
  - kidney and liver damage
  - somnolence
  - convulsions
  - stiffness of extremities
  - ataxia
  - paralysis
  - coma
  - muscular weakness
  - suspected human carcinogen
  - mutagenic and teratogenic effects

HAZARD RISK
◆ combustible
◆ poisonous gases and vapors such as hydrogen chloride and carbon monoxide produced in fire
◆ incompatible with strong oxidizers
◆ capable of creating a dust explosion
◆ Decomposition emits toxic fumes of Cl^-.

EXPOSURE ROUTES
◆ primarily in soil
◆ residue in or on apples, apricots, citrus fruits, pears, potatoes, quinces, and other crops

REGULATORY STATUS/INFORMATION
◆ Resource Conservation and Recovery Act
  - list of hazardous constituents. 40CFR258

ADDITIONAL/OTHER COMMENTS
◆ Symptoms
  - ingestion: somnolence, convulsions, coma, nausea, vomiting
  - inhalation: irritates skin and eyes
◆ First Aid
  - wash eyes immediately with large amounts of water
  - wash skin immediately with soap and water
  - induce vomiting and follow with gastric lavage and supportive therapy

- ◆ fire fighting: use dry chemical, foam, carbon dioxide, and water
- ◆ personal protection: protective clothing, goggles, filter mask, and respiratory apparatus if necessary
- ◆ storage: keep away from contact with oxidizing materials

SOURCES
- ◆ *Material Safety Data Sheets (MSDS)*
- ◆ *Sax's Dangerous Properties of Industrial Materials*, 9th edition, 1996
- ◆ *Aldrich Chemical Company Catalog*, 1996–1997
- ◆ *Environmental Law Index to Chemicals*, 1996
- ◆ *Merck Index*, 12th edition, 1996
- ◆ *Environmental Contaminant Reference Databook*, volume 1, 1995

DDE (yf)

CAS #: 72-55-9
MOLECULAR FORMULA: $C_{14}H_8Cl_4$
FORMULA WEIGHT: 318.03
SYNONYMS: 2,2-bis(p-chlorophenyl)-1,1-dichloroethylene, bis(4-chlorophenyl)-1,1-dichloroethylene(2,2), bis(chloro-4-phenyl)-2,2-dichloro-1,1-ethylene, dichloro-2,2-bis(p-chlorophenyl ethylene-1), dichloro-diphenyldichloro-ethylene, (p,p-dichlorodiphenyldichloroethylene), dichloro-ethenylidene bis (4-chlorobenzene)(1,1), 1,1'-dichloroethenylidene-bis(4-chloro)benzene

USES
- ◆ pesticides
- ◆ insecticide
- ◆ DDE is a breakdown product of DDT and has no uses.

PHYSICAL PROPERTIES
- ◆ white, crystalline solid
- ◆ melting point 85°C
- ◆ vapor pressure 6.5×10^{-6} torr at 20°C
- ◆ log octanol/water partition coefficient (log K_{ow}) 7.0

CHEMICAL PROPERTIES
- ◆ sensitive to light
- ◆ major metabolite of DDT

HEALTH RISK
- ◆ Toxicity Data
  - oral-rat 880 mg/kg
  - ingestion-duck 45 mg/kg/45d. Effect: cracks in the eggshells
  - oral-mouse 250 mg/kg. Effect: hepatomas, liver cell tumors
  - oral-mouse 148 and 261 mg/kg (male and female). Effect: hepatocellular carcinomas
  - oral-hamster 500 and 1000 mg/kg (male and female). Effect: liver cell tumors
  - Studies show increased thyroid tumors in female rats from oral exposure to DDE.
  - EPA has not established a RfC or a RfD for DDE.

- ◆ Other Toxicity Indicators
  - • Traces of DDE have been found in human blood, placental tissue, fat, urine, semen, breast milk, and umbilical cord blood.
  - • EPA classified group B2: probable human carcinogen of medium carcinogenic hazard
  - • $1/Ed_{10}$ value of 1.9 per (mg/kg)/d
  - • oral risk estimate 9.7×10^{-6} $(\mu/l)^{-1}$
- ◆ Acute Risks
  - • no acute risk data available
- ◆ Chronic Risks
  - • Oral animal studies have linked reduced fertility, reduced spermatogenesis, decreased testicular and ovarian weights, embryo toxicity, and fetotoxicity to exposure to DDT, which, when broken down, is DDE.

HAZARD RISK
- ◆ no data available

EXPOSURE ROUTES
- ◆ primarily through food
- ◆ release from hazardous waste sites
- ◆ ingestion

REGULATORY STATUS/INFORMATION
- ◆ CERCLA: medium carcinogenic hazard
- ◆ Clean Air Act Section 304 Water Quality Substances 33USC1251 et seq; Section 307 Priority Pollutants 40CFR423.17 Appendix A
- ◆ Resource Conservation and Recovery Act Hazardous Substances 40CFR part 261; Hazardous Constraints for Groundwater Monitoring 40CFR264 Appendix 1X; Land Disposal Restrictions Halogenated Organic Compounds 40CFR268

ADDITIONAL/OTHER COMMENTS
- ◆ DDE is listed as a pollutant of concern to EPA's great waters program because of its
  - • persistence in the environment
  - • potential to bioaccumulate
  - • toxicity to humans and the environment
- ◆ suspect mutagen
- ◆ DDT, the parent chemical of DDE, has been banned from the United States since the 1970s. Thus, the only remaining sources of DDE are traces left over from the days when DDT was allowed in the country.

SOURCES
- ◆ *Aldrich Chemical Company Catalog*, 1996–1997
- ◆ *Environmental Law Index to Chemicals*, 1996 edition
- ◆ EPA Health Effects Notebook
  http://www.epa.gov/oar/oaqps/airtox/hapindex.html
- ◆ University of Kentucky web page
  http://www.chem.uky.edu/resources/msds.html

DIAZOMETHANE

CAS #: 334-88-3
MOLECULAR FORMULA: CH_2N_2
FORMULA WEIGHT: 42.05
SYNONYMS: azimethylene, diazirine

USES
- ◆ Used as a Methylating agent for Acidic Compounds
  - • carboxylic acids
  - • phenols
  - • enols
- ◆ used as an analytical reagent

PHYSICAL PROPERTIES
- ◆ yellow gas
- ◆ musty odor
- ◆ melting point −145°C
- ◆ boiling point −23°C
- ◆ density 1.45
- ◆ soluble in ether and dioxane

CHEMICAL PROPERTIES
- ◆ forms yellow solutions in ethereal solvents
- ◆ very toxic and explosive
- ◆ Decomposition is rapid if alcohols or water are present.

HEALTH RISK
- ◆ Toxic Exposure Guidelines
  - • OSHA PEL TWA 0.2 ppm
  - • ACGIH TLV TWA 0.2 ppm
  - • DFG MAK animal carcinogen, suspected human carcinogen
- ◆ Toxicity Data
  - • mmo-nsc 250 mmol/L
  - • inhalation-rat TCLo 272 mg/m³/26W-I:ETA
  - • inhalation-mouse TCLo 272 mg/m³/26W-I:ETA
  - • subcutaneous-mouse TDLo 48 g/kg/52W-I:ETA
- ◆ Acute Risks
  - • strong respiratory irritation
  - • irritation of eyes
  - • denudation of mucous membranes
  - • cough
  - • wheezing
  - • asthmatic symptoms
  - • pulmonary edema
  - • fulminating pneumonia
  - • dizziness
  - • weakness
  - • headache
  - • chest pains
  - • fever
  - • moderate cyanosis
  - • malaise
  - • tremors
  - • hepatic enlargement
  - • hypersensitivity
  - • shock
  - • death
  - • hemorrhage emphysema in animals
  - • bronchopneumonia in animals
- ◆ Chronic Risks
  - • no information on the chronic effects in humans
  - • pulmonary adenomas in rats and mice
  - • IRAC Group 3: not classifiable as to its carcinogenicity to humans

HAZARD RISK
- explosive
- highly explosive when shocked or exposed to heat or by chemical reaction
- Undiluted liquid and concentrated solutions may explode violently, especially if impurities are present.
- Gaseous diazomethane may explode on heating to 100°C or on tough glass surfaces.
- Alkali metals produce explosions with diazomethane.
- Copper powder causes active decomposition with the evolution of nitrogen and the formation of insoluble white flakes of polymethylene.
- incompatible with alkali metals and calcium sulfate
- Decomposition emits highly toxic fumes of NO_x.

EXPOSURE ROUTES
- occupational exposure in the workplace

REGULATORY STATUS/INFORMATION
- Clean Air Act Amendment, Title III: hazardous air pollutants
- CERCLA and Superfund
  - list of chemicals. 40CFR372.65
- Occupational Safety and Health Act
  - 29CFR1910.1000 Table Z1 and list of highly hazardous chemicals, toxics, and reactives
- chemical regulated by the State of California

ADDITIONAL/OTHER COMMENTS
- Symptoms
  - labored breath, dizziness, vomiting, coughing, headache, nausea, prickling, diarrhea
- First Aid
  - wash eyes immediately with large amounts of water
  - wash skin immediately with soap and water
  - remove to fresh air immediately or provide respiratory apparatus if necessary
  - drink water and induce vomiting
- personal protection: compressed air/oxygen apparatus, gas-tight suit
- storage: use safety screen and a well-ventilated hood when in use

SOURCES
- *Health Effects Notebook for Hazardous Air Pollutants*, U.S. Environmental Protection Agency, December 1994
- *Material Safety Data Sheets (MSDS)*
- *Sax's Dangerous Properties of Industrial Materials*, 9th edition, 1996
- *Environmental Law Index to Chemicals*, 1996
- *Merck Index*, 12th edition, 1996
- *Hawley's Condensed Chemical Dictionary*, 12th edition, 1993

DIBENZOFURANS

CAS #: 132-64-9
MOLECULAR FORMULA: $C_{12}H_8O$
FORMULA WEIGHT: 168.20
SYNONYMS: dibenzofuranne, dibenzofurano

USES
- insecticide

PHYSICAL PROPERTIES
- white crystals or crystalline solid
- odor threshold 1 mg/m³
- melting point 87°C
- boiling point 288°C
- density 1.089 g/cm³ at 99°C
- vapor pressure 0.0175 mm Hg at 25°C
- solubility in water 3 mg/L at 25°C
- slightly soluble in alcohol, ether, and benzene
- log K_{ow} 3.18 − 4.12

CHEMICAL PROPERTIES
- No information is available on the chemical properties of dibenzofuran.

HEALTH RISK
- Toxic Exposure Guidelines
  - No information is available on the toxic exposure guidelines of dibenzofuran.
- Toxicity Data
  - No information is available on the toxicity data of dibenzofuran.
- Acute Risks
  - No information is available on the acute risks of dibenzofuran for humans or animals.
- Chronic Risks
  - EPA Group D: not classifiable as to human carcinogenicity

HAZARD RISK
- No information is available on the hazard risk of dibenzofuran

EXPOSURE ROUTES
- inhalation and dermal contact
- combustion/carbonization processes
- coal tar and coal gasification operations
- ambient air
- coke dust
- grate ash
- fly ash
- flame soot
- inhalation of contaminated air
- consumption of contaminated drinking water or food
- tobacco smoke

REGULATORY STATUS/INFORMATION
- Clean Air Act Amendment, Title III: hazardous air pollutants
- Resource Conservation and Recovery Act
  - list of hazardous organic and inorganic constituents. 40CFR258
  - groundwater monitoring chemicals list. 40CFR264

- ◆ CERCLA and Superfund
  - • list of chemicals. 40CFR372.65
- ◆ chemical regulated by the State of California

ADDITIONAL/OTHER COMMENTS
- ◆ There are no additional comments on dibenzofuran.

SOURCES
- ◆ *Health Effects Notebook for Hazardous Air Pollutants*, U.S. Environmental Protection Agency, December 1994
- ◆ *Material Safety Data Sheets (MSDS)*
- ◆ *Environmental Law Index to Chemicals*, 1996
- ◆ *Hawley's Condensed Chemical Dictionary*, 12th edition, 1993
- ◆ *Aldrich Chemical Company Catalog*, 1996–1997

1,2-DIBROMO-3-CHLOROPROPANE (pf)

CAS #: 96-12-8
MOLECULAR FORMULA: $C_3H_5Br_2Cl$
FORMULA WEIGHT: 236.35
SYNONYMS: BBC 12, 1-chloro-2,3-dibromopropane, 3-chloro-1,2-di-bromopropane, DBCP, Dibromochloropropan (German), dibromo-chloropropane, dibromochloropropane (DOT), 1,2-dibromo-3-chloro-propane (DOT:OSHA), 1,2-Dibromo-3-chlor-propan (German), 1,2-dibromo-3-cloro-propano (Italian), 1,2-dibroom-3-chloorpropaan (Dutch), fumagon, fumazone, fumazone 86, fumazone 86E, NCI-C00500, nemabrom, nemafume, nemagon, nemagon 20, nemagon 20G, nemagon 90, nemagon 206, nemagone, nemagon soil, fumigant, nemanax, nemapaz, nemaset, nematocide, nematox, nemazon, OS 1897, OXY DBCP, propane 1-chloro-2,3-dibromo-, RCRA waste number U066, SD 1897, UN2872 (DOT), many others

USES
- ◆ pesticide
- ◆ nematocide
- ◆ soil fumigant
- ◆ used as chemical intermediate in the production of organic chemicals
- ◆ All other uses were canceled by the EPA in 1979.

PHYSICAL PROPERTIES
- ◆ colorless liquid when pure, while commercial grades are dark-amber to dark-brown
- ◆ melting point 6.7°C (44.06°F)
- ◆ boiling point 195.5°C (383.9°F) at 760 mm Hg
- ◆ density 2.05 g/cm³ at 20°C
- ◆ specific gravity 2.05 (water = 1.00)
- ◆ vapor density not available (air = 1)
- ◆ surface tension not available
- ◆ viscosity not available
- ◆ heat capacity not available
- ◆ heat of vaporization not available
- ◆ pungent odor
- ◆ odor threshold 0.3 mg/m³

- ◆ vapor pressure 0.8 mm at 21°C
- ◆ slightly soluble in water
- ◆ miscible with oils, dichloropropane, and isopropyl alcohol

CHEMICAL PROPERTIES
- ◆ combustible liquid
- ◆ stable
- ◆ Hazardous polymerization will not occur.
- ◆ combustion and heating to decomposition yields toxic hydrogen chloride, hydrogen bromide, carbon monoxide, and carbon dioxide fumes
- ◆ incompatible with strong oxidizing agents, bases, aluminum, magnesium, tin, and their alloys
- ◆ flash point 76.6°C (170°F)
- ◆ explosion limits in air not available
- ◆ autoignition temperature not available
- ◆ heat of combustion not available
- ◆ enthalpy (heat) of formation not available

HEALTH RISK
- ◆ Toxic Exposure Guidelines
  - • ACGIH TLV TWA not available
  - • OSHA PEL TWA 0.001 ppm (cancer hazard)
  - • NIOSH REL TWA 0.01 ppm/30M
- ◆ Toxicity Data
  - • oral-rat LD_{50} 170 mg/kg
  - • inhalation-rat LC_{50} 103 ppm/8H
  - • subcutaneous-rat LD_{50} 100 mg/kg
  - • oral-mouse LD_{50} 257 mg/kg
  - • intraperitoneal-mouse LD_{50} 123 mg/kg
  - • oral-rabbit LD_{50} 180 mg/kg
  - • skin-rabbit LD_{50} 1400 mg/kg
  - • oral-guinea pig LD_{50} 150 mg/kg
  - • oral-chicken LD_{50} 60 mg/kg
- ◆ Acute Risks
  - • harmful if swallowed, inhaled, or absorbed through the skin
  - • Vapor or mist is irritating to the eyes, mucous membranes, and upper respiratory tract.
  - • causes skin irritation and central nervous system depression
  - • effects may vary from mild irritation to severe destruction to tissue depending on the intensity and duration of exposure.
  - • Exposure can cause gastrointestinal disturbances, nausea, headaches, vomiting, pulmonary congestion and edema, and damage to the eyes, liver, and kidneys.
- ◆ Chronic Risks
  - • confirmed human carcinogen
  - • Effects on the liver, eyes, kidneys, central nervous system, and immune system have been recorded.
  - • mutagen
  - • may alter genetic material
  - • primarily causes male reproductive effects

- causes decreased sperm counts in men
- High incidences of tumors in the nasal cavities of rats and mice and in the lungs of mice have been observed due to inhalation.
- OSHA-regulated carcinogen
- Reference concentration is 0.0002 mg/m³ based on testicular effects in rabbits.
- The reference dose has not been established.

HAZARD RISK
- hazard toxic by ingestion, inhalation, and absorption
- confirmed human carcinogen
- mutagen
- combustible liquid
- may alter genetic material
- primarily causes male reproductive effects
- causes decreased sperm counts in men
- Heating to decomposition yields toxic hydrogen chloride, hydrogen bromide, carbon monoxide, and carbon dioxide fumes.
- incompatible with strong oxidizing agents, bases, aluminum, magnesium, tin, and their alloys
- no longer made in the United States

EXPOSURE ROUTES
- inhalation
- absorption through skin or eyes
- ingestion
- most likely results from the inhalation of contaminated air in the workplace and ingestion of contaminated drinking water
- In the past exposure occurred primarily from its fumigant and nematocide uses.

REGULATORY STATUS/INFORMATION
- NTP 7th Annual Report on Carcinogens
- IARC Cancer Review: Group 2B IMEMDT 7,191,87
- Animal Sufficient Evidence IMEMDT 15,139,77
- Human Limited Evidence IMEMDT 20,83,79
- Animal Sufficient Evidence IMEMDT 20,83,79
- NCI Carcinogenesis Bioassay Completed
- Results Positive: mouse, rat NCITR* NCI-CG-TR-28,78
- EPA Genetic Toxicology Program
- Reported in EPA TSCA Inventory
- Community Right-to-Know List
- Clean Air Act of 1970
  - Chapter I, Subchapter C: Air Programs 40CFR50-99
  - Title III: hazardous air pollutants, originally in 42USC7412
- Resource Conservation Recovery Act of 1976
  - Chapter I, Subchapter I: Solid Wastes 40CFR240-299
  - constituents for detection monitoring (for MSWLF) 40CFR258-Appendix 1
  - list of hazardous inorganic and organic constituents 40CFR258-Appendix 2

- hazardous constituents 40CFR261 Appendix 8, 40CFR261.11
- groundwater monitoring chemicals list 40CFR264 Appendix 9
- risk specific doses 40CFR266 Appendix 5
- health based limits for exclusion of waste-derived residues 40CFR266 Appendix 7
- discarded commercial chemical products, off-specification species, container residues, and spill residues thereof 40CFR261.33, 261.33.f. (U waste)
◆ CERCLA and Superfund
- designation of hazardous substances. 40CFR302.4
- Appendix A: sequential CAS registry # list of CERCLA hazardous substances
- Appendix B: radionuclides
- chemicals and chemical categories to which this part applies 40CFR372.65
◆ EPA Office of Air Quality Planning and Standards, for a hazard ranking under section 112(g) of the Clean Air Act Amendments, has ranked the contaminant in the nonthreshold category.
◆ OSHA, 29CFR1910.1000 Table Z1: limits for air contaminants
◆ OSHA 29CFR1910.1001-1048: specially regulated substances
◆ chemical regulated by the State of California

ADDITIONAL/OTHER COMMENTS
◆ Symptoms
- harmful if swallowed, inhaled, or absorbed through the skin
- Vapor or mist is irritating to the eyes, mucous membranes, and upper respiratory tract.
- causes skin irritation and central nervous system depression
- Effects may vary from mild irritation to severe destruction to tissue depending on the intensity and duration of exposure.
- Exposure can cause gastrointestinal disturbances, nausea, headaches, vomiting, pulmonary congestion and edema, and damage to the eyes, liver, and kidneys.
◆ First Aid
- if inhaled, remove to fresh air, give artificial respiration if not breathing, or oxygen if breathing is difficult, and get medical attention
- if contacted with the eyes, flush with water for 15 minutes and get medical attention
- if contacted with skin, wash with water for at least 15 minutes, while removing clothing; get medical attention
- if swallowed drink water and get medical attention
◆ fire fighting: use carbon dioxide, dry chemical powder, or appropriate foam

- ◆ personal protection: use chemical-resistant gloves, boots, apron, goggles or a face shield, and a NIOSH-approved vapor respirator if needed due to poor ventilation
- ◆ storage: keep the storage container tightly closed and store in a cool, dry place away from heat and open flames

SOURCES
- ◆ *Material Safety Data Sheets (MSDS)*
- ◆ *Sax's Dangerous Properties of Industrial Materials*, 9th edition, 1996
- ◆ *Merck Index*, 12th edition, 1996
- ◆ *Hawley's Condensed Chemical Dictionary*, 12th edition, 1993
- ◆ *Environmental Law Index to Chemicals*, 1996 edition
- ◆ *Health Effects Notebook for Hazardous Air Pollutants*, U.S. Environmental Protection Agency, December 1994

DIBUTYLPHTHALATE

CAS #: 84-74-2
MOLECULAR FORMULA: $C_6H_4(COOC_4H_9)_2$
FORMULA WEIGHT: 278.38
SYNONYMS: benzene-o-dicarboxylic acid di-n-butyl ester, o-benzenedicarboxylic acid, dibutyl ester, n-butyl phthalate (DOT), celluflex DPB, dibutyl-1,2-benzenedicarboxylate, di-n-butyl phthalate, elaol, hexaplas M/B, palatinol C, polycizer DBP, PX 104, staflex DBP, witcizer 300

USES
- ◆ Used as a Plasticizer
  - nitrocellulose lacquers
  - elastomers
  - explosives
  - nail polish
  - solid rocket propellants
  - makes plastics soft and flexible
- ◆ Used in Consumer Products
  - shower curtains
  - raincoats
  - food wraps
  - bowls
  - car interiors
  - vinyl fabrics
  - floor tiles
  - perfume fixative
  - textile lubricating agent
  - safety glass
  - insecticides
  - printing inks
  - paper coatings
  - adhesives

PHYSICAL PROPERTIES
- ◆ colorless, oily liquid
- ◆ melting point −35°C
- ◆ boiling point 340°C
- ◆ density 1.0484 at 20°/20°C
- ◆ vapor density 9.58
- ◆ viscosity 0.203 poise at 20°C
- ◆ distillation range 227–335 at 37 mm Hg
- ◆ vapor pressure 1.0×10^{-5} mm Hg at 25°C
- ◆ critical temperature 513.2°C
- ◆ miscible with common organic solvents
- ◆ insoluble with water
- ◆ log K_{ow} 5.60

CHEMICAL PROPERTIES
- stable
- flash point 312 °F

- autoignition temperature 750 °F
- combustible

- can react with oxidizing materials

HEALTH RISK

◆ Toxic Exposure Guidelines
  - OSHA PEL TWA 5 mg/m³
  - ACGIH TLV TWA 5 mg/m³
  - NIOSH IDLH 9,300 mg/m³
◆ Toxicity Data
  - mmo-sat 100 μg/plate
  - cyt-ham:fbr 30 mg/L/24H
  - intraperitoneal-rat TDLo 6 g/kg (female 3–9D post):REP
  - oral-rat TDLo 2520 mg/kg (1–21D preg):TER
  - oral-rat TDLo 12,600 mg/kg (female 1–21D post):TER
  - oral-mouse TDLo 20 g/kg (female 6–13D post):REP
  - intraperitoneal-rat TDLo 1017 mg/kg (female 5–15D post):REP
  - oral-mouse TDLo 7200 mg/kg (female 1–18D post):REP
  - oral-mouse TDLo 16,800 mg/kg (male 7D pre):REP
  - oral-rat TDLo 16,800 mg/kg (male 7D pre):REP
  - oral-rat TDLo 8400 μg/kg (7D male):REP
  - oral-mouse TDLo 7200 mg/kg (female 1–18D post):TER
  - intraperitoneal-rat TDLo 305 mg/kg (female 5–15D post):TER
  - oral-human TDLo 140 mg/kg:CNS,GI,KID
  - oral-rat LD$_{50}$ 8000 mg/kg
  - inhalation-rat LC$_{50}$ 4250 mg/m³
  - skin-rat LDLo 6 g/kg
  - intraperitoneal-rat: LD$_{50}$ 3050 mg/kg
  - oral-mouse LD$_{50}$ 5289 mg/kg
  - inhalation-mouse LC$_{50}$ 25 g/m³/2H
  - intravenous-mouse LD$_{50}$ 720 mg/kg
◆ Acute Risks
  - eye effects by ingestion
  - hallucinations
  - distorted perceptions
  - nausea
  - vomiting
  - kidney changes
  - ureter changes
  - bladder changes
  - irritation of upper respiratory tract
  - irritation of stomach
◆ Chronic Risks
  - No information is available for chronic effects in humans.
  - effects on liver in animals
  - reduced fetal weight in animals
  - decreased number of viable litters
  - birth defects in mice
  - decreased spermatogenesis and testes weight in animals
  - EPA Group D: not classifiable as to human carcinogenicity

HAZARD RISK

◆ combustible when exposed to heat or flame

- ◆ can react with oxidizing materials
- ◆ violent reaction with chlorine
- ◆ incompatible with chlorine
- ◆ Decomposition emits acrid smoke and fumes.

EXPOSURE ROUTES

- ◆ primarily through food
- ◆ ambient air
- ◆ air of new cars and inside homes
- ◆ vinyl floors
- ◆ drinking water supplies

REGULATORY STATUS/INFORMATION

- ◆ Clean Air Act Amendment, Title III: hazardous air pollutants
- ◆ Resource Conservation and Recovery Act
  - list of hazardous constituents. 40CFR261, 40CFR261.11
  - U waste # 069
- ◆ CERCLA and Superfund
  - designation of hazardous substances. 40CFR302.4
  - list of chemicals. 40CFR372.65
- ◆ OSHA, 29CFR1910.1000 Table Z1
- ◆ chemical regulated by the State of California

ADDITIONAL/OTHER COMMENTS

- ◆ Symptoms
  - inhalation: irritation of the nose, throat, and upper respiratory tract
  - ingestion: nausea, dizziness, lacrimation, photophobia, conjunctivitis
  - eye contact: immediate, severe, stinging pain, profuse tears
- ◆ First Aid
  - wash eyes immediately with large amounts of water
  - wash skin immediately with soap and water
  - remove to fresh air immediately or provide respiratory apparatus if necessary
  - drink water and induce vomiting
- ◆ fire fighting: use carbon dioxide, dry chemical, or foam
- ◆ personal protection: compressed air/oxygen apparatus, gloves

SOURCES

- ◆ *Health Effects Notebook for Hazardous Air Pollutants*, U.S. Environmental Protection Agency, December 1994
- ◆ *Material Safety Data Sheets (MSDS)*
- ◆ *Sax's Dangerous Properties of Industrial Materials*, 9th edition, 1996
- ◆ *Environmental Law Index to Chemicals*, 1996
- ◆ *Merck Index*, 12th edition, 1996
- ◆ *Hawley's Condensed Chemical Dictionary*, 12th edition, 1993
- ◆ *Aldrich Chemical Company Catalog*, 1996-1997

1,4-DICHLOROBENZENE (kjc)

CAS #: 106-46-7
MOLECULAR FORMULA: $C_6H_4Cl_2$
FORMULA WEIGHT: 147.02 g/mol
SYNONYMS: 1,4-dichloorbenzeen (Dutch), 1,4-Dichlorbenzol (German), di-chloricide, 1,4-dichlorobenzene, 1,4-diclorobenzene (Italian), evola, globol, NCI-C54955, p-chlorophenyl chloride, p-dichloorbenzeen (Dutch), p-Dichlorbenzol (German), p-dichlorobenzene, p-dichlorobenzene (ACGIH:DOT:OSHA), p-dichlorobenzol, p-diclorobenzene (Italian), para crystals, paracide, paradi, Paradichlorbenzol (German), paradichlorobenzene, paradichlorobenzol, paradow, paramoth, paranuggets, parazene, pdb, pdcb, persia-perazol, santochlor

USES
- ◆ fumigant
  - moths
  - molds
  - mildews
  - tree-boring insects
  - fruitborers
  - ants
  - soil
  - space deodorizer
- ◆ germicide
- ◆ Used as a Component or in the Manufacturing Process
  - 2,5-dichloroaniline
  - dyes
  - additive to resin-bonded abrasive wheels
- ◆ pharmacy
- ◆ space odorant
  - pig pens
  - refuse containers
  - mothballs
  - toilet deodorizer blocks

PHYSICAL PROPERTIES
- ◆ colorless crystals
- ◆ melting point 53°C (127°F)
- ◆ boiling point 174°C (345°F)
- ◆ density 1.2475 g/mL at 20°C
- ◆ specific gravity 1.241 (water = 1.00)
- ◆ vapor density 5.07 (air = 1.00)
- ◆ surface tension 34.66 dynes/cm at 20°C
- ◆ viscosity 0.839 mN/m^{-2} at 55°C
- ◆ heat capacity not available
- ◆ heat of vaporization 17,260.5 cal/g
- ◆ aromatic, mothball-like odor; sweet taste
- ◆ odor threshold 0.18 ppm
- ◆ vapor pressure 1.03 mm Hg at 25°C, 10 mm Hg at 54.8°C
- ◆ solubility 65.3 mg/L at 25°C
- ◆ soluble in chloroform, carbon disulfide, alcohol, ether, acetone, benzene
- ◆ insoluble in water
- ◆ log octanol/water partition coefficient 3.52

CHEMICAL PROPERTIES
- ◆ volatile (sublimes readily)
- ◆ combustible
- ◆ nonstaining
- ◆ noncorrosive
- ◆ reacts strongly with oxidizing agents, aluminum and its alloys
- ◆ flash point 65°C (150°F)
- ◆ flammability limits 1.8% lower and 7.8% upper
- ◆ autoignition temperature 647°C
- ◆ heat of combustion not available

◆ enthalpy of formation not available

HEALTH RISK
 ◆ Toxic Exposure Guidelines
  • ACGIH TLV (450 mg/m³)
  • ACGIH TWA 10 ppm (60 mg/m³)
  • ACGIH TLV TWA 75 ppm
  • ACGIH TLV STEL 110 ppm
  • ACGIH STEL (675 mg/m³)
  • OSHA PEL (450 mg/m³)
  • OSHA PEL TWA 75 ppm (450 mg/m³)
  • OSHA PEL STEL 110 ppm (675 mg/m³)
  • OSHA STEL (675 mg/m³)
  • DFG MAK 50 ppm (300 mg/m³)
  • NIOSH IDLH 150 ppm (6000 mg/m³)
 ◆ Toxicity Data
  • oral-human LDLo 857 mg/kg
  • unreported-human LDLo 357 mg/kg
  • unreported-man LDLo 221 mg/kg
  • oral-rat LD_{50} 500 mg/kg
  • skin-rat LD_{50} >6 g/kg
  • intraperitoneal-rat LD_{50} 2562 mg/kg
  • oral-mouse LD_{50} 2950 mg/kg
  • oral-guinea pig LDLo 2800 mg/kg
  • intraperitoneal-mouse LD_{50} 2 g/kg
  • subcutaneous-mouse LD_{50} 5145 mg/kg
  • oral-rabbit LD_{50} 2830 mg/kg
  • skin-rabbit LD_{50} >2 g/kg
  • oral-mammal LD_{50} 2600 mg/kg
  • inhalation-mammal LC_{50} 12 g/m³
  • eye-human 80 ppm
 ◆ Other Toxicity Indicators
  • California no-significant risk level: 20 mcg/d
 ◆ Acute Risks
  • irritation of skin, eyes, mucous membranes, and upper respiratory tract
 ◆ Chronic Risks

  • damage to the central nervous system
  • blood disorders
  • EPA Group B2: probable human carcinogen
  • target organs: liver, kidneys, and lungs

HAZARD RISK
 ◆ NFPA rating Health 2, Flammability 2, Reactivity 0, Special Hazards none
 ◆ combustible
 ◆ poisonous gases produced when heated
 ◆ combustion or decomposition products: carbon monoxide, carbon dioxide, hydrogen chloride gas

EXPOSURE ROUTES
◆ primarily by inhalation of mothballs and toilet deodorizer blocks
◆ absorption through skin and eyes possible
◆ ingestion
◆ emissions from burning
◆ eating and drinking of contaminated water and foods
◆ occupational exposure in factories that produce or process 1,4-dichlorobenzene

REGULATORY STATUS/INFORMATION
◆ Clean Air Act Amendments of 1990 Section 111 Volatile Organic Compounds
  • list of hazardous air pollutants. 42USC7412
◆ Safe Drinking Water Act
  • one of 83 contaminants required to be regulated under the 1986 act
  • maximum contaminant level (MCL) for organic contaminants: 0.075 mg/L
  • maximum contaminant level goal (MCLG) for organic contaminants. 40CFR141.50(a): 0.075 mg/L
  • Safe Drinking Water Act Statutory Contaminants
◆ Resource Conservation and Recovery Act
  • RCRA Hazardous Substances
  • RCRA Hazardous Constituents for Groundwater Monitoring
  • RCRA Land Disposal Prohibitions Halogenated Organic Compounds (HOCs)
  • RCRA Land Disposal Prohibitions: Scheduled Third Wastes
  • Design Criteria for Municipal Solid Waste Landfills (MSWLF) under RCRA. 40CFR258.40
  • Constituents for Detection Monitoring for Municipal Solid Waste Landfills. 40CFR258-Appendix 1
  • list of hazardous inorganic and organic constituents. 40CFR258-Appendix 2
  • D waste #: D027
  • U waste #: U070, U071, U072
◆ CERCLA and Superfund
  • designation of hazardous substances. 40CFR302.4
  • list of chemicals. 40CFR372.65
  • Reportable Quantity (RQ): 100 lb
  • SARA Section 110 Priority List of CERCLA Hazardous Substances
  • SARA Title III Section 313 Toxic Chemicals
  • CERCLA Hazardous Substances
◆ Clean Water Act
  • total toxic organics (TTOs). 40CFR413.02
  • 126 priority pollutants. 40CFR423
  • Clean Water Act Section 304 Water Quality Criteria Substances

- Clean Water Act Section 307 Priority Pollutants
- Clean Water Act Section 311 Hazardous Substances
◆ OSHA Table Z-1, Table Z-1-A
◆ Department of Transportation
  - hazard class/division: 3
  - identification number: UN1592
  - labels: combustible
  - emergency response guide #: 58

ADDITIONAL/OTHER COMMENTS

◆ Symptoms
  - inhalation: eye, skin, and throat irritation
  - skin absorption: profuse rhinitis, headaches
  - ingestion: nausea, vomiting, decreased weight
◆ First Aid
  - wash eyes immediately with large amounts of water
  - wash skin immediately with soap and water
  - provide respiratory support
◆ fire fighting: use dry chemical, foam, carbon dioxide, or water spray
◆ Personal Protection
  - wear self-contained breathing apparatus
  - do not wear contact lenses when working with this chemical
  - wear protective clothing
  - use only in chemical fume hood
◆ Storage
  - keep tightly closed
  - keep away from heat and open flame
  - store in cool, dry place

SOURCES

◆ *Environmental Containment Reference Databook*, volume 1, 1995
◆ *Environmental Law Index to Chemicals*, 1996 edition
◆ *Fire Protection Guide to Hazardous Materials*, 11th edition, 1994
◆ *Material Safety Data Sheets (MSDS)*
◆ *NIOSH Pocket Guide to Chemical Hazards*, June 1996
◆ *Sax's Dangerous Properties of Industrial Materials*, 9th edition, volume 2, 1996

3,3'-DICHLOROBENZIDINE

CAS #: 91-94-1
MOLECULAR FORMULA: $C_6H_3ClNH_2C_6H_3ClNH_2$
FORMULA WEIGHT: 253.14
SYNONYMS: C.I. 23060, curithane C126, DCB, 4,4'-diamino-3,3'-dichlorobiphenyl, 4,4'-diamino-3,3'-dichlorodiphenyl, 3,3'-dichlorobenzidin (Czech), 3,3'-dichlorobenzidina (Spanish), dichlorobenzidine, 3,3'-dichlorobenzidene, o,o'-dichlorobenzidine, dichlorobenzidine base, 3,3'-dichloro-4,4'-biphenyldiamine, 3,3'-dichlorobiphenyl-

4,4'-diamine, 3,3'-dichloro-4,4'-diaminobiphenyl, 3,3'-dichloro-4,4'-diamino(1,1-biphenyl)

USES
◆ Used in the Manufacturing Process
  - azo dyes
  - rubber and plastic
  - printing ink
  - textiles, plastics, crayons
◆ Used as a Curing Agent
  - blends with 4,4'-methylenebis for polyurethane elastomers
  - isocyanate-containing polymers
  - solid urethane plastics
◆ Used as an Intermediate
  - detection of gold
  - production of pigments
  - substitutes for lead chromate pigments

PHYSICAL PROPERTIES
◆ gray or purple crystalline solid
◆ melting point 132–133°C
◆ boiling point 402°C
◆ vapor pressure 1.15×10^{-7} mm Hg at 25°C
◆ insoluble in water
◆ soluble in benzene, diethyl ether, ethanol, and glacial acetic acid
◆ log K_{ow} 3.64

CHEMICAL PROPERTIES
◆ nonflammable
◆ decomposes on heating
◆ formation of diazonium salts and alkyl derivatives

HEALTH RISK
◆ Toxic Exposure Guidelines
  - ACGIH TLV suspected human carcinogen
  - NIOSH REL reduce to lowest feasible level
  - DFG MAK 0.1 mg/m³
  - OSHA PEL cancer suspect agent
◆ Toxicity Data
  - mma-sat-5 µg/plate
  - dns-human:hla 100 nmol/L
  - bfa-rat/sat 40 mg/kg
  - otr-hamster:kdy 80 µg/L
  - dnd-mammal:lym 25,500 nmol/L
  - oral-rat TDLo 17 g/kg/50W-C:CAR
  - subcutaneous-rat TDLo 7 g/kg/43W-I:ETA
  - subcutaneous-mouse TDLo 320 mg/kg
  - oral-rat TD 21 g/kg/50W-C:CAR
  - oral-rat LD$_{50}$ 5250 mg/kg
◆ Other Toxicity Indicators
  - EPA Cancer Risk Level (1 in a million excess lifetime risk) 8.0×10^{-5} mg/m³
◆ Acute Risks
  - labored breath
  - unconsciousness
  - blue color
  - low to moderate toxicity
◆ Chronic Risks
  - dermatitis
  - gastrointestinal upset

- upper respiratory tract infections
- liver injury
- abnormal growth in kidneys of fetuses in mice
- increased tumors in animals
- a tumorigen and carcinogen
- EPA Group B2: probable human carcinogen

HAZARD RISK
- decomposes on heating
- Decomposition emits toxic fumes of NO_x and Cl^-.

EXPOSURE ROUTES
- low levels in water
- pressurized spray containers of paints
- lacquers
- enamels
- azo dyes
- synthesis of azo dyes
- workers in garment and leather industries
- workers in printing and paper industries
- homecraft industries

REGULATORY STATUS/INFORMATION
- Clean Air Act Amendment, Title III: hazardous air pollutants
- Resource Conservation and Recovery Act
  - list of hazardous constituents. 40CFR258, 40CFR261, 40CFR261.11
  - groundwater monitoring list. 40CFR264
  - U waste #: U073
- CERCLA and Superfund
  - designation of hazardous substances. 40CFR302.4
  - list of chemicals. 40CFR372.65
- Clean Water Act
  - total toxic organics. 40CFR413.02
  - 126 priority pollutants. 40CFR423
- OSHA 29CFR1910.1000 Table Z1 and specially regulated substances

ADDITIONAL/OTHER COMMENTS
- Symptoms
  - ingestion: labored breath, unconsciousness, blue color
- First Aid
  - wash eyes immediately with large amounts of water
  - wash skin with soap and water
  - remove to fresh air or provide respiratory support
  - induce vomiting
- personal protection: wear self-contained breathing apparatus and protective clothing and goggles

SOURCES
- *Health Effects Notebook for Hazardous Air Pollutants*, U.S. Environmental Protection Agency, December 1994
- *Material Safety Data Sheets (MSDS)*
- *Sax's Dangerous Properties of Industrial Materials*, 9th edition, 1996
- *Hawley's Condensed Chemical Dictionary*, 12th edition, 1993
- *Environmental Law Index to Chemicals*, 1996
- *Merck Index*, 12th edition, 1996

DICHLOROETHYL ETHER

CAS #: 111-44-4
MOLECULAR FORMULA: $ClCH_2CH_2OCH_2CH_2Cl$
FORMULA WEIGHT: 143.02
SYNONYMS: sym-dichloroethyl ether, bis (beta-chloroethyl) ether, bis (2-chloroethyl) ether, chlorex, 1-chloro-1-(beta-chloroethoxy) ethane, chloroethyl ether, clorex, DCEE, 2,2′-Dichlor-diaethylaether (German), beta,beta′-dichlorodiethyl ether, dichloroether, 2,2′-dichloroethyl ether, beta,beta′-dichloroethyl ether (MAK), di (beta-chloroethyl) ether, dichloroethyl oxide, 2,2′-dichloroetiletere (Italian), 2,2′-dichloroorethylether (Dutch), dwuchlorodwuetylowy eter (Polish), ENT 4,504, ether dichlore (French), 1,1′-oxybis (2-chloro) ethane, oxyde de chloroethyle (French)

USES
◆ Used in Industry
  • manufacture of pesticides
  • paints
  • varnishes
  • cleaning fluid for textiles
  • purification of oils and gasoline
◆ Used as a Solvent
  • fats
  • waxes
  • greases
  • esters

PHYSICAL PROPERTIES
◆ colorless liquid
◆ strong, unpleasant odor
◆ odor threshold 0.049 ppm
◆ melting point −51.9°C
◆ boiling point 178.5°C
◆ density 1.2220 at 20°/20°C
◆ vapor pressure 0.71 mm Hg at 20°C
◆ vapor density 4.93
◆ insoluble in water
◆ soluble in Et_2O, MeOH, and C_6H_6
◆ dissolves oils, fats, and greases
◆ log K_{ow} 1.58

CHEMICAL PROPERTIES
◆ nonflammable
◆ flash point 131°F
◆ autoignition temperature 696°F

HEALTH RISK
◆ Toxic Exposure Guidelines
  • OSHA PEL TWA 5 ppm
  • OSHA PEL STEL 10 ppm (skin)
  • ACGIH TLV TWA 5 ppm
  • ACGIH TLV STEL 10 ppm (skin)
  • DFG MAK 10 ppm
◆ Toxicity Data
  • skin-rabbit 10 mg/24H open
  • skin-rabbit 500 mg open MLD
  • eye-rabbit 20 mg
  • mmo-sat 1 mL/plate/2H

- mma-sat 1 mg/plate
- oral-mouse TDLo 33 g/kg/79-C:CAR
- subcutaneous-mouse TDLo 2400 mg/kg/60W-I:ETA
- oral-rat LD_{50} 75 mg/kg
- inhalation-rat LC_{50} 330 mg/m^3/4H
- oral-mouse LD_{50} 112 mg/kg
- inhalation-mouse LC_{50} 650 mg/m^3/2H
- skin-rabbit LD_{50} 720 mg/kg
- skin-guinea pig LD_{50} 300 mg/kg

◆ Other Toxicity Indicators
- EPA Cancer Risk Level (1 in a million excess lifetime risk) 3×10^{-6} mg/m^3

◆ Acute Risks
- extreme irritation of the respiratory tract and skin
- irritation of nose and eyes in animals
- congestion
- edema
- hemorrhage of the lung
- congestion of the brain, liver, and kidneys in animals
- central nervous system effects
- lacrimation
- coughing
- nausea
- vomiting

◆ Chronic Risks
- decreased body weights in rats
- liver tumors in mice
- EPA Group B2: probable human carcinogen

HAZARD RISK
◆ flammable liquid when exposed to heat, flame, or oxidants
◆ dangerous explosion hazard
◆ reacts vigorously with oleum and chlorosulfonic acid
◆ reacts with water or steam to evolve toxic and corrosive fumes
◆ can react vigorously with oxidizing materials
◆ Decomposition emits toxic fumes of Cl^-.

EXPOSURE ROUTES
◆ ambient air
◆ drinking water supplies
◆ groundwater near waste disposal sites
◆ chemical plants where sym-dichloroethyl ether is manufactured or used

REGULATORY STATUS/INFORMATION
◆ Resource Conservation and Recovery Act
- list of hazardous constituents. 40CFR258
- groundwater monitoring chemicals list. 40CFR264
- risk specific doses. 40CFR266
- health based limits for exclusion of waste-derived residues. 40CFR266

◆ CERCLA and Superfund
- designation of hazardous substances. 40CFR302.4
- list of chemicals. 40CFR372.65

◆ Clean Water Act
  • total toxic organics (TTOs). 40CFR413.02
  • 126 priority pollutants. 40CFR423
◆ chemical regulated by the State of California

ADDITIONAL/OTHER COMMENTS
◆ Symptoms
  • labored breath, headache, cough, irritation, stupefying, vomiting, abdominal pain, nausea, diarrhea, dizziness
◆ First Aid
  • wash eyes and skin immediately with large amounts of water
  • remove to fresh air immediately or provide respiratory apparatus if necessary
  • rinse out mouth with water and induce vomiting
◆ fire fighting: use water, foam, mist, fog, spray, or dry chemical
◆ personal protection: compressed air/oxygen apparatus, gastight suit

SOURCES
◆ *Health Effects Notebook for Hazardous Air Pollutants*, U.S. Environmental Protection Agency, December 1994
◆ *Material Safety Data Sheets (MSDS)*
◆ *Sax's Dangerous Properties of Industrial Materials*, 9th edition, 1996
◆ *Environmental Law Index to Chemicals*, 1996
◆ *Merck Index*, 12th edition, 1996
◆ *Hawley's Condensed Chemical Dictionary*, 12th edition, 1993
◆ *Aldrich Chemical Company Catalog*, 1996-1997

1,3-DICHLOROPROPENE (kjc)

CAS #: 542-75-6
MOLECULAR FORMULA: $C_3H_4Cl_2$
FORMULA WEIGHT: 110.97 g/mol
SYNONYMS: 3-chloroallyl chloride, alpha-chloroallyl chloride, gamma-chloroallyl chloride, 3-chloropropenyl chloride, DCP, D-D92, dichloropropene 1,3-dichloropropene (ACGIH), 1,3-dichloropropene-1, 1,3-dichloro-2-propene, 1,3-dichloropropylene, alpha, gamma-dichloropropylene, dorlone II, NCI-C03985, telone, telone II, telone II soil fumigant, vidden D

USES
◆ soil fumigant
◆ organic synthesis

PHYSICAL PROPERTIES
◆ clear, colorless liquid
◆ melting point 48°C (119°F)
◆ boiling point 104°C (219°F) at 730 mm Hg
◆ density not available
◆ specific gravity 1.198 (water = 1.00)
◆ vapor density 3.80 (air = 1.00)
◆ surface tension not available
◆ viscosity not available
◆ heat capacity not available
◆ heat of vaporization not available
◆ sweet chloroform-like odor

- odor threshold 1.0 ppm
- vapor pressure 34–43 mm Hg at 25°C
- soluble in acetone, toluene, and octane
- insoluble in water
- log octanol/water partition coefficient 1.60

CHEMICAL PROPERTIES

- oxidizable
- reacts strongly with active metals, aluminum, magnesium, halogens, oxidizers
- flash point 25°C (78°F)
- flammability limits 5.3% lower and 14.5% upper
- autoignition temperature not available
- heat of combustion not available
- enthalpy of formation not available

HEALTH RISK

- Toxic Exposure Guidelines
  - ACGIH TLV TWA 1 ppm (4.5 mg/m^3)
  - OSHA PEL TWA 1 ppm (5 mg/m^3)
  - NIOSH REL TWA 5 mg/m^3
- Toxicity Data
  - oral-rat LD$_{50}$ 470 mg/kg
  - skin-rat LD$_{50}$ 775 mg/kg
  - intraperitoneal-rat LD$_{50}$ 175 mg/kg
  - oral-mouse LD$_{50}$ 640 mg/kg
  - inhalation-mouse LC$_{50}$ 4650 mg/m^3/2H
  - skin-rabbit LD$_{50}$ 504 mg/kg
  - oral-rat TDL$_o$ 15,600 mg/kg/2Y-I:CAR
  - inhalation-mouse TCLo 60 ppm/6H/5D/2Y:NEO
  - inhalation-guinea pig LCLo 400 ppm/7H
- Other Toxicity Indicators
  - California no-significant risk level: 4 mcg/d
- Acute Risks
  - hemorrhage of the lungs, small intestine, and liver
  - mucous membrane irritation
  - nausea
  - vomiting
  - malaise
  - chest pain
  - breathing difficulty
  - emphysema and edema
- Chronic Risks
  - EPA Group B2: probable human carcinogen
  - skin sensitization
  - damage to the nasal mucosa and the urinary bladder
  - target organs: eyes, skin, respiratory system, liver, kidneys
  - alters genetic material

HAZARD RISK

- NFPA rating Health 2, Flammability 3, Reactivity 0, Special Hazards none
- combustible
- moderate fire risk
- combustion or decomposition products: carbon monoxide, carbon dioxide, hydrogen chloride gas

EXPOSURE ROUTES
- ◆ primarily by inhalation
- ◆ absorption through skin and eyes possible
- ◆ ingestion
- ◆ drinking of contaminated water
- ◆ occupational exposure in industries during its manufacture, formulation, or application as a soil fumigant

REGULATORY STATUS/INFORMATION
- ◆ Clean Air Act Amendments of 1990 Section 112 Statutory Air Pollutants
- ◆ Resource Conservation and Recovery Act
  - • RCRA Hazardous Substances
  - • U waste #: U084
- ◆ Safe Drinking Water Act
  - • one of 83 contaminants required to be regulated under the 1986 act
  - • Safe Drinking Water Act Drinking Water Priority List Substances
  - • Safe Drinking Water Act Synthetic Organic Chemical Monitoring Contaminants
- ◆ CERCLA and Superfund
  - • list of chemicals. 40CFR372.65
  - • Reportable Quantity (RQ): 100 lb
  - • CERCLA Hazardous Substances
  - • SARA Title III Section 313 Toxic Chemicals
- ◆ OSHA Table Z-1, Table Z-1-A
- ◆ Clean Water Act
  - • total toxic organics (TTOs). 40CFR413.02
  - • Section 304 Water Quality Criteria Substances
  - • Section 307 Priority Pollutants
  - • Section 311 Hazardous Substances
- ◆ National Toxicology Program (NTP) Annual Report on Carcinogens
- ◆ Department of Transportation
  - • DOT Hazardous Substances and Radionuclides
  - • hazard class/division: 3
  - • identification number: UN2047
  - • labels: flammable liquid
  - • emergency response guide #: 29
- ◆ Emergency Planning and Community Right-to-Know Act
  - • section 313 de minimis concentration: 0.1%

ADDITIONAL/OTHER COMMENTS
- ◆ Symptoms
  - • inhalation: eye, skin, and throat irritation
  - • skin absorption: respiratory system, irritated eyes, skin burns, lacerations
  - • ingestion: headaches, dizziness
- ◆ First Aid
  - • wash eyes immediately with large amounts of water
  - • wash skin immediately with soap and water

- provide respiratory support
◆ fire fighting: use dry chemical, foam, carbon dioxide, or water spray
◆ Personal Protection
  - wear self-contained breathing apparatus
  - use only in chemical fume hood
◆ Storage
  - keep tightly closed
  - keep away from heat and sparks
  - store in cool, dry place

SOURCES
◆ *Environmental Containment Reference Databook*, volume 1, 1995
◆ *Environmental Law Index to Chemicals*, 1996 edition
◆ *Fire Protection Guide to Hazardous Materials*, 11th edition, 1994
◆ *Material Safety Data Sheets (MSDS)*
◆ *NIOSH Pocket Guide to Chemical Hazards*, June 1996
◆ *Sax's Dangerous Properties of Industrial Materials*, 9th edition, volume 2, 1996

DICHLORVOS

CAS #: 62-73-7
MOLECULAR FORMULA: $(CH_3O)_2O(O))CH:CCL_2$
FORMULA WEIGHT: 220.98
SYNONYMS: apavap, astrobot, atgard, bay 19149, benfos, bibesol, brevinyl, canogard, cekusan, chlorovinphos, cyanophos, cypona, DDVF, DDVP, dedevap, deriban, derribante, devikol, (2,2-dichloor-vinyl)-dimethyl-fosfaat (Dutch), dichloorovo (Dutch), dichlorfos (Polish), 2,2-dichloroethanol dimethyl phosphate, 2,2-dichloroethenyl dimethyl phosphate, 2,2-dichloroethenyl phosphoric acid dimethyl ester, dichlorophos, dichlorovas, 2,2-dichlorovinyl alcohol, dimethyl phosphate, 2,2-dichlorovinyl dimethyl phosphate, 2,2-dichlorovinyl dimethyl phosphoric acid ester, dichlorphos, (2,2-Dichlorvinyl)-dimethyl-phosphat (German), o-(2,2-dicloro-vinil) dimetilfosfato (Italian), dimethyl-2,2-dichloroethenyl phosphate, dimethyl dichlorovinyl phosphate, o,o-dimethyl dichlorovinyl phosphate, dimethyl-2,2-dichlorovinyl phosphate, o,o-dimethyl-o-2,2-dichlorovinyl phosphate, o,o-Di-methyl-o-(2,2-dichlor-vinyl)-phosphate (German), divipan, dquigard, duo-kill, duravos, equigel, estrosel, estrosol, fecama, fly-die, fly fighter, herkal, krecalvin, lindan, mafu, marvex, mopari, nerkol, nogos, nopest, no-pest strip, nuva, oko, oms 14, phosphate de dimethyl et de 2,2-dichlorovinyle (French), phosphoric acid-2,2-dichloroethenyl dimethyl ester, phosvit, szklarniak, task, tenac, tetravos, vapona, vaponite, verdican, verdipor, vinylofos, vinylophos

USES
◆ Used as an Agricultural Insecticide
  - on crops
  - animals
  - stored products

- slow release on pest strips in homes
- mushrooms
- flies
- aphids
- spider mites
- caterpillars
- thrips
- outdoor fruit
- vegetables

◆ Used as an Anthelmintic (Worming Agent)
  - for dogs
  - for swine
  - for horses
◆ Used as a Botacide (Agent That Kills Larvae)
  - for horses
  - flea collars for dogs

PHYSICAL PROPERTIES

◆ oily colorless to amber liquid
◆ aromatic chemical odor
◆ boiling point 140°C at 20 mm Hg
◆ density 1.415 at 25°C
◆ vapor pressure 0.012 mm Hg at 20°C
◆ slightly soluble in water and glycerin
◆ miscible with aromatic and
◆ chlorinated hydrocarbon solvents and alcohols
◆ log K_{ow} 2.03

CHEMICAL PROPERTIES

◆ stable to heat
◆ corrosive to iron and milled steel

HEALTH RISK

◆ Toxic Exposure Guidelines
  - OSHA PEL TWA 1 mg/m³ (skin)
  - ACGIH TLV TWA 0.1 ppm (skin)
  - DFG MAK 0.1 ppm (1 mg/m³)
  - NIOSH REL 1 mg/m³
  - MSHA standard 1 mg/m³
◆ Toxicity Data
  - dns-human:oth 65 mmol/L
  - dni-human:lym 62 mg/L
  - cyt-hamster-intraperitoneal 3 mg/kg
  - oral-rat TDLo 39,200 µg/kg (14–21D preg):REP
  - inhalation-rabbit TCLo 4 mg/m³/23H
  - oral-rabbit TDLo 65 mg/kg
  - unr-rat TDLo 40 mg/kg
  - oral-pig TDLo 255 mg/kg
  - intraperitoneal-rat TDLo 15 mg/kg
  - oral-rat TDLo 4120 mg/kg/2Y-C:NEO
  - oral-rat TDLo 2060 mg/kg/2Y-C:CAR
  - oral-mouse TDLo 20,600 mg/kg/2Y-C:CAR
  - oral-rat LD$_{50}$ 17 mg/kg
  - inhalation-rat LC$_{50}$ 15 mg/m³/4H
  - skin-rat LD$_{50}$ 70,400 µg/kg
  - intraperitoneal-rat LD$_{50}$ 23,300 µg/kg
  - subcutaneous-rat LD$_{50}$ 10,800 µg/kg
  - oral-mouse LD$_{50}$ 61 mg/kg
  - inhalation-mouse LC$_{50}$ 13 mg/m³/4H

- skin-mouse LD_{50} 206 mg/kg
- intraperitoneal-mouse LD_{50} 22 mg/kg
- subcutaneous-mouse LD_{50} 24 mg/kg
- intravenous-mouse LD_{50} 18 mg/kg
- oral-dog LD_{50} 100 mg/kg
- oral-rabbit LD_{50} 10 mg/kg
- skin-rabbit LD_{50} 107 mg/kg

◆ Acute Risks
- reduction in cholinesterase levels in blood
- tightness of chest
- wheezing
- pupil constriction
- blurred vision
- tearing
- headaches
- nausea
- vomiting
- cramps
- diarrhea
- sweating
- twitching
- paralysis
- respiratory failure
- miosis
- aching eyes
- rhinorrhea
- anorexia
- giddiness
- ataxia
- low blood pressure
- cardiac irregularities
- irritation of skin

◆ Chronic Risks
- decreased plasma and red blood cholinesterase levels
- increases in forestomach tumors in mice
- leukemia in rats
- pancreatic adenomas in rats
- lung tumors in male rats
- mammary tumors in female rats
- EPA Group B2: probable human carcinogen

HAZARD RISK
◆ Toxic gases and vapors (HCl gas, phosphoric acid mist, CO) may be released in a fire.
◆ Decomposition emits toxic fumes of Cl^- and PO_x.

EXPOSURE ROUTES
◆ manufacture and formulation of dichlorovos
◆ application in agriculture, household, and public health uses
◆ indoor air buildings (pest strips or sprays for insect control)
◆ small amount in food

REGULATORY STATUS/INFORMATION
◆ Clean Air Act Amendment, Title III: hazardous air pollutants
◆ CERCLA and Superfund
  - designation of hazardous substances. 40CFR302.4
  - list of extremely hazardous substances and their threshold planning quantities. 40CFR355-AB
  - list of chemicals. 40CFR372.65
◆ Clean Water Act
  - designation of hazardous substances. 40CFR116.4
  - determination of reportable quantities. 40CFR117.3
◆ OSHA, 29CFR1910.1000 Table Z1
◆ chemical regulated by the State of California

ADDITIONAL/OTHER COMMENTS
- ◆ Symptoms
  - • headache, fatigue, dizziness, blurred vision, excessive sweating, nausea, vomiting, stomach cramps, diarrhea, salivation, muscular twitching
- ◆ First Aid
  - • wash eyes and skin immediately with large amounts of water
  - • remove to fresh air immediately or provide respiratory apparatus if necessary
  - • drink water or milk and induce vomiting
- ◆ fire fighting: use carbon dioxide, dry chemical, or foam
- ◆ personal protection: compressed air/oxygen apparatus, gastight suit

SOURCES
- ◆ *Health Effects Notebook for Hazardous Air Pollutants*, U.S. Environmental Protection Agency, December 1994
- ◆ *Material Safety Data Sheets (MSDS)*
- ◆ *Sax's Dangerous Properties of Industrial Materials*, 9th edition, 1996
- ◆ *Environmental Law Index to Chemicals*, 1996
- ◆ *Merck Index*, 12th edition, 1996
- ◆ *Hawley's Condensed Chemical Dictionary*, 12th edition, 1993
- ◆ *Environmental Contaminant Reference Databook*, volume 1, 1995

DIETHANOLAMINE

CAS #: 111-42-2
MOLECULAR FORMULA: $(HOCH_2CH_2)_2NH$
FORMULA WEIGHT: 105.16
SYNONYMS: bis (2-hydroxyethyl) amine, DEA, Diaethanolamin (German), diethanolamin (Czech), diethylolamine, 2,2'-dihydroxydiethylamine, di (2-hydroxyethyl) amine, diolamine, 2,2'-iminobisethanol, 2,2'-iminodiethanol, NCI-C55174

USES
- ◆ Used in the Production Process
  - • lubricants for the textile industry
  - • organic synthesis
  - • cosmetics
  - • pharmaceuticals
  - • cutting oils
  - • shampoos
  - • cleaners
  - • polishes
  - • solubilizer for 2,4-D
  - • gas conditioning agent
- ◆ Used as a Chemical Intermediate
  - • fatty alkanolamides
  - • for liquid detergents
  - • textile chemicals
  - • morpholine
  - • fatty acid salt emulsifiers
  - • alkyl sulfate salt emulsifiers
  - • rubber chemicals intermediate
  - • humectant and softening agent

- emulsifier
- dispersing agent in agricultural chemicals

PHYSICAL PROPERTIES
- faintly colored
- viscous liquid or deliquescent prisms
- slight ammonia-like odor
- odor threshold 0.27 ppm
- melting point 28°C
- boiling point 270°C
- density 1.0966 at 20°C
- viscosity 351.9 cP at 30°C
- vapor pressure 0.577 mm Hg at 25°C
- vapor density 3.65
- very soluble in water and alcohol
- insoluble in ether and benzene
- log K_{ow} −1.46

CHEMICAL PROPERTIES
- combustible
- flash point 305°F
- autoignition temperature 1224°F
- explosion limits in air 10.6% upper and 1.6% lower

HEALTH RISK
- Toxic Exposure Guidelines
  - ACGIH TLV TWA 0.46 ppm (skin)
  - OSHA PEL TWA 3 ppm
- Toxicity Data
  - skin-rabbit 50 mg open MLD
  - skin-rabbit 500 mg/24H MLD
  - eye-rabbit 5500 mg SEV
  - eye-rabbit 750 µg/24H SEV
  - oral-rat TDLo 18,382 mg/kg
  - oral-rat LD_{50} 710 mg/kg
  - intraperitoneal-rat LD_{50} 120 mg/kg
  - subcutaneous-rat LD_{50} 2200 mg/kg
  - intravenous-rat LD_{50} 778 mg/kg
  - ims-rat LD_{50} 1500 mg/kg
  - oral-mouse LD_{50} 3300 mg/kg
  - intraperitoneal-mouse LD_{50} 2300 mg/kg
  - subcutaneous-mouse LD_{50} 3553 mg/kg
  - oral-guinea pig LD_{50} 2 g/kg
- Acute Risks
  - irritation of nose and throat
  - irritation of skin
  - increased blood pressure
  - pupillary dilatation
  - salivation
  - sedation
  - coma
  - death
  - burn skin
  - impair vision on eye contact
  - inflammation and edema of larynx and bronchi
  - chemical pneumonitis
  - pulmonary edema
- Chronic Risks
  - No information is available on chronic effects in humans.
  - effects on liver, kidney, bone marrow, brain, spinal cord, and skin in animals

- testicular degeneration
- reduced sperm motility
- No information is available on the carcinogenic effects on humans or animals.

HAZARD RISK
◆ combustible when exposed to heat or flame
◆ incompatible with oxidizing agents, copper, copper alloys, zinc, galvanized iron
◆ hazardous decomposition products: carbon monoxide, carbon dioxide, and nitrogen oxides

EXPOSURE ROUTES
◆ primarily dermal exposure
◆ soaps
◆ shampoos
◆ cosmetics
◆ detergents

◆ inhalation during use of lubricating liquids in machine building and metallurgy

REGULATORY STATUS/INFORMATION
◆ Clean Air Act Amendment, Title III: hazardous air pollutants
◆ CERCLA and Superfund
  • list of chemicals. 40CFR372.65
◆ Toxic Substance Control Act
  • list of substances. 40CFR716.120.a
◆ chemical regulated by the State of California

ADDITIONAL/OTHER COMMENTS
◆ Symptoms
  • exposure: irritation to eyes and skin
  • inhalation: coughing, smothering sensation, nausea, headache
◆ First Aid
  • wash eyes and skin immediately with large amounts of water
  • remove to fresh air immediately or provide respiratory apparatus if necessary
  • wash out mouth with water
◆ fire fighting: use carbon dioxide, dry chemical powder, or appropriate foam
◆ personal protection: self-contained breathing apparatus and protective clothing
◆ storage: keep tightly closed in a cool, dry place

SOURCES
◆ *Health Effects Notebook for Hazardous Air Pollutants*, U.S. Environmental Protection Agency, December 1994
◆ *Material Safety Data Sheets (MSDS)*
◆ *Sax's Dangerous Properties of Industrial Materials*, 9th edition, 1996
◆ *Hawley's Condensed Chemical Dictionary*, 12th edition, 1993
◆ *Aldrich Chemical Company Catalog*, 1996-1997
◆ *Environmental Law Index to Chemicals*, 1996
◆ *Environmental Contaminant Reference Databook*, volume 1, 1995

N,N-DIETHYL ANILINE(sf)

CAS #: 121-69-7
MOLECULAR FORMULA: $C_8H_{11}N$
FORMULA WEIGHT: 121.20
SYNONYMS: benzenamine, N,N-dimethyl-(9CI), (dimethylamino) benzene, N,N-dimethylaniline, N-dimethyl-aniline (OSHA), N,N-dimethylbenzeneamine, dimethylphenylamine, N,N-dimethylphenylamine, NCI-C56428, NL 63/10, Versneller

USES
◆ Manufacture of Other Chemicals
  • dyes
  • vanillin
  • Michler's ketone
  • solvent
  • stabilizer
  • alkylating agent
  • methyl violet

PHYSICAL PROPERTIES
◆ yellow, oily liquid
◆ density 0.956 g/mL
◆ no particular odor listed
◆ odor threshold 0.013 ppm (mg/m^3)
◆ melting point 1.5–2.5°C (34.4–35.9°F)
◆ boiling point 193–194°C (379.8–380.6°F) 1 atm
◆ specific gravity 0.956 (water = 1)
◆ vapor density 4.2 (air = 1)
◆ surface tension 35.52 mN/m
◆ viscosity 1.295×10^{-3} Pa s at 25°C
◆ heat capacity 199.1 J/gmol-K (liquid at 25°C)
◆ heat of vaporization 52.83 kJ/gmol
◆ insoluble in water
◆ miscible with alcohol and ether
◆ vapor pressure 1 mm Hg at 30°C, 10 mm Hg at 70°C

CHEMICAL PROPERTIES
◆ flash point 145°F (62.7°C)
◆ autoignition temperature 370°C (700°F)
◆ flammability limits 1.0% lower and 6.4% upper
◆ enthalpy (heat) of formation at 25°C (liquid) 100.5 kJ/gmol
◆ heat of combustion −4525 kJ/mol
◆ generally stable
◆ will not polymerize
◆ reacts with acids, acid chlorides, acid anhydrides, oxidizing agents,
◆ chloroforms, and halogens
◆ hazardous combustion or decomposition products. Thermal decomposition may produce CO, CO_2, and NO_x.

HEALTH RISK
- ◆ Toxic Exposure Guidelines
  - • AGCIH TLV TWA 5 ppm; STEL 10 ppm (skin)
  - • OSHA PEL TWA 5 ppm; STEL 10 ppm (skin)
  - • NIOSH REL TWA 5 ppm; STEL 10 ppm (skin)
- ◆ Toxicity Data
  - • oral-human LDLo 50 mg/kg
  - • oral-rat LD_{50} 1410 mg/kg
  - • TDLo 16,250 mg/kg/13W-I
  - • inhalation-rat LCLo 250 mg/m^3/4H
  - • TCLo 10,700 mcg/m^3/5H/17W-I
  - • TCLo 300 mcg/m^3/24H/14W-C
  - • skin-rabbit LD_{50} 1770 mg/kg
- ◆ Other Toxicity Indicators
  - • EPA Cancer Risk Level: not classified
- ◆ Acute Risks
  - • irritant to skin, eyes, mucous membranes, and upper respiratory tract
  - • serious damage to eyes
  - • inhibitor of central nervous and circulatory systems
  - • headaches
  - • cyanosis
  - • dizziness
  - • labored breathing
  - • paralysis
  - • convulsions, tonic and clonic
  - • tremors
- ◆ Chronic Risks
  - • blood disorders (increased methemoglobin, decreased hemoglobin concentration)
  - • significant changes in central nervous system, blood, and liver
  - • splenomegaly (enlargement of spleen)
  - • hemosiderosis of liver, kidneys, and testes
  - • suspected carcinogen

HAZARD RISK
- ◆ NFPA rating Health 3, Flammability 2, Reactivity 0
- ◆ dangerous fire hazard
- ◆ explodes on contact with benzoyl peroxide or diisopropyl peroxycarbonate
- ◆ incompatible or reacts strongly with acids, acid chlorides, acid anhydrides, chloroformates, halogens, and oxidizing agents

EXPOSURE ROUTES
- ◆ primarily by inhalation
- ◆ absorption through skin and eyes
- ◆ ingestion
- ◆ exposure in workplace
- ◆ present as an impurity in β-lactam antibiotics (penicillin and cephalosporin)

REGULATORY STATUS/INFORMATION

◆ Clean Air Act Amendments of 1990 Section 112 Hazardous Air Pollutant
◆ Superfund Amendments and Reauthorization Act
  • subject to section 313 reporting requirements
◆ CERCLA and Superfund
  • subject to listing under 40CFR372.65
  • Reportable quantity is 1 lb.
◆ OSHA Table Z1 and Table Z1-A
◆ Department of Transportation
  • identification number UN2253
  • hazard class/division 6.1
  • label: poison
◆ Community Right-to-Know Threshold Planning Quantity: not listed

ADDITIONAL/OTHER COMMENTS

◆ Symptoms
  • Inhalation and contact with vapor irritates skin, eyes, mucous membranes, and upper respiratory tract and causes headaches, dizziness, labored breathing, and in extreme cases convulsions, tremors, and paralysis.
  • Skin absorption may also cause many of these symptoms.
  • Effects may be delayed 2 to 4 hours or longer.
◆ First Aid
  • flush eyes and/or skin with large amounts of water for at least 15 minutes while removing contaminated garments and shoes
  • separate eyelids with fingers while flushing
  • seek fresh air
  • provide respiratory support and oxygen
  • If swallowed wash out mouth with water if conscious
  • seek medical help immediately
◆ fire fighting: use water, carbon dioxide, dry chemical powder, or foam; use water spray to cool hot containers
◆ Personal Protection
  • wear appropriate NIOSH/MSHA approved respirator
  • safety goggles, face shield (8 in. minimum)
  • long rubber or neoprene gauntlet gloves
  • rubber apron, sleeves, and other protective clothing
◆ Storage
  • keep container closed
  • cool, dry place away from heat and open flame

SOURCES

◆ *AGCIH Threshold Limit Values (TLVs) for 1996–1997*
◆ *Aldrich Chemical Company Catalog*, 1996–1997
◆ *CRC Handbook of Chemistry and Physics*, 77th edition, 1996–1997
◆ *CRC Handbook of Thermophysical and Thermochemical Data*, 1994
◆ *Environmental Law Index to Chemicals*, 1996 edition

◆ *Hawley's Condensed Chemical Dictionary*, 12th edition, 1993
◆ *MSDS*, Aldrich Chemical Company, Inc.
◆ *Physical and Thermodynamic Properties of Pure Chemicals: Data Compilation*, AiChE and NSRSD, 1996
◆ *Sax's Dangerous Properties of Industrial Materials*, 9th edition, 1996
◆ *Suspect Chemicals Sourcebook*, 1996 edition
◆ *Health Effects Notebook for Hazardous Air Pollutants*, U.S. Environmental Protection Agency, 1994

DIETHYL SULFATE

CAS #: 64-67-5
MOLECULAR FORMULA: $(C_2H_5O)_2SO_2$
FORMULA WEIGHT: 154.20
SYNONYMS: Diaethylsulfat (German), diethylester kyseliny sirove, diethyl ester sulfuric acid, diethyl tetraoxosulfate, DS, ethyl sulfate

USES
◆ Used in Industry
  • ethylating agent
  • accelerator in sulfation of ethylene
  • accelerator in sulfonations
  • alkylating agent
◆ Used as a Chemical Intermediate
  • ethyl derivatives of phenols
  • ethyl derivatives of amines
  • ethyl derivatives of thiols

PHYSICAL PROPERTIES
◆ colorless, oily liquid
◆ faint ethereal or peppermint odor
◆ melting point −25°C
◆ boiling point 209.5°C
◆ density 1.19 at 18°C
◆ vapor pressure 0.29 mm Hg at 25°C
◆ vapor density 5.31
◆ viscosity 1.79 cP at 20°C
◆ insoluble in water
◆ miscible with alcohol and ether
◆ log K_{ow} 1.14

CHEMICAL PROPERTIES
◆ combustible
◆ flash point 104.4°C
◆ autoignition temperature 436°C
◆ decomposes by hot water
◆ explosion limits in air 4.1% lower

HEALTH RISK
◆ Toxic Exposure Guidelines
  • DFG TRK 0.03 ppm, animal carcinogen, possible human carcinogen
◆ Toxicity Data
  • skin-rabbit 10 mg/24H SEV
  • skin-rabbit 500 mg open MLD
  • eye-rabbit 2 mg SEV
  • mmo-sat 5 mg/plate

- mma-sat 5 mg/plate
- slt-mouse-intraperitoneal 100 mg/kg/10W
- cct-mouse-intraperitoneal 150 mg/kg
- msc-hamster:ovr 1 mmol/L
- intravenous-rat TDLo 340 mg/kg
- oral-rat TDLo 3700 mg/kg
- subcutaneous-rat TDLo 800 mg/kg
- intravenous-rat TDLo 85 mg/kg
- oral-rat LD_{50} 880 mg/kg
- inhalation-rat LCLo 250 ppm/4H
- subcutaneous-rat LD_{50} 350 mg/kg
- oral-mouse LD_{50} 647 mg/kg
- skin-rabbit LD_{50} 600 mg/kg

◆ Acute Risks
- extremely destructive to tissue of the mucous membranes
- destructive to upper respiratory tract, eyes, and skin
- spasm of the larynx and bronchi
- inflammation and edema of the larynx and bronchi
- chemical pneumonitis
- pulmonary edema
- burning sensation
- coughing
- wheezing
- laryngitis
- shortness of breath
- headache
- nausea
- vomiting
- skin cells killed in rabbits on contact

◆ Chronic Risks
- laryngeal cancer
- excess mortality rate
- tumors in forestomach in rats
- increased incidence of skin cancer in mice
- IARC Group 2A: probable human carcinogen

HAZARD RISK
◆ combustible when exposed to heat or flame
◆ reacts with oxidizing materials
◆ Moisture causes liberation of H_2SO_4.
◆ violent reaction with potassium-tert-butoxide
◆ reacts violently with 3,8-dinitro-6-phenylphenanthridine + water
◆ Reaction with iron + water forms explosive hydrogen gas.
◆ rapid decomposition when heated above 100°C
◆ vigorous exothermic hydrolysis when reacts with water in presence of caustic catalysts at temperatures above 50°C
◆ Decomposition emits toxic fumes of SO_x.

EXPOSURE ROUTES
◆ dermal contact during production or use
◆ inhalation during production or use
◆ ambient environment
◆ fugitive emissions

REGULATORY STATUS/INFORMATION
◆ Clean Air Act Amendment, Title III: hazardous air pollutants

- ◆ CERCLA and Superfund
  - list of chemicals. 40CFR372.65
- ◆ chemical regulated by the State of California

ADDITIONAL/OTHER COMMENTS
- ◆ Symptoms
  - inhalation: spasm, inflammation, and edema of larynx and bronchi, burning sensation, coughing, wheezing, laryngitis, shortness of breath, headache, nausea, vomiting
- ◆ First Aid
  - wash eyes and skin immediately with large amounts of water
  - remove to fresh air immediately or provide respiratory apparatus if necessary
  - wash mouth out with water
- ◆ fire fighting: use carbon dioxide, dry chemical powder, or appropriate foam
- ◆ personal protection: self-contained breathing apparatus and protective clothing
- ◆ storage: keep tightly closed away from heat and open flame in a cool, dry place

SOURCES
- ◆ *Health Effects Notebook for Hazardous Air Pollutants*, U.S. Environmental Protection Agency, December 1994
- ◆ *Material Safety Data Sheets (MSDS)*
- ◆ *Sax's Dangerous Properties of Industrial Materials*, 9th edition, 1996
- ◆ *Hawley's Condensed Chemical Dictionary*, 12th edition, 1993
- ◆ *Aldrich Chemical Company Catalog*, 1996–1997
- ◆ *Environmental Law Index to Chemicals*, 1996
- ◆ *Merck Index*, 12th edition, 1996

3,3′-DIMETHOXYBENZIDINE

CAS #: 119-90-4
MOLECULAR FORMULA: $[C_6H_3(OCH_3)NH_2]_2$
FORMULA WEIGHT: 244.32
SYNONYMS: acetamine diazo black RD, amacel developed navy SD, azoene fast blue base, azofix blue B salt, azogine fast blue B, blue BN base, brentamine fast blue B base, cellitazol B, C.I. 24110, C.I. azoic diazo component 48, cibacete diazo navy blue 2B, C.I. disperse black 6, diacelliton fast grey G, diacel navy DC, o-Dianisidin (Czech, German), o-dianisidina (Italian), 3,3′-dianisidine, o-dianisidine, o,o′-dianisidine, diato blue base B, 3,3′-dimethoxybenzidin (Czech), 3,3′-dimetossibenzodina (Italian), fast blue B base, hiltonil fast blue B base, hiltosal fast blue B salt, hindasol blue B salt, kako blue B salt, kayaku blue B base, lake blue B base, meisei teryl diazo blue HR, mitsui blue B base, naphthanil blue B base, neutrosel navy BN, sanyo fast blue salt B, setacyl diazo navy R, spectrolene blue B

USES
- ◆ Used as an Intermediate

- production of dyes
- pigments
- manufacture of azo dyes

PHYSICAL PROPERTIES

- colorless crystalline solid
- turns violet when exposed to air
- melting point 137°C
- vapor pressure 8.8×10^{-9} mm Hg at 25°C
- vapor density 8.5
- soluble in alcohol and ether
- insoluble in water
- log K_{ow} 1.81

CHEMICAL PROPERTIES

- combustible
- flash point 206°C
- incompatible with strong oxidizers
- will not polymerize
- decomposes above 300°C

HEALTH RISK

- Toxic Exposure Guidelines
  - NIOSH REL reduce to lowest feasible level
  - DFK MAK animal carcinogen, suspected human carcinogen
- Toxicity Data
  - mmo-sat 333 µg/plate
  - mma-sat 1 µg/plate
  - sce-hamster:ovr 500 µg/L
  - oms-dog:oth 100 µmol/L
  - dnd-dog:oth 100 µmol/L
  - oral-rat TDLo 12 g/kg/56W-I:ETA
  - oral-rat LD_{50} 1920 mg/kg
  - oral-dog LDLo 600 mg/kg
- Acute Risks
  - no information available for humans
  - irritation to skin
  - allergic skin reaction
  - irritation to eyes
  - moderate acute toxicity in animals
- Chronic Risks
  - no information available for humans
  - effects on liver
  - effects on kidney
  - effects on bladder and gastritis
  - intestinal hemorrhage
  - weight loss
  - EPA Group 2B: probable human carcinogen
  - tumors of mammary gland, ovary, bladder, intestine, skin, and stomach

HAZARD RISK

- combustible when exposed to heat or flame
- Combustion will produce carbon dioxide, carbon monoxide, and oxides of nitrogen.
- capable of creating dust explosions
- incompatible with strong oxidizers
- Decomposition emits toxic fumes of NO_x.

EXPOSURE ROUTES
- ◆ dye manufacturing and processing plants
- ◆ use of packaged dyes
- ◆ use of pigments for home use

REGULATORY STATUS/INFORMATION
- ◆ Clean Air Act Amendment, Title III: hazardous air pollutants
- ◆ Resource Conservation and Recovery Act
  - • list of hazardous constituents. 40CFR261, 40CFR261.11
  - • U waste #: U091
- ◆ CERCLA and Superfund
  - • designation of hazardous substances. 40CFR302.4
  - • list of chemicals. 40CFR372.65
- ◆ chemical regulated by the State of California

ADDITIONAL/OTHER COMMENTS
- ◆ Symptoms
  - • contact: irritation or allergic reaction on skin
  - • inhalation: irritation to eyes
- ◆ First Aid
  - • wash eyes and skin immediately with large amounts of water
  - • remove to fresh air or provide respiratory support
  - • induce vomiting
- ◆ fire fighting: use dry chemical, water spray or mist, or carbon dioxide
- ◆ personal protection: wear self-contained breathing apparatus and protective clothing and goggles
- ◆ storage: keep away from contact with oxidizing materials

SOURCES
- ◆ *Health Effects Notebook for Hazardous Air Pollutants*, U.S. Environmental Protection Agency, December 1994
- ◆ *Material Safety Data Sheets (MSDS)*
- ◆ *Sax's Dangerous Properties of Industrial Materials*, 9th edition, 1996
- ◆ *Hawley's Condensed Chemical Dictionary*, 12th edition, 1993
- ◆ *Aldrich Chemical Company Catalog*, 1996–1997
- ◆ *Environmental Law Index to Chemicals*, 1996
- ◆ *Merck Index*, 12th edition, 1996

DIMETHYL AMINOAZOBENZENE (jl)

CAS #: 60-11-7
MOLECULAR FORMULA: $C_{14}H_{15}N_3$
FORMULA WEIGHT: 25.295
SYNONYMS: atul fast yellow, benzeneamine, benzeneazodimethylaniline, brilliant fast oil yellow, brilliant fast spirit yellow, cerasine yellow gg, C.I. 11020, C.I. solvent yellow 2, dab (carcinogen), dimethylaminoazobenzene, N,N-dimethyl-4-aminoazobenzene, 4-(N,N-dimethylamino)azobenzene, para-dimethylaminoazobenzene, dimethylaminoazobenzol, 4-dimethylaminoazobenzol, 4-dimethylaminophenylazobenzene, N,N-dimethyl-p-azoaniline, N,N-dimethyl-4-(phenylazo),

N,N-dimethyl-4-(phenylazo)benzamine, N,N-dimethyl-4-(phenylazo)-benzenamine, N,N-dimethyl-4-(phenylazo)-benzeneamine, di-
methyl yellow, dimethyl yellow analar, dimethyl yellow N,N-dimethyl-
aniline, dmab, enial yellow 2g, fast oil yellow b, fast yellow, fat yellow,
fat yellow a, fat yellow ad oo, fat yellow es, fat yellow es extra, fat
yellow extra conc., fat yellow r, fat yellow r (8186), grasal brilliant
yellow, Methyl yellow, oil yellow, cil yellow 20, oil yellow 2625, oil
yellow 7463, oil yellow bb, oil yellow d, oil yellow dn, oil yellow ff, oil
yellow fn, oil yellow g, oil yellow 2g, oil yellow g-2, oil yellow gg, oil
yellow gr, oil yellow ii, oil yellow n, oil yellow pel, oleal yellow 2g,
organol yellow adm, orient oil yellow gg, p.d.a.b., petrol yellow wt,
resinol yellow gr, resoform yellow gga, silotras yellow t2g, somalia
yellow a, stear yellow jb, sudan gg, sudan yellow, sudan yellow gg,
sudan yellow gga, toyo oil yellow g, waxoline yellow ad, waxoline
yellow ads, yellow g soluble in grease

USES
◆ coloring polishes
◆ wax products
◆ polystyrene
◆ soap
◆ indicators

PHYSICAL PROPERTIES
◆ yellowish, crystalline leaflets
◆ melting point 111°C
◆ boiling point not avilable
◆ density 1050 kg/m^3
◆ specific gravity 1.05 (water = 1)
◆ vapor density not available
◆ surface tension not available
◆ viscosity not available
◆ heat capacity not available
◆ heat of vaporization not available
◆ aromatic odor; chloroform-like odor
◆ vapor pressure 3.3×10^{-7} mm Hg at 25°C
◆ insoluble in water

CHEMICAL PROPERTIES
◆ flash point not available
◆ flammability limits not available
◆ autoignition temperature not available
◆ heat of combustion not available
◆ enthalpy (heat of formation) not available
◆ will not polymerize
◆ Incompatible with
  • strong acids
  • strong oxidizing agents

HEALTH RISK
◆ Toxic Exposure Guidelines
  • U.S. EPA has not established a reference dose
◆ Toxicity Data
  • LD_{50} (oral-rat)(mg/kg) 200

- LD_{50} (intraperitoneal-rat)(mg/kg) 230
- LD_{50} (oral-mus) (mg/kg) 300
- LD_{50} (intraperitoneal-mus) (mg/kg) 230
- TDLo (intraperitoneal-mus) 3353 µg/kg
- TD (intraperitoneal-mus) 10 mg/kg
- TDLo (intraperitoneal-frog) 12 mg/kg
- TDLo (oral-rat) 89 g/kg/57W-C
- TDLo (scu-mus) 296 mg/kg (15–19D preg)
- TDLo (skin-rat) 1965 mg/kg/2Y-I

◆ Acute Risks
- The solid or solutions of this chemical may irritate the skin, causing a rash or burning feeling on contact.
- Tests involving acute exposure of animals, such as the LD_{50} test in rats, have shown 4-dimethylaminoazobenzene to have high acute toxicity from oral exposure.
- may be fatal if inhaled, swallowed, or absorbed through skin

◆ Chronic Risks
- Cancer Hazard
  - a probable cancer causing agent in humans
  - has been shown to cause liver, bladder, lung, and skin cancer in animals
- Reproductive Hazard
  - may damage the developing fetus
- Other Long-Term Effects
  - Repeated exposure may cause a skin rash.
- Target Organs
  - liver
  - bladder

HAZARD RISK

◆ Unusual Fire and Explosion Hazards
- a noncombustible solid
- extinguish fire using an agent suitable for type of surrounding fire
- Contact with strong oxidizers may cause fire.

◆ Hazardous Decomposition Products
- carbon monoxide
- carbon dioxide
- nitrogen oxides

EXPOSURE ROUTES

◆ ingestion
◆ inhalation
◆ absorption
◆ eye contact
◆ skin contact

REGULATORY STATUS/INFORMATION

◆ SARA Title III Section 313 Toxic Chemicals (40CFR372)
◆ OSHA Air Contaminants (20CFR1910)
◆ OSHA Specifically Regulated Substances
◆ NTP Annual Report on Carcinogens

◆ chemical regulated by the State of California
◆ Canadian Workplace Hazardous Materials Information System (WHMIS) Ingredient Disclosure List: CN2—Ingredient must be disclosed at a concentration of 0.1%.
◆ Massachusetts Substance List: MA1CE—Carcinogen, extraordinarily hazardous substance
◆ New Jersey Right to Know Hazardous Substances List: special health hazard

ADDITIONAL/OTHER COMMENTS
◆ First Aid
  • eye contact: immediately flush with large amounts of water for at least 15 minutes, occasionally lifting upper and lower lids
  • skin contact: quickly remove contaminated clothing. Immediately wash area with large amounts of soap and water.
◆ Personal Protection
  • Respiratory and Eye: At any exposure level, use an MSHA/NIOSH approved supplied air respirator with a full facepiece operated in the positive pressure mode or with a full facepiece, hood, or helmet operated in the continuous flow mode. An MSHA/NIOSH approved self-contained breathing apparatus with a full facepiece operated in pressure demand or other positive pressure mode is also recommended.
  • Skin: uniform, protective suit; rubber gloves are recommended
  • use only in a chemical fume hood
  • avoid all contact
◆ fire fighting: use water spray, alcohol foam, dry chemical, or carbon dioxide
◆ Spill or Leak
  • restrict persons not wearing protective equipment from area of spill or leak until cleanup is complete
  • ventilate the area of spill or leak
  • collect spilled material in the most convenient and safe manner and deposit in sealed containers for reclamation or for disposal in an approved facility
  • Liquid containing dimethylaminoazobenzene should be absorbed in vermiculite, dry sand, earth or a similar material.
  • It may be necessary to contain and dispose of dimethylaminoazobenzene as a hazardous waste. Contact your environmental program for specific recommendations.
◆ Storage
  • keep container tightly closed
  • store in a cool, dry, well-ventilated area

SOURCES
◆ *Aldrich Chemical Company Catalog,* 1996
◆ *MSDS,* Aldrich Chemical Company, Inc.
◆ *Suspect Chemicals Sourcebook,* 1990 edition

♦ *Dangerous Properties of Industrial Materials*, 7th edition, volume 3, 1989
♦ *Merck Index*, 10th edition, 1983

3,3'-DIMETHYL BENZIDINE

CAS #: 119-93-7
MOLECULAR FORMULA: [-C₆H₃(CH₃)-4-NH₂]₂
FORMULA WEIGHT: 212.32
SYNONYMS: bianizidine, 4,4'-bi-o-toluidine, C.I. 37230, C.I. azoic diazo component 113, (4,4'-diamine-3,3'-dimethyl(1,1'-biphenyl), 4,4'-diamino-3,3'-dimethylbiphenyl, 4,4'-diamino-3,3'-dimethyldiphenyl, diaminoditolyl, 3,3'-dimethylbenzidin, 3,3'-dimethyl-4,4-biphenyldiamine, 3,3'-dimethylbiphenyl-4,4'-diamine, 3,3'-dimethyl-(1,1'-biphenyl)-4,4'-diamine, 3,3-dimethyl-4,4'-diphenyldiamine, 3,3'-dimethyldiphenyl-4,4'-diamine, 4,4-di-o-toluidine, fast dark blue base R, 2-Tolidin (German), 2-tolidina (Italian), tolidine, 3,3'-tolidine, o-tolidine, o,o'-tolidine, 2-tolidine, o-toluidine

USES
♦ Used as an Intermediate
  • dyes
  • pigments
♦ Used as a Sensitive Reagent
  • gold
  • free chlorine in water
  • curing agent for urethane resins

PHYSICAL PROPERTIES
♦ white to red crystalline solid
♦ melting point 130°C
♦ vapor pressure 1.7×10^{-4} mm Hg at 25°C
♦ slightly soluble in water
♦ soluble in alcohol and ether
♦ log K_{ow} 4.59

CHEMICAL PROPERTIES
♦ combustible but stable
♦ Polymerization will not occur.
♦ incompatible with strong oxidizers

HEALTH RISK
♦ Toxic Exposure Guidelines
♦ ACGIH TLV suspected human carcinogen
♦ NIOSH REL 0.02 mg/m³, avoid skin contact
♦ DFG MAK animal carcinogen, suspected human carcinogen
♦ Toxicity Data
  • dns-human:hla 1 μmol/L
  • dns-rat:lvr 1 μmol/L
  • oral-rat TDLo 4500 mg/kg/27D-I:CAR
  • subcutaneous-rat TDLo 1650 mg/kg/33W-I:ETA
  • oral-rat LD₅₀ 404 mg/kg
  • intraperitoneal-rat LDLo 125 mg/kg
  • intraperitoneal-mouse LDLo 125 mg/kg
  • oral-dog LDLo 600 mg/kg
♦ Acute Risks
  • irritation of nose
  • irritation of throat
  • irritation of eyes

◆ Chronic Risks
  • tumors of intestines
  • tumors of skin
  • central nervous system ef-
    fects
  • tumors of lung and liver
  • confusion
  • loss of balance
  • dizziness
  • EPA Group B2: probable
    human carcinogen

HAZARD RISK
  ◆ combustible
  ◆ Combustion will produce carbon dioxide, carbon monoxide,
    and oxides of nitrogen.
  ◆ capable of creating dust explosions
  ◆ incompatible with strong oxidizers
  ◆ Decomposition emits toxic fumes of NO_x.
  ◆ affected by light

EXPOSURE ROUTES
  ◆ dye manufacturing and processing plants
  ◆ use of packaged dyes
  ◆ use of pigments for home use

REGULATORY STATUS/INFORMATION
  ◆ Clean Air Act Amendment, Title III: hazardous air pollutants
  ◆ Resource Conservation and Recovery Act
    • list of hazardous constituents. 40CFR258, 40CFR261,
      40CFR261.11
    • groundwater monitoring list. 40CFR264
    • U waste #: U095
  ◆ CERCLA and Superfund
    • designation of hazardous substances. 40CFR302.4
    • list of chemicals. 40CFR372.65
  ◆ chemical regulated by the State of California

ADDITIONAL/OTHER COMMENTS
  ◆ Symptoms
    • inhalation: irritation to eyes, nose, and throat
  ◆ First Aid
    • wash eyes and skin immediately with large amounts
      of water
    • remove to fresh air or provide respiratory support
    • induce vomiting
  ◆ fire fighting: use dry chemical, water spray or mist, or carbon
    dioxide
  ◆ personal protection: wear self-contained breathing apparatus
    and protective clothing and goggles
  ◆ storage: keep away from contact with oxidizing materials

SOURCES
  ◆ *Health Effects Notebook for Hazardous Air Pollutants*, U.S.
    Environmental Protection Agency, December 1994
  ◆ *Material Safety Data Sheets (MSDS)*
  ◆ *Sax's Dangerous Properties of Industrial Materials*, 9th edi-
    tion, 1996
  ◆ *Hawley's Condensed Chemical Dictionary*, 12th edition, 1993

♦ *Aldrich Chemical Company Catalog*, 1996–1997
♦ *Environmental Law Index to Chemicals*, 1996
♦ *Merck Index*, 12th edition, 1996

DIMETHYL CARBAMOYL CHLORIDE (kjc)

CAS #: 79-44-7
MOLECULAR FORMULA: C_3H_6ClNO
FORMULA WEIGHT: 107.54 g/mol
SYNONYMS: carbamic acid, carbamyl chloride, chlorid kyseliny, chloroformic acid dimethylamide, ddc, dimethyl- (9CI), dimethylamid kyseliny chlormravenci (Czech), (dimethylamino) carbonyl chloride, N,N-dimethylaminocarbonyl chloride, dimethylcarbamic acid chloride, N,N-dimethylcarbamic acid chloride, dimethylcarbamic chloride, dimethylcarbamidoyl chloride, N,N-dimethylcarbamidoyl chloride, dimethylcarbamoyl chloride, dimethyl carbamoyl chloride (ACGIH:DOT), N,N-dimethylcarbamoyl chloride, dimethylcarbamyl chloride, N,N-dimethylcarbamyl chloride, dimethylchloroformamide, dimethylkarbaminove (Czech), dimethylkarbamoylchlorid (Czech), DMCC, TL 389

USES
♦ Chemical Intermediate
  • pharmaceuticals
  • pesticides
  • dye synthesis

PHYSICAL PROPERTIES
♦ colorless liquid
♦ melting point −33°C (−27°F)
♦ boiling point 167°C (333°F) at 775 mm Hg
♦ specific gravity 1.168 (water = 1.00)
♦ vapor density 3.73 (air = 1.00)
♦ surface tension not available
♦ viscosity not available
♦ heat capacity not available
♦ heat of vaporization not available
♦ odor threshold not available
♦ vapor pressure not available
♦ solubility 1.4540 mg/L at 20°C
♦ log octanol/water partition coefficient not available

CHEMICAL PROPERTIES
♦ combustible
♦ reacts strongly with strong oxidizing agents, strong bases
♦ decomposes on exposure to moist air or water
♦ flash point 68°C (155°F)
♦ flammability limits: not available
♦ autoignition temperature not available
♦ heat of combustion not available
♦ enthalpy of formation not available

HEALTH RISK
♦ Toxic Exposure Guidelines
  • ACGIH TLV TWA 0.1 ppm (40 mg/m³)
♦ Toxicity Data
  • oral-rat LD_{50} 1 g/kg

- inhalation-rat LC_{50} 180 ppm/6H/6W-I:CAR
- intraperitoneal-mouse LD_{50} 300 mg/kg
- skin-mouse TDLo 17,280 mg/kg/72W-I:CAR
- skin-mouse LD 13 g/kg/55W-I:NEO
- intraperitoneal-mouse TDLo 2560 mg/kg/64W-I:NEO
- subcutaneous-mouse TDLo 894 mg/kg/52W-I:CAR
- subcutaneous-mouse LD 8 g/kg/40W-I:NEO
- inhalation-hamster TCLo 1 ppm/6H/89W-I:CAR
- inhalation-rat 111 ppm/6H/6W-I:ETA
- inhalation-rat 111 ppm:ETA
- subcutaneous-mouse LD 5200 mg/kg/26W-I:NEO

◆ Other Toxicity Indicators
  - California no-significant risk level: 0.05 mcg/d
◆ Acute Risks
  - skin irritation
  - damage to mucous membranes of nose, throat, and lungs
  - skin tumors
  - local sarcomas
  - liver disturbance
  - conjunctivitis
  - keratitis
◆ Chronic Risks
  - EPA Group B2: probable human carcinogen
  - alters genetic material

HAZARD RISK
◆ NFPA rating: none

EXPOSURE ROUTES
◆ primarily by inhalation
◆ ingestion
◆ emissions from burning
◆ absorption through skin and eyes possible
◆ occupational exposure during manufacture of pharmaceuticals and pesticides

REGULATORY STATUS/INFORMATION
◆ Clean Air Act Amendments of 1990 Section 112 Statutory Air Pollutants
◆ Clean Air Act Amendments of 1990 Section 112 High-Risk Pollutants
◆ Resource Conservation and Recovery Act
  - RCRA Hazardous Substances
  - RCRA Land Disposal Prohibitions: Scheduled Second Third Wastes
  - U waste #: U097
◆ CERCLA and Superfund
  - CERCLA Hazardous Substances
  - designation of hazardous substances. 40CFR302.4
  - SARA Title III Section 313 Toxic Chemicals
  - list of chemicals. 40CFR372.65
  - Reportable Quantity (RQ): 1 lb
◆ Department of Transportation
  - DOT Hazardous Materials

- DOT Hazardous Substances and Radionuclides
- hazard class/division: 8
- identification number: UN2262
- labels: corrosive
- emergency response guide #: 60
◆ Emergency Planning and Community Right-to-Know Act
  - section 313 de minimis concentration: 0.1%

ADDITIONAL/OTHER COMMENTS
◆ Symptoms
  - inhalation: eye irritation, damage to the mucous membranes of the nose, throat, and lungs
  - skin absorption: skin irritation
  - ingestion: no tests performed
◆ First Aid
  - wash eyes immediately with large amounts of water
  - wash skin immediately with large amounts of water
  - provide respiratory support
◆ fire fighting: use dry chemical, foam, carbon dioxide
◆ Personal Protection
  - wear self-contained breathing apparatus
  - wear protective clothing
  - use only in chemical fume hood
  - emits toxic fumes under fire conditions
◆ Storage
  - keep tightly closed
  - keep away from heat and open flame
  - moisture sensitive
  - store in cool, dry place

SOURCES
◆ *Environmental Containment Reference Databook*, volume 1, 1995
◆ *Environmental Law Index to Chemicals*, 1996 edition
◆ *Fire Protection Guide to Hazardous Materials*, 11th edition, 1994
◆ *Material Safety Data Sheets (MSDS)*
◆ *NIOSH Pocket Guide to Chemical Hazards*, June 1996
◆ *Sax's Dangerous Properties of Industrial Materials*, 9th edition, volume 2, 1996

DIMETHYL FORMAMIDE

CAS #: 68-12-2
MOLECULAR FORMULA: $HOON(CH_3)_2$
FORMULA WEIGHT: 73.11
SYNONYMS: Dimethylformamid (German), dimethylformamide, N,N-dimethyl formamide, dimetilformamide (Italian), dimetylformamidu (Czech), DMF, DMFA, dwumethyloformamid (Polish), N-formyldimethylamine, NSC 5356, U-424

USES
◆ Primarily Used as an Industrial Solvent

- vinyl resins
- butadiene
- Used in the Production Process
  - polymer fibers
  - films
  - surface coatings
  - to permit easy spinning of acrylic fibers

- acid gases
- polyacrylic fibers

- wire enamels
- crystallization medium in pharmaceutical industry
- catalyst in carboxylation reactions

PHYSICAL PROPERTIES
- colorless to slightly yellow hygroscopic liquid
- fishy, unpleasant odor
- odor threshold 2.2 ppm
- melting point $-61°C$
- boiling point 152.8°C
- density 0.945 at 22.4'/4'

- vapor pressure 3.7 mm Hg at 25°C
- vapor density 2.51
- miscible in water, EtOH, Et_2O, C_6H_6, and $CHCl_3$
- critical temperature 323.4°C
- critical pressure 52.5 atm
- log K_{ow} -1.01

CHEMICAL PROPERTIES
- combustible
- lel 2.2% at 100°C
- uel 15.2% at 100°C

- flash point 136°C
- autoignition temperature 833°F

HEALTH RISK
- Toxic Exposure Guidelines
  - OSHA PEL TWA 10 ppm (skin)
  - ACGIH TLV TWA 10 ppm (skin)
  - DFG MAK 10 ppm
  - MSHA standard 30 mg/m^3
- Toxicity Data
  - skin-human 100%/24H MLD
  - skin-rabbit 10 mg/24H open
  - eye-rabbit 100 mg RNS SEV
  - mma-sat 600 µg/plate
  - cyt-human:lym 100 nmol/L
  - inhalation-rat TDLo 600 mg/m^3/24H
  - inhalation-rat TCLo 4 mg/m^3/4H
  - skin-rat TDLo 3600 mg/kg
  - intraperitoneal-mouse TDLo 2100 mg/kg
  - oral-rabbit TDLo 2600 mg/kg
  - oral-rat LD$_{50}$ 2800 mg/kg
  - intraperitoneal-rat LD$_{50}$ 1400 mg/kg
  - subcutaneous-rat LD$_{50}$ 3800 mg/kg
  - intravenous-rat LD$_{50}$ 2000 mg/kg
  - oral-mouse LD$_{50}$ 3750 mg/kg
  - inhalation-mouse LC$_{50}$ 9400 mg/m^3/2H
  - intraperitoneal-mouse LD$_{50}$ 650 mg/kg
  - subcutaneous-mouse LD$_{50}$ 4500 mg/kg
  - intravenous-mouse LD$_{50}$ 2500 mg/kg
  - ims-mouse LD$_{50}$ 3800 mg/kg
  - intravenous-dog LD$_{50}$ 470 mg/kg
  - intraperitoneal-cat LD$_{50}$ 500 mg/kg

- skin-rabbit LD$_{50}$ 4720 mg/kg
- ◆ Acute Risks
  - damage of liver
  - abdominal pain
  - nausea
  - vomiting
  - jaundice
  - alcohol intolerance
  - rashes
  - damage to liver, kidneys, and lungs in animals
  - dermatitis
  - skin and severe eye irritation
  - high blood pressure
- ◆ Chronic Risks
  - effects on liver
  - digestive disturbances
  - minimal hepatic changes
  - liver abnormalities
  - increased rate of spontaneous abortion
  - embryotoxic in animals
  - reduced implantation efficiency in rats
  - decreased mean fetal weight in rats
  - malformed fetuses in rabbits
  - increase in testicular germ-cell tumors
  - cancers of the buccal cavity or pharynx
  - IARC Group 2B: possibly carcinogenic to humans

HAZARD RISK
- ◆ flammable liquid when exposed to heat or flame
- ◆ can react with oxidizing materials
- ◆ explosion hazard when exposed to flame
- ◆ explosive reaction with bromine, potassium permanganate, triethylaluminum + heat
- ◆ forms explosive mixtures with lithium azide and uranium perchlorate
- ◆ ignition on contact with chromium trioxide
- ◆ violent reactions with chlorine, sodium hydroborate + heat, diisocyanatomethane, carbon tetrachloride + iron, 1,2,3,4,5-hexachlorocyclohexane + iron
- ◆ vigorous exothermic reaction with magnesium nitrate, sodium + heat, sodium hydroxide + heat, sulfinyl chloride + traces of iron or zinc, 2,4,6-trichloro-1,3,5-triazine, and many other materials
- ◆ avoid contact with halogenated hydrocarbons, inorganic and organic nitrates, 2,5-dimethyl pyrrole + P(OCl)$_3$, C$_6$Cl$_6$, methylene diisocyanates, and P$_2$O$_3$
- ◆ Decomposition emits toxic fumes of NO$_x$.

EXPOSURE ROUTES
- ◆ most likely exposed in the workplace
- ◆ effluents from sewage treatment plants and industrial plants

REGULATORY STATUS/INFORMATION
- ◆ CERCLA and Superfund
  - list of chemicals. 40CFR372.65

ADDITIONAL/OTHER COMMENTS
- ◆ Symptoms

- skin contact: drying and defatting of skin, itching, scaling, abdominal distress, loss of appetite, nausea, vomiting, constipation or diarrhea, headache, hepatomegaly, facial flushing
  - eye contact: irritation, redness, burning, tears
- ◆ First Aid
  - wash eyes immediately with large amounts of water
  - wash skin immediately with soap
  - remove to fresh air immediately or provide respiratory apparatus if necessary
  - drink water and induce vomiting
- ◆ fire fighting: use foam, carbon dioxide, or dry chemical
- ◆ personal protection: compressed air/oxygen apparatus, gastight suit
- ◆ storage: take precautions necessary for a flammable material

SOURCES
- ◆ *Health Effects Notebook for Hazardous Air Pollutants*, U.S. Environmental Protection Agency, December 1994
- ◆ *Material Safety Data Sheets (MSDS)*
- ◆ *Sax's Dangerous Properties of Industrial Materials*, 9th edition, 1996
- ◆ *Environmental Law Index to Chemicals*, 1996
- ◆ *Merck Index*, 12th edition, 1996
- ◆ *Hawley's Condensed Chemical Dictionary*, 12th edition, 1993
- ◆ *Aldrich Chemical Company Catalog*, 1996–1997

1,1-DIMETHYL HYDRAZINE (pf)

CAS #: 57-14-7
MOLECULAR FORMULA: $C_2H_8N_2$
FORMULA WEIGHT: 60.12
SYNONYMS: dimazine, dimethylhydrazine, asym-dimethylhydrazine, N,N-dimethylhydrazine, uns-dimethylhydrazine, unsym-dimethylhydrazine, 1,1-Dimethylhydrazine (German), dimethylhydrazine unsymmetrical (DOT), DMH, niesymetryczna dwu methlohydrazyna (Polish), RCRA waste number U098, UMDH (DOT), many others

USES
- ◆ plant growth control agent
- ◆ Basis as a fuel
  - high-energy fuel in military applications
  - component of jet and rocket fuels
  - component in small electrical power generating units
- ◆ absorbent for acid gases
- ◆ photography

PHYSICAL PROPERTIES
- ◆ fuming, colorless liquid that gradually turns yellow upon contact with air
- ◆ melting point 58°C (136.4°F)
- ◆ boiling point 63.9°C (147.0°F) at 760 mm Hg
- ◆ density 0.782 g/cm^3 at 25°C
- ◆ specific gravity 0.78 (water = 1.00)

- vapor density 1.94 (air = 1)
- surface tension not available
- viscosity not available
- heat capacity not available
- heat of vaporization not available
- ammonia-like fishy odor
- odor threshold 1.7 ppm

CHEMICAL PROPERTIES

- flammable
- corrosive
- stable
- Hazardous polymerization will not occur.
- hygroscopic liquid
- a powerful reducing agent: will ignite spontaneously (hypergolic) with many oxidants (dinitrogen tetroxide, hydrogen peroxide, and nitric acid); reacts vigorously with oxidizing materials (air and fuming nitric acid)

- vapor pressure 103 mm at 20°C, 566 mm at 55°C
- completely soluble in water, liberating heat
- miscible with ethanol, dimethylformamide, and hydrocarbons, though hardly with ether

- incompatible with copper, copper alloys, brass, iron, and iron salts
- flash point −1.1°C (34°F)
- explosion limits in air 2% lower and 95% upper
- autoignition temperature 247°C (476.6°F)
- heat of combustion not available
- enthalpy (heat) of formation not available

HEALTH RISK

- Toxic Exposure Guidelines
  - ACGIH TLV TWA 0.5 ppm (0.025 mg/m ^{3}) (skin)
  - OSHA PEL 0.5 ppm (1 mg/m ^{3}) (skin), (proposed TWA 0.01 ppm (skin))
  - NIOSH REL 0.15 mg/m ^{3}/2H
  - DFG MAK animal carcinogen, suspected human carcinogen
- Toxicity Data
  - otr-human:fbr 167 μmol/L
  - dnd-human:fbr 300 μmol/L
  - intraperitoneal-rat TDLo 600 mg/kg (6–15 days pregnant): Teratogen
  - intraperitoneal-rat TDLo 600 mg/kg (6–15 days pregnant): REP
  - oral-rat TDLo 150 mg/kg/7W-intermittent:ETA
  - subcutaneous-rat TDLo 21 mg/kg:ETA
  - oral-mouse TDLo 5880 mg/kg/42W-C:Carcinogen
  - intraperitoneal-rat TDLo 144 mg/kg 8W-intermittent:ETA;TER
  - subcutaneous-mouse TDLo 420 mg/kg/21W-intermittent:NEO
  - oral-hamster TDLo 228 g/kg/48W-intermittent:Carcinogen
  - subcutaneous-hamster TDLo 2686 mg/kg/72W-intermittent:Carcinogen
  - oral-mouse TD 288 mg/kg/8W-intermittent:ETA
  - subcutaneous-rat TD 400 mg/kg/20W-intermittent:ETA
  - oral-rat TD 300 mg/kg/14W-intermittent:ETA

- subcutaneous-mouse LD 200 mg/kg/25W-intermittent:ETA
- oral-rat LD $_{50}$ 122 mg/kg
- inhalation-rat LC $_{50}$ 252 ppm/4H
- intraperitoneal-rat LD $_{50}$ 102 mg/kg
- intravenous-rat LD $_{50}$ 119 mg/kg
- intracerebral-rat LDLo 27 mg/kg
- oral-mouse LD $_{50}$ 265 mg/kg
- inhalation-mouse LC $_{50}$ 172 ppm/4H
- intraperitoneal-mouse LD $_{50}$ 113 mg/kg
- intravenous-mouse LD $_{50}$ 250 mg/kg
- inhalation-dog LC $_{50}$ 3580 ppm/15M

◆ Acute Risks
  - nose and throat irritation, burning sensation, coughing, laryngitis, shortness of breath, headaches, mild conjunctivitis, nausea, and vomiting caused in humans due to inhalation
  - highly corrosive and irritating to the skin, eyes, and mucous membranes
  - Neurological symptoms have been observed due to burns from the substance.
  - central nervous system stimulation and convulsions observed in animals
  - high acute toxicity from inhalation, oral, and dermal exposures in rats, mice, hamsters, rabbits, and guinea pigs
  - causes burns and corrosive to skin
  - may be fatal if inhaled, swallowed, or absorbed through the skin

◆ Chronic Risks
  - liver damage in humans
  - hemolytic anemia and central nervous system effects in animals such as convulsive seizures caused by chronic inhalation
  - mild kidney damage in animals
  - reference concentration under EPA review
  - reference dose not established by the EPA
  - no information on reproductive or developmental effects on humans or animals
  - carcinogenic effects observed in lungs and liver of animals upon inhalation and oral exposure
  - probable human carcinogen
  - may alter genetic material
  - can cause gastrointestinal disturbances

HAZARD RISK
◆ hazard rating of 3 indicating severe hazard level for toxicity, fire, and reactivity
◆ dangerous fire hazard
◆ Vapor may cause flashback by traveling to the ignition source.
◆ Decomposition results in flammable and/or explosive mixtures in air.

◆ Heating to decomposition also results in production of toxic substances CO, CO_2, and nitrogen oxides.
◆ flammable upon exposure to air
◆ forms the carcinogen 1,1-dimethylnitrosamine upon exposure to the atmosphere
◆ powerful reducing agent that can react vigorously with oxidizing materials
◆ spontaneously ignites (hypergolic) with many oxidants (N_2O_4, H_2O_2, and HNO_3)
◆ incompatible with copper, copper alloys, brass, iron, and iron salts
◆ avoid contact with water due to liberation of heat

EXPOSURE ROUTES
◆ inhalation
◆ absorption through skin or eyes
◆ ingestion
◆ due to spills, leaks, and venting during loading, transfer, and storage
◆ from ambient atmosphere when used as jet or rocket fuel
◆ intraperitoneal, intravenous, intracerebral routes

REGULATORY STATUS/INFORMATION
◆ NTP 7th Annual Report on Carcinogens
◆ IARC Cancer Review: Group 2B IMEMDT 7,56,87
◆ Animal Sufficient Evidence IMEMDT 4,137,74
◆ EPA Genetic Toxicology Program
◆ Community Right-to-Know List
◆ EPA Extremely Hazardous Substance List
◆ reported in EPA TSCA Inventory
◆ EPA Office of Air Quality Planning and Standards, for a hazard ranking under section 112(g) of the 1990 Clean Air Amendments, has not yet assigned the contaminant for ranking as a carcinogenic hazard.
◆ OSHA, 29CFR1910.119 Appendix A: list of highly hazardous chemicals, toxics, and reactives
◆ chemical regulated by the State of California
◆ subject to Superfund Amendments and Reauthorization Act (SARA) of 1986, Section 313 reporting requirements
◆ Department of Transportation
  • hazard class division 6.1
  • identification number UN 1163
  • labels: poison, flammable liquid, corrosive

ADDITIONAL/OTHER COMMENTS
◆ Symptoms
  • inhalation: nose and throat irritation, burning sensation, coughing, laryngitis, shortness of breath, headaches, mild conjunctivitis, nausea, and vomiting caused in humans
  • absorption: highly corrosive and irritating to the skin, eyes, and mucous membranes; causes burns and corrosive to skin
  • Neurological symptoms have been observed due to burns from the substance.

- may be fatal if inhaled, swallowed, or absorbed through the skin
◆ First Aid
  - if inhaled, remove to fresh air, give artificial respiration if not breathing or oxygen if breathing is difficult, and get medical attention
  - if contacted with the eyes, flush with water for 15 minutes and get medical attention
  - if contacted with the skin, wash with water for at least 15 minutes, while removing clothing; get medical attention if symptoms persist after washing
  - if swallowed, do not induce vomiting; take in water unless unconscious and call poison control center immediately
◆ fire fighting: use alcohol foam, carbon dioxide, dry chemical
◆ personal protection: use chemical-resistant gloves, boots, apron, goggles or a face shield and a NIOSH approved vapor respirator if needed due to poor ventillation

Note: use of a respirator requires professional expertise to select and fit the respirator and a medical exam to determine if a respirator can be worn without causing heart and/or other problems.

◆ storage: store under nitrogen in a tightly sealed container away from combustible materials and ignition sources

SOURCES
◆ *Material Safety Data Sheets (MSDS)*
◆ *Sax's Dangerous Properties of Industrial Materials*, 9th edition, 1996
◆ *Merck Index*, 12th edition, 1996
◆ *Hawley's Condensed Chemical Dictionary*, 12th edition, 1993
◆ *Environmental Law Index to Chemicals*, 1996 edition
◆ *Health Effects Notebook for Hazardous Air Pollutants*, U.S. Environmental Protection Agency, December 1994
◆ http://hazard.com:80/fish/acros/5/1480.html

DIMETHYL PHTHALATE

CAS #: 131-11-3
MOLECULAR FORMULA: C_6H_4-1,2-$(CO_2CH_3)_2$
FORMULA WEIGHT: 194.20
SYNONYMS: avolin, 1,2-benzenedicarboxylic acid, dimethyl-1,2-benzenedicarboxylate, dimethyl benzeneorthodicarboxylate, dimethyl ester, DMP, ENT 262, fermine, methyl phthalate, mipax, NTM, palatinol M, phthalic acid methyl ester, Phthalsaeuredimethylester (German), solvanom, solvarone

USES
◆ Used in Industry
  - solid rocket propellants
  - lacquers
  - plastics
  - resins
  - safety glasses
  - rubber coating agents
  - molding powders
  - insect repellents
  - pesticides

- solvent and plasticizer for cellulose acetate and cellulose acetate–butyrate compositions

PHYSICAL PROPERTIES
- colorless, oily liquid
- slightly sweet odor
- melting point 0°C
- boiling point 282.4°C
- specific gravity 1.190
- vapor pressure 4.19×10^{-3} mm Hg at 20°C
- vapor density 6.69
- bulk density 9.93 lb/gal
- viscosity 17.2 cP at 25°C
- miscible with alcohol and ether
- insoluble in water and paraffinic hydrocarbons
- slightly soluble in mineral oil
- log K_{ow} 1.56

CHEMICAL PROPERTIES
- combustible
- flash point 149°C
- autoignition temperature 555°C
- heat of vaporization 93.1 gcal/g
- heat of combustion 119.7 kcal/mole
- explosion limits in air 0.94% lower and 8.03% upper
- incompatible with oxidizing agents and acids

HEALTH RISK
- Toxic Exposure Guidelines
  - ACGIH TLV TWA 5 mg/m^3
  - OSHA PEL TWA 5 mg/m^3
- Toxicity Data
  - eye-rabbit 119 mg
  - mmo-sat 200 µg/plate
  - cyt-rat-skin 25 g/kg/4W-I
  - intraperitoneal-rat TDLo 1125 mg/kg
  - oral-rat LD$_{50}$ 6800 mg/kg
  - intraperitoneal-rat LD$_{50}$ 3375 mg/kg
  - oral-mouse LD$_{50}$ 6800 mg/kg
  - intraperitoneal-mouse LD$_{50}$ 1380 mg/kg
  - subcutaneous-mouse LDLo 6500 mg/kg
  - inhalation-cat LCLo 9630 mg/m^3/6H
  - oral-rabbit LD$_{50}$ 4400 mg/kg
  - oral-guinea pig LD$_{50}$ 2400 mg/kg
  - oral-chicken 8500 mg/kg
- Acute Risks
  - irritation of eyes
  - irritation of nose
  - irritation of throat
  - central nervous system depression
  - not absorbed through skin
- Chronic Risks
  - No information is available on the chronic effects in humans.
  - slight effects on growth and on the kidney in animals
  - may cause respiratory disorders
  - EPA Group D: not classifiable as to human carcinogenicity

HAZARD RISK
- combustible when exposed to heat or flame
- can react with oxidizing materials

- ◆ hazardous decomposition products: carbon monoxide and carbon dioxide
- ◆ Decombustion emits acrid smoke and irritating fumes.

EXPOSURE ROUTES
- ◆ food
- ◆ use of hemodialysis tubing
- ◆ polyvinylchloride bags containing intravenous solutions
- ◆ drinking water
- ◆ factories that manufacture or use dimethyl phthalate

REGULATORY STATUS/INFORMATION
- ◆ Clean Air Act Amendment, Title III: hazardous air pollutants
- ◆ Resource Conservation and Recovery Act
  - • list of hazardous constituents. 40CFR258, 40CFR261, 40CFR261.11
  - • groundwater monitoring list. 40CFR264
  - • PICs found in stack effluents. 40CFR266
  - • U waste #: U102
- ◆ CERCLA and Superfund
  - • designation of hazardous substances. 40CFR302.4
  - • list of chemicals. 40CFR372.65
  - • Clean Water Act
  - • total toxic organics (TTOs). 40CFR413.02
  - • 126 priority pollutants. 40CFR423
- ◆ Toxic Substance Control Act
  - • testing consent orders for substances and mixtures with CAS registry #s
  - • OSHA 29CFR1910.1000 Table Z1
  - • chemical regulated by the State of California

ADDITIONAL/OTHER COMMENTS
- ◆ Symptoms
  - • exposure: burning sensation, coughing, wheezing, laryngitis, shortness of breath, headache, nausea, vomiting
- ◆ First Aid
  - • wash eyes and skin immediately with large amounts of water
  - • remove to fresh air immediately or provide respiratory apparatus if necessary
  - • wash out mouth with water
- ◆ fire fighting: use carbon dioxide and dry chemical
- ◆ personal protection: self-contained breathing apparatus, protective clothing, and goggles
- ◆ storage: keep tightly closed in a cool, dry place

SOURCES
- ◆ *Health Effects Notebook for Hazardous Air Pollutants*, U.S. Environmental Protection Agency, December 1994
- ◆ *Material Safety Data Sheets (MSDS)*
- ◆ *Sax's Dangerous Properties of Industrial Materials*, 9th edition, 1996
- ◆ *Hawley's Condensed Chemical Dictionary*, 12th edition, 1993
- ◆ *Aldrich Chemical Company Catalog*, 1996–1997

♦ *Environmental Law Index to Chemicals*, 1996
♦ *Merck Index*, 12th edition, 1996

DIMETHYL SULFATE

CAS #: 77-78-1
MOLECULAR FORMULA: $(CH_3O)_2SO_2$
FORMULA WEIGHT: 126.14
SYNONYMS: cwumetylowy siarczan (Polish), dimethylester kyseliny sirove (Czech), dimethylmonosulfate, dimethylsulfaat (Dutch), dimethylsulfat (Czech), dimetilsolfato (Italian), DMS, DMS (methyl sulfate), methyle (sulfate de), (French), methyl sulfate (DOT), sulfate de methyle (French), sulfate dimethylique (French), sulfuric acid, dimethyl ester

USES
♦ Used in the Manufacturing Process
  • many organic chemicals
  • dyes
  • perfumes
  • separation of mineral oils
  • analysis of auto fluids
  • war gas
  • methylating agent for amines and phenols
  • polyurethane based adhesives

PHYSICAL PROPERTIES
♦ colorless, oily liquid
♦ faint, onion-like odor
♦ melting point −31.8°C
♦ boiling point 188°C
♦ density 1.322 at 15°C
♦ vapor pressure 0.5 mm Hg at 20°C
♦ vapor density 4.35
♦ slightly soluble in water
♦ slightly soluble in hexane, EtOH, and C_6H_6
♦ soluble in Et_2O and Me_2CO
♦ log K_{ow} 0.032

CHEMICAL PROPERTIES
♦ combustible
♦ flash point 83.3°C
♦ autoignition temperature 494°C
♦ incompatible with strong oxidizing agents, strong bases, and ammonia

HEALTH RISK
♦ Toxic Exposure Guidelines
  • ACHIH TLV TWA 0.1 ppm (skin)
  • OSHA PEL TWA 0.1 ppm (skin)
  • DFG MAK production 0.02 ppm, use 0.04 ppm
  • MSHA standard 5 mg/m³
♦ Toxicity Data
  • skin-rabbit 10 mg/24K open SEV
  • eye-rabbit 100 mg/4S rns SEV
  • eye-rabbit 50 µg/24H SEV
  • mma-sat 4300 nmol/L/1H
  • dnr-omi 640 µg/plate
  • dnd-human:lym 1 mmol/L
  • intravenous-rat TDLo 100 mg/kg
  • inhalation-rat TCLo 17 mg/m³/19W-I:ETA

- subcutaneous-rat TDLo 50 mg/kg:ETA
- intravenous-rat TDLo 20 mg/kg
- inhalation-human LCLo 97 ppm/10M
- oral-rat LD_{50} 205 mg/kg
- inhalation-rat LC_{50} 45 mg/m³/4H
- subcutaneous-rat LD_{50} 100 mg/kg
- oral-mouse LD_{50} 140 mg/kg
- inhalation-mouse LC_{50} 280 mg/m³
- oral-rabbit LDLo 45 mg/kg
- subcutaneous-rabbit LDLo 53 mg/kg
- intravenous-rabbit LDLo 50 mg/kg
- inhalation-guinea pig LC_{50} 32 ppm/1H

◆ Acute Risks

- severe inflammation and necrosis of the eyes, mouth, and respiratory tract
- severe and fatal damage to the lungs
- primarily damages the lungs
- injures the liver
- injures the kidneys
- injures the heart and central nervous system
- convulsions
- delirium
- paralysis
- coma
- death
- severe blistering
- conjunctivitis
- catarrhal inflammation of the mucous membranes of the nose, throat, larynx, and trachea
- ulceration
- local necrosis

◆ Chronic Risks

- Cornea shows clouding.
- pulmonary edema
- serious injury to liver and kidneys
- prostration
- suppression of urine
- jaundice
- albuminuria
- hematuria
- decreased body weight in rats and mice
- increased mortality in rats and mice
- produces tumors in offspring of rats
- tumors in nasal passages, lungs, and thorax of animals
- EPA Group B2: probable human carcinogen

HAZARD RISK
◆ flammable when exposed to heat, flame, or oxidizers
◆ can react with oxidizing materials
◆ violent reaction with NH_4OH and NaN_3
◆ may decompose on exposure to moist air or water
◆ hazardous decomposition products: carbon monoxide, carbon dioxide, sulfur oxides
◆ Decomposition emits toxic fumes of SO_x.

EXPOSURE ROUTES
◆ most likely in the workplace
◆ fly ash

- ◆ airborne particulate matter generated by coal combustion processes
- ◆ breathing ambient air near coal-fired power generating plants

REGULATORY STATUS/INFORMATION

- ◆ Clean Air Act Amendment, Title III: hazardous air pollutants
- ◆ Resource Conservation and Recovery Act
  - list of hazardous constituents. 40CFR261, 40CFR261.11
  - U waste #: U103
- ◆ CERCLA and Superfund
  - designation of hazardous substances. 40CFR302.4
  - list of extremely hazardous substances and their threshold planning quantities. 40CFR355-AB
  - list of chemicals. 40CFR372.65
- ◆ Toxic Substance Control Act
  - OHSA chemicals in need of dermal absorption testing 40CFR712.30.e10, 40CFR716.120.d10
- ◆ chemical regulated by the State of California

ADDITIONAL/OTHER COMMENTS

- ◆ Symptoms
  - exposure: irritation to skin, eyes, and mucous membranes, conjunctivitis, catarrhal inflammation of mucous membranes of nose, throat, larynx, and trachea, reddening of skin
- ◆ First Aid
  - wash eyes and skin immediately with large amounts of water
  - remove to fresh air immediately or provide respiratory apparatus if necessary
  - wash out mouth with water
- ◆ fire fighting: use water spray, carbon dioxide, dry chemical powder, or appropriate foam
- ◆ personal protection: self-contained breathing apparatus and protective clothing
- ◆ storage: keep tightly closed away from heat and open flame

SOURCES

- ◆ *Health Effects Notebook for Hazardous Air Pollutants*, U.S. Environmental Protection Agency, December 1994
- ◆ *Material Safety Data Sheets (MSDS)*
- ◆ *Sax's Dangerous Properties of Industrial Materials*, 9th edition, 1996
- ◆ *Hawley's Condensed Chemical Dictionary*, 12th edition, 1993
- ◆ *Aldrich Chemical Company Catalog*, 1996–1997
- ◆ *Environmental Law Index to Chemicals*, 1996
- ◆ *Merck Index*, 12th edition, 1996

4,6-DINITRO-o-CRESOL and SALTS

CAS #: 534-52-1
MOLECULAR FORMULA: $CH_3C_6H_2(NO_2)_2OH$
FORMULA WEIGHT: 198.13
SYNONYMS: antinonin, antinonnin, arborol, cresol(o-), degrassan, dekrysil, detal, dillex, dinitro, 4,6-dinitro-, dinitro-cresol, dinitro-o-cresol, 4,6-dinitro-o-cresol, dinitro-o-cresol(2,4-), dinitro-o-cresol (3,5-), dinitro-dendtroxal, dinitro-2-hydroxytoluene(3,5-), dinitro-2-methylphenol(4,6-), dinitro-6-methylphenol(2,4-), dinitrol, dinoc, dinurania, ditrosol, DNOC, effusan, effusan-3436, elgetol, elgetol-30, elipol, extrar, flavin-sandoz, hedolit, K-III, K-IV, kreozan, krezotol-50, lipan, methyl-2,4-dinitrophenol(6-), methyl-4,6-dinitrophenol(2-), neudoriff-DN 50, nitro-fan, phenol, 2-methyl-4,6-dinitro, phenol, 2-methyl-4,6-dinitro- and salts, prokarbol, rafex, rafex-35, raphatox, sandolin, sandolin-A, selinon, sinox, winterwash

USES
- ◆ Used in Industry
  - herbicide
  - defoliant
  - dyestuff industry
  - control of broad-leaved weeds in cereals
  - emulsifiable concentrate formulations
  - preharvest desiccation of potatoes
  - leguminous seed crops
  - weight reducing agent (previously)
  - free-radical polymerization inhibitor
- ◆ Used as an Insecticide
  - toxic to eggs of certain insects
  - dormant spray
  - thinning blossoms of fruit trees
  - to kill locusts
- ◆ Used as a Fungicide
  - against overwintering venturia inaequalis
  - against foreign fungi in mushroom houses

PHYSICAL PROPERTIES
- ◆ yellow prismatic solid
- ◆ Sodium salt is a red powder.
- ◆ melting point 87.5°C
- ◆ boiling point 312°C
- ◆ vapor pressure 1.05×10^{-4} at 25°C
- ◆ vapor density 6.82
- ◆ log K_{ow} 2.564
- ◆ pK_a 4.4
- ◆ sparingly soluble in water
- ◆ NH_4, K, Ca salts are soluble in water.
- ◆ soluble in alkaline aqueous solutions

CHEMICAL PROPERTIES
- ◆ heat of combustion −3920 cal/g
- ◆ stable during transport
- ◆ heat of decomposition −1.17 kcal/g
- ◆ autoignition temperature 435°C
- ◆ incompatibility with strong oxidizers
- ◆ will not polymerize

HEALTH RISK
- ◆ Toxic Exposure Guidelines
  - OSHA PEL TWA 0.2 mg/m³
  - ACGIH TLV TWA 0.2 mg/m³

- DFG MAK 0.2 mg/m³
- NIOSH REL TWA 0.2 mg/m³
- MSHA standard 0.2 mg/m³
◆ Toxicity Data
  - skin-rabbit 105 mg/9D-I MLD
  - eye-rabbit 20 mg/24H MOD
  - mma-sat 1 μmol/plate
  - sln-dmg-oral 250 μmol/L
  - oral-man TDLo 7500 μg/kg/7D:CNS
  - inhalation-human TCLo 1 mg/m³
  - unr-man LDLo 29 mg/kg
  - oral-rat LD_{50} 10 mg/kg
  - skin-rat LD_{50} 200 mg/kg
  - intraperitoneal-rat LDLo 28 mg/kg
  - subcutaneous-rat LD_{50} 25,600 μg/kg
  - oral-mouse LD_{50} 47 mg/kg
  - intraperitoneal-mouse LD_{50} 19 mg/kg
  - intravenous-dog LDLo 15 mg/kg
  - inhalation-cat LCLo 40 mg/m³
◆ Acute Risks
  - profuse sweating
  - deep, rapid respiration
  - thirst
  - fatigue
  - lethargy
  - headache
  - nausea
  - appetite loss
  - malaise
  - collapse
  - coma
  - greenish-yellow pigmentation of the conjunctiva
  - yellow coloring of nails, hands, and hair
  - damage to liver and kidneys
  - damage to nervous system
  - local necrosis
  - eye and skin irritant
  - severe increase in basal metabolic rate
  - death

◆ Chronic Risks
  - fatigue
  - chest pain
  - dyspnea
  - flushing
  - diarrhea
  - fever
  - jaundice
  - vomiting
  - restlessness
  - anxiety
  - excessive sweating
  - unusual thirst
  - loss of weight
  - lethargy
  - headache
  - malaise
  - collapse
  - coma
  - yellow staining of conjunctiva
  - cataract formation
  - blindness
  - cumulative poison
  - effects to cardiovascular, gastrointestinal, and central nervous system
  - changes in blood count
  - kidney damage
  - hyperpyrexia
  - hyperglycemia
  - glycosuria
  - decreased liver enzyme activity

- changes in absolute and relative organ weights
- dermal necrosis

HAZARD RISK
- ◆ may detonate in a fire
- ◆ usually moistened with 10% water to reduce risk of explosion
- ◆ Combustion will produce carbon dioxide, carbon monoxide, and oxides of nitrogen.
- ◆ incompatibility with strong oxidizers

EXPOSURE ROUTES
- ◆ inhalation or dermal contact
- ◆ manufacture, formulation, application of the pesticide
- ◆ in water in industrial effluents
- ◆ direct pesticidal applications
- ◆ pesticide leaching or runoff

REGULATORY STATUS/INFORMATION
- ◆ Clean Air Act Amendment, Title III: hazardous air pollutants
- ◆ Federal Insecticide, Fungicide, and Rodenticide Act: class of dinitrophenols. 180.3(6)
- ◆ Resource Conservation and Recovery Act
  - list of hazardous constituents. 40CFR261, 40CFR261.11
  - P waste #: P047
- ◆ CERCLA and Superfund
  - designation of hazardous substances. 40CFR302.4
  - list of extremely hazardous substances. 40CFR355-AB

ADDITIONAL/OTHER COMMENTS
- ◆ Symptoms
  - short-term exposure: rapid respiration, sweating, thirst, heat sensation, motor weakness, collapse, coma, death
- ◆ First Aid
  - wash eyes and skin immediately with large amounts of water
  - remove to fresh air or provide respiratory support
  - induce vomiting by touching back of throat with finger or blunt object
- ◆ fire fighting: use water, foam, dry chemical, or carbon dioxide
- ◆ personal protection: wear self-contained breathing apparatus and protective clothing and goggles
- ◆ storage: keep away from contact with oxidizing materials and away from fire

SOURCES
- ◆ *Health Effects Notebook for Hazardous Air Pollutants*, U.S. Environmental Protection Agency, December 1994
- ◆ *Material Safety Data Sheets (MSDS)*
- ◆ *Sax's Dangerous Properties of Industrial Materials*, 9th edition, 1996
- ◆ *Hawley's Condensed Chemical Dictionary*, 12th edition, 1993
- ◆ *Aldrich Chemical Company Catalog*, 1996–1997
- ◆ *Environmental Law Index to Chemicals*, 1996
- ◆ *Merck Index*, 12th edition, 1996

2,4-DINITROPHENOL

CAS #: 51-28-5
MOLECULAR FORMULA: $C_6H_4N_2O_5$
FORMULA WEIGHT: 184.12
SYNONYMS: aldifen, chemox Pe, 2,4-dinitrofenol (Dutch), dinitrofenolo (Italian), alpha-dinitrophenol, 2,4-DNP, fenoxyl carbon N, 1-hydroxy-2,4-dinitrobenzene, maroxol-50, nitro kleenup, NSC 1532, solfo black B, solfo black BB, solfo black 2B supra, solfo black G, solfo black SB, tertrosulphur black PB, tertrosulphur PBR

USES
♦ Used as an Intermediate in the Manufacture of Products
  • wood preservatives
  • dyes
  • pesticides
  • fungicides
  • miticides
♦ used as an indicator for the detection of potassium and ammonium ions

PHYSICAL PROPERTIES
♦ yellow crystals
♦ melting point 112°C
♦ density 1.683 at 24°C, 1.4281 g/cm³ at 121°C
♦ vapor density 6.35
♦ vapor pressure 1.42×10^{-7} mm Hg at 25°C
♦ log K_{ow} 1.91
♦ slightly soluble in water
♦ volatile with steam
♦ very soluble in benzene

CHEMICAL PROPERTIES
♦ sublimes when heated
♦ forms explosive salts with alkalies and ammonia

HEALTH RISK
♦ Toxic Exposure Guidelines
  • EPA RfD 0.002 mg/kg/d
♦ Toxicity Data
  • skin-rabbit 300 mg/4W-I MLD
  • oms-ofs:oth 100 μmol/L
  • cyt-mouse-intraperitoneal 10 g/kg
  • oral-rat TDLo 2040 mg/kg
  • intraperitoneal-mouse TDLo 40,800 μg/kg
  • oral-human LDLo 4300 μg/kg
  • oral-rat LD_{50} 30 mg/kg
  • intraperitoneal-rat LD_{50} 20 mg/kg
  • subcutaneous-rat LD_{50} 25 mg/kg
  • oral-mouse LD_{50} 45 mg/kg
  • intraperitoneal-mouse LD_{50} 26 mg/kg
  • oral-rabbit LD_{50} 30 mg/kg
  • oral-guinea pig LD_{50} 81 mg/kg
  • oral-dog LDLo 30 mg/kg
  • inhalation-dog LCLo 300 mg/m³/30M
  • subcutaneous-guinea pig LDLo 25 mg/kg

◆ Acute Risks
- nausea
- vomiting
- sweating
- dizziness
- headache
- loss of weight
- excessive sweating
- feeling of warmth

- weakness and fatigue
- rapid respiration
- tachycardia
- rise in body temperature
- rapid metabolism and excretion
- death

◆ Chronic Risks
- cataracts
- skin lesions
- effects on bone marrow
- effects on central nervous system

- effects on cardiovascular system
- fetal growth inhibition
- polyneuropathy
- exfoliative dermatitis

HAZARD RISK
◆ an explosive
◆ forms explosive salts with alkalies and ammonia
◆ upon decomposition, emits toxic fumes of NO_x
◆ may explode when shocked or exposed to heat or chemical
◆ reaction with reducing materials

EXPOSURE ROUTES
◆ primarily from pesticide runoff to water
◆ readily absorbed through intact skin
◆ vapors absorbed through respiratory tract

REGULATORY STATUS/INFORMATION
◆ Clean Air Act Amendment, Title III: hazardous air pollutants
◆ Safe Drinking Water Act: priority list of drinking water contaminants. 55FR1470
◆ Resource Conservation and Recovery Act
- list of hazardous constituents. 40CFR258, 40CFR261, 40CFR261.11
- groundwater monitoring list. 40CFR264
- reference air concentrations. 40CFR266
- P waste #: P048
◆ CERCLA and Superfund
- designation of hazardous substances. 40CFR302.4
- list of chemicals. 40CFR372.65
◆ Clean Water Act
- designation of hazardous substances. 40CFR116.4
- total toxic organics (TTOs). 40CFR413.02
- 126 priority pollutants. 40CFR423
◆ Toxic Substance Control Act
- IRIS chemicals. 40CFR716.120.d7
◆ chemical regulated by the State of California

ADDITIONAL/OTHER COMMENTS
◆ Symptoms
- ingestion: nausea, vomiting, sweating, dizziness, headache, weight loss
- skin contact: irritation

228

◆ First Aid
  • wash eyes and skin immediately with large amounts of water
  • provide respiratory protection
◆ personal protection: protective clothing, butyl rubber gloves, face shield, goggles, and respiratory protection
◆ storage: keep in a cool, well-ventilated place away from heat, ignition sources, and sun rays

SOURCES
◆ *Health Effects Notebook for Hazardous Air Pollutants*, U.S. Environmental Protection Agency, December 1994
◆ *Material Safety Data Sheets (MSDS)*
◆ *Sax's Dangerous Properties of Industrial Materials*, 9th edition, 1996
◆ *Aldrich Chemical Company Catalog*, 1996–1997
◆ *Environmental Law Index to Chemicals*, 1996
◆ *Merck Index*, 12th edition, 1996

2,4-DINITROTOLUENE

CAS #: 121-14-2
MOLECULAR FORMULA: $C_7H_6N_2O_4$
FORMULA WEIGHT: 182.15
SYNONYMS: benzene, 1-methyl-2,4-dinitro, benzene, 2-methyl-1,3-dinitro-, dichloro-4-hydroxybenzene(1,3-), dinitro-toluene(2,4-), dinitro-toluene(2,6-), dinitro-toluene mixture, 2,4-/2,6-, dinitro-toluene mixture(2,4/2,6-), dinitro-toluol(2,4-), DNT(2,4-), methyl-1,3-dinitro-benzene(2-), methyl-2,4-dinitrobenzene(1-), toluene, 2,4-dinitro-, toluene, 2,6-dinitro

USES
◆ Used as an Intermediate in the Manufacture of Chemicals and Products
  • polyurethanes
  • organic synthesis
  • toluidines
  • explosives
  • dyes
  • smokeless gunpowders
◆ used as a gelatinizing and waterproofing agent

PHYSICAL PROPERTIES
  • slightly yellow, sharp crystals
  • melting point 66–68°C
  • boiling point 300°C
  • vapor pressure 0.1 hPa at 50°C, 0.0051 mm Hg at 20°C
  • solubility 0.3 g/L at 20°C
  • solubility in water 270 mg/L at 22°C
  • soluble in alcohol and ether
  • specific gravity 1.3208
  • slight odor
  • log K_{ow} 2.0

CHEMICAL PROPERTIES
◆ generally stable under normal temperature and pressure
◆ decomposition temperature 250°C
◆ flash point 404°C

- ◆ autoignition temperature 420°C
- ◆ combustible when exposed to heat or flame
- ◆ reacts with oxidizing agents, reducing agents, and strong bases
- ◆ ignites on contact with sodium oxide

HEALTH RISK

- ◆ Toxic Exposure Guidelines
  - ACGIH TLV 1.5 mg/m³
  - NIOSH REL 1.5 mg/m³
  - OSHA PEL TWA 1.5 mg/m³
- ◆ Toxicity Data
  - mma-sat 125 μg/plate
  - dnd-rat:lvr 3 mmol/L
  - oms-rat-oral 10 mg/kg
  - mmo-sat 10 μg/plate
  - cyt-mouse-oral 840 μg/kg
  - dlt-mouse-oral 840 μg/kg
  - oral-rat TDLo 3094 mg/kg
  - oral-rat TDLo 2620 mg/kg/78W-C:NEO
  - oral-mouse TDLo 10,080 mg/kg/2Y-C:CAR
  - oral-rat TD 28 g/kg/2Y-C:CAR
  - oral-rat TD 12,775 mg/kg/2Y-C:ETA
  - oral-rat LD_{50} 268 mg/kg
  - oral-mouse LD_{50} 790 mg/kg
  - subcutaneous-cat LDLo 25 mg/kg
  - oral-guinea pig LD_{20} 1300 mg/kg
- ◆ Other Toxicity Indicators
  - EPA has classified 2,4-dinitrotoluene as a Group B2, probable human carcinogen of medium carcinogenic hazard, $1/ED_{10}$ value of 3.8 per (mg/kg)/d.
- ◆ Acute Risks
  - Information on humans is not available.
  - cyanosis and ataxia in animals
  - moderate to high acute toxicity from oral exposure in animals
- ◆ Chronic Risks (Noncancer)
  - metallic taste in mouth
  - muscular weakness
  - headache
  - appetite loss
  - giddiness
  - dizziness
  - nausea
  - insomnia
  - tingling pains in the extremities in humans
  - effects to the blood
  - pallor
  - methemoglobinemia
  - cyanosis
  - anemia
  - elevated mortality from heart disease
  - muscular incoordination
  - weakness
  - tremors
  - convulsions
  - ataxia
  - paralysis
  - effects on liver and kidneys in animals
  - reduction in sperm counts and normal sperm morphology

- increase in spontaneous abortions
- decreased fertility in animals
- testicular atrophy
- degenerated seminiferous tubules in male rats
- ovarian atrophy and dysfunction in female rats

◆ Chronic Risks (Cancer)
  - renal tumors and liver tumors
  - probable human carcinogen

HAZARD RISK
◆ combustible when exposed to heat or flame
◆ reacts with oxidizing materials, reducing agents, and strong bases
◆ instances of explosion during manufacture or storage
◆ Mixture with nitric acid is a high explosive.
◆ Mixture with sodium carbonate decomposes with significant pressure increase at 210°C.
◆ ignites on contact with sodium oxide
◆ When heated to decomposition it emits toxic fumes of NO_x.

EXPOSURE ROUTES
◆ primarily occupational by inhalation or dermal contact
◆ ingestion and subcutaneous routes
◆ absorbed by skin

REGULATORY STATUS/INFORMATION
◆ Clean Air Act Amendment, Title III: hazardous air pollutants
◆ Safe Drinking Water Act: priority list of drinking water contaminants. 55FR1470
◆ Resource Conservation and Recovery Act
  - list of hazardous constituents. 40CFR258, 40CFR261, 40CFR261.11
  - groundwater monitoring list. 40CFR264
  - risk specific doses. 40CFR266
  - health based limits for exclusion of waste-derived residues. 40CFR266
  - D waste #: D030
  - U waste #: U105
◆ CERCLA and Superfund
  - designation of hazardous substances. 40CFR302.4
  - list of chemicals. 40CFR372.65
◆ Clean Water Act
  - total toxic organics (TTOs). 40CFR413.02
  - 126 priority pollutants. 40CFR423
◆ Toxic Substance Control Act
  - OHSA chemicals in need of dermal absorption testing. 40CFR712.30.e10, 40CFR716.120.d10
◆ chemical regulated by the State of California

ADDITIONAL/OTHER COMMENTS
◆ Symptoms
  - an irritant and an allergen
  - inhalation: muscular weakness, headache, appetite loss, giddiness, dizziness, nausea, insomnia, tingling pains

- ◆ First Aid
  - • wash eyes immediately with large amounts of water
  - • flush skin with plenty of soap and water
  - • remove to fresh air immediately
- ◆ fire fighting: use dry chemical, water spray or mist, chemical foam, or alcohol-resistant foam
- ◆ personal protection: wear self-contained breathing apparatus and chemical goggles

SOURCES

- ◆ *Health Effects Notebook for Hazardous Air Pollutants*, U.S. Environmental Protection Agency, December 1994
- ◆ *Material Safety Data Sheets (MSDS)*
- ◆ *Sax's Dangerous Properties of Industrial Materials*, 9th edition, 1996
- ◆ *Hawley's Condensed Chemical Dictionary*, 12th edition, 1993
- ◆ *Aldrich Chemical Company Catalog*, 1996–1997
- ◆ *Environmental Law Index to Chemicals*, 1996

1,4-DIOXANE (bh)

CAS #: 123-91-1
MOLECULAR FORMULA: $C_4H_8O_2$
FORMULA WEIGHT: 88.11
SYNONYMS: diaksan (Polish), diethylene dioxide, 1,4-diethylene dioxide, diethylene dioxide (OSHA), diethylene ether, di(ethylene oxide), 1,4-diethyleneoxide, diokane, diossano-1,4 (Italian), dioxaan-1,4 (Dutch), 1,4- dioxacyclohexane, dioxan, Dioxan-1,4 (German), P-dioxan (Czech), dioxane, dioxane-1,4, 1,4-dioxane, dioxane (ACGIH:DOT:OSHA), dioxanne (French), P-dioxin, tetrahydro-, di-oxyethylene ether, glycol ethylene ether, NCI-C03689, RCRA waste number U108, tetrahydro-P-dioxin, tetrahydro-1,4-dioxin, UN1165

USES

- ◆ Used as an Industrial Solvent for
  - • cellulose acetate
  - • ethyl cellulose
  - • benzyl cellulose
  - • oils
  - • resins
  - • waxes
  - • some dyes
  - • other organic and inorganic compounds
  - • stabilizer in chlorinated solvents

PHYSICAL PROPERTIES

- ◆ colorless liquid
- ◆ faint, pleasant odor
- ◆ melting point 11.80°C (53.24°F)
- ◆ boiling point 101.1°C (214°F) at 1 atm
- ◆ specific gravity 1.0329
- ◆ vapor density 3 kg/m³ at 20°C
- ◆ vapor pressure 27 mm Hg at 20°C and 40 mm Hg at 25.2°C

- ◆ soluble in water
- ◆ soluble in organic solvents
- ◆ odor threshold 24 ppm (86.4 mg/m^3)
- ◆ specific heat 0.0370 kcal/mol/°C at 20°C
- ◆ viscosity 0.0120 cP at 25°C

CHEMICAL PROPERTIES

- ◆ flammable liquid
- ◆ explosion limits 2.0% lower and 22% upper by volume in air
- ◆ autoignition temperature 356°C (673°F)
- ◆ flash point 12°C (54°F)
- ◆ heat of combustion 581 kcal/mol
- ◆ heat of fusion 2.98 kcal/mol

HEALTH RISK

- ◆ Toxic Exposure Guidelines
  - • ACGIH TLV TWA 25 ppm (90 mg/m^3) (skin)
  - • OSHA PEL TWA 100 ppm (360 mg/m^3) (skin)
  - • NIOSH REL TWA 1 ppm/30M (3.6 mg/m^3)
  - • DFG MAK 50 ppm (180 mg/m^3) Suspected Human Carcinogen
  - • EPA Cancer Risk Level (1 in a million excess lifetime risk) 3.0E-6 g/L
- ◆ Toxicity Data
  - • eye-human 300 ppm/15M
  - • skin-rabbit 515 mg open MLD
  - • eye-rabbit 21 mg
  - • eye-guinea pig 10E-6 g MOD
  - • oral-rat TDLo 185 g/kg/2Y-C:CAR
  - • inhalation-human TCLo 470 ppm:CNS,CVS,GIT
  - • inhalation-human TCLo 5500 ppm/1M:EYE, PUL
  - • inhalation-human LCLo 470 ppm/3d
  - • inhalation-rat LC$_{50}$ 46 g/m^3/2H
  - • inhalation-mouse LC$_{50}$ 37 g/m^3/2H
- ◆ Acute Risks
  - • harmful or fatal if swallowed, inhaled, or absorbed through skin
  - • irritation of eyes, nasal passages, throat, and lungs
  - • dizziness
  - • headaches
  - • damage to kidney and liver
  - • skin irritation
  - • nausea
- ◆ Chronic Risks
  - • target organs: liver, kidneys, and central nervous system
  - • may alter genetic material
  - • EPA Group B2: probable human carcinogen
  - • hepatic and renal lesions

HAZARD RISK

- ◆ dangerous fire hazard
- ◆ incompatible with strong oxidizers, halogens, oxygens, reducing agents, heat, moisture

- ◆ forms explosive peroxides
- ◆ forms carbon monoxide and carbon dioxide upon combustion
- ◆ toxic

EXPOSURE ROUTES
- ◆ primarily through occupation
- ◆ ingestion: found in drinking water: both surface and groundwater

REGULATORY STATUS/INFORMATION
- ◆ Clean Air Act
  - • Title III: hazardous air pollutant
- ◆ Resource Conservation Recovery Act
  - • Appendix 9: groundwater monitoring. 40CFR264
  - • Appendix 5: risk specific doses. 40CFR266
  - • discarded commercial chemical products, off-specification species, container residues, and spill residues thereof. 40CFR261.33
- ◆ CERCLA and Superfund
  - • Designation of Hazardous Substances. 40CFR302.4
  - • Appendix A: sequential CAS registry # list of CERCLA hazardous substances
  - • Appendix B: radionuclides
  - • chemicals and chemical categories to which this part applies (CAS # listing) 40CFR372.65
- ◆ Occupational Safety and Health Act of 1970
  - • limits for air contaminants applicable for the transitional period and to the extent set forth in paragraph 1910.1000.f under OSHA. 29CFR1910.1000
- ◆ chemical regulated by the State of California

ADDITIONAL/OTHER COMMENTS
- ◆ First Aid
  - • flush eyes or skin with copious amounts of water for at least 15 minutes
  - • remove contaminated clothing
  - • if inhaled, remove to fresh air
  - • if not breathing, give artificial respiration
  - • if breathing is difficult, give oxygen
  - • if ingested, wash out mouth with water and call physician
- ◆ Spill or Leak
  - • evacuate area and keep personnel upwind
  - • shut off all sources of ignition
  - • wear self-contained breathing apparatus, rubber boots, and heavy rubber gloves
  - • shut off leak if there is no risk
  - • ventilate area and wash spill site after material pickup is complete
  - • cover with activated carbon adsorbent

SOURCES
- ◆ *Encyclopedia of Chemical Technology*, 4th edition, 1991
- ◆ *Environmental Law Index to Chemicals*, 1996 edition

◆ Health Effects Notebook for Hazardous Air Pollutants, U.S. Environmental Protection Agency, 1996
◆ *Handbook of Environmental Data on Organic Chemicals*, 2nd edition, 1983
◆ *Material Safety Data Sheets (MSDS)*
◆ *Merck Index*, 12th edition, 1996
◆ *Sax's Dangerous Properties of Industrial Materials*, 9th edition, 1996

1,2-DIPHENYLHYDRAZINE (bh)

CAS #: 122-66-7
MOLECULAR FORMULA: $C_{12}H_{12}N_2$
FORMULA WEIGHT: 184.26
SYNONYMS: N,N'-bianiline, N,N'-diphenylhydrazine, (sym)-diphenylhyrazine, 1,2-diphenylhydrazine (9CI), hydrazobenzen (Czech), hydrozodi-benzene, NCI-C01854, RCRA waste number U109

USES
◆ Used as a Starting material in the Manufacture of
  • benzidine
  • benzidine-based dyes
  • anti-inflammatory drugs
◆ 1,2-diphenylhydrazine is no longer produced in the United States.

PHYSICAL PROPERTIES
◆ orange-yellow crystalline powder
◆ melting point 126–127°C (260°F)
◆ boiling point decomposes
◆ specific gravity 1.580
◆ very slightly soluble in water
◆ insoluble in acetylene
◆ odor threshold not established
◆ specific heat not available

CHEMICAL PROPERTIES
◆ nonflammable solid
◆ explosion limits not available
◆ autoignition temperature not available
◆ flash point not available

HEALTH RISK
◆ Toxic Exposure Guidelines
  • no data available
◆ Toxicity Data
  • dni-mouse-ipr 100 mg/kg
  • oral-rat TDLo 2620 mg/kg/78W-C:CAR
  • scu-rat TDLo 6 g/kg/27W-I: ETA, REP
  • oral-mouse TDLo 26 g/kg/78W-C:CAR
  • skin-mouse TDLo 5280 mg/kg/26W-I: ETA
  • oral-rat TD 36 g/kg/53W-I: ETA, TER
◆ Acute Risks
  • harmful if swallowed, inhaled, or absorbed through skin

- may cause irritation
- There is little information available on the short-term effects of 1,2-diphenylhydrazine.
◆ Chronic Risks
  - NCI Carcinogenesis Bioassay (feed): Clear Evidence: mouse, rat
  - NTP 7th Annual Report on Carcinogens, 1992: Anticipated to be a carcinogen
  - EPA Cancer Risk Level (1 in a million excess lifetime risk) 5.0E-06 mg/m^3
  - degenerative alterations in the liver of mice and rats
  - EPA Group B2: probable human carcinogen
  - may alter genetic material
  - intestinal hemorrhage in mice
  - hepatocellular carcinomas in rats and mice

HAZARD RISK

◆ incompatible with strong oxidizers, strong acids, acid chlorides, acid anhydrides
◆ forms carbon monoxide, carbon dioxide, or nitrogen oxide upon combustion
◆ toxic

EXPOSURE ROUTES

◆ primarily occupational
◆ ingestion: found in drinking water

REGULATORY STATUS/INFORMATION

◆ Clean Air Act
  - Title III: hazardous air pollutant
◆ Safe Drinking Water Act
  - SDWA priority list of drinking water contaminants
◆ Resource Conservation Recovery Act
  - Appendix 8: hazardous constituents. 40CFR261, also 40CFR261.11
  - Appendix 9: groundwater monitoring. 40CFR264
  - Appendix 7: health based limits for exclusion of waste-derived residues. 40CFR266
  - UX109: discarded commercial chemical products, off-specification species, container residues, and spill residues thereof. 40CFR261.33
◆ CERCLA and Superfund
  - designation of hazardous substances. 40CFR302.4
  - Appendix A: sequential CAS registry # list of CERCLA hazardous substances
  - Appendix B: radionuclides
  - chemicals and chemical categories to which this part applies (CAS # listing). 40CFR372.65
◆ Clean Water Act
  - total toxic organics (TTOS). 40CFR413.02
  - Appendix A: 126 priority pollutants. 40CFR423
◆ chemical regulated by the State of California

ADDITIONAL/OTHER COMMENTS
◆ First Aid
  • flush eyes or skin with copious amounts of water for at least 15 minutes
  • remove contaminated clothing
  • if inhaled, remove to fresh air
  • if not breathing, give artificial respiration
  • if breathing is difficult, give oxygen
  • if ingested, wash out mouth with water and call physician
◆ Spill or Leak
  • evacuate area and keep personnel upwind
  • wear self-contained breathing apparatus, rubber boots, and heavy rubber gloves
  • ventilate area and wash spill site after material pickup is complete
  • sweep up, place in a bag, and hold for waste disposal
SOURCES
◆ *Environmental Law Index to Chemicals*, 1996 edition
◆ *Health Effects Notebook for Hazardous Air Pollution*, U.S. Environmental Protection Agency, 1994
◆ *Material Safety Data Sheets (MSDS)*
◆ *Sax's Dangerous Properties of Industrial Materials*, 9th edition, 1996

EPICHLOROHYDRIN (jl)

CAS #: 106-89-8
MOLECULAR FORMULA: C_3H_5 OCl
FORMULA WEIGHT: 92.525
SYNONYMS: alpha-epichlorohydrin; alpha-epichorohydrin; chloro-2,3-epoxypropane; 1-chloro-2,3-epoxypropane; 3-chloro-1,2-epoxypropane; 1-chloro-2,3-epoxy propane; chloromethyl; (chloromethyl) ethylene oxide; (chloromethyl) oxirane; 2-(chloromethyl) oxirane; chloropropylene, chloropropylene oxide; 3-chloro-1,2-propylene oxide; ech; epichlorohydrin; epichlorohydrine, epichlorophydrin; 1,2-epoxy-3-chloropropane; 2,3-epoxypropyl chloride; gamma-chloropropylene oxide; glycerol epichlorohydrin; oxirane, chloromethyl; skekhg

USES
◆ epoxy resins
◆ glycerol
◆ solvent for natural and synthetic resins, gums, cellulose esters and ethers, paints, varnishes, nail enamels and lacquers, and cements for celluloid
◆ a stabilizer in chlorine-containing materials and an intermediate in the preparation of condensates with polyfunctional substances
◆ pharmaceuticals
◆ agricultural chemicals
◆ coatings
◆ ion-exchange resins

- glycidyl esters
- surface-active agents
- insecticides
- textile chemicals
- adhesives
- plasticizers
- ethymyl-ethylenic alcohol and fatty acid derivatives

PHYSICAL PROPERTIES
- colorless liquid with an irritating, chloroform-like odor
- melting point 57°C
- boiling point 117.9°C at 1 atm
- density 1180 kg/m³
- specific gravity 1.18 (water = 1)
- vapor density 3.2 (air = 1)
- surface tension 37.0 dynes/cm
- viscosity not available
- heat capacity not available
- heat of vaporization 97.9 cal/g
- aromatic odor; chloroform-like odor
- vapor pressure 13 mm Hg at 20°C
- negligible solubility in water
- Highly soluble in water. Concentrations of 1000 milligrams and more will mix with a liter of water.

CHEMICAL PROPERTIES
- flash point −33°C
- flammability limits 3.8% lower and 21% upper
- autoignition temperature 779°F
- heat of combustion −4524 cal/g
- enthalpy (heat of formation) not available
- will polymerize at temperatures >325°C
- Incompatible with
  - strong acids
  - strong bases
  - strong oxidizing agents
  - metallic salts
  - amines
  - aluminum
  - chlorine and chlorine compounds
  - most common metals

HEALTH RISK
- Toxic Exposure Guidelines
  - This substance is listed as an IARC probable human carcinogen (groups 2a and 2b).
  - threshold limit value (TLV/TWA) 10 mg/m³ (2 ppm)
  - short-term exposure limit (STEL) 20 mg/m³ (5 ppm)
  - permissible exposure limit (PEL) 19 mg/m³ (5 ppm)
- Toxicity Data
  - TCLo (inhalation-human) 20 ppm
  - LD_{50} (oral-rat) 90 mg/kg
  - LC_{50} (inhalation-rat) 250 ppm

- LDLo (skin-rat) 1000 mg/kg
- LD_{50} (oral-mus) 236 mg/kg
- LD_{50} (oral-rabbit) 345 mg/kg
- LD_{50} (skin-rabbit) 515 mg/kg
- LD_{50} (oral-guinea pig) 280 mg/kg
- TCLo (inhalation-human) 40 ppm
- LD_{50} (skin-mus) 250 mg/kg
- LD_{50} (intraperitoneal-rat) 133 mg/kg
- LD_{50} (subcutaneous-rat) 150 mg/kg
- LDLo (parenteral-mus) 250 mg/kg
- LD_{50} (intraperitoneal-rabbit) 118 mg/kg
- LD_{50} (intraperitoneal-guinea pig) 118 mg/kg
- LD_{50} (intravenous-rat) 154 mg/kg

◆ Acute Risks
  - Inhalation may cause inflammation of the lungs.
  - asthmatic bronchitis
  - central nervous system depression
  - Inhalation and ingestion are harmful and may be fatal.
  - Contact with skin or eyes may cause severe irritation or burns.
  - Liquid may cause permanent eye damage.
  - Ingestion may cause headache, nausea, vomiting, gastrointestinal irritation, unconsciousness, convulsions.

◆ Chronic Risks
  - Laboratory experiments have shown mutagenic effects.
  - target organs: liver, kidney

HAZARD RISK
◆ Unusual Fire & Explosion Hazards
  - Vapors may flow along surfaces to distant ignition sources and flash back.
  - Closed containers exposed to heat may explode.
  - Contact with strong oxidizers may cause fire.
◆ Toxic Gases Produced
  - hydrogen chloride
  - carbon monoxide
  - carbon dioxide
  - phosgene

EXPOSURE ROUTES
◆ ingestion
◆ inhalation
◆ absorption
◆ eye contact
◆ skin contact

REGULATORY STATUS/INFORMATION
◆ Clean Water Act Section 311 Hazardous Substances
◆ Safe Drinking Water Act Maximum Contaminant Level Goal (MCLG) Substances (40 CFR 141)
◆ Safe Drinking Water Act Statutory Contaminants
◆ Clean Air Act Section 111 Volatile Organic Compounds (40CFR60)

◆ Clean Air Act Section 112 Statutory Air Pollutants (42USC4212)
◆ Clean Air Act Section 112(r) Accidental Release Prevention Substances
◆ NIOSH Recommendation Substances
◆ RCRA Hazardous Substances (40CFR241)
◆ RCRA Land Disposal: Scheduled First Third Wastes
◆ CERCLA Hazardous Substances (40CFR302)
◆ SARA Title III Section 302 Extremely Hazardous Substances (40CFR355.50)
◆ SARA Title III Section 313 Toxic Chemicals (40CFR372)
◆ OSHA Air Contaminants (29CFR1910)
◆ NTP Annual Report on Carcinogens
◆ IARC Human Carcinogens
◆ DOT Hazardous Materials (40CFR171)
◆ DOT Hazardous Substances and Radionuclides (49CFR172.101)
◆ DOT Marine Pollutants
◆ Chemical regulated by the State of California
◆ Canadian Workplace Hazardous Materials Information System (WHMIS) Ingredient Disclosure List: CN2—Ingredient must be disclosed at a concentration of 0.1%
◆ Massachusetts Substance List: MA1CE—Carcinogen, extraordinarily hazardous substance
◆ New Jersey Right to Know Hazardous Substances List: special health hazard

ADDITIONAL/OTHER COMMENTS
◆ First Aid
  • call a physician
  • if swallowed, if conscious, give large amounts of water; induce vomiting
  • if inhaled, remove to fresh air
  • if not breathing, give artificial respiration
  • if breathing is difficult, give oxygen
  • in case of contact, immediately flush eyes or skin with plenty of water for at least 15 minutes while removing contaminated clothing and shoes
  • wash clothing before reuse
◆ Fire Fighting
  • use water spray, alcohol foam, dry chemical, or carbon dioxide
  • Firefighters should wear proper protective equipment and self-contained breathing apparatus with full facepiece operated in positive pressure mode.
  • move containers from fire area if it can be done without risk
  • use water to keep fire-exposed containers cool
◆ Personal Protection
  • Respiratory
    • respiratory protection required if airborne concentration exceeds TLV

240

- At concentrations above 2 ppm, self-contained breathing apparatus is advised.
- Eye/Skin
  - safety goggles and face shield
  - uniform
  - protective suit
  - Rubber gloves are recommended.
- ◆ Storage
  - bond and ground containers when transferring liquid
  - keep container tightly closed
  - store in a cool, dry, well-ventilated, flammable liquid storage area

SOURCES
- ◆ *Aldrich Chemical Company Catalog*, 1996–1997
- ◆ *MSDS*, Aldrich Chemical Company, Inc.
- ◆ *Suspect Chemicals Sourcebook*, 1996
- ◆ *NIOSH Pocket Guide to Chemical Hazards*, 1994
- ◆ *Merck Index*, 10th edition, 1983.

1,2-EPOXYBUTANE (bh)

CAS #: 106-88-7
MOLECULAR FORMULA: C_4H_8O
FORMULA WEIGHT: 72.12
SYNONYMS: 1-butene oxide, 1,2-butene oxide, butylene oxide, 1,2-butylene oxide, 1,2-butylene oxide-stabilzed (DOT), epoxybutane, ethyl ethylene oxide, ethylene oxide-ethyl-, ethyloxirane, NCI-C55527, UN3022 (DOT)

USES
- ◆ Used as a Component in the Manufacture of
  - butylene glycol and its derivatives
  - butanolamines
  - surface-active agents
  - gasoline additives
- ◆ Used as a stabilizer in chlorinated hydrocarbon solvents

PHYSICAL PROPERTIES
- watery, white liquid
- sweet and disagreeable odor
- boiling point 63°C (145°F) at 1 atm
- specific gravity 0.837
- vapor density 2.2 kg/m³
- vapor pressure 140 mm Hg at 20°C
- soluble in water
- miscible in organic solvents
- Odor threshold has not been determined.

CHEMICAL PROPERTIES
- extremely flammable liquid
- explosion limits 1.7% lower and 19% upper by volume in air
- autoignition temperature 369°C (698°F)
- flash point 17°C (1°F)

- may decompose on exposure to moisture
- corrosive

HEALTH RISK

◆ Toxic Exposure Guidelines
  - no data available
◆ Toxicity Data
  - skin-rabbit 500 mg/24H maximum lethal dosage (MLD)
  - eye-rabbit 100 mg/24H maximum oral dosage (MOD)
  - inhalation-rabbit TCLo 1000 ppm/7H (10–24D preg): reproductive (REP)
  - inhalation-rat TCLo 400 ppm/6H/5D/2Y-C:cardiovascular (CAR)
  - oral-rat LD_{50} 500 mg/kg
  - inhalation-rat LCLo 4000 ppm/4H
◆ Acute Risks
  - harmful or fatal if swallowed, inhaled, or absorbed through skin
  - irritation of eyes, nasal passages, throat, and lungs
  - Material is extremely destructive to tissue of the mucous membranes and upper respiratory tract.
  - Inhalation may cause spasms, inflammation, and edema of larynx and bronchi.
  - chemical pneumonitis and pulmonary edema
  - headache
  - nausea and vomiting
◆ Chronic Risks
  - atrophy and necrosis of the spleen and thymus, and renal necrosis in mice
  - may alter genetic material
  - Repeated dermal exposure may cause blistering and necrosis in humans.
  - EPA Cancer Risk: not classified

HAZARD RISK

◆ dangerous fire hazard
◆ incompatible with acids, bases, heat, moisture
◆ forms carbon monoxide and carbon dioxide upon combustion
◆ toxic
◆ mutagen

EXPOSURE ROUTES

◆ primarily occupational by inhalation or dermal contact

REGULATORY STATUS/INFORMATION

◆ Clean Air Act
  - Title III: hazardous air pollutant
◆ chemical regulated by the State of California

ADDITIONAL/OTHER COMMENTS

◆ First Aid
  - flush eyes or skin with copious amounts of water for at least 15 minutes
  - remove contaminated clothing
  - if inhaled, remove to fresh air

- if not breathing, give artificial respiration
- if breathing is difficult, give oxygen
- if ingested, wash out mouth with water and call physician
◆ Spill or Leak
  - evacuate area and keep personnel upwind
  - shut off all sources of ignition
  - wear self-contained breathing apparatus, rubber boots, and heavy rubber gloves
  - absorb on sand or vermiculite and place in closed container for disposal
  - ventilate area and wash spill site after material pickup is complete

SOURCES

◆ *Environmental Law Index to Chemicals*, 1996 edition
◆ *Health Effects Notebook for Hazardous Air Pollutants*, U.S. Environmental Protection Agency, 1994
◆ *Material Safety Data Sheets* (*MSDS*)
◆ *Sax's Dangerous Properties of Industrial Materials*, 9th edition, 1996

ETHYL ACRYLATE (sm)

CAS #: 140-88-5
MOLECULAR FORMULA: $C_5H_8O_2$
FORMULA WEIGHT: 100.12
SYNONYMS: acrylic acid ethyl ester, ethoxycarbonylethylene, ethyl propenoate, ethyl-2-propenoate, FEMA No. 2418, NCI-C50384, 2-propenoic acid, propenoic acid ethyl ester 1050, RCRA waste # U113

USES
◆ Used in the Production of Emulsion Based Polymers Used as
  - fabric finishes
  - dirt release agents
  - specialty plastics
  - latex paints
  - textiles
  - paper coatings
◆ monomer in the manufacture of water emulsion paint vehicles
◆ imparts flexibility to hard films

PHYSICAL PROPERITES
◆ colorless liquid
◆ melting point −71°C at 760 mm Hg
◆ boiling point 99.4°C at 760 mm Hg
◆ heat of vaporization 8.27 kcal/mol
◆ surface tension 0.025 dyne/cm at 20°C
◆ vapor pressure 29.5 mm at 20°C
◆ vapor density 3.45 (air = 1.00)
◆ specific gravity 0.924
◆ miscible with alcohol and ether
◆ slightly soluble in alcohol, chloroform, ether, and water

CHEMICAL PROPERTIES
◆ flash point 15°C (60°F)
◆ may polymerize on exposure to light
◆ autoignition temperature 382°C (721°F)

- slightly soluble in alcohol, chloroform, ether, and water
- heat of combustion 655.49 kcal/mol
- explosion limits 1.8% lower and 12.1% upper (by volume in air)
- readily polymerized

HEALTH RISK
- Toxic Exposure Guidelines
  - ACGIH TLV TWA 5 ppm (20 mg/m^3)
  - OSHA PEL TWA 5 ppm (20 mg/m^3)
- Toxicity Data
  - oral-human TDLo 130 ppm
  - inhalation-human LCLo 50 ppm
  - oral-rat LD$_{50}$ 760–1020 mg/kg
  - inhalation-rat LC$_{50}$ 1000–2000 ppm/4h
- Acute Risks
  - destructive of eyes, skin, and upper respiratory tract
  - shortness of breath
  - wheezing
  - inflammation and edema of the larynx and bronchi
  - nausea and vomiting
  - Inhalation may be fatal.
  - headaches
  - laryngitis
  - chemical pneumonitis and pulmonary edema
- Chronic Risks
  - known EPA carcinogen
  - difficulty in breathing
  - severe local irritation of gastroenteric tract
  - coughing
  - chest pains
  - lung, liver, and kidney damage

HAZARD RISK
- NFPA rating Health 2, Flammability 3, Reactivity 2, Special Hazards P
- highly flammable
- reacts violently with chlorosulfonic acid
- reacts vigorously with oxidizing materials

EXPOSURE ROUTES
- primarily by inhalation
- absorption through skin and eye
- migrates to food from packaging materials

REGULATORY STATUS/INFORMATION
- Clean Air Act, Title III: hazardous air pollutants
- Resource Recovery and Conservation Act, 40CFR261.33
  - U waste #: U113
- CERCLA and Superfund
  - 40CFR355 Appendix B
  - 40CFR372.65
  - Superfund Amendments and Reauthorization Act of 1986 (SARA) Title III Section 313: Toxic chemicals
  - Reportable quantity is 1000 lb.
- OSHA 29CFR1910.1000 Table Z1
- Toxic Substance Control Act (TSCA), 40CFR716.120
- Department of Transportation (DOT) Hazardous Substances and Radionuclides

ADDITIONAL/OTHER COMMENTS
◆ Symptoms
  • inhalation: irritates eyes, skin, and respiratory system
  • skin absorption: potential occupational carcinogen
◆ First Aid
  • wash eyes immediately with copious amounts of water
  • wash skin immediately with copious amounts of water
  • provide respiratory support
◆ fire fighting: use carbon dioxide, dry chemical powder, or appropriate foam

SOURCES
◆ *ACGIH Threshold Limit Values (TLVs) for 1996*
◆ *Aldrich Chemical Company Catalog*, 1996–1997
◆ *CRC Handbook of Chemistry and Physics*, 77th edition, 1996–1997
◆ *Environmental Contaminant Reference Databook*, volume 1, 1995
◆ *Hawley's Condensed Chemical Dictionary*, 12th edition, 1993
◆ *Health Effects Notebook for Hazardous Air Pollutants*, U.S. Environmental Protection Agency, December 1994
◆ *Material Safety Data Sheets* (*MSDS*)
◆ *Merck Index*, 12th edition, 1996
◆ *NIOSH Pocket Guide to Chemical Hazards*, June 1994
◆ *Suspect Chemicals Sourcebook*, 1996 edition

ETHYL BENZENE (sm)

CAS #: 100-41-4
MOLECULAR FORMULA: C_8H_{10}
FORMULA WEIGHT: 106.18
SYNONYMS: benzene ethyl, eb, ethyl-benzene, ethylbenzol, NCI-C56393, phenylethane, UN1175, NCI-C56393

USES
◆ Intermediate in Production of Styrene
  • production of synthetic rubber as a solvent or diluent
  • a component of automotive and aviation fuels
  • manufacture of cellulose acetate
◆ Solvent for
  • alkyd surface coatings
  • propylene oxide
  • chemical intermediate for diethylbenzene and acetophenone
  • unrecovered component of gasoline
  • ethyl anthraquinone
  • ethylbenzene sulfonic acids
  • alpha-methyl-benzene alcohol

PHYSICAL PROPERTIES
◆ colorless liquid
◆ boiling point 136.25°C at 760 mm Hg
◆ melting point −95.01°C at 760 mm Hg
◆ viscosity 0.64 cP at 25°C

- ◆ negligibly soluble in water
- ◆ miscible in alcohol and ether
- ◆ surface tension 31.50 dynes/cm
- ◆ density 0.867 g/mL at 20°C
- ◆ aromatic odor
- ◆ vapor density 3.66 (air = 1.00)
- ◆ soluble in ethyl alcohol, ethyl ether, water, benzene, carbon tetrachloride, and SO_2
- ◆ heat of vaporization 9301.3 g/cal/gmole

CHEMICAL PROPERTIES
- ◆ flash point 18°C (64°F)
- ◆ reacts strongly with oxidizing agents
- ◆ soluble in ethyl alcohol, ethyl ether, and water
- ◆ flammability limits 1.0% lower and 6.7% upper (in air by volume)
- ◆ heat of combustion −9877 cal/g (−413.5×10^5 J/kg)

HEALTH RISK
- ◆ Toxic Exposure Guidelines
  - • ACGIH TLV TWA 100 ppm (434 mg/m^3)
  - • OSHA PEL TWA 100 ppm (435 mg/m^3)
  - • NIOSH REL TWA 100 ppm (434 mg/m^3)
  - • IDLH 800 ppm
- ◆ Toxicity Data
  - • inhalation-human TCLo 100 ppm/8h Toxic effect: eye, central nervous system, pulmonary system
  - • oral-rat LD_{50} 3500 mg/kg
  - • intraperitoneal-mouse LD_{50} 2272 mg/kg
- ◆ Acute Risks
  - • irritation of skin, eyes, and upper respiratory tract
  - • vomiting
  - • nausea
  - • headache
- ◆ Chronic Risks
  - • central nervous system depression
  - • death in high exposures

HAZARD RISK
- ◆ NFPA rating Health 2, Flammability 3, Reactivity 0, Special Hazards none

EXPOSURE ROUTES
- ◆ primarily by inhalation
- ◆ occupational exposure in industries that use ethyl benzene
- ◆ Absorption through skin and eyes is possible.

REGULATORY STATUS/INFORMATION
- ◆ Clean Air Act Title III: hazardous air pollutants
- ◆ Safe Drinking Water Act
  - • maximum contaminant level goal (MCLG) for organic contaminants 40CFR141.50(b)
  - • variances and exemptions from maximum contaminant levels (MCLs) 40CFR141.62
  - • waste specific prohibitions: solvent wastes 40CFR148.10
- ◆ Resource Recovery and Conservation Act
  - • constituents for detection monitoring 40CFR258-Appendix 1

- list of hazardous inorganic and organic constituents 40CFR258-Appendix 1
- groundwater monitoring chemicals 40CFR264-Appendix 9
◆ CERCLA and Superfund
  - designation of hazardous substances 40CFR302.4
  - sequential CAS registry # list of CERCLA hazardous substances 40CFR302.4-Appendix A
  - chemicals and chemical categories (CAS # listing) 40CFR372.65
◆ Clean Water Act
  - designation of hazardous substances 40CFR116.4
  - list of hazardous substances 40CFR116.4-Table116.4A
  - list of hazardous substances by CAS # 40CFR116.4-Table116.4B
  - reportable quantities of hazardous substances designated pursuant to sec 311 of CWA list of hazardous substances 40CFR117.3-Table 117.3
  - toxic pollutants 40CFR401.15
  - total toxic organics 40CFR413.02
◆ Occupational Safety and Health Act
  - limits for air contaminants OSHA 29CFR1910.1000 Table Z1

ADDITIONAL/OTHER COMMENTS
◆ Symptoms
  - inhalation: irritates eyes, skin, mucous membranes
  - ingestion: headache, dermatitis, narcosis, coma
◆ First Aid
  - eye: wash immediately with large amounts of water
  - skin: flush contaminated skin with water promptly
  - inhalation: provide respiratory support
  - ingestion: get medical attention immediately

SOURCES
◆ *ACGIH Threshold Limit Values (TLVs) for 1996*
◆ *Aldrich Chemical Company Catalog*, 1996–1997
◆ *CRC Handbook of Chemistry and Physics*, 77th edition, 1996–1997
◆ *Environmental Contaminant Reference Databook*, volume 1, 1995
◆ *Environmental Law Index to Chemicals*, 1996 edition
◆ *Hawley's Condensed Chemical Dictionary*, 12th edition, 1993
◆ *Health Effects Notebook for Hazardous Air Pollutants*, U.S. Environmental Protection Agency, December 1994
◆ *Material Safety Data Sheets (MSDSs)*
◆ *Merck Index,* 12th edition, 1996
◆ *NIOSH Pocket Guide to Chemical Hazards*, June 1994

ETHYL CARBAMATE (sm)

CAS #: 51-79-6
MOLECULAR FORMULA: $C_3H_7NO_2$
FORMULA WEIGHT: 89.09
SYNONYMS: A11032, carbamic acid ethyl ester, estane 5703, ethylure-than, ethylurethane, leucethane, leucothane , NSC 746, pracabamin, pracarbamine, RCRA waste number U238, U-compound, urethan, urethane

USES
- ◆ Intermediate for
  - pharmaceuticals
  - fungicides
  - pesticides
  - organic synthesis
- ◆ biochemical research
- ◆ medicine (antineoplastic)
- ◆ solvent for various organic materials

PHYSICAL PROPERTIES
- ◆ melting point 49°C
- ◆ density 0.9862 g/mL
- ◆ vapor pressure 10 mm Hg at 10°C
- ◆ soluble in water, alcohol, ether, glycerol, chloroform
- ◆ boiling point 180°C at 760 mm Hg
- ◆ colorless crystals or white powder
- ◆ vapor density 3.07 (air = 1.00)

CHEMICAL PROPERTIES
- ◆ will react strongly with strong acids and bases and oxidizing agents
- ◆ flash point 92°C (198°F)

HEALTH RISK
- ◆ Toxicity Data
  - oral-rat LD_{50} 1809 mg/kg
  - intraperitoneal-rat LD_{50} 1500 mg/kg
- ◆ Acute Risks
  - irritation of skin, eye, mucous membranes, upper respiratory tract
  - depression
  - nausea
  - vomiting
- ◆ Chronic Risks
  - damage to immune system, liver, bone marrow, central nervous system
  - carcinogen
  - central nervous system depression

HAZARD RISK
- ◆ reacts with phosphorus pentachloride to form an explosive product
- ◆ when heated emits toxic fumes of NO_x

EXPOSURE ROUTES
- primarily by inhalation
- absorption through skin and eyes

REGULATORY STATUS/INFORMATION
- ◆ Clean Air Act Title III
  - • hazardous air pollutants 42USC7412
- ◆ Resource Conservation and Recovery Act
  - • hazardous constituents 40CFR26-Appendix 8
  - • discarded commercial chemical products, off-specification species, container residues, and spills thereof 40CFR261.33
  - • U waste #: 238
- ◆ CERCLA and Superfund
  - • designation of hazardous substances 40CFR302.4
  - • sequential CAS registry # list of CERCLA hazardous substances 40CFR302.4-Appendix A
- ◆ Occupational Safety and Health Act
  - • limits for air contaminants OSHA 29CFR1910.1000 Table Z1

ADDITIONAL/OTHER COMMENTS
- ◆ First Aid
  - • flush eyes with copious amounts of water
  - • wash skin with large amounts of water
  - • provide respiratory support
- ◆ fire fighting: use carbon dioxide, dry chemical, or appropriate foam

SOURCES
- ◆ *Aldrich Chemical Company Catalog*, 1996–1997
- ◆ *CRC Handbook of Chemistry and Physics*, 77th edition, 1996–1997
- ◆ *Environmental Law Index to Chemicals*, 1996 edition
- ◆ *Hawley's Condensed Chemical Dictionary*, 12th edition, 1993
- ◆ *Health Effects Notebook for Hazardous Air Pollutants*, U.S. Environmental Protection Agency, December 1994
- ◆ *Material Safety Data Sheets* (*MSDS*)
- ◆ *Merck Index*, 12th edition, 1996
- ◆ *NIOSH Pocket Guide to Chemical Hazards*, June 1994

ETHYL CHLORIDE (sm)

CAS #: 75-00-3
MOLECULAR FORMULA: C_2H_5Cl
FORMULA WEIGHT: 64.52
SYNONYMS: aethylis, aethylis chloridium, anodynon, chelen, chloridum, chloroethane, chloroethyl, chloryl, chloryl anesthetic, ether chloratus, ether hydrochloric, ether muriatic, hydrochloric ether, kelene, monochlorethane, muriatic ether, narcotile, NCI-C06224

USES
- ◆ Manufacture of Other Chemicals
  - • tetraethyl lead
  - • ethylcellulose
- ◆ Solvent for
  - • phosphorus
  - • fats

- resins
- sulfur

- oils
- waxes

◆ Other Uses
  - anesthetic
  - alkylating agent
  - analytical reagent

  - organic synthesis
  - refrigeration

PHYSICAL PROPERTIES

◆ colorless liquid
◆ ether-like odor
◆ boiling point 12.5°C at 760 mm Hg
◆ vapor pressure 1000 mm at 20°C
◆ specific gravity 0.891 (water = 1)

◆ gas at room temperature
◆ density 0.9214 g/mL
◆ melting point −138.7°C
◆ vapor density 2.22 (air = 1.00)
◆ slightly soluble in water
◆ miscible in alcohol and ether

CHEMICAL PROPERTIES

◆ volatile at room temperature
◆ flash point −50°C (−58°F)
◆ reacts with strong oxidizing agents and sodium, potassium, and their alloys
◆ stable and noncorrosive when dry

◆ flammability limits 3.6% lower and 14.8% upper (by volume in air)
◆ autoignition temperature 518°C (966°F)
◆ hydrolyzes in the presence of water or alkalies

HEALTH RISK

◆ Toxic Exposure Guidelines
  - ACGIH TLV TWA 1000 ppm (2640 mg/m^3)
  - OSHA PEL TWA 1000 ppm (2600 mg/m^3)
  - NIOSH REL TWA
◆ Toxicity Data
  - inhalation-rat LC$_{50}$ 146 g/m^3
◆ Acute Risks
  - eye irritation
  - skin irritation
  - nausea
  - vomiting

  - irritating to mucous membranes and upper respiratory tract
  - headache
  - abdominal cramps

◆ Chronic Risks
  - kidney and liver damage
  - cardiac arrhythmias
  - cardiac arrest

HAZARD RISK

◆ NFPA rating Health 2, Flammability 4, Reactivity 0, Special Hazards none
◆ extremely volatile
◆ severe flammable and explosion risk
◆ reacts strongly with oxidizing materials
◆ reacts with water or steam to produce toxic fumes
◆ incompatible with potassium
◆ When heated to decomposition it emits toxic fumes of phosgene and Cl$^-$.

EXPOSURE ROUTES
◆ primarily through inhalation
◆ absorption through skin

REGULATORY STATUS/INFORMATION
◆ Clean Air Act, Title III: hazardous air pollutants
◆ Resource Conservation and Recovery Act
  • list of hazardous inorganic and organic constituents 40CFR258-Appendix 2
◆ CERCLA and Superfund
  • designation of hazardous substances 40CFR302.4
  • sequential CAS registry # list of CERCLA hazardous substances 40CFR302.4-Appendix A
◆ Occupational Safety and Health Act
  • limits for air contaminants OSHA 29CFR1910.1000 Table Z1

ADDITIONAL/OTHER COMMENTS
◆ Symptoms
  • inhalation: irritates eyes, skin, and respiratory system
◆ First Aid
  • wash eyes immediately with large amounts of water
  • wash contaminated area with water
  • provide respiratory support
◆ fire fighting: use carbon dioxide

SOURCES
◆ *Aldrich Chemical Company Catalog*, 1996–1997
◆ *CRC Handbook of Chemistry and Physics*, 77th edition, 1996–1997
◆ *Environmental Law Index to Chemicals*, 1996 edition
◆ *Hawley's Condensed Chemical Dictionary*, 12th edition, 1993
◆ *Health Effects Notebook for Hazardous Air Pollutants*, U.S. Environmental Protection Agency, December 1994
◆ *Material Safety Data Sheets (MSDS)*
◆ *Merck Index*, 12th edition, 1996
◆ *NIOSH Pocket Guide to Chemical Hazards*, June 1994

ETHYLENE DIBROMIDE (rp)

CAS #: 106-93-4
MOLECULAR FORMULA: $C_2H_4Br_2$
FORMULA WEIGHT: 187.88
SYNONYMS: a,a-dibroom, Aethylenbromid (German), ai3-15349, alpha,beta-dibromoethane, bromofume, bromuro-di-etile (Italian), 1,2-dibromoetano (Italian), dibromoethane, 1,2-dibromoethane, dibromure-d'ethylene (French), 1,2-dibroomethaan (Dutch), dowfume w 85, dwubromoetan (Polish), edb, edb-85, ent-15349, ethylene bromide, glycol-bromide, glycol-dibromide, iscobrome-d, nci-coo522, nefis, nephis, pestmaster, sanhyuum, soilfume, solibrom, sym-dibromoethane, unifume

USES
◆ Manufacture of Other Chemicals

- dyes
- resins
- Grignard reagents
- vinyle bromide
- pharmaceuticals
- organic synthesis
- vinyl bromide
- ethylene

◆ Used as a Component or in the Manufacturing Process
  - fumigant
  - gums
  - waxes
  - insecticide
  - nematicide
  - scavenger for lead in gasoline
  - waterproofing
  - antiknock gasolines

PHYSICAL PROPERTIES
◆ colorless, heavy liquid
◆ mildly sweet odor
◆ density 2.172 g/mL at 25°C
◆ viscosity 1.727 cP at 20°C
◆ boiling point 10.06°C at 706 mm Hg
◆ freezing point 9.0°C
◆ melting point 9.8°C
◆ heat of vaporization 82.1 Btu/lb = 45.6 cal/g
◆ octanol/water partition coefficient 135
◆ vapor pressure 11.7 mm at 25°C
◆ refractive index 1.5337 at 25°C
◆ bulk density 18.1 lb/gal
◆ vapor pressure 17.4 mm Hg at 30°C

CHEMICAL PROPERTIES
◆ soluble in water
◆ surface tension 38.75 dynes/cm = 0.03875 newtons/m at 20°C
◆ flammability limit none
◆ critical temperature 373°C
◆ critical pressure 52.7 atm
◆ incompatibilities: alkali metals, oxidizing agents, aluminum, magnesium
◆ heat of combustion 6647 J/g (1289 cal/g)
◆ soluble in water

HEALTH RISK
◆ Toxic Exposure Guidelines
  - MSHA TWA
  - 20 ppm (145 mg/m^3)
  - NIOSH TWA 0.045 ppm
  - OSHA PEL 25 ppm (190 mg/m^3)
  - LD_{50} (man) 65 mg/kg
  - NIOSH CL 1.0 mg/m^3
  - NIOSH REL 0.4 mg/m^3

- EPA Cancer Risk Level 5×10^6 mg/m^3
- Inhalation unit risk estimate 2.2×10^{-4} µg/m^3
◆ Acute Risks
  - blistering of skin
  - allergic reactions
  - shortness of breath
  - headache
  - nausea
  - vomiting
  - gastrointestinal disturbances
  - wheezing
  - burning sensation
  - skin irritation
  - laryngitis
◆ Chronic Risks
  - carcinogen
  - genetic alteration
  - reproductive disorders
  - target organs: liver, kidney, lungs, eyes

HAZARD RISK
◆ Decomposition emits toxic fumes of Br_2.

EXPOSURE ROUTES
◆ skin: by contact with contaminated soil
◆ inhalation of ambient air
◆ ingestion of drinking water

REGULATORY STATUS/INFORMATION
◆ Clean Air Act
  - 40CFR261.33
  - 40CFR116.4
  - section 112(g)
◆ Resource Conservation and Recovery Act
  - 47 FR 9352
◆ Federal Insecticide, Fungicide, and Rodenticide Act
  - FIFRA 180.3(3)
◆ CERCLA and Superfund
  - superfund, 40CFR302.4

ADDITIONAL/OTHER COMMENTS
◆ Fire Fighting
  - use dry chemical, carbon dioxide, or foam extinguisher and water to keep fire-exposed containers cool
  - emits toxic fumes under fire conditions
◆ Personal Protection
  - wear a face shield (8 in. minimum)
  - avoid prolonged or repeated exposure
  - wear appropriate NIOSH/MSHA approved respirator
  - wear gloves, safety goggles, and other protective clothing
  - prevent contact with skin or eyes
◆ Spill or Leak
  - evacuate area
  - shut off all sources of ignition

- absorb on sand or vermiculite
- ventilate area and wash spill site after material pickup
- storage: do not use handling equipment or containers composed of magnesium, aluminum or their alloys
◆ storage: do not use handling equipment or containers composed of magnesium or aluminum or their alloys

SOURCES

◆ *Material Safety Data Sheets (MSDS)*
◆ *Sax's Dangerous Properties of Industrial Materials*, 9th edition, 1996
◆ *Merck Index*, 12th edition, 1996
◆ *Fire Protection Guide to Hazardous Materials*, 11th edition, 1996
◆ gopher://ecosys.drdr.Virginia.edu: 70/00/library/gen/toxics
◆ *Health Effects Notebook for Hazardous Air Pollutants*, U.S. Environmental Protection Agency, December 1994

ETHYLENE DICHLORIDE (jm)

CAS #: 107-06-2
MOLECULAR FORMULA: $C_2H_4Cl_2$
FORMULA WEIGHT: 98.96
SYNONYMS: alpha, beta-dichlorethane, 1,2-dichlorethane, 1,2-dichloro, ethane, ethane-dichloride, ethyleendichloride, ethylene chloride, freon-150, glycol dichloride, sym-dichloroethane

USES

◆ Manufacture of Other Chemicals
- acetyl cellulose
- varnish and finish removers
- organic compounds
- pharmaceutical products
- resins
◆ Used as a Component or in the Manufacturing process
- paint
- soaps and scouring compounds
- leather cleaning
- degreasing agent
- rubber cement
- acrylic adhesives
- photography

PHYSICAL PROPERTIES

◆ clear, odorless, oily liquid
◆ pleasant odor and sweet taste
◆ melting point −35°C
◆ boiling point 83°C
◆ vapor pressure 100 mm at 29.4°C
◆ specific gravity 1.256
◆ heat of vaporization 76.4 cal/g
◆ odor threshold between 6 and 40 ppm
◆ solubility 0.869 g/100 mL water at 20°C

- ◆ miscible with chloroform
- ◆ soluble in ordinary organic solvents
- ◆ density 1.2351 g/mL at 20°C
- ◆ vapor density 3.4

CHEMICAL PROPERTIES
- ◆ stable in the presence of alkali, acids
- ◆ flash point 15°C (60°F)
- ◆ flammability limits 6.2% lower and 15.6% upper by volume in air
- ◆ autoignition temperature 413°C (775°F)
- ◆ corrodes iron and other metals at elevated temperatures when in contact with water
- ◆ combustible

HEALTH RISK
- ◆ Toxic Exposure Guidelines
  - OSHA PEL TWA 1 ppm
  - OSHA PEL STEL 2 ppm
  - ACGIH TLV TWA 10 ppm (40 mg/m^3)
  - NIOSH REL TWA 1 ppm
  - NIOSH REL CL 2 ppm/15M
- ◆ Toxicity Data
  - oral-rabbit LD_{50} 860 mg/kg
  - inhalation rabbit LCLo 3000 ppm/7H
  - skin-rabbit 500 mg/24H
  - eye-rabbit 500 mg/24H
  - oral-human LDLo 286 mg/kg
  - oral-human TDLo 428 mg/kg
  - inhalation-human TCLo 4000 ppm/H
  - oral-rat LD_{50} 670 mg/kg
  - oral-rat TDLo 5286 mg/kg
  - inhalation-rat LCLo 300 ppm/7H
  - inhalation-rat LC_{50} 1000 ppm/7H
  - oral-dog LD_{50} 5700 mg/kg
- ◆ Acute Risks
  - harmful if swallowed, inhaled, or absorbed through skin
  - irritation of skin, eyes, mucous membranes, and upper respiratory tract
  - causes dermatitis
  - Prolonged exposure can cause nausea, headache, and vomiting.
  - damage to liver and kidney
- ◆ Chronic Risks
  - carcinogen
  - may alter genetic material
  - target organs: liver and kidneys
  - Chronic poisoning causes loss of appetite, nausea, and vomiting.
  - tremors
  - low blood sugar levels
  - highly toxic with chronic exposure

HAZARD RISK
- explosions when mixed with nitrogen tetroxide, dimethyl amino propyl amine, or liquid ammonia
- vigorous reaction when mixed with propylene dichloride, orthodichlorobenzene, and aluminum
- incompatible with strong oxidizers, strong caustics, and active metals such as aluminum, powdered magnesium, sodium, or potassium
- emits toxic vapors when heated to decomposition
- dangerous fire hazard if exposed to heat, flame, or oxidizers

EXPOSURE ROUTES
- primarily by inhalation
- ingestion
- absorption through skin and eyes

REGULATORY STATUS/INFORMATION
- Toxic Substance Control Act (TSCA) Section 4e Interagency Testing Committee Priority List. 40CFR790, 40CFR796, 40CFR797, and 40CFR798
- Clean Water Act
  - section 304 Water Quality Criteria Substances
  - section 307 Priority Pollutant. 40CFR401.15 and 40CFR423.17
  - section 311 Hazardous Substances. 40CFR116.4
- Safe Drinking Water Act
  - one of 83 contaminants required to be regulated under the 1986 act
- Clean Air Act
  - section 111 Volatile Organic Compound. 40CFR60.489
  - section 112 (1990 amendments) Hazardous Air Pollutant. 42U.S.C.7412
- NIOSH Recommendation Substances
- Resource Conservation and Recovery Act
  - list of hazardous substances. 40CFR261
  - groundwater monitoring list. 40CFR264
  - land disposal prohibition: halogenated compounds. 40CFR268
  - land disposal prohibition: scheduled first third wastes. 40CFR268.10
- CERCLA and Superfund
  - designation of hazardous substances. 40CFR302.4
- OSHA 29CFR1910.1000 Table Z1 and Table Z2
  - ACGIH threshold limit value characteristics
  - IARC human carcinogens (groups 1, 2A, 2B)
  - National Toxicology Program Testing Program Substances
  - Department of Transportation
  - hazard class/division: 3
  - labels: flammable liquid, poison
  - hazardous materials list. 40CFR171
  - hazardous substances list. 49CFR172.101
  - marine pollutants. 49CFR172.101 (appendix B)

ADDITIONAL/OTHER COMMENTS
- ◆ Symptoms
  - inhalation: irritates eyes and skin
  - skin absorption: headache, nausea, respiratory system
- ◆ First Aid
  - wash eyes immediately with large amounts of water
  - if inhaled, remove to fresh air
  - provide respiratory support
  - fire fighting: use carbon dioxide, dry chemical powder, or foam
  - personal protection: wear appropriate NIOSH/MSHA approved respirator, chemical-resistant gloves, safety goggles, and other protective clothing
  - storage: keep tightly closed away from heat, sparks, and open flame; preferably in a cool, dry place

SOURCES
- ◆ *Environmental Contaminant Reference Databook*, volume 1, 1995
- ◆ *Fire Protection Guide to Hazardous Materials*, 11th edition, 1994
- ◆ *Material Safety Data Sheets* (*MSDS*)
- ◆ *Merck Index*, 12th edition, 1996
- ◆ *Sax's Dangerous Properties of Industrial Materials*, 9th edition, volume 2, 1996
- ◆ *Suspect Chemicals Sourcebook*, 1996

ETHYLENE GLYCOL (jm)

CAS #: 107-21-1
MOLECULAR FORMULA: $C_2H_6O_2$
FORMULA WEIGHT: 62.08
SYNONYMS: athylenglycol, 1,2-dihydroxyethane, ethane-1,2-Diol, 1,2-ethendiol, ethylene alcohol, ethylene dihydrate, glycol alcohol, lutrol-9, macrogol 400, M.E.G., monoethylene glycol, norkool, ramp, tescol, ucar 17

USES
- ◆ Manufacture of Other Chemicals
  - organic compounds such as esters and ethers
  - cosmetics
  - lacquers
  - alkyd resins
- ◆ Used as a Component or in the Manufacturing Process
  - paint
  - coolant and antifreeze
  - low-pressure laminates
  - brake fluids
  - printing inks
  - adhesives
  - wood stains
  - leather dyeing

- textile processing
- tobacco
- foam stabilizer

PHYSICAL PROPERTIES
- ◆ clear, colorless, syrupy liquid
- ◆ slightly viscous
- ◆ melting point $-13°C$
- ◆ boiling point 197.6°C at 760 mm Hg
- ◆ vapor pressure 0.06 mm Hg at 20°C
- ◆ density 1.1135 g/mL at 20°C
- ◆ specific gravity 1.113
- ◆ heat of vaporization 191 cal/g; 344 Btu/lb at 760 mm Hg
- ◆ odor threshold 3 ppm at 20°C
- ◆ surface tension 48.4 dynes/cm at 20°C
- ◆ miscible with water, lower aliphatic alcohols, glycerol, acetic acid, acetone and similar ketones, aldehydes, and pyridine
- ◆ slightly soluble in ether
- ◆ viscosity 26 cP at 15°C, 21 cP at 20°C, 17.3 cP at 25°C

CHEMICAL PROPERTIES
- ◆ heat of combustion -7259 Btu/lb $(-4033$ cal/g$)$
- ◆ reacts vigorously with oxidants
- ◆ flammability limits 3.2% by volume in air lower and 15.3% upper
- ◆ autoignition temperature 398°C (748°F)
- ◆ combustible

HEALTH RISK
- ◆ Toxic Exposure Guidelines
  - OSHA PEL 125 mg/m^3
  - ACGIH 125 mg/m^3
  - TLV 125 mg/m^3
  - DFG MAK 10 ppm
  - MSHA standard 10 mg/m^3
- ◆ Toxicity Data
  - skin-rabbit LD_{50} 9530 μL/kg
  - oral-human LDLo 786 mg/kg
  - oral-human LDLo 398 mg/kg
  - oral-rat LD_{50} 4700 mg/kg
  - inhalation-rat LC_{50} 10,876 mg/kg
  - oral-dog LD_{50} 5500 mg/kg
  - oral-mouse LD_{50} 5500 mg/kg
  - intraperitoneal-mouse LD_{50} 5614 mg/kg
  - oral-cat LD_{50} 1650 mg/kg
- ◆ Acute Risks
  - harmful if swallowed, inhaled, or absorbed through skin
  - irritation of eyes, mucous membranes, and upper respiratory tract
  - skin irritation
  - nervous system disturbances
  - Prolonged exposure can cause nausea, headache, and vomiting.

- damage to liver and kidney
◆ Chronic Risks
  - reproduction disorders
  - target organs: central nervous system, liver, and kidneys

HAZARD RISK
◆ moderate explosions when exposed to flame
◆ violent reaction when mixed with chlorosulfonic acid, H_2SO_4, $HClO_4$, and P_2S_5
◆ incompatible with strong oxidizing agents and strong bases
◆ hazardous combustion products: carbon monoxide and carbon dioxide
◆ ignites on contact with chromium trioxide, potassium permanganate and sodium peroxide

EXPOSURE ROUTES
◆ inhalation
◆ ingestion
◆ absorption through skin and eyes
◆ direct contact
◆ drinking contaminated water
◆ subcutaneous, intravenous and intramuscular routes

REGULATORY STATUS/INFORMATION
◆ Clean Air Act
  - section 111 Volatile Organic Compound. 40CFR60.489
◆ CERCLA and Superfund
  - designation of hazardous substances. 40CFR302.4
  - list of chemicals. 40CFR372.65
◆ OSHA 29CFR1910.1000 Table Z1

ADDITIONAL/OTHER COMMENTS
◆ Symptoms
  - inhalation: irritates eyes and skin
  - ingestion: stupor coma which may lead to fatal kidney injury
◆ First Aid
  - wash eyes immediately with large amounts of water
  - if inhaled, remove to fresh air
  - provide respiratory support
  - if swallowed, wash out mouth provided person is conscious
◆ fire fighting: use water spray, carbon dioxide, dry chemical powder, or appropriate foam
◆ personal protection: wear appropriate NIOSH/MSHA approved respirator, chemical-resistant gloves, safety goggles, and other protective clothing
◆ storage: keep tightly closed; store in a cool, dry place

SOURCES
◆ *Health Effects Notebook for Hazardous Air Pollutants*, U.S. Environmental Protection Agency, December 1994
◆ *Environmental Contaminant Reference Databook*, volume 1, 1995
◆ *Fire Protection Guide to Hazardous Materials*, 11th edition, 1994

- *Material Safety Data Sheets(MSDS)*
- *Merck Index*, 12th edition, 1996
- *Sax's Dangerous Properties of Industrial Materials*, 9th edition, volume 2, 1996
- *Suspect Chemicals Sourcebook*, 1996

ETHYLENE IMINE (rp)

CAS #: 151-56-4
MOLECULAR FORMULA: C_2H_5N
FORMULA WEIGHT: 43.08
SYNONYMS: Aethylenimin (German), aminoethylene, azacyclopropane, aziran, azirane, aziridin (German), aziridine, 1h-azirine, dihydro, dihydroazirene, dihydroazirine, dimethyleneimine, dimethylenimine, el, ent-50324, ethyleenimine (Dutch), ethylenimine, 7-ethylimine, etilenimina (Italian), tl-337, vinylamine
USES
- Manufacture of Other Chemicals
  - taurine
  - triethylmelamine
  - ethyleneimine
  - resins
  - alkylazirano
  - fuel oil
- Used as a Component or in the Manufacturing Process
  - polymerization
  - adhesives
  - binders
  - lubricants
  - coating
  - cosmetics
  - photographic chemicals
  - flameproofing
  - pharmaceuticals
  - insect repellents
  - varnishes
  - ion exchange resins
  - protective coating
PHYSICAL PROPERTIES
- colorless fluid
- ammoniacal odor
- melting point $-71.5°C$
- freezing point $-78°C$
- refractive index 1.4123 (25°C)
- boiling point 56–57°C
- density 0.832 g/cm³ at 20°C
- vapor pressure 160 mm Hg at 20°C
- heat of vaporization 333 Btu/lb (-8850 cal/g)
CHEMICAL PROPERTIES
- very corrosive

- ◆ flash point −13°C
- ◆ autoignition temperature 612°F
- ◆ flammability limits: 3.6% lower and 46% upper (by volume in air)
- ◆ heat of combustion −15.930 Btu/lb (−8850 cal/g)
- ◆ surface tension 34.5 dynes/cm = 0.0345 N/m at 20°C
- ◆ very soluble in water and ethanol
- ◆ reacts violently with acids, chlorinating agents, and silver

HEALTH RISK

- ◆ Toxic Exposure Guidelines
  - OSHA TLV (TWA) 0.5 ppm (0.88 mg/m³)
  - OSHA PEL (TWA) 1 mg/m³ (skin)
  - OSHA PEL 0.5 ppm
  - ACGIH TLV (TWA) 0.5 ppm (skin)
  - MSHA 1 mg/m³
  - LC_{50} (cats) 250 ppm
  - odor threshold 1.5 ppm
- ◆ Acute Risks
  - irritation of skin and eyes
  - nausea
  - vomiting
  - headache
  - dizziness
  - irritation of mouth, nose, and upper respiratory tract
  - nasal secretion
  - death in high exposure
- ◆ Chronic Risks
  - mutation of the trachea and bronchi
  - edema of lungs and secondary bronchial pneumonia
  - carcinogenic
  - deep necrosis
  - severe, deep burns
  - genetic damage
  - neoplastigenic
  - pulmonary edema

HAZARD RISK

- ◆ very dangerous fire and explosion hazard when exposed to heat, flame, or oxidizers
- ◆ Reacts Violently with Some Compounds
  - acids
  - aluminum chloride + substituted anilines
  - acetic acid
  - acetic anhydride
  - acrolein
  - acrylic acid
  - allyl chloride
  - CS_2
  - Cl_2
  - chlorosulfonic acid
  - epichlorohydrin
  - glyoxal
  - HCl
  - HF
  - HNO_3
  - oleum
  - β-propiolactone
  - Ag
  - NaOCl
  - H_2SO_4
  - vinyl acetate

◆ reacts with chlorinating agents to form explosive 1-chloro
   aziridine
◆ reacts with silver or its alloys to form explosive silver derivatives
◆ Heat and/or the presence of catalytically active metals or
   chloride ions can cause a violent exothermic reaction.
◆ Decomposition emits acrid smoke and irritating fumes.

EXPOSURE ROUTES
◆ primarily by inhalation
◆ Absorption through skin and eyes is possible.
◆ ingestion
◆ intraperitoneal route

REGULATORY STATUS/INFORMATION
◆ Clean Air Act
   • nonthreshold category
   • section 112(g)
◆ Resource Conservation and Recovery Act
   • waste number P054
   • 40CFR264
◆ CERCLA and Superfund
   • 40CFR355
   • 40CFR372.65
   • 1/ED 340 per (mg/kg)/d
   • PO54

ADDITIONAL/OTHER COMMENTS
◆ Fire Fighting
   • use dry chemical, carbon dioxide, or foam extinguisher and
     water to keep fire-exposed containers cool
   • emits toxic fumes under fire conditions
◆ Personal Protection
   • wear a face shield (8 in. minimum)
   • avoid prolonged or repeated exposure
   • wear appropriate NIOSH/MSHA approved respirator
   • wear gloves, safety goggles, and other protective clothing
   • prevent contact with skin or eyes
◆ Spill or Leak
   • evacuate area
   • shut off all sources of ignition
   • absorb on sand or vermiculite
   • ventilate area and wash spill site after material pickup
◆ storage: do not use handling equipment or containers composed of magnesium or aluminum or their alloys

SOURCES
◆ *Material Safety Data Sheet (MSDS)*
◆ *Sax's Dangerous Properties of Industrial Materials*, 9th edi-
   tion, 1996
◆ *Merck Index*, 12th edition, 1996
◆ *Fire Protection Guide to Hazardous Materials*, 11th edition,
   1996
◆ gopher://ecosys.drdr.Virginia.edu: 70/00/library/gen/toxics

ETHYLENE OXIDE (jm)

CAS #: 75-21-8
MOLECULAR FORMULA: C_2H_4O
FORMULA WEIGHT: 44.06
SYNONYMS: aethylenoxid, amprolene, anprolene, dihydrooxirene, dimethylene oxide, E.O., 1,2-epoxyaethan, epoxyethane, 1,2-epoxyethane, ethene oxide, ethyleenoxide, ethylene, eto, merpol, oxane, oxidoethane, oxirane, oxyfume, RCRA waste number U115, T-gas

USES
  ◆ Manufacture of Other Chemicals
    • acrylonitrate and nonionic surfactants
    • organic compounds
  ◆ Used as a Component or in the Manufacturing Process
    • fumigant for foodstuffs and textiles
    • used to sterilize surgical instruments
    • agricultural fungicide

PHYSICAL PROPERTIES
  ◆ colorless gas at room temperature
  ◆ melting point $-11.3°C$
  ◆ boiling point $10.7°C$
  ◆ vapor pressure 1093 mm Hg at 20°C
  ◆ density 0.8711 g/mL at 20°C
  ◆ miscible in water and alcohols
  ◆ very soluble in ether
  ◆ very flammable
  ◆ vapor density 1.52
  ◆ specific gravity 0.882
  ◆ odor threshold 430 ppm

CHEMICAL PROPERTIES
  ◆ stable
  ◆ flash point $-4°F$
  ◆ will polymerize
  ◆ autoignition temperature 804°F
  ◆ will react violently with acids, alcohols, alkali metals, ammonia, bases, oxidizing agents, chemically active metals and their salts, water

HEALTH RISK
  ◆ Toxic Exposure Guidelines
    • OSHA PEL TWA 1 ppm cancer hazard
    • ACGIH TLV TWA 1 ppm suspected human carcinogen
    • NIOSH REL TWA 0.1 ppm
    • NIOSH REL CL 5 ppm/10 M/D
    • NIOSH PEL < 0.2 mg/m³
  ◆ Toxicity Data
    • oral-guinea pig LD_{50} 270 mg/kg
    • inhalation-guinea pig LC_{50} 1500 mg/m³/4H

- inhalation-mouse LC_{50} 836 ppm/ 4H
- intraperitoneal-mouse LD_{50} 175 mg/kg
- oral-rat LD_{50} 72 mg/kg
- inhalation-rat LC_{50} 800 ppm/4 H
- inhalation-dog LC_{50} 960 ppm/4H
◆ Acute Risks
- harmful if swallowed, inhaled, or absorbed through skin
- causes severe irritation
- may cause allergic skin reaction
- Symptoms of exposure may include burning sensation, coughing, wheezing, laryngitis, shortness of breath, headache, nausea and vomiting.
- High concentrations are extremely destructive to tissues of the mucous membranes and upper respiratory tract, eyes, and skin.
◆ Chronic Risks
- carcinogen
- may alter genetic material
- may cause reproductive disorders
- target organs: nerves and lungs

HAZARD RISK
◆ severe explosion when exposed to flame
◆ highly flammable liquid or gas
◆ incompatible with bases, alcohols, air, copper, iron chlorides, iron oxides, potassium, tin chlorides, alkane thiols, bromoethane
◆ emits acrid smoke and irritating fumes when heated to decomposition
◆ explosive reaction with glycerol at 200°C

EXPOSURE ROUTES
◆ inhalation
◆ ingestion
◆ absorption through skin and eyes

REGULATORY STATUS/INFORMATION
◆ Toxic Substance Control Act
- listed in Chemical Hazard Information Profile Substances
◆ Clean Air Act
- section 111 Volatile Organic Compound. 40CFR60.489
- section 112 (1990 Amendments) Statutory Air Pollutants. 42U.S.C.7412
- section 112 High Risk Pollutants. 40CFR63
- section 112r one of 77 listed in Accidental Release Prevention Substances. 40CFR68
◆ NIOSH Recommendation Substances
◆ Resource Conservation and Recovery Act
- list of hazardous substances. 40CFR261
- land disposal prohibition: scheduled first third wastes. 40CFR268.10
◆ CERCLA and Superfund
- designation of hazardous substances. 40CFR302.4

- listed among extremely hazardous substances. 40CFRCh.1, Part 355.50
- listed among toxic chemicals. 40CFRCh.1, Part 372
◆ OSHA 29CFR1910.1000 Table Z1
  - listed in specifically regulated substances. 29CFR1910.1047
  - listed as National Toxicity Program Carcinogen
  - ACGIH threshold limit value characteristics
  - IARC human carcinogens (groups 1, 2A, 2B)
  - listed in highly hazardous chemicals. 29CFR1910
◆ National Toxicology Program Testing Program Substances
◆ Department of Transportation
  - hazard class/division: 2.3
  - labels: flammable gas, poison gas
  - hazardous substances list. 49CFR172.101

ADDITIONAL/OTHER COMMENTS
◆ First Aid
  - wash eyes immediately with large amounts of water
  - if inhaled, remove to fresh air
  - provide respiratory support
  - if swallowed, wash out mouth with water provided person is conscious
◆ fire fighting: use water spray
◆ personal protection: wear appropriate NIOSH/MSHA approved respirator, chemical-resistant gloves, safety goggles, and other protective clothing
◆ storage: keep tightly closed away from heat, sparks, and open flame; preferably in a cool, dry place

SOURCES
◆ *Agency for Toxic Substances and Disease Registry Public Health Statement* (http://atsdr1.atsdr.cdc.gov:8080), December 1990
◆ *Health Effects Notebook for Hazardous Air Pollutants*, U.S. Environmental Protection Agency, December 1994
◆ *Fire Protection Guide to Hazardous Materials*, 11th edition, 1994
◆ *Material Safety Data Sheets* (*MSDS*)
◆ *Merck Index*, 12th edition, 1996
◆ *Sax's Dangerous Properties of Industrial Materials*, 9th edition, volume 2, 1996
◆ *Suspect Chemicals Sourcebook*, 1996

ETHYLENE THIOUREA (jm)

CAS #: 96-45-7
MOLECULAR FORMULA: $C_3H_6N_2S$
FORMULA WEIGHT: 102.17
SYNONYMS: 4,5-dihydroimadazalene, 1,3-ethylenethiourea, 1,3-ethylene-2-thiourea, ETU, 2-mercaptoimidazoline, 2-mercaptol, imidazo-

lene-2-thiol, NA-22, NA-22-D, sodium-22 neoprene accelerator, 2-thiol-dihydroglyoxaline, urea, warecure C

USES
- ◆ Manufacture of Other Chemicals
  - dyes
  - pharmaceuticals
  - synthetic resins
- ◆ Used as a Component or in the Manufacturing Process
  - electroplating baths
  - insecticides
  - fungicides
  - vulcanization indicators
  - accelerator for neoprene rubbers

PHYSICAL PROPERTIES
- ◆ white crystals, needles, or prisms from pentanol
- ◆ melting point 197–200°C
- ◆ solubility 9 g/100 mL at 30°C
- ◆ faint amine odor
- ◆ moderately soluble in methanol, ethanol, ethylene glycol, pyridene
- ◆ insoluble in acetone, ether, chloroform, benzene, ligroin
- ◆ slightly soluble in acetic acid at room temperature

CHEMICAL PROPERTIES
- ◆ flash point 486°F
- ◆ combustible solid
- ◆ incompatible with acrolein
- ◆ EPA Group B2: probable human carcinogen

HEALTH RISK
- ◆ Toxic Exposure Guidelines
  - NOAEL 0.25 mg/kg/d
  - IARC group 2B: agent is possibly carcinogenic to humans
  - NIOSH REL ETU use encapsulated form; minimize exposure; potential human carcinogen
- ◆ Toxicity Data
  - inhalation-rat TCLo 120 mg/m^3
  - oral-rat LD$_{50}$ 1832 mg/kg
  - oral-rat TDLo 5306 mg/kg
  - oral-mouse LD$_{50}$ 3 GM/kg
  - oral-mouse TDLo 1600 mg/kg
  - intraperitoneal-mouse LD$_{50}$ 200 mg/kg
- ◆ Acute Risks
  - harmful if swallowed, inhaled, or absorbed through skin
  - irritation of skin, eyes, mucous membranes, and upper respiratory tract
  - Prolonged or repeated exposure can cause allergic reactions in certain sensitive individuals.
- ◆ Chronic Risks
  - carcinogen
  - may alter genetic material
  - target organs: liver, thyroid, pituitary

- may cause cancer
- may cause congenital malformation in the fetus

HAZARD RISK
- ◆ hazardous combustion or decomposition products
- ◆ toxic fumes of carbon monoxide, carbon dioxide, nitrogen oxides, sulfur oxides
- ◆ incompatible with strong oxidizing agents
- ◆ emits toxic fumes of NO_x and SO_x when heated to decomposition

EXPOSURE ROUTES
- ◆ primarily by inhalation
- ◆ ingestion
- ◆ absorption through skin and eyes

REGULATORY STATUS/INFORMATION
- ◆ Safe Drinking Water Act
  - listed among water priority list substances
- ◆ Clean Air Act
  - section 112 (1990 amendments) Statutory Air Pollutants. 42 U.S.C. 7412
- ◆ NIOSH Recommendation Substances
- ◆ Resource Conservation and Recovery Act
  - list of hazardous substances. 40 CFR 261
  - land disposal prohibition: scheduled second third wastes. 40 CFR 268.11
- ◆ CERCLA and Superfund
  - designation of hazardous substances. 40 CFR 302.4
  - listed among toxic chemicals. 40 CFR CH.1, Part 355.50
- ◆ Occupational Safety and Health Act
  - listed as National Toxicology Program carcinogen
  - IARC human carcinogens (groups 1, 2A, 2B)
- ◆ Department of Transportation
  - hazardous substances list. 49 CFR 172.101

ADDITIONAL/OTHER COMMENTS
- ◆ Symptoms
  - inhalation: irritates eyes and skin
  - skin absorption: respiratory system
- ◆ First Aid
  - wash eyes immediately with large amounts of water
  - if inhaled, remove to fresh air
  - provide respiratory support
  - if swallowed, wash out mouth with water, provided that the person is conscious
- ◆ fire fighting: use water spray, carbon dioxide, dry chemical powder, or appropriate foam
- ◆ Personal Protection
  - wear appropriate NIOSH/MSHA approved respirator, chemical-resistant gloves, safety goggles, and other protective clothing
  - use only in a chemical fume hood

- do not breathe the dust
- storage: keep tightly closed in a cool, dry place

SOURCES
- *Health Effects Notebook for Hazardous Air Pollutants*, U.S. Environmental Protection Agency, December 1994
- *Environmental Contaminant Reference Databook*, volume 1, 1995
- *Hawley's Condensed Chemical Dictionary*, 12th edition, 1993
- *Material Safety Data Sheets (MSDS)*
- *Merck Index*, 12th edition, 1996
- *NIOSH Pocket Guide to Chemical Hazards*, June 1994
- *Sax's Dangerous Properties of Industrial Materials*, 9th edition, volume 2, 1996
- *Suspect Chemicals Sourcebook*, 1996

ETHYLIDENE DICHLORIDE (rp)

CAS #: 75-34-3
MOLECULAR FORMULA: $C_2H_4Cl_2$
FORMULA WEIGHT: 98.97
SYNONYMS: Aethylidenchlorid (German), chlorinated hydrochloric ether, chloure d'ethylidene (French), cloruro di etilidene (Italian), 1,1-dichloorethaan (Dutch), 1,1-Dichloraethan (German), 1,1-dicloroetano (Italian), 1,1-dichloroethane, ethylidene chloride, ethylidene dichlororide

USES
- Manufacture of Other Chemicals
  - vinyl chloride
  - 1,1,1-trichloroethane
  - solvents
- Used as a Component or in the Manufacturing Process
  - high vacuum rubber
  - plastics
  - oils
  - fats

PHYSICAL PROPERTIES
- colorless, oily liquid
- Odor is similiar to ether.
- saccharine taste
- density 1.174 g/mL at 17°C
- boiling point 57–59°C
- freezing point −98°C
- refractive index 1.4166 (20°C)
- odor threshold 120 ppm

CHEMICAL PROPERTIES
- volatile
- will react violently with strong oxidizing agents
- When heated to decomposition, it will emit highly toxic fumes of phosgene and Cl^-.
- reacts vigorously with oxidizing agents

HEALTH RISK
- ◆ Toxic Exposure Guidelines
  - ACGIH TLV (TWA) 810 mg/m³
  - ACGIH STEL 1010 mg/m³
  - OSHA PEL (TWA) 200 ppm
  - OSHA STEL 250 ppm (310 mg/m³)
  - LC$_{50}$ (rats) 64,765 mg/m³
- ◆ Acute Risks
  - skin burns
  - scaliness
  - rashes
  - central nervous system depression
  - cardiostimulatory effects
  - cardiac arrhythmias
- ◆ Chronic Risks
  - noncancerous
  - body weight depression
  - liver damage

HAZARD RISK
- ◆ very dangerous fire hazard and moderate explosion hazard when exposed to heat or flame
- ◆ can react vigorously with oxidizing materials
- ◆ Decomposition emits highly toxic fumes of phosgene and Cl⁻.

EXPOSURE ROUTES
- ◆ absorption through skin
- ◆ inhalation
- ◆ ingestion

REGULATORY STATUS/INFORMATION
- ◆ Clean Air Act
- ◆ Resource Conservation and Recovery Act
- ◆ waste number U076
  - 40CFR258-32

ADDITIONAL/OTHER COMMENTS
- ◆ Fire Fighting
  - use dry chemical, carbon dioxide, or foam extinguisher and water to keep fire-exposed containers cool
  - emits toxic fumes under fire conditions
- ◆ Personal Protection
  - wear a face shield (8 in. minimum)
  - avoid prolonged or repeated exposure
  - wear appropriate NIOSH/MSHA approved respirator
  - wear gloves, safety goggles, and other protective clothing
  - prevent contact with skin or eyes
- ◆ Spill or Leak
  - evacuate area
  - shut off all sources of ignition
  - absorb on sand or vermiculite
  - ventilate area and wash spill site after material pickup
- ◆ storage: do not use handling equipment or containers composed of magnesium or aluminum or their alloys

269

SOURCES
- *Material Safety Data Sheet* (*MSDS*)
- *Sax's Dangerous Properties of Industrial Materials*, 9th edition, 1996
- *Merck Index*, 12th edition, 1996
- *Fire Protection Guide to Hazardous Materials*, 11th edition, 1994
- *Health Effects Notebook for Hazardous Air Pollutants*, U.S. Environmental Protection Agency, December 1994

FINE MINERAL FIBERS

CAS #: Not applicable to fine mineral fibers
MOLECULAR FORMULA: Not applicable to fine mineral fibers
Note: The 1990 Amendments to the Clean Air Act state that the Hazardous Air Pollutant "Fine Mineral Fibers" "includes mineral fiber emissions from facilities manufacturing or processing glass, rock, or slag fibers (or other mineral derived fibers) of average diameter 1 micrometer or less."
FORMULA WEIGHT: Not applicable to fine mineral fibers
SYNONYMS: Not applicable to fine mineral fibers

USES
◆ Fiberglass is used in insulation.
PHYSICAL PROPERTIES (Fiberglass)
◆ white, off-white solid
◆ odorless
◆ specific gravity 1.96
PHYSICAL PROPERTIES (Mineral Wool, Rockwool Insulation)
◆ gray to brown fibrous appearance
◆ odorless
◆ melting point 982°C
◆ specific gravity >1
CHEMICAL PROPERTIES
◆ Fiberglass and rockwool are both stable.
◆ Polymerization will not occur.
HEALTH RISK
◆ Toxic Exposure Guidelines
  • OSHA has no PEL for airborne mineral fibers.
  • listed as one of 18 Priority Safety and Health Hazards by OSHA
  • EPA or IARC possible human carcinogens
◆ Toxicity Data
  • 42% of hamsters exposed by inhalation after 2 years: mesothelioma from refractory ceramic fibers
◆ Acute Risks (Fiberglass)
  • persistent, dry, hacking cough
  • sore throat
  • blood in sputum
  • sinusitis
  • rhinitis
  • severe respiratory infections
  • headaches
  • nausea
  • dizziness
  • insomnia

- occupational asthma
- allergy-like symptoms
- ear, eye, and skin infections
- extreme sensitivity to ambient pollutants, especially formaldehyde

◆ Chronic Risks (Fiberglass)
  - occupational asthma
  - sick building syndrome
  - multiple chemical sensitivities
  - respiratory disease
  - development of tumors in rats and mice

◆ Chronic Risks (Erionite)
  - high mortality from malignant mesothelioma in animals
  - induced transformation in mice cells
  - unscheduled DNA synthesis in mice and humans
  - high proportions of ferruginous bodies in lung tissue in animals

◆ Chronic Risks (Glasswool)
  - elevated mortality
  - respiratory cancer
  - lung cancer
  - increased incidences of lung tumors in hamsters
  - increased incidences of mesotheliomas in hamsters
  - pleural tumors in rats
  - sarcomas of the peritoneal cavity in rats
  - induced chromosomal alterations in mammalian cells in vitro
  - morphological transformations in rodent cells in vitro

◆ Chronic Risks (Rockwool and Slagwool)
  - increased mortality from lung cancer
  - increased incidence of pleural mesotheliomas in rats
  - abdominal cavity tumors in rats

HAZARD RISK
◆ hazardous decomposition products of fiberglass: carbon monoxide, carbon dioxide, hydrocarbons, and water
◆ hazardous decomposition product of rockwool: dust suppressant oil additive
◆ Dust suppressant oil additive may emit carbon monoxide as a decomposition product.

EXPOSURE ROUTES
◆ inhalation
◆ dermal exposure in industry
◆ in home insulation
◆ installation of fiberglass insulation
◆ ambient air

REGULATORY STATUS/INFORMATION
◆ Clean Air Act Amendment, Title III: hazardous air pollutants

ADDITIONAL/OTHER COMMENTS
◆ Symptoms
  - fiberglass overdose: dry cough, sore throat, blood in sputum, severe sinusitis, rhinitis, respiratory infections, headaches, dizziness, nausea, insomnia, asthma, ear, eye, and skin infections, sensitivity to ambient pollutants

- ◆ First Aid (Fiberglass)
  - flush eyes with water
  - wash skin with mild soap and water; use washcloth to remove fibers
  - do not rub or scratch irritated area
  - move to fresh air
- ◆ First Aid (Rockwool)
  - wash skin with mild soap and water; use washcloth to remove fibers
  - move to fresh air
- ◆ fire fighting: use water, foam, carbon dioxide, and dry chemical
- ◆ personal protection: self-contained breathing apparatus, protective clothing, and goggles

SOURCES
- *Material Safety Data Sheets* (*MSDS*) for fiberglass and rockwool
- *Environmental Law Index to Chemicals*, 1996
- *Hawley's Condensed Chemical Dictionary*, 12th edition, 1993
- Excerpt from EPA Report: *Documentation of de minimus Emission Rates—Proposed Background Document*
- *Answers to Selected Questions on Environmental Health Issues, Man-Made Mineral Fibers and Cancer*
- The Fiberglass Information Network
- Report by Robert Horowitz (off the Internet)

FORMALDEHYDE (rp)

CAS #: 50-00-0
MOLECULAR FORMULA: CH_2O
FORMULA WEIGHT: 30.03
SYNONYMS: BFV, dormol, fannoform, formaldehyde-solution, formalin, formalith, formic aldehyde, formol, fyde, ivalon, lysoform, methanal, methyl aldehyde, methylene oxide, morbicid, oxomethane, oxymethylene, superlysoform

USES
- ◆ Manufacture of Other Chemicals
  - dyes
  - phenolic resins
  - artificial silk
  - cellulose esters
  - explosives
  - glass mirrors
  - fatty amides
  - resins
  - casein
  - albumin
  - gelatin
- ◆ Used as a Component or in the Manufacturing Process
  - disinfecting
  - germicide
  - fungicide
  - waterproofing
  - preserving
  - coatings
  - molding compound
  - adhesives
  - dry cleaning
  - finishes

PHYSICAL PROPERTIES
◆ clear, water-white
◆ pungent odor
◆ pH range of 2.8 to 4.0
◆ density 1.067 (air = 1.00)
◆ boiling point −19.5°C
◆ melting point −92°C
◆ odor threshold 0.5 to 1.0 ppm
◆ heat of vaporization 5917.9 gcal/gmole
◆ specific gravity 1.083
◆ vapor density 1.03
◆ vapor pressure 52 mm at 37°C

CHEMICAL PROPERTIES
◆ corrosive to carbon steel
◆ volatile
◆ slowly oxidizes to formic acid
◆ flash point 133°F
◆ flammability limits 7.0% lower and 73% upper (by volume in air)
◆ autoignition temperature 300°C (formaldehyde gas)
◆ heat of combustion −136.42 kcal/gmole at 25°C (gas)
◆ will react violently with strong oxidizing agents
◆ flammable when exposed to heat

HEALTH RISK
◆ Toxic Exposure Guidelines
  • ACGIH TLV TWA 1 ppm
  • OSHA PEL TWA 0.75 ppm
  • OSHA STEL 2 ppm
  • LD_{50} (rats) oral 800 mg/kg
  • NIOSH REL limit to lowest feasible level
  • USEPA Cancer Risk Level 8×10^{-5} mg/m^3
  • inhalation unit risk 1.3×10^{-5} µg/m^3
◆ Acute Risks
  • harmful if swallowed, inhaled, or absorbed through skin
  • destructive to tissue of the mucous membranes and upper respiratory tract
  • wheezing
  • shortness of breath
  • coughing
  • headache
  • nausea
  • vomiting
  • chest pains
  • gastrointestinal disturbances
  • skin reactions
◆ Chronic Risks
  • alter genetic material
  • carcinogen
  • pulmonary edema
  • respiratory edema
  • coma
  • hemorrhage

HAZARD RISK
◆ combustible liquid when exposed to heat or flame
◆ can react vigorously with oxidizers
◆ moderate explosion hazard when exposed to heat or flame
◆ Gas is a more dangerous fire hazard than the vapor.
◆ If formaldehyde is involved in fire, irritating gaseous formaldehyde may evolve.
◆ When aqueous formaldehyde solutions are heated above their flash points, a potential for an explosion hazard exists.

- reaction with NO_x at 180°C which becomes explosive
- reacts violently with perchloric acid + aniline, performic acid, nitromethane, magnesium carbonate, and H_2O_2
- If storage tank is ruptured, irritating fumes which may exist in toxic concentrations locally evolve.
- Decomposition emits acrid smoke and fumes.

EXPOSURE ROUTES
- absorption through skin
- inhalation of indoor air and ambient air
- ingestion of food
- smoking

REGULATORY STATUS/INFORMATION
- Clean Water Act
  - 40USC7412
  - 40CFR116.4
- Clean Air Act
  - section 112(g)
- Resource Conservation and Recovery Act
  - waste number 4122
  - 40CFR261
- Federal Insecticide, Fungicide, and Rodenticide Act
  - 40CFR261
- CERCLA and Superfund
  - superfund, 40CFR302.4
  - 1/ED 3 per (mg/kg)/d

ADDITIONAL/OTHER COMMENTS
- Fire Fighting
  - use dry chemical, carbon dioxide, or foam extinguisher and water to keep fire-exposed containers cool
  - emits toxic fumes under fire conditions
- Personal Protection
  - wear a face shield (8 in. minimum)
  - avoid prolonged or repeated exposure
  - wear appropriate NIOSH/MSHA approved respirator
  - wear gloves, safety goggles, and other protective clothing
  - prevent contact with skin or eyes
- Spill or Leak
  - evacuate area
  - shut off all sources of ignition
  - absorb on sand or vermiculite
  - ventilate area and wash spill site after material pickup
- storage: do not use handling equipment or containers composed of magnesium or aluminum or their alloys

SOURCES
- *Material Safety Data Sheet (MSDS)*
- *Sax's Dangerous Properties of Industrial Materials*, 9th edition, 1996
- *Merck Index*, 12th edition, 1996
- *Fire Protection Guide to Hazardous Materials*, 11th edition, 1994

GLYCOL ETHERS

CAS #: Glycol ethers have variable CAS #s. The CAS # for 2-ethoxyethanol is 110-80-5. The CAS # for 2-methoxyethanol is 109-86-4.

MOLECULAR FORMULA: Glycol ethers have variable molecular formulas. The molecular formula for 2-ethoxyethanol is $C_4H_{10}O_2$. The molecular formula for 2-methoxyethanol is $C_3H_8O_2$.

Note 1: For all hazardous air pollutants (HAPs) that contain the word "compounds" in the name, and for glycol ethers, the following applies: Unless otherwise specified, these HAPs are defined by the 1990 Amendments to the Clean Air Act as including any unique chemical substance that contains the named chemical (i.e., antimony, arsenic, etc.) as part of that chemical's infrastructure.

Note 2: The 1990 Amendments to the Clean Air Act also state the following with regard to the HAP glycol ethers: "includes mono- and di-ethers of ethylene glycol, diethylene glycol, and triethylene glycol R-$(OCH_2CH_2)_n$-OR' where

n = 1, 2, or 3
R = alkyl or aryl groups
R' = R, H, or groups which, when removed, yield glycol ethers with the structure: R-$(OCH_2CH)_n$-OH. Polymers are excluded from the glycol category."

FORMULA WEIGHT: Glycol ethers have variable formula weights. The formula weight for 2-ethoxyethanol is 90.10. The formula weight for 2-methoxyethanol is 76.1.

SYNONYMS: cellosolve (DOT), cellosolve solvent, dowanol EE, ekta-solve EE, ether monoethylique de l'ethylene-glycol (French), ethyl cellosolve, ethylene glycol ethyl ether, ethylene glycol monoethyl ether, ethylene glycol monoethyl ether (DOT), 2-Ethoxyethanol: Athylen-glykol-monoathylather (German), etoksyetylowy alkohol (Polish), glycol ether EE, glycol ethyl ether, glycol monoethyl ether, hydroxy ether, jeffersol EE, oxitol, poly-solv EE

USES
◆ Glycol Ethers Used as a Solvent
  • resins
  • lacquers
  • paints
  • varnishes
  • gum
  • perfume
  • dyes
  • inks
◆ Glycol Ethers Used as a Constituent
  • paints and pastes
  • cleaning compounds
  • liquid soaps
  • cosmetics
  • hydraulic fluids
◆ Glycol Ethers used in Industry
  • semiconductors
  • removal of flux
  • removal of solder paste
  • removal of greases, inks, and oils

275

PHYSICAL PROPERTIES
- ◆ Glycol ethers are colorless liquids with a slight odor.

PHYSICAL PROPERTIES (2-Methoxyethanol)
- ◆ melting point $-87°C$
- ◆ boiling point 124°C
- ◆ specific gravity 1.0
- ◆ vapor pressure 9.5 mm Hg at 25°C
- ◆ vapor density 2.6
- ◆ soluble in water
- ◆ log K_{ow} -0.74

PHYSICAL PROPERTIES (2-Ethoxyethanol)
- ◆ boiling point 135.1°C
- ◆ specific gravity 0.93
- ◆ vapor pressure 5.5 mm Hg at 25°C
- ◆ vapor density 3.0
- ◆ miscible in water
- ◆ log K_{ow} -1.0

CHEMICAL PROPERTIES (2-Methoxyethanol)
- ◆ flash point 39°C
- ◆ explosion limits 2.3% lower and 24.5% upper
- ◆ can react with strong oxidizing materials

CHEMICAL PROPERTIES (2-Ethoxyethanol)
- ◆ flash point 202°F
- ◆ autoignition temperature 455°F
- ◆ explosion limits 1.7% lower and 15.6% upper

HEALTH RISK
- ◆ Toxic Exposure Guidelines
  - OSHA PEL (2-ethoxyethanol) 740 mg/m³
  - OSHA PEL (2-methoxyethanol) 80 mg/m³
  - ACGIH TLV (2-ethoxyethanol) 18 mg/m³
  - ACGIH TLV (2-methoxyethanol) 16 mg/m³
  - NOAEL (2-ethoxyethanol) 68 mg/m³
  - NOAEL (2-methoxyethanol) 17 mg/m³
  - RfC (2-ethoxyethanol) 0.2 mg/m³
  - RfC (2-methoxyethanol) 0.02 mg/m³
- ◆ Toxicity Data
  - LC_{50} (2-methoxyethanol) 4668.7 mg/m³
- ◆ Acute Risks
  - narcosis
  - pulmonary edema
  - severe liver damage
  - severe kidney damage
  - conjunctivitis
  - upper respiratory tract irritation
  - temporary corneal clouding
  - adverse effects on weight gain in animals
  - adverse effects on peripheral blood counts in animals
  - bone marrow effects in animals
  - lymphoid tissue effects in animals
- ◆ Acute Risks (2-Methoxyethanol)
  - irritation of eyes, nose, and throat
  - drowsiness
  - weakness
  - shaking
  - headache
  - fatigue
  - personality change
  - decreased mental ability

◆ Chronic Risks
- fatigue
- lethargy
- nausea
- anorexia
- tremor
- anemia
- reduced body weight gain in animals
- irritation of eyes and nose
- red blood cell depression
- slight kidney damage in animals
- encephalopathy (degenerative brain disease)
- bone marrow depression
- pancytopenia
- potential to cause adverse reproductive effects
- embryotoxicity in animals
- increased incidences of embryonic death in animals
- teratogenesis or growth
- retardation in animals
- testicular atrophy and sterility in male animals

HAZARD RISK
◆ Glycol ethers are flammable and combustible when exposed to heat or flame.
◆ Glycol ethers can react vigorously with oxidizing materials.
◆ 2-ethoxyethanol is a moderate explosion hazard in the form of vapor when exposed to heat or flame.
◆ 2-ethoxyethanol mixture with hydrogen peroxide + polyacrylamide gel + toluene is explosive when dry.

EXPOSURE ROUTES
◆ use of cleaning compounds
◆ liquid soaps
◆ cosmetics
◆ other consumer products
◆ workers in the chemical industry

REGULATORY STATUS/INFORMATION
◆ Clean Air Act Amendment, Title III: hazardous air pollutants
◆ CERCLA and Superfund
- designation of hazardous substances. 40CFR302.4
◆ chemical regulated by the State of California

ADDITIONAL/OTHER COMMENTS
◆ Symptoms
- exposure: headaches, drowsiness, weakness, fatigue, staggering, personality change, decreased mental ability
◆ First Aid
- wash eyes and skin immediately with large amounts of water
- remove to fresh air immediately or provide respiratory apparatus if necessary
- for 2-ethoxyethanol, induce vomiting
◆ fire fighting: use alcohol foam, dry chemical, and carbon dioxide
◆ personal protection: self-contained breathing apparatus, protective clothing, and goggles
◆ storage (2-methoxyethanol): outside or detached storage; inside in standard flammable liquid storage room or cabinet

◆ storage (2-ethoxyethanol): keep in closed container in cool, well-ventilated area away from heat and sources of ignition, separate from oxidizers and alkalies

SOURCES

• *Health Effects Notebook for Hazardous Air Pollutants*, U.S. Environmental Protection Agency, December 1994
• *Material Safety Data Sheets (MSDS)* for 2-methoxyethanol and 2-ethoxyethanol
• *Sax's Dangerous Properties of Industrial Materials*, 9th edition, 1996
• *Environmental Law Index to Chemicals*, 1996
• SAGE: Solvent Alternatives Guide: Glycol Ethers, Process Unique Information (from Internet)

HEPTACHLOR (cp)

CAS #: 76-44-8
MOLECULAR FORMULA: $C_{10}H_5Cl_7$
FORMULA WEIGHT: 373.32
SYNONYMS: agroceres, amatin, anticarie, bunt-cure, bunt-no-more, 3-chlorochlordene, co-op hexa, drinox, ent-15,152, esaclorobenzene, goldcrest H-60, heptox, many more

USES
◆ pesticides (EPA canceled registration for this use)
◆ insecticides

PHYSICAL PROPERTIES
◆ manmade
◆ white powder
◆ melting point 95°C (203°F) to 96°C (205°F) at 1.5 mm Hg
◆ boiling point 135°C (275°F) to 145°C (293°F) at 1.5 mm Hg
◆ density 1.57 g/cm³
◆ specific gravity 1.66
◆ flash point noncombustible
◆ stable under normal temperatures and pressures
◆ does not ignite readily
◆ water partition coefficient (log) 3.87 to 5.44
◆ camphor-like odor
◆ odor threshold 0.02 ppm (in water)
◆ vapor pressure 3.0 mm Hg at 25°C
◆ solubility 0.0006% (by weight) in water at 20°C

CHEMICAL PROPERTIES
◆ nearly insoluble in water
◆ soluble in organic solvents
◆ soluble in flammable liquids
◆ noncombustible as a solid
◆ when heated to decomposition emits hydrogen-chlorine gas, toxic oxides of carbon
◆ stable to light, moisture, and air
◆ organochlorine insecticide

HEALTH RISK
- ◆ Toxic Exposure Guidelines
  - ACGIH TLV TWA 0.05 (skin)
  - OSHA PEL TWA 0.5 (skin)
  - NIOSH REL TWA 0.5 (skin)
  - NIOSH REL TWA 0.5 mg/m^3 (inhale)
- ◆ Toxicity Data
  - oral-rat LD$_{50}$ 40–220 mg/kg
  - skin-rat LD$_{50}$ 119–320 mg/kg
  - oral-mice LD$_{50}$ 30–68 mg/kg
  - oral-guinea pigs LD$_{50}$ 116 mg/kg
  - skin-guinea pigs LD$_{50}$ 1000 mg/kg
  - oral-hamster LD$_{50}$ 100 mg/kg
  - oral-chicken LD$_{50}$ 62 mg/kg
  - skin-rabbit LD$_{50}$ >2000 mg/kg
  - inhale-cat LCLo 150 mg/m^3
- ◆ Other Toxicity Indicators
  - EPA Cancer Risk Level (1 in a million excess lifetime risk) 8×10^{-7} mg/m^3
- ◆ Acute Risks
  - liver damage
  - tremors
  - vomiting
  - convulsions
  - kidney damage
  - respiratory collapse
  - death
- ◆ Chronic risks
  - Oral Effects
    - personality changes
    - brain effects such as loss of memory and concentration
    - neurological effects such as dizziness, muscle tremors, and convulsions
    - liver damage
    - kidney damage
  - Inhalation Effects
    - blood dyscrasias
    - potential for causing reproductive damage

HAZARD RISK
- ◆ NFPA Hazard Rating: Flammability not rated, Reactivity not rated
- ◆ effects due to inhalation and passing through skin
- ◆ Exposure causes feeling of anxiety and irritability, headache, dizziness, muscle twitching.
- ◆ High or repeated exposure may cause liver and kidney damage.
- ◆ confirmed carcinogen to be handled with caution
- ◆ Containers may explode if heated.

EXPOSURE ROUTES
- ◆ ingestion

- ◆ skin contact
- ◆ inhalation
- ◆ intravenous
- ◆ subsurface ground injection (insecticide)

REGULATORY STATUS/INFORMATION

- ◆ American Conference of Governmental and Industrial Hygienists and Occupational Safety and Health Act
  - max concentration in workplace for an 8 hour day, 40 hour workweek 0.5 mg/m^3
- ◆ Environmental Protection Agency
  - banned sales and restricted use of products containing chemical
  - commercial use permitted for fire ant control
  - recommended maximum level for drinking water or seafood 2.78 ppt
  - drinking water exposure for children: short term 10,000 ppt, long term 5000 ppt
  - Quantities greater than one pound entering the environment must be reported to The National Response Center.
  - inhalation unit risk: 1.3×10^{-3} (g/m^3)
- ◆ Food and Drug Administration
  - Maximum Concentration
    - raw food crops 0–10 ppb
    - edible seafood 300 ppb
    - fat of food producing animal 200 ppb
- ◆ National Institute for Occupational Safety and Health
  - Airborne concentration of 100 mg/m^3 or higher is immediately threatening to health and life

ADDITIONAL/OTHER COMMENTS

- ◆ First Aid
  - eye contact: wash out eyes immediately with large amounts of water for 15 minutes, occasionally lifting lids
  - skin contact: remove contaminated clothes immediately; wash with large amounts of soap and water
  - inhalation: remove person from exposure; apply resuscitation if breathing has stopped, and CPR if heart failure. Transfer to medical facility.
- ◆ fire fighting: use dry chemical, carbon dioxide, water spray, or foam extinguisher
- ◆ personal protection: use local exhaust when working with contaminant; wear protective clothing, breathing apparatus, and dustproof goggles. Wash thoroughly after exposure.
- ◆ storage: in a tightly closed container in a cool, well-ventilated area, protected from any physical damage, strong oxidizers, excessive heat, sparks, open flames

SOURCES

- ◆ *Hawley's Condensed Chemical Dictionary*, 12th edition, 1993
- ◆ *Fire Protection Guide to Hazardous Materials*, 11th edition, 1994
- ◆ *Merck Index*, 12th edition, 1996

◆ http://atsdr1.cdc.gov:8080/mrls.html
◆ http://www.epa.gov/oar/oaqps/airtox/hapindex.html
◆ http://mail.odsnet.com/TRIFacts

HEXACHLOROBENZENE (cp)

CAS #: 118-74-1
MOLECULAR FORMULA: C_6Cl_6
FORMULA WEIGHT: 285.2
SYNONYMS: amatin, anticarie, bunt-cure, bunt-no-more, ceku c.b., co-op hexa, granox nm, hexa c.b., julin's carbon chloride, no bunt, no bunt 40, no bunt 80, no bunt liquid, pentachlorophenyl chloride, perchlorobenzene, sanocid, others

USES
◆ Direct Uses
  • pesticide
  • fungicide
◆ Indirect Uses
  • intermediate for synthesis of wood preservatives
  • intermediate for synthesis of rubber
  • intermediate for synthesis of dyes
PHYSICAL PROPERTIES
◆ white needles
◆ melting point 227°C (440.6°F) to 229°C (444.2°F)
◆ boiling point 323°C (613.4°F) to 326°C (618.8°F)
◆ density 2.044 at 23°C
◆ flash point 241°C (open cup)
◆ vapor pressure 0.00001 mm Hg at 20°C
◆ octanol/water partition coefficient 6.18
◆ soluble in benzene and boiling alcohol
◆ insoluble in water
◆ odor
◆ odor threshold 3.0 ppm
◆ vapor density 9.8 g/m³
CHEMICAL PROPERTIES
◆ incompatible with strong oxidizing agents
◆ autoignition temperature 596.2°C
HEALTH RISK
◆ Toxic Exposure Guidelines
  • ACGIH TLV 0.025 mg/m³
◆ Toxicity Data
  • oral-rat LD_{50} 10 g/kg
  • inhale-rat LC_{50} 3600 mg/m³
  • oral-mouse LD_{50} 4 g/kg
  • inhale-mouse LC_{50} 4 g/m³
  • oral-cat LD_{50} 1700 mg/kg
  • inhale-cat LC_{50} 1600 mg/m³
  • oral-rabbit LD_{50} 2600 mg/kg
  • inhale-rabbit LC_{50} 1800 mg/m³
  • oral-guinea pig LD_{50} >3 g/kg

- oral-quail $LD_{50} > 6400$ mg/kg
- ◆ Other Toxicity Indicators
  - EPA Cancer Risk Level (1 in a million excess lifetime risk) 2×10^{-6} mg/m^3
- ◆ Acute Risks
  - irritation of eye membranes
  - irritation of the skin
  - irritation of mucous membranes and upper respiratory tract
  - may cause photosensitivity, resulting in skin lesions when exposed to the sun
  - possible cutaneous, hepatic, arthritic, urinary, and neurological effects
  - drowsiness
  - incoordination
  - headache
  - vomiting
  - numbness of extremities
- ◆ Chronic Risks
  - carcinogen
  - may alter genetic material
  - cancer of the liver
  - development of porphyria cutanea tarda
  - bullous lesions
  - cancer of thyroid
  - cancer of kidneys

HAZARD RISK
- ◆ hazardous combustion and decomposition products
- ◆ toxic fume formation of carbon monoxide, carbon dioxide, hydrogen chloride gas
- ◆ skin blistering after 60 minutes
- ◆ asphyxiating
- ◆ corrosive
- ◆ vapor ignitable by electric spark

EXPOSURE ROUTES
- ◆ ingestion
- ◆ inhalation of vapor
- ◆ absorption through skin

REGULATORY STATUS/INFORMATION
- ◆ Clean Water Act
  - listed on Water Quality Criteria List
  - listed as a Priority Pollutant 40CFR423.17
- ◆ Environmental Protection Agency
  - Reference Dose (RfD) of 0.0008 mg/kg/d
  - carcinogenic potency: 1/ED10 value of 13 per (mg/kg)/d[4]
  - Inhalation Unit Risk: 4.6×10^{-4} (μg/m^3)$^{-1}$
  - listed as probable human carcinogenic medium
- ◆ Safe Drinking Water Act
  - given a maximum contaminant level 40CFR141
  - maximum contaminant level: 0.001 mg/L

- listed as maximum contaminant goal substance 40CFR141
- maximum contaminant level goal: 0 mg/L
- listed as a Special Monitoring Substance 40CFR141.40
- ◆ Clean Air Act
  - section 111 as a Volatile Organic Compound 40CFR60
  - section 112 Statutory Air Pollutant
  - section 112 High Risk Pollutant 40CFR63
- ◆ Resource Conservation and Recovery Act
  - listed as a Hazardous Substance 40CFR261
  - RCRA Hazardous Constituents for Groundwater Monitoring 40CFR264
  - RCRA Land Disposal Restrictions: Halogenated Organic Compound 40CFR268
  - RCRA Land Disposal Restrictions: Second Third Wastes
  - Hazardous Waste # U127
- ◆ CERCLA
  - listed as a Hazardous Substance 40CFR302
- ◆ Superfund Amendment and Reauthorization Act
  - section 313 Toxic Chemicals 40CFR372
  - SARA Section 110 Priority List of CERCLA Hazardous Substances
  - Health Assessment Risk: 92
- ◆ National Toxicology Program
  - National Toxicology Program Carcinogen
- ◆ Department of Transportation
  - Hazardous Material Table 49CFR171
  - Hazardous Class: 6.1
  - DOT Identification # UN2729
  - Hazardous Substance Other than Radionuclides 49CFR172.101
  - Statutory RQ of "1*"

ADDITIONAL/OTHER COMMENTS

- ◆ First Aid
  - skin contact: immediately remove contaminated clothing; wash with large amounts of water for 15 minutes
  - inhalation: remove from exposure; if breathing has stopped, give artificial respiration; if breathing is difficult supply oxygen
  - eye contact: immediately wash out eyes for 15 minutes with water
- ◆ fire fighting: use a water spray, carbon dioxide, dry chemical powder, or foam
- ◆ Personal Protection
  - wear appropriate NIOSH/MSHA approved respirator
  - do not breathe dust
  - wash thoroughly after handling or exposure
  - use in a chemical fume hood or a well-ventilated workplace only
  - wear protective clothing such as safety goggles, and chemical-resistant gloves

◆ storage: store in a tightly sealed container in a cool, dry place
SOURCES
- ◆ *Hawley's Condensed Chemical Dictionary*, 12th edition, 1993
- ◆ *Fire Protection Guide to Hazardous Materials*, 11th edition, 1994
- ◆ *Merck Index*, 12th edition, 1996
- ◆ *NIOSH Pocket Guide to Chemical Hazards*, 1994
- ◆ *Sax's Dangerous Properties of Industrial Materials*, 9th edition, 1996
- ◆ *Suspect Chemicals Sourcebook*, 1996
- ◆ http://atsdr1.atsdr.cdc.gov:8080/toxfaq.html
- ◆ http://www.epa.gov/oar/oaqps/airtox/hapindex.html
- ◆ http://mail.odsnet.com/TRIFacts
- ◆ http://sulaco.oes.orst.edu:70/1s/ext/extoxnet/pips

HEXACHLOROBUTADIENE (cp)

CAS #: 87-68-3
MOLECULAR FORMULA: C_4Cl_6
FORMULA WEIGHT: 260.76
SYNONYMS: dolen-pur, HCBD, hexachlor-1,3-butadien (Czech), per-chlorobutadiene, and others

USES
- ◆ Direct Uses
  - • solvent
  - • pesticide (fumigant)
  - • lubricants
  - • heat transfer fluid
  - • hydraulic fluid
  - • isolating agent for transformers
- ◆ Indirect Uses
  - • intermediate for lubricants
  - • synthetic rubber

PHYSICAL PROPERTIES
- ◆ slightly yellowish to clear liquid
- ◆ melting point −21°C at 760 mm Hg
- ◆ boiling point 210°C at 760 mm Hg
- ◆ density 1.675 at 16°C (reference 16°C)
- ◆ specific gravity 1.665
- ◆ flash point none
- ◆ octanol/water partition coefficient 4.78
- ◆ viscosity 2.447 cP at 37.7°C
- ◆ vapor density 8.99 (air = 1)
- ◆ faint turpentine odor
- ◆ odor threshold 1.0 ppm (12 mg/m³)
- ◆ vapor pressure 0.3 mm Hg at 77°C
- ◆ solubility 9.77 mmol/L

CHEMICAL PROPERTIES
- ◆ insoluble in water
- ◆ soluble in alcohol and ether

- ◆ autoignition temperature 610°C (1130°F)
- ◆ incompatible with strong oxidizing agents
- ◆ compatible with numerous resins
- ◆ Decomposition by-products include carbon monoxide, carbon dioxide, and hydrogen chloride.

HEALTH RISK
- ◆ Toxic Exposure Guidelines
  - OSHA PEL 0.24 mg/m^3
  - ACGIH TLV 0.24 mg/m^3
  - NIOSH REL 0.24 mg/m^3
- ◆ Toxicity Data
  - oral-rat LD$_{50}$ 82 mg/kg
  - oral-mouse LD$_{50}$ 87 mg/kg
  - inhale-mouse LD$_{50}$ 370 mg/m^3
  - skin-mouse LD$_{50}$ 150 mg/kg
  - oral-cat LD$_{50}$ 187 mg/kg
  - skin-rabbit LD$_{50}$ 100 mg/kg
  - oral-guinea pig LD$_{50}$ 90 mg/kg
  - oral-hamster LD$_{50}$ 90 mg/kg
  - inhale-mouse LC$_{50}$ 370 mg/m^3
- ◆ Other Toxicity Indicators
  - EPA Cancer Risk Level (1 in a million excess lifetime risk) 5×10^{-5} mg/m^3
- ◆ Acute Risks
  - skin and eye irritation or burning
  - Inhalation may cause death due to respiratory spasm.
  - High concentration may cause kidney damage.
  - coughing
  - wheezing
  - laryngitis
  - shortness of breath
  - headache
  - nausea
  - vomiting
- ◆ Chronic Risks
  - cancer of kidneys
  - cancer of liver
  - damage to developing fetus

HAZARD RISK
- ◆ Department of Health rating: Flammability 1, Reactivity 1, Health 2
- ◆ flammable
- ◆ risk of explosion above flash point temperature
- ◆ toxic
- ◆ carcinogenic to be handled with caution
- ◆ gas heavier than air
- ◆ formation of toxic vapors in fire
- ◆ formation of corrosive vapors
- ◆ Minimal Risk Level: oral 0.0002 mg/kg/d

EXPOSURE ROUTES
- ◆ primarily inhalation
- ◆ skin absorption
- ◆ ingestion (least likely)

REGULATORY STATUS/INFORMATION
- ◆ American Conference of Governmental and Industrial Hygienists
  - • workplace limit: airborne exposure for 8 hour work day 0.02 ppp; if skin contact involved, this level is considered overexposure
- ◆ Environmental Protection Agency
  - • no reference concentration (RfC) established
  - • Reference Dose (RfD) under consideration
  - • carcinogenic potency: 1/ED10 value of 0.36 per (mg/kg)/d[4]
  - • inhalation unit risk: 2.2×10^{-5} (μg/m³)$^{-1}$
- ◆ Clean Water Act
  - • listed under section 304: Water Quality Criteria Substances
  - • listed under section 307: Priority Pollutant 40CFR423.17
- ◆ Clean Air Act
  - • listed under section 112: Statutory Air Pollutants 42USC7412
- ◆ Safe Drinking Water Act
  - • listed as Special Monitoring Substance 40CFR141
  - • listed on Water Priority List 56FedReg1470
- ◆ Resource Conservation and Recovery Act
  - • listed as Hazardous Substance 40CFR261
  - • Hazardous Waste # U128
  - • listed as a Hazardous Constituent for Groundwater Monitoring 40CFR264
  - • has Land Disposal Restrictions as a Halogenated Organic Compound (HOC) 40CFR268
- ◆ CERCLA
  - • listed as Hazardous Substance 40CFR302
  - • on Priority List of Hazardous Substances
  - • Statutory RQ of 1 lb
  - • Final RQ of 1 lb
- ◆ Occupational Safety and Health Act
  - • listed as Air Contaminant 29CFR1910
  - • Table Z-1-A
- ◆ Department of Transportation
  - • listed as a Hazardous Material 49CFR171
  - • Hazardous Class 6.1: keep away from food
  - • ID# UN2279
  - • listed as Hazardous Substance Other than Radionuclides 49CFR172.101
  - • listed as Severe Marine Pollutant
- ◆ Superfund Amendment and Reauthorization Act
  - • Health Assessment Rank: 14
- ◆ Internal Agency for Research on Contaminants (IARC)
- ◆ New Jersey Right-to-Know (Listed but no value)

ADDITIONAL/OTHER COMMENTS
◆ First Aid
  • inhalation: remove person from exposure; apply resuscitation if breathing stopped, CPR if heart stopped; transfer to medical facility immediately
  • eye contact: flush with large amounts of water for 15 minutes, separating eyelids
  • skin contact: remove contaminated clothes; wash with soap and water immediately
  • for all cases, seek medical attention immediately
◆ fire fighting: use a dry chemical, carbon dioxide, water spray, alcohol foam extinguisher
◆ Personal Protection
  • use personal NIOSH/MSHA approved respirator
  • ventilation of contaminated workplace
  • wear protective clothing when working with or near chemical, including splashproof goggles/face mask and solvent-resistant gloves
  • wash immediately after exposure
◆ storage: tightly sealed container in a cool, well-ventilated place

SOURCES
◆ *Hawley's Condensed Chemical Dictionary*, 12th edition, 1993
◆ *Fire Protection Guide to Hazardous Materials*, 11th edition, 1994
◆ *Merck Index*, 12th edition, 1996
◆ *NIOSH Pocket Guide to Chemical Hazards*, 1994
◆ *Sax's Dangerous Properties of Industrial Materials*, 9th edition, 1996
◆ *Suspect Chemicals Sourcebook*, 1996
◆ http://atsdr1.atsdr.cdc.gov:8080/toxfaq.html
◆ http://chemfinder.camsoft.com
◆ http://www.epa.gov/oar/oaqps/airtox/hapindex.html
◆ http://mail.odsnet.com/TRIFacts
◆ http://sulaco.oes.orst.edu:70/1s/ext/extoxnet/pips

HEXACHLOROCYCLOPENTADIENE (cp)

CAS #: 77-47-4
MOLECULAR FORMULA: C_5C_{l6}
FORMULA WEIGHT: 272.29
SYNONYMS: C-56, graphlox, HCCPD, hexachlorocyklopentadien, HRS 1655, NCI-C55607, PCL, perchlorocyclopentadiene, others

USES
  • intermediate for dyes
  • intermediate for pesticides, such as heptachlor, aldrin, and endrin
  • intermediate for fungicides
  • intermediate for pharmaceuticals
  • intermediate for flame retardant materials

PHYSICAL PROPERTIES
- pale-yellow liquid
- melting point −10°C at 753 mm Hg
- boiling point 239°C at 753 mm Hg
- density 1.717 at 15°C
- specific gravity 1.702
- flash point none
- noncombustible
- vapor density 9.42
- flammability limits none
- bulk density 14.3 lb/gal at 15.5°C
- pungent odor
- odor threshold 0.03 ppm
- octanol/water partition coefficient 5.04
- vapor pressure 0.13 psi at 20°C
- solubility

CHEMICAL PROPERTIES
- incompatible with strong oxidizing agents
- decomposes when exposed to moist air or water
- autoignition temperature none

HEALTH RISK
- Toxic Exposure Guidelines
  - ACGIH TWA 0.01 ppm
  - OSHA TWA 0.01 ppm
  - NIOSH TWA 0.01 ppm
- Toxicity Data
  - inhale-rabbit LC_{50} 58 mg/m^3
  - inhale-rat LC_{50} 18 mg/m^3
- Acute Risks
  - fatal if inhaled
  - fatal if swallowed
  - fatal if absorbed through skin
  - very destructive to mucous membranes of upper respiratory tract
  - irritates eye mucous membranes
  - spasm and inflammation of larynx
  - inflammation and edema of the bronchi
  - chemical pneumonitis
  - pulmonary edema
  - burning sensation to exposed area
  - coughing
  - wheezing
  - laryngitis
  - shortness of breath
  - headache
  - nausea
  - vomiting
  - possible blistering of exposed skin
- Chronic Risks
  - damage to liver

- damage to kidneys
- degeneration of lung tissue

HAZARD RISK
- ◆ highly toxic if inhaled
- ◆ readily absorbed through skin
- ◆ causes skin burns
- ◆ corrosive
- ◆ emits toxic fumes of carbon monoxide, carbon dioxide, hydrogen chloride gas
- ◆ emits toxic fumes on exposure to moist air or water

EXPOSURE ROUTES
- ◆ inhalation of vapors
- ◆ absorption through the skin
- ◆ ingestion

REGULATORY STATUS/INFORMATION
- ◆ Toxic Substance Control Act
  - listed as a Chemical Hazardous Information Profile Substance
- ◆ Clean Water Act
  - section 304 Water Quality Criteria Substance
  - section 307 Priority Pollutants 40CFR423.17
  - section 307 Hazardous Substance 40CFR141.40
- ◆ Safe Drinking Water Act
  - Maximum Contaminant Level Substance
  - 40CFR141
  - MCL of 0.05 mg/L
  - Maximum Contaminant Level Goal Substance
  - 40CFR141
  - MCLG of 0.05 mg/L
  - Special Monitoring Substance
  - 40CFR141.40
  - SDWA Statutory Contaminant
- ◆ Environmental Protection Agency
  - An RfC has not been established.
  - considered a group D contaminant: not classifiable as to human carcinogenicity
- ◆ Clean Air Act
  - section 112 Statutory Air Pollutant
  - High Risk Pollutant 40CFR63
- ◆ Resource Conservation and Recovery Act
  - Hazardous Substance 40CFR261
  - Hazardous Constituent for Groundwater Monitoring 40CFR264
  - Land Disposal Restrictions: solvents, dioxins
  - Land Disposal Restrictions: Halogenated Organic Compounds 40CFR268
  - Land Disposal Restrictions: First Third Wastes
  - RCRA Designation #U130
- ◆ CERCLA
  - Hazardous Substance 40CFR302

- Statutory RQ of 1 lb
- Final RQ of 10 lb
◆ Superfund Amendment and Reauthorization Act
- Extremely Hazardous Substance 40CFR355.50
- Toxic Chemical 40CFR372
- SARA Priority List of CERCLA Hazardous Substances
- Health Assessment Risk: 126
◆ Occupational Safety and Health Act
- Air Contaminant 29CFR1910
- Table Z-1-A
◆ American Conference of Governmental Industrial Hygienists
- Threshold Limit Value Chemical
- National Toxic Program
- Testing Program Substance
◆ Department of Transportation
- Hazardous Material Table 49CFR171
- Hazardous Class 6.1
- ID# UN2646
- Hazardous Substance Other than Radionuclides 49CFR172.101

ADDITIONAL/OTHER COMMENTS

◆ First Aid
- eye contact: immediately flush eyes with water for 15 minutes, separating eyelids occasionally
- skin contact: remove contaminated clothing; rinse exposed area with water for 15 minutes
- inhalation: remove to fresh air; if breathing stopped give artificial resuscitation
- discard any contaminated clothing
◆ fire fighting: use carbon dioxide, dry chemical powder, or appropriate foam
◆ Personal Protection
- wear appropriate NIOSH/MSHA approved breathing apparatus
- wear chemical-resistant gloves
- use a face shield
- avoid prolonged exposure
- wash thoroughly after handling
- wear protective clothing
◆ Storage
- keep in a tightly sealed container, in a cool dry place
- moisture sensitive

SOURCES

◆ *Hawley's Condensed Chemical Dictionary*, 12th edition, 1993
◆ *Fire Protection Guide to Hazardous Materials*, 11th edition, 1994
◆ *NIOSH Pocket Guide to Chemical Hazards*, June 1994
◆ *Sax's Dangerous Properties of Industrial Materials*, 9th edition, 1996
◆ *Suspect Chemicals Sourcebook*, 1996

- http://atsdr1.atsdr.cdc.gov:8080/toxfaq.html
- http://chemfinder.camsoft.com
- http://www.epa.gov/oar/oaqps/airtox/hapindex.html
- http://sulco.oea.orst.edu:70/1s/ext/extoxnet/pips

HEXACHLOROETHANE (ns)

CAS #: 67-72-1
MOLECULAR FORMULA: C_2Cl_6
FORMULA WEIGHT: 236.72
SYNONYMS: alvotane, avolthane, carbon hexachloride, distokal, distopan, distopin, egitol, ethane hexachloride, ethylene hexachloride, falkitol, fasciolin, hexachlorethane, hexachloroethane, 1,1,1,2,2,2-hexachloroethane, hexachloroethane (ACGIH:OSHA), hexachloroethylene, mottenhexe, NCI-C04604, perchloroethane, phenohep, RCRA waste number U131

USES
- organic synthesis
- retarding agent in fermentation; rubber accelerator
- ocamphor substitute in nitrocellulose
- pyrotechnics and smoke devices
- added to the feed of ruminants to prevent methanogenesis and increase feed efficiency
- solvent
- explosives
- insecticide
- veterinary antihelmintic to destroy tapeworms
- ignition suppressant
- lube oils additive
- pesticide, acaricide
- intermediate for pharmaceuticals, moth repellents, pyrotechnic materials, and smoke generators
- used in metal and alloy production
- polymer additive

PHYSICAL PROPERTIES
- white crystalline powder
- melting point 186°C at 760 mm Hg
- boiling point sublimes at 185°C
- no flash point
- density 2.091 g/cm³ at 20°C
- critical temperature 442°C
- critical pressure 64.79 atm
- vapor pressure 1 mm Hg at 33°C
- specific gravity 2.091
- vapor density 8.16
- not soluble in water
- soluble in alcohol, ether, hot fluoric acid, benzene, and chloroform
- camphor-like odor

CHEMICAL PROPERTIES
- nonflammable
- carcinogenic
- Water promotes corrosion of metals.
- reacts with metals

HEALTH RISK
- ◆ Toxic Exposure Guidelines
  - OSHA PEL TWA 1 ppm (skin)
  - ACGIH TLV TWA 1 ppm ; suspected human carcinogen
  - DFG MAK 1 ppm (10 mg/m^3)
  - NIOSH REL (hexachloroethane) reduce to lowest level
- ◆ Toxicity Data
  - oral-rat TDLo 5500 mg/kg (6–16D preg) REP
  - oral-mouse TDLo 230 g/kg/78W-ICAR
  - oral-mouse TD 460 g/kg/78W-ICAR
  - oral-rat LD$_{50}$ 4460 mg/kg
  - inhalation-rat LCLo 5900 ppm/8H
  - intraperitoneal-rat LDLo 2900 mg/kg
  - intraperitoneal-mouse LD$_{50}$ 4500 mg/kg
  - intravenous-dog LDLo 325 mg/kg
  - skin-rabbit LD$_{50}$ 32 g/kg
  - subcutaneous-rabbit LDLo 4000 mg/kg
  - oral-guinea pig LD$_{50}$ 4970 mg/kg
  - RfD 0.001 mg/kg/d
  - Group C, possible human carcinogen of low carcinogenic hazard
  - 1/ED$_{10}$ 0.051 per (mg/kg)/d
  - RfD 0.001 mg/kg/d
  - inhalation unit risk estimate 4.0×10^{-6} (μg/m^3)$^{-1}$
- ◆ Acute Risks
  - eye and skin irritation
  - irritating to mucous membranes and upper respiratory tract
  - dizziness
  - headache
  - sleepiness
  - sore throat
  - coughing
  - stupefying
  - prickling
  - pain
  - redness
  - vomiting
  - diarrhea
  - abdominal pain
  - narcosis
  - central nervous system depressant in humans
- ◆ Chronic Risks
  - Prolonged exposure may cause dermatitis.
  - blepharospasm
  - photophobia

- lacrimation
- damage to liver
- may cause cancer

HAZARD RISK
- ◆ produces toxic vapors upon combustion
- ◆ slightly explosive by spontaneous chemical reaction

EXPOSURE ROUTES
- inhalation of contaminated air
- skin absorption
- eye contact
- ingestion of contaminated drinking water

REGULATORY STATUS/INFORMATION
- Clean Air Act
- Chapter 1, Subchapter C—Air Programs, 40CFR51.390-51.464
- Title III: Hazardous Air Pollutants
- ◆ Safe Drinking Water Act
- Chapter 1, Subchapter D: Water Programs, 40CFR141-149
- SDWA, 55FR1470: SDWA priority list of drinking water contaminants
- ◆ Resource Conservation and Recovery Act
- Chapter 1, Subchapter J: Superfund, Emergency Planning and Community Right-to-Know Programs, 40CFR400-699
- RCRA, 40CFR258; Appendix 2: List of Hazardous Inorganic and Organic constituents
- RCRA, 40CFR261; Appendix 8: Hazardous Constituents
- RCRA, 40CFR264; Appendix 9: Groundwater Monitoring Chemical List
- RCRA, 40CFR266; Appendix 5: Risk Specific Doses
- RCRA, 40CFR266; Appendix 7: Health Based Limits for Exclusion of Waste-Derived Residues
- Unlisted Hazardous Wastes Characteristic of Extraction Procedure (EP)
- Toxicity (40CFR261.24 and 302.4)
- RCRA, 40CFR261.33: Discarded Commercial Chemical Products Off-Specification Species, Container Residues, and Spill Residues thereof
- ◆ CERCLA and Superfund
- Superfund, 40CFR302.4: Designation of Hazardous Substances
- Appendix A: Sequential CAS Registry
- Appendix B: Radionuclides
- Superfund, 40CFR372.65
- ◆ Clean Water Act
- CWA, 40CFR413.02; Total Toxic Organics (TTOs)
- CWA, 40CFR423; Appendix A: 126 Priority Pollutants
- ◆ Occupational Safety and Health Act
- Chapter XVII—OSHA Administration Department of Labor, 29CFR1900-1910

- OSHA, 29CFR1910.119; Appendix Z1: Limits for Air Contaminants under Occupational Safety and Health Act
- chemical regulated by the State of California

ADDITIONAL/OTHER COMMENTS

◆ Personal Protection
  - wear appropriate NIOSH/MSHA approved respirator
  - chemical-resistant gloves
  - safety gloves
  - protective clothing
  - safety shower and eye bath
  - use only in chemical fume hood
  - do not breathe dust
  - avoid all contact
  - do not get in eyes, on skin, or on clothing
  - avoid prolonged or repeated exposure
  - wash thoroughly after handling
  - keep tightly closed
  - store in cool, dry place

◆ fire fighting: wear self-contained breathing apparatus and protective clothing to prevent contact with skin and eyes; extinguish with carbon dioxide, dry chemical powder, or appropriate foam

◆ spill or leak: wear self-contained breathing apparatus, rubber boots and heavy rubber gloves, disposable coveralls (discard them after use); sweep up and place in a bag and hold for waste disposal (avoid raising dust); ventilate area and wash spill site after material pickup is complete

◆ First Aid
  - eyes and skin: flush eyes (separate eyelids with fingers) or skin with copious amounts of water for 15 minutes; remove contaminated clothing
  - inhalation: remove to fresh air; if not breathing, give artificial respiration
  - swallowing: wash out mouth with water provided person is conscious; call a physician
  - discard contaminated clothing and shoes

SOURCES

- *Material Safety Data Sheets (MSDS)*
- *Environmental Law Index to Chemicals*, 1996 edition
- *Sax's Dangerous Properties of Industrial Materials*, 9th edition, volume 2, 1996
- ECDIN Databank
- *Hawley's Condensed Chemical Dictionary*, 12th edition, 1993
- *Health Effects Notebook for Hazardous Air Pollutants*, U.S. Environmental Protection Agency, December 1994

HEXAMETHYLENE-1,6-DIISOCYANATE (kjc)

CAS #: 822-06-0
MOLECULAR FORMULA: $C_8H_{12}N_2O_2$
FORMULA WEIGHT: 168.22
SYNONYMS: desmodur h, desmodur n, 1,6-diisocyanatohexane, 1,6-hexanediol diisocyanate, hexamethylendiisokyanat (czech), 1,6-hexamethylene diisocyanate, hexamethylene diisocyanate (DOT), hexamethylene ester, hexamethylene-1,6-diisocyanate, HMDI, isocyanic acid, isocyanic acid, diester with 1,6-hexanediol, metyleno-bis-fenyloizocyjanian (polish), szesciomethylenodwuizocyjanian (polish), TL 78

USES
- production of dental materials, contact lenses, and medical absorbents
- production of polyurethane

PHYSICAL PROPERTIES
- clear, colorless liquid
- melting point 25°C (77°F)
- boiling point 255°C (491°F)
- density 1.053 g/mL at 20°C
- specific gravity 1.040 (water = 1.00)
- vapor density not available
- surface tension not available
- viscosity not available
- heat capacity not available
- heat of vaporization not available
- sharp, pungent odor
- odor threshold 0.001 ppm
- vapor pressure 0.05 mm Hg at 25°C
- solubility 1.4585 mg/L

CHEMICAL PROPERTIES
- combustible
- polymerizes above 392°F
- reacts strongly with water, strong acids, strong bases, amines, carboxylic acids, organotin, heat, strong oxidizing agents
- flash point 140°C (284°F)
- flammability limits not available
- autoignition temperature not available
- heat of combustion not available
- enthalpy of formation not available

HEALTH RISK
- Toxic Exposure Guidelines
  - ACGIH TLV TWA 0.005 ppm (0.034 mg/m³)
  - NIOSH REL TWA 0.005 ppm (35 µg/m³)
- Toxicity Data
  - oral-rat LD_{50} 710 µL/kg
  - oral-mouse LD_{50} 350 mg/kg
  - inhalation-mouse LC_{50} 30 mg/m³
  - intravenous-mouse LD_{50} 5600 µg/kg
  - skin-rabbit LD_{50} 570 µL/kg
- Other Toxicity Indicators
  - California no-significant risk level not available
- Acute Risks
  - irritation of eyes, nose, and throat
  - pulmonary edema
  - shortness of breath

◆ Chronic Risks
  • chronic lung problems
  • depressed weight gain
                                    • target organs: eyes, skin,
                                      respiratory system

HAZARD RISK
◆ NFPA rating none
◆ poisonous gases produced under fire conditions

EXPOSURE ROUTES
◆ inhalation
◆ absorption through skin and eyes possible
◆ ingestion
◆ emissions from burning
◆ occupational exposure when produced or used as a chemical
  intermediate for polyurethane

REGULATORY STATUS/INFORMATION
◆ Clean Air Act Amendments of 1990 Section 112 Statutory Air
  Pollutants
  • list of hazardous air pollutants. 42USC7412
◆ CERCLA and Superfund
  • designation of hazardous substances. 40CFR302.4
  • list of chemicals. 40CFR372.65
  • Reportable Quantity (RQ): 1 lb
  • SARA Title III Section 313 Toxic Chemicals
  • CERCLA Hazardous Substances
◆ Department of Transportation
  • hazard class/division: 6.1
  • identification number: UN2281
  • labels: poison
  • emergency response guide #: 55
◆ Emergency Planning and Community Right-to-Know Act
  • section 313 de minimis concentration: 1.0%

ADDITIONAL/OTHER COMMENTS
◆ Symptoms
  • inhalation: irritates eyes, skin, respiratory system
  • ingestion: coughing, wheezing
◆ First Aid
  • wash eyes immediately with large amounts of water
  • wash skin immediately with soap and water
  • provide respiratory support
◆ fire fighting: use dry chemical or carbon dioxide
◆ Personal Protection
  • wear self-contained breathing apparatus
  • wear protective clothing
  • use only in chemical fume hood
◆ Storage
  • keep tightly closed
  • moisture sensitive
  • store in cool, dry place

SOURCES
◆ *Environmental Containment Reference Databook*, volume 1,
  1994
◆ *Environmental Law Index to Chemicals*, 1996 edition

- *Fire Protection Guide to Hazardous Materials*, 11th edition, 1994
- *Material Safety Data Sheets* (*MSDS*)
- *NIOSH Pocket Guide to Chemical Hazards*, June 1996
- *Sax's Dangerous Properties of Industrial Materials*, 9th edition, volume 2, 1996

HEXAMETHYLPHOSPHORAMIDE (ns)

CAS #: 680-31-9
MOLECULAR FORMULA: $C_6H_{18}N_3OP$
FORMULA WEIGHT: 179.24
SYNONYMS: eastman inhibitor hpt, ent 50882, hempa, hexametapol, hexamethyl phosphoramide (ACGIH), hexamethylphosphoramide, hexamethylphosphoric acid triamide, hexamethylphosphoric triamide, N,N,N,N,N,N-hexamethylphosphoric triamide, hexamethylphosphorotriamide, hexamethylphosphotriamide, hmpa, hmpt, hpt, phosphoric acid hexamethyltriamide, phosphoric tris (dimethylamide), phosphoryl hexamethyltriamide, tri(dimethylamino) phosphineoxide, tris-(dimethylamino) phosphine oxide, tris(dimethylamino) phosphorus oxide

USES
- UV inhibitor in polyvinyl chloride
- chemosterilant for insects
- promoting stereospecific reactions
- specialty solvent
- polymerization catalyst and stabilizer
- antistatic agent
- flame retardant
- intermediate for aramid polyamide fibers, kevlar fabrics, and functional fluids

PHYSICAL PROPERTIES
- water-white liquid
- melting point 7°C
- boiling point 230°C at 760 mm Hg
- flash point 222°F
- vapor pressure 0.07 mm at 25°C
- density 1.024 g/mL at 25°C
- specific gravity 1.030
- vapor density 6.18
- soluble in water and in polar and nonpolar solvents
- mild amine odor
- corrosive

CHEMICAL PROPERTIES
- produces toxic fumes upon combustion
- highly toxic
- combustible
- decomposes on heating
- carcinogenic

HEALTH RISK

◆ Toxic Exposure Guidelines
  • ACGIH TLV suspected human carcinogen—no TWA
  • DFG MAK animal carcinogen, suspected human carcinogen

◆ Toxicity Data
  • oral-rat TDLo 2430 mg/kg (MGN)REP
  • inhalation-rat TCLo 550 ppb/52W-CCAR
  • inhalation-rat TC 400 ppb/35W-ICAR
  • inhalation-rat TD 50 ppb/6H/52W-IETA
  • oral-rat LD_{50} 2650 mg/kg
  • skin-rat LDLo 3500 mg/kg
  • oral-mouse LD_{50} 2400 mg/kg
  • intraperitoneal-mouse LD_{50} 1600 mg/kg
  • intravenous-mouse LD_{50} 800 mg/kg
  • oral-rabbit LDLo 1500 mg/kg
  • skin-rabbit LD_{50} 2600 mg/kg
  • oral-guinea pig LD_{50} 1600 mg/kg
  • skin-guinea pig LD_{50} 1175 mg/kg
  • oral-chicken LD_{50} 835 mg/kg
  • oral-quail LD_{50} 1 g/kg

◆ Acute Risks
  • eye and skin irritation
  • irritating to mucous membranes and upper respiratory tract
  • burning sensation of the throat
  • labored breathing
  • shortness of breath

◆ Chronic Risks
  • abdominal pain
  • may cause bronchopneumonia
  • long-term exposure may affect central nervous and gastro-intestinal systems
  • may alter genetic material
  • may impair male fertility
  • damage to kidneys and lungs
  • may cause cancer

HAZARD RISK

◆ toxic by skin contact, ingestion, intraperitoneal, and intravenous routes
◆ if temperature exceeds flash point then risk explosion
◆ experimentally carcinogenic

EXPOSURE ROUTES

◆ inhalation
◆ ingestion
◆ skin absorption
◆ eye contact

REGULATORY STATUS/INFORMATION

◆ Clean Air Act
  • Chapter 1, Subchapter C—Air Programs, 40CFR51.390-51.464

- Title III: Hazardous Air Pollutants
◆ CERCLA and Superfund
  - Superfund, 40CFR372.65: Chemicals and chemical categories to which this part applies (CAS # listing)
◆ Toxic Substance Control Act
  - Chapter I, Subchapter R—TSCA, 40CFR1500-1517
  - TSCA, 40CFR721.225—9975: Subpart E: Significant new uses of chemical substances

ADDITIONAL/OTHER COMMENTS

◆ First Aid
  - eyes and skin: flush eyes with copious amounts of water for at least 15 minutes and remove contaminated clothing
  - inhalation: remove to fresh air; if not breathing give artificial respiration
  - ingestion: wash out mouth with water provided person is conscious; call a physician
◆ fire fighting: wear self-contained breathing apparatus and protective clothing to prevent contact with skin and eyes; extinguish with water spray, carbon dioxide, dry chemical powder, or appropriate foam
◆ Personal Protection
  - wear appropriate NIOSH/MSHA approved respirator
  - chemical-resistant gloves
  - safety goggles
  - wear protective clothing
  - safety shower and eye bath
  - use only in a chemical fume hood
  - avoid inhalation
  - do not get in eyes, on skin, or on clothing
  - avoid prolonged or repeated exposure
  - wash thoroughly after handling
  - keep tightly closed
  - discard contaminated clothing and shoes
◆ spill or leak: wear self-contained breathing apparatus, rubber boots and heavy rubber gloves, disposable coveralls (discard them after use); cover with dry lime or soda ash, pick up, keep in a closed container and hold for waste disposal (keep out of sewer and surface water); ventilate area and wash spill site after material pickup is completed
◆ storage: store in a cool, dry place

SOURCES

◆ *Environmental Law Index to Chemicals*, 1996 edition
◆ ECDIN Databank
◆ *Sax's Dangerous Properties of Industrial Materials*, 9th edition, volume 2, 1996
◆ *Material Safety Data Sheets (MSDS)*
◆ *Hawley's Condensed Chemical Dictionary*, 12th edition, 1993

HEXANE

CAS #: 110-54-3
MOLECULAR FORMULA: $CH_3(CH_2)_4CH_3$
FORMULA WEIGHT: 86.20
SYNONYMS: esani (Italian), gettysolve-B, heksan (Polish), hexane (DOT), hexanen (Dutch), hexanes (FCC)

USES
- ◆ Used as a Solvent
  - glues
  - varnishes
  - cements
  - inks
  - extraction of edible fats and oils
- ◆ Used in Industry
  - dry cleaners
  - pesticides
  - low temperature thermometers
  - fuel component
  - alcohol denaturant
  - intermediate for pharmaceuticals
  - determination of refractive index of minerals

PHYSICAL PROPERTIES
- ◆ colorless, clear liquid
- ◆ faint odor
- ◆ odor threshold 130 ppm
- ◆ melting point −95.6°C
- ◆ boiling point 69°C
- ◆ density 0.655 at 25°/4°
- ◆ vapor pressure 150 mm Hg at 25°C
- ◆ vapor density 2.97
- ◆ critical temperature 234.3°C
- ◆ critical pressure 29.73 atm
- ◆ insoluble in water
- ◆ miscible in chloroform, ether, and alcohol

CHEMICAL PROPERTIES
- ◆ highly flammable
- ◆ flash point −9.4°F
- ◆ autoignition temperature 437°F
- ◆ lower explosive limit 1.2%
- ◆ upper explosive limit 7.5%

HEALTH RISK
- ◆ Toxic Exposure Guidelines
  - OSHA PEL TWA 50 ppm
  - ACGIH TLV TWA 50 ppm
  - DFG MAK 50 ppm
  - NIOSH REL TWA (alkanes) 350 mg/m^3
- ◆ Toxicity Data
  - eye-rabbit 10 mg MLD
  - cyt-hamster:fbr 500 mg/L
  - inhalation-rat TCLo 10,000 ppm/7H
  - inhalation-human TCLo 190 ppm/8W:PNS
  - oral-rat LD_{50} 28,710 mg/kg
  - intraperitoneal-rat LDLo 9100 mg/kg
  - inhalation-mouse LCLo 120 g/m^3
- ◆ Acute Risks
  - mild central nervous system depression
  - irritation to mucous membranes

- dizziness
- giddiness
- slight nausea
- headache
- dermatitis
- irritation of eyes
◆ Chronic Risks
  - polyneuropathy
  - numbness in extremities
  - muscular weakness
  - blurred vision
  - headache
  - fatigue
  - neurotoxic effects in rats
  - mild inflammatory, erosive, and degenerative lesions in the olfactory and respiratory epithelium of the nasal cavity in mice

- irritation of throat
- light-headedness
- numbness of extremities
- muscle weakness
- chemical pneumonia

- pulmonary lesions in rabbits
- testicular damage in male rats
- teratogenic effects in offspring of rats
- EPA Group D: not classifiable as to human carcinogenicity

HAZARD RISK
◆ flammable liquid
◆ very dangerous fire and explosion hazard when exposed to heat or flame
◆ can react vigorously with oxidizing materials
◆ Mixtures with dinitrogen tetraoxide may explode at 28°C.
◆ formation of toxic vapors
◆ may generate electrostatic charges

EXPOSURE ROUTES
◆ primarily by inhalation
◆ primarily in workplace
◆ Hexane is a widely occurring atmospheric pollutant.

REGULATORY STATUS/INFORMATION
◆ CERCLA and Superfund
  - list of chemicals. 40CFR372.65
◆ TSCA: identification of specific chemical substance and mixture testing requirements. 40CFR799

ADDITIONAL/OTHER COMMENTS
◆ Symptoms
  - unconsciousness, dizziness, headache, sleepiness, nausea, irritation to skin, eyes, nose, and throat
  - ingestion: vomiting, nausea, diarrhea, abdominal pain, drowsiness, labored breath, prickling
◆ First Aid
  - wash eyes immediately with large amounts of water
  - wash skin immediately with soap
  - remove to fresh air immediately or provide respiratory apparatus if necessary
  - drink water or milk
◆ fire fighting: use carbon dioxide or dry chemical

◆ personal protection: compressed air/oxygen apparatus, gas-tight suit, fireproof suit
◆ storage: storage necessary for flammable substances
SOURCES
◆ *Health Effects Notebook for Hazardous Air Pollutants*, U.S. Environmental Protection Agency, December 1994
◆ *Material Safety Data Sheets (MSDS)*
◆ *Sax's Dangerous Properties of Industrial Materials*, 9th edition, 1996
◆ *Environmental Law Index to Chemicals*, 1996
◆ *Merck Index*, 12th edition, 1996
◆ *Hawley's Condensed Chemical Dictionary*, 12th edition, 1993
◆ *Aldrich Chemical Company Catalog*, 1996–1997

HYDRAZINE (ns)

CAS #: 302-01-2
MOLECULAR FORMULA: H_4N_2
FORMULA WEIGHT: 32.06
SYNONYMS: diamide, diamine, hydrazine (ACGIH, OSHA), hydrazine, anhydrous (DOT), hydrazine aqueous solutions with >64% hydrazine by weight (DOT), levoxine, nitrogen hydride, oxytreat 35, RCRA waste number U133, UN2029 (DOT)
USES
◆ chemical blowing agent
◆ agricultural pesticides
◆ water treatment for corrosion protection
◆ pharmaceuticals—the tuberculostatic drug isoniazid
◆ explosives
◆ polymers and polymer additives
◆ antioxidants
◆ metal reductants
◆ hydrogenation of organic groups
◆ photography chemicals
◆ xerography
◆ textile dyes
◆ reactant in fuel cells in military uses; nuclear fuel processing
◆ jet and rocket fuel
◆ electrolytic plating of metals on glass and plastics
◆ water additive for cooling systems and boiler feeders
◆ water treatment for corrosion protection
PHYSICAL PROPERTIES
◆ clear, colorless, oily liquid
◆ freezing point 2.0°C
◆ boiling point 113.5°C
◆ vapor pressure 1.92 kPa/14.4 mm Hg at 25°C
◆ liquid density 1.004 g/mL at 25°C
◆ surface tension 66.45 dynes/cm at 25°C
◆ log K_{ow} 0.08
◆ liquid viscosity 0.913 cP at 25°C
◆ heat of vaporization 39.079 kJ/mol

- ◆ flash point 52°C at COC
- ◆ very soluble in water
- ◆ insoluble in chloroform and ether

CHEMICAL PROPERTIES

- ◆ heat of fusion 12.66 kJ/mol
- ◆ heat capacity 3.0778 J/(g-°K) at 25°C
- ◆ heat of combustion −622.1 kJ/mol
- ◆ heat of formation 50.434 kJ/mol
- ◆ Gibbs free energy of formation 149.24 kJ/mol
- ◆ entropy of formation 121.21 J/(mol-°K)
- ◆ combustible—highly exothermic
- ◆ flammable
- ◆ explodes during distillation if traces of air are present
- ◆ turns salts with inorganic acids
- ◆ highly toxic
- ◆ carcinogenic

HEALTH RISK

- ◆ Toxic Exposure Guidelines
  - OSHA PEL TWA 1 ppm (0.1 mg/m^3)
  - ACGIH TLV TWA 0.1 ppm (air) (0.01 mg/m^3)
  - MSHA 1.3 mg/m^3
  - DFG TRK 0.1 ppm
  - NIOSH REL CL 0.04 mg/m^3/2H
  - DOT Classification 3; Label; Flammable Liquid, Poison, Corrosive
  - odor threshold 3.7 ppm
- ◆ Toxicity Data
  - subcutaneous-rat TDLo 80 mg/kg (female 11–20D post)
  - inhalation-rat TCLo 1 mg/m^3/24H (female 1–11D post)
  - oral-rat TDLo 900 mg/kg/2Y-C:NEO
  - inhalation-rat TCLo 5 ppm/6H/1Y-I:CAR
  - inhalation-mouse TCLo 1 ppm/6H/1Y-I:ETA
  - intraperitoneal-mouse TDLo 400 mg/kg/5W-ICAR
  - inhalation-rat LC$_{50}$ 570 ppm/4H
  - intravenous-rat LD$_{50}$ 59 mg/kg
  - intravenous-rat LD$_{50}$ 55 mg/kg
  - oral-mouse LD$_{50}$ 59 mg/kg
  - inhalation-mouse LC$_{50}$ 252 ppm/4H
  - intraperitoneal-mouse LD$_{50}$ 62 mg/kg
  - intravenous-mouse LD$_{50}$ 57 mg/kg
  - skin-dog LDLo 96 mg/kg
  - intravenous-dog LD$_{50}$ 25 mg/kg
  - skin-rabbit LD$_{50}$ 91 mg/kg
  - intravenous-rabbit LD$_{50}$ 20 mg/kg
  - skin-guinea pig LD$_{50}$ 190 mg/kg
  - EPA Group B2: probable human carcinogen
  - inhalation unit risk estimate 4.9×10^{-3} (µg/m^3)$^{-1}$
  - 1/ED$_{10}$ 107 per (mg/kg)/d
- ◆ Acute Risks
  - itching

- swelling
- blistering
- temporary blindness
- causes burns
- nausea
- dizziness
- headaches
- pulmonary edema
- seizures
◆ Chronic Risks
  - damage to liver, lungs, and central nervous system
  - coma
  - may cause cancer
  - may cause heritable genetic damage
  - Severe exposures may cause death.

HAZARD RISK

◆ decomposes at or above room temperature
◆ Hydrazine vapor in air is flammable at 4.7 to 100% hydrazine by volume—handle with nitrogen.
◆ Monel, bronze, brass, cadmium, gold, molybdenum, and stainless steel with more than 0.5% molybdenum or rust cause decomposition of hydrazine.

EXPOSURE ROUTES

◆ occupational exposure in workplace
◆ inhalation
◆ Accidental discharge into water, air, and soil may occur during storage, handling, transport, and improper waste disposal.
◆ ingestion
◆ skin absorption
◆ eye contact
◆ has been detected in tobacco smoke

REGULATORY STATUS/INFORMATION

◆ Clean Air Act
  - Chapter 1, Subchapter C – Air Programs, 40CFR51.390-51.464
  - Title III: Hazardous Air Pollutants
◆ Resource Conservation and Recovery Act
  - Chapter 1, Subchapter J: Superfund, Emergency Planning and Community Right-to-Know Programs, 40CFR400-699
  - RCRA, 40CFR261.8; Appendix 8: Hazardous Constituents
  - RCRA, 40CFR266; Appendix 5: Risk Specific Doses
  - RCRA, 40CFR266; Appendix 7: Health based limits for exclusion of waste-derived residues
  - RCRA, 40CFR261.33: Discarded commercial chemical products off-specification species, container residues, and spill residues thereof
◆ CERCLA and Superfund
  - Superfund, 40CFR302.4: Designation of Hazardous Substances

- Appendix A: Sequential CAS Registry
- Appendix B: Radionuclides
- Superfund, 40CFR355-Appendix B: List of extremely hazardous substances and their threshold planning quantities (CAS # order)
  - Superfund, 40CFR372.65: Chemicals and chemical categories to which this part applies (CAS # listing)
◆ Occupational Safety and Health Act
  - Chapter XVII—OSH Administration Department of Labor, 29CFR1900-1910
  - OSHA, 29CFR19100.1000; Table Z1: Limits for Air Contaminants under Occupational Safety and Health Act
  - OSHA, 29CFR1910.119; Appendix A: List of Highly Hazardous Chemicals, Toxics and Reactives

ADDITIONAL/OTHER COMMENTS
◆ Personal Protection
  - wear appropriate NIOSH/MSHA approved respirator
  - chemical-resistant gloves
  - safety goggles
  - protective clothing
  - safety shower and eye bath
  - face shield (8-in. minimum)
  - do not breathe vapor
  - do not get in eyes, on skin, or on clothing
  - avoid prolonged or repeated exposure
  - wash thoroughly after handling
  - keep tightly closed
  - keep away from open flame and heat
  - store in a cool, dry place
  - wash contaminated clothing before reuse and discard contaminated shoes
◆ fire fighting: extinguish with water spray, carbon dioxide, or dry chemical powder; wear self-contained breathing apparatus and protective clothing to prevent contact with skin and eyes
  - spill or leak: eliminate all ignition sources; use water spray to cool and disperse vapors, protect personnel and dilute spills to form nonflammable mixtures; control runoff and isolate discharged material for proper disposal
◆ First Aid
  - eyes and skin: immediately flush eyes or skin (separate the eyelids with fingers) with copious amounts of water for 15 minutes and remove contaminated clothing
  - inhalation: remove to fresh air and observe for delayed development of symptoms
  - ingestion: do not induce vomiting; give egg whites or other emollient; wash out mouth with water provided person is conscious and call a physician immediately

SOURCES
◆ *Environmental Contaminant Reference Databook*, volume 1, 1995
◆ *Environmental Law Index to Chemicals*, 1996 edition
◆ *Fire Protection Guide to Hazardous Materials*, 11th edition, 1994
◆ *Material Safety Data Sheets, (MSDS)*
◆ *Sax's Dangerous Properties of Industrial Materials*, 9th edition, volume 2, 1996
◆ ECDIN Databank
◆ *Encyclopedia of Chemical Technology*, 4th edition, volume 13
◆ *Hawley's Condensed Chemical Dictionary*, 12th edition, 1993
◆ *Suspect Chemicals Sourcebook*, volume 1, 1996
◆ *Health Effects Notebook for Hazardous Air Pollutants*, U.S. Environmental Protection Agency, December 1994

HYDROCHLORIC ACID (ns)

CAS #: 7647-01-0
MOLECULAR FORMULA: HCl
FORMULA WEIGHT: 36.5
SYNONYMS: anhydrous hydrochloric acid (UN 1050), aqueous hydrogen chloride, chlorohydric acid, EPA pesticide chemical code 045901, hydrochloric acid solution (UN 1789), hydrochloride, hydrogen chloride (HCl), hydrogen chloride—refrigerated liquid (UN 2186), muriatic acid, spirits of salt, toilet bowl cleaner

USES
◆ Used in Environmental Applications
  • ore treatment including extraction, separation, purification and water
  • treatment
  • recovery of semiprecious metals
  • pH control
  • neutralization of waste streams
◆ Used in Chemical Processes
  • chemical intermediate
  • brine acidification prior to electrolysis in chlorine/caustic cells
  • acidizing of petroleum wells
◆ Used in the Production of
  • calcium chloride
  • phosphoric acid
  • chlorine dioxide
  • chloride chemicals
  • vinyl chloride from acetylene
  • alkyl chlorides from olefins
◆ converts ethanol to ethyl chloride
◆ polymerization, isomerization, alkylation, and nitration reactions
◆ Used in Industrial Processes

- steel pickling and metal cleaning
- industrial acidizing
- boiler and heat exchanger scale removal
- electroplating
- leather tanning
- photographic industry
- textile industry
- prepare synthetic rubber product
- clean and prepare other metal coatings
- recovery of zinc from galvanized iron scrap

◆ General Cleaning
- cleaning of membrane in desalination plants
- alcohol denaturant
- laboratory reagent

◆ Used as a Component in Food Processing
- starch modifier
- sugar refining
- manufacture of high fructose corn syrup for sweetening soft drinks
- hydrolyzed vegetable protein and soy sauce
- manufacture of gelatin and sodium glutamate
- acidification of vegetable juices, canned foods, and artificial sweeteners
- brewing industry

◆ veterinary antiseptic
◆ in toilet bowls and urinals against animal pathogenic bacteria
◆ swimming pool disinfectant
◆ rodenticide
◆ catalyst
◆ pesticides
◆ pigments

PHYSICAL PROPERTIES

◆ clear, colorless liquid or slightly yellow
◆ vapor pressure 3.23 psi at 21.1°C and 7.93 psi at 37.7°C
◆ specific gravity 1.2 at −85°C
◆ melting point −114.2°C (−173.6°F) at 760 mm Hg
◆ boiling point −85°C (−121.0°F) at 760 mm Hg
◆ critical temperature 51.6°C
◆ critical pressure 82.6 atm or 8.32 MPa
◆ critical volume 0.069 L/mol
◆ critical density 424 g/L
◆ irritating, pungent odor
◆ surface tension 23 dynes/cm at 118.16°K
◆ critical compressibility factor 0.117
◆ density 1.045 g/cm^3 at 118.16K and 0.630 g/cm^3 at 319.15K
◆ heat of vaporization 16.14 kJ/mol at −85.05°C
◆ viscosity 0.405 cP at 118.16K
◆ thermal conductivity 335 mW/mol-K at 118.16K
◆ soluble in water, alcohol, and benzene

CHEMICAL PROPERTIES
- enthalpy −92.31 kJ/mol at 198K
- Gibbs free energy −95.30 kJ/mol at 298K
- entropy 186.79 J/mol-K at 298K
- dissociation energy 431.62 kJ at 298K
- highly corrosive
- noncombustible
- usually fuming
- emits toxic vapors when heated to decomposition

HEALTH RISK
- Toxic Exposure Guidelines
  - OSHA PEL CL 5 ppm
  - ACGIH TLV CL 5 ppm
  - DFG MAK 5ppm (7 mg/m^3)
  - DOT classification 8; Label Corrosive (UN 1789); 2,3; Label Poison Gas, Corrosive (UN 1050, UN 2186)
- Toxicity Data
  - eye-rabbit 100 mg rns MLD
  - inhalation-rat TCLo 450 mg/m^3/1H (1D pre):TRE
  - inhalation-human LCLo 1300 ppm/30M
  - inhalation-human LCLo 3000 ppm/5M
  - unreported-man LDLo 81 mg/kg
  - inhalation-rat LC_{50} 3124 ppm/H
  - inhalation-mouse LC_{50} 1108 ppm/H
  - intraperitoneal-rat LDLo 40,142 g/kg
  - intraperitoneal-mouse LD_{50} 1449 mg/kg
  - oral-rabbit LD_{50} 900 mg/kg
  - inhalation-rabbit LCLo 4416 ppm/30M
  - RfC 0.007 mg/m^3
  - not classified with respect to potential carcinogenicity
- Acute Risks
  - hoarseness
  - burning sensation
  - coughing
  - wheezing
  - laryngitis
  - shortness of breath
  - dental discoloration and erosion
  - headache
  - nausea
  - vomiting
  - inflammation and ulceration of the respiratory tract
- Chronic Risks
  - may be fatal if inhaled, swallowed, or absorbed through skin
  - causes burns
  - destructive to tissue of the mucous membranes, upper respiratory tract, eyes, and skin
  - Inhalation may lead to spasms, inflammation of the larynx and bronchi, chemical pneumonitis, and pulmonary edema.

- can damage vision
- gastritis
- dermatitis

HAZARD RISK
- ◆ emits highly toxic vapors when heated to decomposition
- ◆ Aqueous hydrochloric acid solutions react with most metals, forming inflammable H_2 gas.
- ◆ reacts vigorously with alkalies and with many organic materials
- ◆ ignites on contact with fluorine; hexalithium disilicide; metal acetylides or carbides

EXPOSURE ROUTES
- ◆ occupational exposure via inhalation or dermal contact during its production and use
- ◆ inhalation
- ◆ ingestion
- ◆ skin absorption
- ◆ eye contact

REGULATORY STATUS/INFORMATION
- ◆ Clean Air Act
  - Chapter 1, Subchapter C—Air Programs, 40CFR51.390-51.464
  - Title III: Hazardous Air Pollutants
  - CERCLA and Superfund
  - Superfund, 40CFR302.4: Designation of Hazardous Substances
  - Appendix A: Sequential CAS Registry
  - Appendix B: Radionuclides
  - Superfund, 40CFR372.65
- ◆ Clean Water Act
  - CWA, 40CFR116.4: Designation of Hazardous Substances
  - CWA, 40CFR117.3: Determination of Hazardous Substances
- ◆ Occupational Safety and Health Act
  - Chapter XVII—OSHA Administration Department of Labor, 29CFR1900-1910
  - OSHA, 29CFR1910.1000; Table Z1: Limits for Air Contaminants under Occupational Safety and Health Act
  - OSHA, 29CFR1910.119; Appendix A: List of Highly Hazardous Chemicals, Toxics and Reactives
- ◆ FIFRA 40CFR180.1001; HCl is exempted from the requirement of a tolerance when used in accordance with good agricultural practice as inert ingredients on pesticide formulations applied to growing crops only.

ADDITIONAL/OTHER COMMENTS
- ◆ should be manufactured in closed systems
- ◆ First Aid
  - eyes and skin: flush eyes (separate the eyelids with fingers) or skin with water for 15 minutes; remove contaminated clothing and shoes

- inhalation: remove to fresh air or apply artificial respiration if not breathing
- ingestion: wash out mouth with water if person is conscious; call a physican immediately; have person drink milk and do not induce vomiting

◆ fire fighting: wear special protective clothing and positive pressure self-contained breathing apparatus; use water; neutralize with chemically basic substances such as soda ash or slaked lime

◆ disposal considerations: dilute through addition of ice water, quench and neutralize with lime or caustic solution; route to sewage plant

◆ Personal Protection
- chemical safety goggles
- face shield (8-in. minimum)
- NIOSH/MSHA approved respirator in nonventilated area
- rubber gloves
- avoid breathing vapor
- mechanical exhaust required
- do not get in eyes, on skin, or on clothing
- avoid prolonged repeated exposure
- wash thoroughly after handling
- safety shower and eye bath (if needed)
- keep tightly closed
- store in a cool, dry place
- wash contaminated clothing, before reuse and discard contaminated shoes

◆ spill or leak: stop or control leak; use water fog or spray to knock down and absorb vapors; control runoff and isolate discharged material for proper disposal

SOURCES

◆ *Environmental Contaminant Reference Databook*, volume 1, 1995

◆ *Environmental Law Index to Chemicals*, 1996 edition

◆ *Fire Protection Guide to Hazardous Materials*, 11th edition, 1994

◆ *Material Safety Data Sheets (MSDS)*

◆ *Sax's Dangerous Properties of Industrial Materials*, 9th edition, volume 2, 1996

◆ ECDIN Databank

◆ *Encyclopedia of Chemical Technology*, 4th edition, volume 13

◆ *Hawley's Condensed Chemical Dictionary*, 12th edition, 1993

◆ *Suspect Chemicals Sourcebook*, volume 1, 1996

◆ *Health Effects Notebook for Hazardous Air Pollutants*, U.S. Environmental Protection Agency, December 1994

HYDROGEN FLUORIDE (dt)

CAS #: 7664-39-3
MOLECULAR FORMULA: HF
FORMULA WEIGHT: 20.01
SYNONYMS: acide fluorhydrique (French), acido fluoridrico (Italian), fluorowodor (Polish), fluorwasserstoff (German), fluorwaserstof (Dutch), hydrofluoric acid, hydrogen fluoride (ACGIH, OSHA), hydrogen fluoride anhydrous (UN1052) (DOT), RCRA waste number U134, rubigine, UN1052 (DOT), UN1790 (DOT)

USES
- ◆ Manufacture of Other Chemicals
  - decomposition of cellulose
  - fluorination of aluminum
  - separating uranium isotopes
  - making plastic
  - metals
  - gasoline
- ◆ Used as a Component or in the Manufacturing Process
  - cleaning cast iron, copper, brass
  - removing efflorescence from brick, stone, or sand particle
  - working over heavily weighted silks
  - frosting
  - polishing crystal glass
  - etching glass and enamel
  - dye chemistry
  - laundry mixture in place of oxalic acid
  - clouding electrical bulbs
  - active constituent in drinking water
  - catalyst in allylation
  - isomerization, condensation, and dehydration

PHYSICAL PROPERTIES
- ◆ clear and colorless
- ◆ odor threshold 0.042 ppm
- ◆ irritating odor
- ◆ melting point −83.1°C
- ◆ boiling point 19.54°C at 1 atm
- ◆ liquid density 0.699 g/L at 22°C
- ◆ gas density 0.901 g/L
- ◆ vapor pressure 400 mm Hg at 2.5°C
- ◆ specific volume 17 cu ft/lb at 21.1°C, 1 atm
- ◆ very soluble in water and alcohol; slightly soluble in ether
- ◆ specific gravity 1.150 (water 1.00)

CHEMICAL PROPERTIES
- ◆ heat of formation −71.65 kcal/mol at 25°C
- ◆ dissociation constant $Ka = 6.46 \times 10^{-4}$ mol/L
- ◆ heat of vaporization 1.8 kcal/mol
- ◆ heat of fusion 1094 cal/gmole
- ◆ when heated, emits highly corrosive fumes

HEALTH RISK
- ◆ Toxic Exposure Guidelines
  - NIOSH 3 ppm
  - OSHA PEL TWA 3 ppm
  - IDLH 30 ppm
  - STEL 6 ppm
  - ACGIH TLV CL 3 ppm
  - DFG MAK 3 ppm
  - DOT label corrosive and poison
- ◆ Toxicity Data
  - inhalation-rat TCLo 0.470 mg/m³/4H (1–22 D Preg): REP
  - inhalation-human LCLo 50 ppm/30M
  - ipr-rat LDLo 25 mg/kg
  - inhalation-mouse LC_{50} 342 ppm/1H
  - skin-mouse LDLo 500 mg/kg
  - inhalation-monkey LC_{50} 1774 ppm/1H
  - inhalation-rabbit LCLo 260 mg/m³/7H
  - inhalation-gpg LC_{50} 4327 ppm/15M
  - dnd-dmg-inh 1300 ppb/6W
  - sln-dmg-ihl 2900 ppb
  - inhalation-rat TCLo 4980 µg/m³/4H (1–22D Preg): TER
  - inhalation-man TCLo 100 mg/m³/1M
  - scu-frg LDLo 112 mg/kg
- ◆ Acute Risks
  - highly toxic via ingestion or inhalation or when absorbed through skin
  - extremely destructive to tissue of the mucous membranes, upper respiratory tract, eyes, and skin
  - Symptoms of exposure may include burning sensation, coughing, wheezing, laryngitis, shortness of breath, headache, nausea and vomiting.
  - Concentrations of 50–250 ppm are dangerous even for brief exposure.
  - bleaching hair
- ◆ Chronic Risks
  - target organs: liver, kidney, eyes, skin, respiratory system

HAZARD RISK
- ◆ severe health hazard
- ◆ not combustible, but if involved in a fire it is extremely irritating
- ◆ evolves heat when combined with water
- ◆ corrosive
- ◆ may produce pain or visible damage if in contact with dilute solution (< 20% in water)
- ◆ reacts to form hydrogen gas on contact with metals

EXPOSURE ROUTES
- ◆ inhalation: emissions from industrial processes, coal combustion, volcanic activity and dust, weathering of fluoride-containing rocks and soil, cigarette smoking
- ◆ ingestion: drinking water dosed with fluoride; food

◆ skin/eye contact

REGULATORY STATUS/INFORMATION

◆ Clean Air Act hazardous air pollutant, original list published in 42USC7412
◆ Resource Conservation and Recovery Act
  • hazardous constituents 40CFR261, 40CFR261.11
  • U waste # U134
◆ CERCLA and Superfund
  • designation of hazardous substance 40CFR302.4
  • list of extremely hazardous substances and their threshold planning quantities
  • 40CFR355-AB
  • chemicals and chemical categories 40CFR372.65
  • Occupational Safety and Health Act
  • limits for air contaminants applicable for the transitional period 29CFR1910.1000, Table Z2
  • list of highly hazardous chemicals, toxics and reactives 29CFR1910.119 appendix A

ADDITIONAL/OTHER COMMENTS

◆ personal protection: wear plastic lensed eye goggles, self-contained air-line or canister-type respiratory apparatus, rubberized acid protective garment
◆ cleanup: ventilate the spillage or leakage area
◆ Symptoms
  • serious and painful burns to eyes and skin
  • irritation in eyes, nose, and throat
  • corneal ulcer
  • bleaching hair
◆ First Aid
  • skin: flush with water immediately
  • eye: irrigate and seek medical attention
  • inhalation: respiratory support
  • ingestion: medical attention
  • storage: Store in cool, dry, well-ventilated location separate from silica, incompatible metals, concrete, glass, ceramics, and oxidizing metals. Do not put even dilute solution in a glass container.

SOURCES

◆ *Sax's Dangerous Properties of Industrial Materials*, 9th edition, volume 3, 1996
◆ *Material Safety Data Sheet (MSDS)*
◆ *Environmental Contaminants Reference Databook*, volume 1, 1995
◆ *Merck Index*, 12th edition, 1996
◆ *Fire Protection Guide to Hazardous Materials*, 11th edition, 1994
◆ *NIOSH Pocket Guide to Chemical Hazards*, June 1994
◆ *Environmental Law Index to Chemicals*, 1996 edition
◆ *CRC Handbook of Chemistry and Physics*, 70th edition, 1989–1990

HYDROQUINONE (dt)

CAS #: 123-31-9
MOLECULAR FORMULA: $C_6H_6O_2$
FORMULA WEIGHT: 110.12
SYNONYMS: arctuvin, benzene p-dihydroxy-, p-benzenediol, 1,4-benzenediol, benzohydroquinone, benzoquinol, black and white bleaching cream, 1,4-dihydroxy-benzen (Dutch), 1,4-dihydroxybenzen, dihydroxybenzene (OSHA), 1,4-Dihydroxy-benzol (German), 1,4-diidrobenzene (Italian), p-dioxobenzene, p-dioxybenzene, eldopaque, eldoquin, hydrochinon (Czech, Polish), hydroquinol, hydroquinole, alphahydroquinone, p-hydroquinone, hydroquinone (ACGIH:OSHA), hydroquinone (liquid or solid) (DOT), p-hydroxyphenol, idrochinone (Italian), NCI-C55834, phiaquin, pyrogentistic acid, quinol, betaquinol, tecquinol, tenox HQ, tequinol, UN2662 (DOT), USAF EK-356

USES
- ◆ Manufacture of Other Chemicals
  - chemical intermediate for dyes
  - food and rubber antioxidant
  - depigmenter used as polymerization inhibitor for vinyl acetate and acrylic monomers
  - stabilizer in paints and varnishes
  - motor fuels and oils
  - antioxidant for fats and oils
- ◆ Used as a Component or in the Manufacturing Process
  - photographic reducer and developer
  - reagent in determining small quantities of phosphate
  - polymerization inhibitor
  - in human medicine for skin blemishes

PHYSICAL PROPERTIES
- ◆ colorless, hexagonal prisms, white crystals, monoclinic prisms
- ◆ melting point 170–171°C
- ◆ boiling point 285–287°C at 730 mm Hg
- ◆ liquid density 1.358 g/L at 20°C
- ◆ vapor density 3.81 g/L
- ◆ vapor pressure 4 mm Hg at 150°C
- ◆ very soluble in ether and alcohol; slightly soluble in benzene
- ◆ specific gravity 1.3 (water = 1.00)
- ◆ dissociation constant 9.96
- ◆ not soluble in water
- ◆ octanol/water partition coefficient 0.590

CHEMICAL PROPERTIES
- autoignition temperature 960°F (498°C)
- combustible
- rapid oxidation in presence of alkali and light
- air oxidizes to quinone and produces brown solution
- flash point 329°F

HEALTH RISK
- ◆ Toxic Exposure Guidelines

- NIOSH REL CL 2 mg/m^3/15M
- OSHA PEL TWA 2 mg/m^3
- IDLH 50 ppm
- ACGIH TLV TWA 2 mg/m^3
- DFG MAK 2 mg/m^3
- DOT: classification 6.1; container should be labeled "keep away from food"
- MSHA standard-air TWA 2 mg/m^3

◆ Toxicity Data
  - skin-human 2% MLD
  - skin-human 5% SEV
  - oral-rat LD$_{50}$ 320 mg/kg
  - ipr-rat LD$_{50}$ 170 mg/kg
  - oms-human-lym 5 µmol/L
  - sce-hmn-lym 5 µmol/L
  - mnt-mus-oral 200 mg/kg
  - oral-human LDLo 29 mg/kg
  - ihv-rat LDL$_{50}$ 245 mg/kg
  - ipr-mouse LD$_{50}$ 100 mg/kg
  - ipr-rabbit LD$_{50}$ 125 mg/L
  - scu-mouse LD$_{50}$ 182 mg/kg

◆ Acute Risks
  - moderately toxic if swallowed or inhaled
  - irritation to tissue of the mucous membranes, upper respiratory tract, eyes, and skin
  - may cause allergic reaction
  - skin burn
  - increased pulse rate without fall in blood pressure

◆ Chronic Risks
  - carcinogen
  - target organs: kidney, eyes
  - Poisoning causes anemia and a general loss of pigment.

HAZARD RISK
◆ human poison by ingestion
◆ strong oxidizing agent
◆ air and light sensitive
◆ allergen and irritant
◆ combustible if exposed to heat or flame
◆ explosive reaction with oxygen

EXPOSURE ROUTES
◆ inhalation: cigarette smoke and diesel exhaust
◆ ingestion: food
◆ skin/eye contact: individuals who develop black-and-white film
◆ skin absorption

REGULATORY STATUS/INFORMATION
◆ Clean Air Act, hazardous air pollutant, original list published in 42USC7412
◆ Toxic Substance Control Act
  - list of substances: 40CFR716.120.a

- chemicals subject to test rules or consent orders for which the testing reimbursement period has passed: 40CFR799.18
- ◆ CERCLA and Superfund
  - list of extremely hazardous substances and their threshold planning quantities 40CFR355-AB
  - chemicals and chemical categories 40CFR372.65
- ◆ Occupational Safety and Health Act
  - limits for air contaminants under OSHA 29 CFR1910.1000 Table Z1

ADDITIONAL/OTHER COMMENTS
- ◆ Symptoms
  - Ingestion may cause ringing in the ears, nausea, dizziness, suffocation, vomiting, muscular delirium, and collapse.
  - Direct particulate contamination of eye can cause irritation and may result in ulcer of cornea.
- ◆ Cleanup
  - ventilate the spillage or leakage area after pickup is completed
  - evacuate area and avoid raising dust
  - sweep up, place in a bag, and hold for waste disposal
- ◆ First Aid
  - skin: wash with soap and flush immediately
  - eye: irrigate and seek medical attention
  - inhalation: respiratory support
  - ingestion: medical attention
- ◆ personal protection: wear NIOSH/MSHA approved respirator, chemical-resistant gloves, safety goggles, and protective clothing. Do not wear contact lenses. Use only in fume hood.
- ◆ storage: Store in cool, dry place, under nitrogen. Keep it tightly closed.

SOURCES
- ◆ *Sax's Dangerous Properties of Industrial Materials*, 9th edition, volume 3, 1996
- ◆ *Material Safety Data Sheet* (*MSDS*)
- ◆ *Environmental Contaminants Reference Databook*, volume 1, 1995
- ◆ *Merck Index*, 12th edition, 1996
- ◆ *Fire Protection Guide to Hazardous Materials*, 11th edition, 1994
- ◆ *NIOSH Pocket Guide to Chemical Hazards*, June 1994
- ◆ *Environmental Law Index to Chemicals*, 1996 edition

ISOPHORONE (dt)

CAS #: 78-59-1
MOLECULAR FORMULA: $C_9H_{14}O$
FORMULA WEIGHT: 138.2
SYNONYMS: isoacetophorone, isoforon, isoforone (Italian), isophoron, isophorone (ACGIH, OSHA), izoforon (Polish), NCI-C55618,

1,1,3-trimethyl-3-cyclohexene-5-one-1-one, 3,3,5-trimethyl-5-3,3,5-tri-methyl-5-cyclohexen-1-one, 3,5,5-trimethyl-2-cyclohexene-1-one, 3,5,5-Trimethyl-2-cyclohexen-1-on (German, Dutch), 3,5,5-trimetil-2-cicloesen-1-one (Italian).

USES
- ◆ Manufacture of Other Chemicals
  - pesticides
  - for polyvinyl and nitrocellulose resins
  - in solvent mixture for finishes
  - solvent in lacquers, printing ink, paints, adhesives
  - metals
  - gasoline

PHYSICAL PROPERTIES
- ◆ water-white-liquid with a camphor-like odor
- ◆ bulk density 7.7 lb/gal at 20°C
- ◆ melting point −8°C
- ◆ boiling point 213–214°C at 760 mm Hg
- ◆ liquid density 0.923g/L at −8°C
- ◆ index of refraction for the sodium D line at 20°C
- ◆ vapor pressure 0.2 mm Hg at 20°C
- ◆ viscosity 2.62 cP at 20°C
- ◆ soluble in vinyl resins, cellulose ester, ether, and acetone alcohol; not soluble in water
- ◆ specific gravity 0.92

CHEMICAL PROPERTIES
- ◆ flash point 84°C
- ◆ autoignition temperature 864°F
- ◆ flammability limit lower 0.8%, upper 3.8% (by volume in air)
- ◆ reacts with oxidizing materials, strong alkalies, and amines
- ◆ combustible
- ◆ octanol/water partition coefficient 1.67
- ◆ odor threshold 0.2 ppm

HEALTH RISK
- ◆ Toxic Exposure Guidelines
  - NIOSH REL TWA (ketone) 23 mg/m^3
  - OSHA PEL TWA 25 ppm
  - IDLH 200 ppm
  - ACGIH TLV CL 5 ppm
  - DFG MAK 5 ppm
  - MSHA standard-air TWA 25 ppm
- ◆ Toxicity Data
  - eye-human 25 ppm/15M
  - skin-rabbit 100 mg/24H MLD
  - eye-rabbit 920 µg SEV
  - eye-guinea pig 840 ppm 14H SEV
  - mutation in mamalian somatic cells-mouse lymphocyte 1 g/L
  - sister chromatid exchange hamster-ovary 1 g/L
  - oral-mouse TDL 258 g/kg/2Y-I, CAR
  - inhalation-hamster TCLo 25 ppm nose, eye, pul

- oral-rat LD_{50} 1870 mg/kg
- inhalation-rat LCLo 1840 ppm/4H
- oral-mouse LD_{50} 2690 mg/kg
◆ Acute Risks
  - severe irritant to skin and eyes
  - irritating to mucous membranes and upper respiratory tract
  - chest tight, sore throat, asthma, possible respiratory sensitization, breathing difficulty, cough, bronchitis
◆ Chronic Risks
  - target organs: lungs, kidney, eyes, skin, central nervous system
  - carcinogen
  - Exposure can cause central nervous system depression.

HAZARD RISK
◆ toxic
◆ combustible liquid
◆ risk of serious damage to eyes
◆ Breathing vapor can be hazardous.
◆ moderately toxic if ingested
◆ low volatility
◆ flammable and exposive when exposed to heat or flame

EXPOSURE ROUTES
◆ inhalation: airborne in printing, coal-fired power plants, exposure to ink, paints, lacquers, and adhesives, and metal coating industries
◆ ingestion: drinking water
◆ skin/eye contact
◆ skin absorption: ink

REGULATORY STATUS/INFORMATION
◆ Clean Air Act hazardous air pollutant, original list published in 42USC7412
◆ Resource Conservation and Recovery Act
  - list of hazardous inorganic and organic constituents RCRA 40CFR258-A2
  - groundwater monitoring chemical list: RCRA, 40 CFR264-A9
◆ CERCLA and Superfund
  - designation of hazardous substance suuperfund 40CFR302.4
◆ Occupational Safety and Health Act
  - limits for air contaminants under the OSHA 29CFR1910.1000, Table Z2
◆ Clean Water Act
  - toxic pollutant: CWA, 40CFR401.15
  - total organics: CWA, 40CFR413.02
  - 126 priority pollutants: CWA, 40CFR423 Appendix A
◆ TSCA: list of substances: TSCA, 40CFR716.120a

ADDITIONAL/OTHER COMMENTS

◆ symptoms: Symptoms of exposure may include burning sensation, coughing, wheezing, laryngitis, shortness of breath, headache, nausea and vomiting.

◆ cleanup: stop or control leak if this can be done without undue risk; use water spray to cool and disperse vapors; control runoff and isolate discharge material for proper disposal

◆ First Aid
  • skin: flush with water immediately
  • eye: irrigate and seek medical attention
  • inhalation: respiratory support
  • ingestion: medical attention

◆ personal protection: wear safety goggles, appropriate respirator, chemical-resistant gloves; avoid contact and do not breathe vapor

◆ storage: Store in cool, dry, well-ventilated location separate from oxidizing agents. Outside or detached storage is preferred.

SOURCES

◆ *Sax's Dangerous Properties of Industrial Materials*, 9th edition, volume 3, 1996
◆ *Material Safety Data Sheet (MSDS)*
◆ *Merck Index*, 12th edition, 1996
◆ *Fire Protection Guide to Hazardous Materials*, 11th edition, 1994
◆ *NIOSH Pocket Guide to Chemical Hazards*, June 1994
◆ *Environmental Law Index to Chemicals*, 1996 edition
◆ *Hawley's Condensed Chemical Dictionary*, 12th edition, 1993

LEAD COMPOUNDS

CAS #: Lead compounds have variable CAS #s. The CAS # for lead is 7439-92-1.

MOLECULAR FORMULA: Lead compounds have variable molecular formulas. The molecular formula for lead is Pb.

Note: For all hazardous air pollutants (HAPs) that contain the word "compounds" in the name, and for glycol ethers, the following applies: Unless otherwise specified, these HAPs are defined by the 1990 Amendments to the Clean Air Act as "including any unique chemical substance that contains the named chemical (i.e., antimony, arsenic, etc.) as part of that chemical's infrastructure."

FORMULA WEIGHT: Lead compounds have variable formula weights. The formula weight for lead is 207.2.

SYNONYMS: glover, lead: C.I. pigment metal, lead flake, lead S2, olow (Polish), omaha, omaha & grant, SI, SO

USES

◆ Used in Industry Production
  • batteries
  • metal products
  • sheet lead
  • solder
  • pipes
  • gasoline additives

- ammunition
- cable covering
- roofing
- devices to shield X-rays
- tank linings
- X-ray and atomic radiation protection
- manufacture of tetraethyllead

- pigments for paints
- ceramics
- plastics
- electronic devices
- building construction
- Babbitt metal alloys
- foil

PHYSICAL PROPERTIES

- bluish-gray, soft metal
- present as a variety of compounds
- melting point 327.43°C
- boiling point 1740°C
- specific gravity 11.3
- vapor pressure 1.0 mm Hg at 980°C
- Pure lead is insoluble in water.

- Lead compounds vary in solubility from insoluble to water soluble
- viscosity 3.2 cP at 327.4°C (molten lead)
- Brinell hardness 4.0 (high purity lead)
- poor electrical conductor
- good sound and vibration absorber

CHEMICAL PROPERTIES

- noncombustible
- heat of vaporization 206 cal/g at 1760°C
- heat capacity 0.031 cal/g/C at 20°C
- resistivity 20.65 μ-ohm-cm at 20°C
- Thermal conductivity varies from 0.083 at 50°C to 0.077 at 225°C.
- reacts with hot concentrated nitric acid
- reacts with boiling hydrochloric or sulfuric acid
- attacked by pure water
- attacked by weak organic acids in presence of O_2
- resistant to tap water, hydrofluoric acid, brine, and solvents

HEALTH RISK

- Toxic Exposure Guidelines
  - ACGIH TLV TWA 0.15 mg/m^3
  - NIOSH REL TWA 0.10 mg/m^3
  - OSHA PEL TWA 0.05 mg/m^3
  - NAAQS 0.0015 mg/m^3
- Toxicity Data
  - LC_{50} (tetramethyl lead) 8870 mg/m^3
  - LC_{50} (tetraethyl lead) 850 mg/m^3
- Acute Risks
  - death from lead poisoning
  - brain and kidney damage
  - gastrointestinal symptoms
  - effects on central nervous system, especially in children
  - colic
  - anorexia

  - vomiting
  - malaise
  - convulsions due to increased intracranial pressure
  - loss of appetite
  - insomnia
  - headache

- irritability
- muscle and joint pains
- tremors
◆ Chronic Risks
  - affects the blood
  - anemia
  - affects central nervous system
  - neurological symptoms
  - slowed nerve conduction in peripheral nerves
  - affects hearing threshold and growth in children
  - effects on blood pressure
  - effects on kidney function
  - interference with vitamin D metabolism
  - severe depression of sperm count
  - decreased function of the prostate
  - decreased function of the seminal vesicles
  - spontaneous abortion
- flaccid paralysis
- hallucinations and distorted perceptions

- toxic effects on the human fetus
- increased risk of preterm delivery
- low birth weight
- impaired mental development
- decreased IQ scores in children
- kidney cancer in rats and mice
- The carbonate, the monoxide, and the sulfate of lead are more toxic than metallic lead or other lead compounds.
- Lead arsenate is very toxic.
- EPA Group B2: probable human carcinogen

HAZARD RISK
◆ flammable in the form of dust when exposed to heat or flame
◆ moderately explosive in the form of dust when exposed to heat or flame
◆ Mixtures of hydrogen peroxide + trioxane explode on contact with lead.
◆ Rubber gloves containing lead may ignite in nitric acid.
◆ violent reaction on ignition with chlorine trifluoride, concentrated hydrogen peroxide, ammonium nitrate, and sodium acetylide
◆ incompatible with NaN_3, Zr, disodium acetylide, and oxidants
◆ can react vigorously with oxidizing materials
◆ common air contaminant
◆ Decomposition emits highly toxic fumes of lead.

EXPOSURE ROUTES
◆ leaded gasoline combustion
◆ combustion of solid waste, coal, and oils
◆ emissions from iron and steel production
◆ lead smelters
◆ tobacco smoke
◆ food and soil
◆ lead-based paints
◆ flaking paint
◆ paint chips
◆ weathered paint powder
◆ drinking water
◆ pipes, solder, and fixtures
◆ lead smelting and refining industries
◆ steel and iron factories
◆ gasoline stations
◆ battery manufacturing plants

REGULATORY STATUS/INFORMATION
- ◆ CERCLA and Superfund
  - designation of hazardous substances. 40CFR302.4
- ◆ chemical regulated by the State of California

ADDITIONAL/OTHER COMMENTS
- ◆ Symptoms
  - ingestion and inhalation: loss of appetite, anemia, malaise, insomnia, headache, irritability, muscle and joint pains, tremors, flaccid paralysis, hallucinations, muscle weakness, gastric and liver changes
- ◆ First Aid
  - wash eyes immediately with large amounts of water
  - for hot metal burns, cool exposed area with water
  - remove to fresh air immediately
- ◆ fire fighting: dry chemicals or sand used with molten metals
- ◆ personal protection: self-contained breathing apparatus, protective clothing, and goggles

SOURCES
- ◆ *Health Effects Notebook for Hazardous Air Pollutants*, U.S. Environmental Protection Agency, December 1994
- ◆ *Material Safety Data Sheets (MSDS)*
- ◆ *Sax's Dangerous Properties of Industrial Materials*, 9th edition, 1996
- ◆ *Hawley's Condensed Chemical Dictionary*, 12th edition, 1993
- ◆ *Environmental Law Index to Chemicals*, 1996
- ◆ *Merck Index*, 12th edition, 1996
- ◆ Agency for Toxicological Substances and Disease Registry

LINDANE (ALL ISOMERS)

CAS #: 58-89-9
MOLECULAR FORMULA: $C_6H_6Cl_6$
FORMULA WEIGHT: 290.82
SYNONYMS: aalindan, aficide, agrisol G-20, agrocide, agronexit, ameisenatod, ameisenmittel merck, aparsin, aphtiria, aplidal, arbitex, BBH, ben-hex, bentox 10, gamma-benzene hexachloride, bexol, BHC, gamma-BHC, celanex, chloresene, codechine, DBH, detmol-extrakt, detox 25, devoran, dol granule, drill tox-spezial aglukon, ENT 7796, entomoxan, exagama, forlin, gallogama, gamacid, gamaphex, gamene, gamiso, gamma-col, gammahexa, gammahexane, gammalin, gammo-paz, HCCH, HCH, gamma-HCH, heclotox, hexachloran, gamma-hexachloran, gamma-hexachlorane, gamma-hexachlorobenzene, 1-alpha, 2-alpha, 3-beta, 4-alpha, 5-alpha, 6-beta-hexachlorocyclohexane, gamma-hexachlorocyclohexane (MAK), 1,2,3,4,5,6-hexachlorocyclo-hexane, gamma-isomer, hexatox, hexicide, HGI, inexit, isotox, jacutin, kokotine, kwell, lendine, lentox, lidenal, lindagrain, lintox, milbol 49, mszycol, neo-scabicidol, nexit, novigam, ovadziak, pedraczak, quella-da, sang gamma, streunex, tap 85, viton

USES
- ◆ Used as an Insecticide

- field crops
- corn
- wheat
- ornamentals
- pasture
- forage crops
- forestry
- timber protection
- livestock
- soil and seed treatment
- viticulture
◆ Used as a Medication
  - pediculicide
  - scabicide
  - ectoparasiticide (vet)
  - treatment of head and body lice and scabies

PHYSICAL PROPERTIES

◆ white powder
◆ slight musty odor
◆ odor threshold 12 ppm
◆ melting point 112.5°C
◆ boiling point 323.4°C at 760 mm Hg
◆ density 1.85
◆ vapor pressure 9.4×10^{-6} mm Hg at 20°C
◆ insoluble in water
◆ log K_{ow} 3.72

CHEMICAL PROPERTIES

◆ stable to heat, light, and oxidation
◆ corrosive to metals
◆ volatile in air
◆ incompatible with strong oxidizing agents

HEALTH RISK

◆ Toxic Exposure Guidelines
  - OSHA PEL TWA 0.5 mg/m³ (skin)
  - ACGIH TLV TWA 0.5 mg/m³ (skin)
  - NIOSH REL 0.5 mg/m³ (skin)
  - DFG MAK 0.5 mg/m³
◆ Toxicity Data
  - dns-ofs:lvr 45 μmol/L
  - msc-hamster:lng 200 mg/L
  - oral-rabbit TDLo 60 mg/kg
  - oral-dog TDLo 473 mg/kg
  - oral-rat TDLo 100 mg/kg
  - oral-rabbit TDLo 260 mg/kg
  - intratesticular-rat TDLo 10 mg/kg
  - oral-rat TDLo 200 mg/kg
  - oral-mouse TDLo 14 g/kg/2Y-C:CAR
  - oral-mouse TD 25 g/kg/73W-C:NEO
  - oral-child LDLo 180 mg/kg:CNS,PUL
  - oral-child TDLo 111 mg/kg:CNS
  - skin-man TDLo 20 mg/kg/6W
  - oral-rat LD$_{50}$ 76 mg/kg
  - skin-rat LD$_{50}$ 500 mg/kg
  - intraperitoneal-rat LDLo 35 mg/kg
  - oral-mouse LD$_{50}$ 44 mg/kg
  - intraperitoneal-mouse LD$_{50}$ 125 mg/kg
  - oral-dog LD$_{50}$ 40 mg/kg
  - intravenous-dog LDLo 8 mg/kg
  - oral-rabbit LD$_{50}$ 50 mg/kg

- skin-rabbit LD_{50} 50 mg/kg
- intravenous-rabbit LDLo 4500 μg/kg
- oral-guinea pig LD_{50} 127 mg/kg
- oral-hamster LD_{50} 360 mg/kg
- intraperitoneal-hamster LD_{50} 640 mg/kg
- intramuscular-wild bird LDLo 26 mg/kg

◆ Acute Risks
- irritation of the nose and throat
- effects on the blood
- anemia
- skin effects
- elevated, itchy patches of skin
- effects on the nervous system
- seizures
- convulsions
- dyspnea
- cyanosis
- dizziness
- headache
- nausea
- vomiting
- diarrhea
- tremors
- weakness
- effects on the cardiovascular system
- effects on the gastrointestinal system
- effects on the musculoskeletal system
- effects on liver and kidney in animals
- effects on immune system in animals

◆ Chronic Risks
- effects on the liver, blood, cardiovascular, and immune systems
- hepatic damage in animals
- local sensitivity reactions
- decreased sperm count in animals
- increased testicular weight in animals
- disruption of spermatogenesis in animals
- liver carcinogen in rats and mice
- EPA Group B2/C: possible human carcinogen

HAZARD RISK
◆ incompatible with strong oxidizing agents
◆ hazardous decomposition products: carbon monoxide, carbon dioxide, and hydrogen chloride gas
◆ Decomposition emits toxic fumes of Cl^-, HCl, and phosgene.

EXPOSURE ROUTES
◆ oral ingestion of food
◆ released in air during formulation
◆ wind erosion of contaminated soil
◆ release from hazardous waste sites
◆ groundwater
◆ surface water in hazardous waste sites

REGULATORY STATUS/INFORMATION
◆ CERCLA and Superfund
- designation of hazardous substances. 40CFR302.4

ADDITIONAL/OTHER COMMENTS
◆ Symptoms

- exposure: hyperirritability, central nervous system excitation, vomiting, restlessness, muscle spasms, ataxia, clonic and tonic convulsions
◆ First Aid
  - wash eyes and skin immediately with large amounts of water
  - remove to fresh air immediately or provide respiratory apparatus if necessary
  - wash out mouth with water
◆ fire fighting: use carbon dioxide, dry chemical, water spray, or foam
◆ personal protection: self-contained breathing apparatus, protective clothing, and goggles
◆ storage: keep tightly closed in a cool, dry place

SOURCES
◆ *Health Effects Notebook for Hazardous Air Pollutants*, U.S. Environmental Protection Agency, December 1994
◆ *Material Safety Data Sheets (MSDS)*
◆ *Sax's Dangerous Properties of Industrial Materials*, 9th edition, 1996
◆ *Hawley's Condensed Chemical Dictionary*, 12th edition, 1993
◆ *Aldrich Chemical Company Catalog*, 1996–1997
◆ *Environmental Law Index to Chemicals*, 1996
◆ *Environmental Contaminant Reference Databook*, volume 1, 1995
◆ *Merck Index*, 12th edition, 1996

MALEIC ANHYDRIDE (mz)

CAS #: 108-31-6
MOLECULAR FORMULA: $C_4H_2O_3$
FORMULA WEIGHT: 98.06
SYNONYMS: anhydrid kyseliny (Czech), cis-butenedioic anhydride, dihydro-2,5-dioxofuran, 2,5-furandione, maleic-acid-anhydride, maleinanhydride, toxilic-anhydride,

USES
◆ Manufacture of Other Chemicals
  - intermediate for unsaturated polyester resins
  - lubricating oil additives
  - fumaric and tartaric acid
  - captan
  - malathion
  - maleic hydride
  - endothall
  - pesticides
  - copolymers
  - surfactants
  - alkyd coating resins
  - plasticizers
  - preservative for oils and fats
  - permanent-press resins (textiles)
  - Diels-Alder synthesis

PHYSICAL PROPERTIES
◆ colorless to white solid that forms orthorhombic crystals
◆ boiling point 202°C
◆ melting point 53°C

- density 1.48 g/mL at 20°/4°
- flash point 103°C (218°F)
- autoignition temperature 465°C (870°F)
- vapor pressure 0.1 to 0.12 mm Hg at 25°C
- vapor density (air = 1) 3.4 g/cm^3
- flammability limits 1.4% lower and 7.1% upper (by volume)
- soluble in water, alcohol, and dioxane
- acrid odor threshold 1.84 mg/m (low) to 1.96 mg/L (high)
- partially soluble in chloroform and benzene
- specific heat 0.285 (solids); 0.396 (liquids)
- irritating and choking odor
- specific gravity (water = 1) 1.43
- grade: technical; rods, flakes, lumps, briquettes, and molten

CHEMICAL PROPERTIES

- corrosive
- stable under normal laboratory conditions
- heat of combustion −3298 cal/g (or) −138×10^{-5} J/kg (or) −5936 Btu/lb
- incompatible with strong oxidizing agents, strong bases, strong reducing agents, alkali metals, amines
- may decompose on exposure to moist air or water
- hazardous combustion or decomposition products: toxic fumes of CO and CO_2

HEALTH RISK

- Toxic Exposure Guidelines
  - ACGIH TLV TWA 0.25 ppm
  - MSHA STANDARD-AIR TWA 0.25 ppm (1 mg/m^3)
  - OSHA PEL (GEN INDUS) 8H TWA 0.25 (1 mg/m^3)
  - OSHA PEL (CONSTRUC) 8H TWA 0.25 ppm (1 mg/m^3)
  - OSHA PEL (SHIPYARD) 8H TWA 0.25 ppm (1 mg/m^3)
  - OSHA PEL (FED CONT) 8H TWA 0.25 ppm (1 mg/m^3)
  - NIOSH REL TO MALEIC ANHYDRIDE-AIR 10H TWA 0.25 ppm
  - DFG MAK 0.1 ppm (0.4 mg/m^3)
- Toxicity Data
  - eye-rabbit 1% SEV
  - oral-rat TDLo 4060 mg/kg
  - subcutaneous-rat TDLo 1220 mg/kg/61W-I:ETA
  - oral-rat LD$_{50}$ 400 mg/kg
  - intraperitoneal-rat LD$_{50}$ 97 mg/kg
  - oral-mouse LD$_{50}$ 465 mg/kg
  - oral-rabbit LD$_{50}$ 875 mg/kg
  - skin-rabbit LD$_{50}$ 2620 mg/kg
  - oral-guinea pig LD$_{50}$ 390 mg/kg
  - inhalation-rat TCLo 9800 μg/m^3/6H/26W-I
  - cyt-hamster:ingest 230 mg/L
- Acute Risks
  - irritation of the respiratory tract
  - burning in the larynx
  - reflex cough
  - lacrimation
  - nosebleeds

- headaches
- eye irritation
- impairment of vision

- Prolonged or repeated exposure may cause allergic reactions in certain sensitive individuals.

◆ Chronic Risks
  - asthma
  - dermatitis
  - chronic bronchitis
  - upper respiratory tract infection
  - eye irritation
  - renal lesions (observed in rats)
  - pulmonary edema (effects may be delayed)

HAZARD RISK
◆ moderate poison
◆ flammable
◆ caustic with contact to water
◆ corrosives
◆ emits toxic vapors when heated
◆ acidic
◆ Explosion may occur from a dust cloud by a flame or spark.
◆ may generate electrostatic charges

EXPOSURE ROUTES
◆ Environmental/Occupational
  - spills
  - fugitive emissions
  - vent gases
  - manufacturing
  - transporting
◆ Personal
  - sensitization by inhalation
  - ingestion
  - skin-absorbable poison
  - eye irritant from vapors
  - breathing in dust

REGULATORY STATUS/INFORMATION
◆ Superfund
  - designation of hazardous substances, 40CFR302.4
  - chemicals and chemical categories (CAS # list), 40CFR372.65
◆ Federal Water Pollution Control Act under section 311(b)(2)(A); designated hazardous substance
◆ Clean Water Act
  - designation of hazardous substances, 40CFR116.4
  - determination of reportable quantities, 40CFR117.3
◆ TSCA, 716.120a: list of substances
◆ OSHA, 29CFR1910.1000 Table Z1
◆ Resource Conservation Recovery Act
  - U waste # 261.33.f
  - hazardous constituents, 40CFR261
  - reference air concentration, 40CFR266
◆ Clean Air Act, Title III: hazardous air pollutants

ADDITIONAL/OTHER COMMENTS
◆ First Aid
  - wash out eyes immediately with large amounts of water
  - remove contaminated clothing

- if inhaled, exit to fresh air
- if not breathing, give artificial respiration
- if breathing difficulty, give oxygen
- if swallowed, wash out mouth and call physician
◆ fire fighting: use water spray or carbon dioxide but not a dry chemical powder extinguisher
◆ personal protection: wear appropriate NIOSH/MSHA approved respirator chemical-resistant gloves, safety goggles, and other protective clothing

SOURCES
◆ *Health Effects Notebook for Hazardous Air Pollutants*, U.S. Environmental Protection Agency December 1994
◆ *Material Safety Data Sheets (MSDS)*
◆ *Sax's Dangerous Properties of Industrial Materials*, 9th edition, volume 3, 1996
◆ *Environmental Law Index to Chemicals*, 1996
◆ *Hawley's Condensed Chemical Dictionary*, 12th edition, 1993
◆ *Environmental Contaminant Reference Databook*, volume 1, 1995
◆ *NIOSH Pocket Guide to Chemical Hazards*, June 1994

MANGANESE COMPOUNDS

CAS #: Manganese compounds have variable CAS #s. The CAS # for manganese is 7439-96-5.
MOLECULAR FORMULA: Manganese compounds have variable molecular formulas. The molecular formula for manganese is Mn.
Note: For all hazardous air pollutants (HAPs) that contain the word "compounds" in the name, and for glycol ethers, the following applies: Unless otherwise specified, these HAPs are defined by the 1990 Amendments to the Clean Air Act as "including any unique chemical substance that contains the named chemical (i.e., antimony, arsenic, etc.) as part of that chemical's infrastructure."
FORMULA WEIGHT: Manganese compounds have variable formula weights. The formula weight for manganese is 64.94.
SYNONYMS: colloidal manganese, magnacat, mangan (Polish), mangan nitrodovany (Czech), tronamang

USES
◆ Metallic Manganese
  - steel production
  - improves hardness
  - improves stiffness
  - improves strength
  - carbon steel
  - stainless steel
  - high-purity salt for chemical uses
  - purifying and scavenging agent in metal production
  - manufacture of aluminum by Toth process
  - railway points and crossings
◆ Manganese Dioxide
  - production of dry-cell batteries
  - matches
  - fireworks

- production of other manganese compounds
- ◆ Manganese Chloride
  - precursor for other manganese compounds
  - animal feed
- ◆ Manganese Sulfate
  - glazes
  - varnishes
  - ceramics
  - nutritional supplement

PHYSICAL PROPERTIES
- ◆ silver-colored metal
- ◆ forms compounds in the environment with chemicals such as oxygen, sulfur, and chlorine
- ◆ Manganese compounds are solids that do not evaporate.
- ◆ Small dust particles can become suspended in air.
- ◆ melting point 1245°C
- ◆ boiling point 2150°C
- ◆ specific gravity 7.43 g/m³
- ◆ vapor pressure 1 mm Hg at 1292°C

CHEMICAL PROPERTIES
- ◆ specific heat 0.115 cal/g-°C
- ◆ latent heat of fusion 3.5 kcal/g-atom
- ◆ superficially oxidized on exposure to air
- ◆ burns with an intense white light when heated in air
- ◆ decomposes water slowly in the cold, rapidly on heating
- ◆ Pure electrolytic manganese is not attacked by water at ordinary temperatures.
- ◆ slightly attacked by steam
- ◆ reacts with diluted mineral acids with evolution of hydrogen and formation of divalent manganous salts
- ◆ reacts with aqueous solutions of sodium or potassium bicarbonate
- ◆ in form of powder, reduces most metallic oxides on heating
- ◆ on heating, reacts directly with carbon, phosphorus, antimony, or arsenic

HEALTH RISK
- ◆ Toxic Exposure Guidelines
  - ACGIH TLV (dust and compounds) 5 mg/m³
  - OSHA PEL (dust and compounds) 5 mg/m³
  - ACGIH TLV STEL (fume) 3 mg/m³
  - OSHA PEL STEL (fume) 3 mg/m³
  - ACGIH TLV (fume) 1 mg/m³
  - OSHA PEL TWA (fume) 1 mg/m³
  - DFG MAK 5 mg/m³
- ◆ Toxicity Data (Metallic Manganese Only)
  - skin-rabbit 500 mg/24H MLD
  - eye-rabbit 500 mg/24H MLD
  - mrc-smc 8 mmol/L/18H
  - intramuscular-rat TDLo 400 mg/kg/1Y-I:ETA
  - inhalation-man TCLo 2300 µg/m³:BRN, CNS
- ◆ Acute Risks
  - No information is available on the acute effects in humans.
  - Symptoms usually take from 1 to 3 years to develop.

- in animals: impaired growth, skeletal abnormalities, impaired reproductive function in females, testicular degeneration in males, and ataxia
- ◆ Chronic Risks
  - effects on central nervous system
  - weakness
  - lethargy
  - speech disturbances
  - mask-like face
  - tremors
  - psychological disturbances
  - increased incidence of cough
  - bronchitis
  - increased susceptibility to infectious lung disease
  - increased risk of pneumonia
  - languor
  - sleepiness
  - speak with slow monotonous voice
  - muscular twitching
  - nocturnal cramps of legs
  - slight increase in tendon reflexes, ankle and patellar clonus
  - typical Parkinsonian slapping gait
  - minute handwriting
  - Symptoms may simulate bulbar paralysis, postencephalitic Parkinsonism, multiple sclerosis, amyotrophic lateral sclerosis, and progressive lenticular degeneration (Wilson's disease).
  - impotence and loss of libido in males
  - severe degenerative changes in the seminiferous tubules leading to sterility
  - decreased testosterone production in animals
  - decrease on average pup weight in offspring of mice
  - manganese deficiency
  - EPA Group D: not classifiable as to carcinogenicity in humans

HAZARD RISK
- ◆ flammable and moderately explosive in the form of dust or powder when exposed to flame
- ◆ Dust may be pyrophoric in air.
- ◆ Dust may explode when heated in carbon dioxide.
- ◆ Mixtures of aluminum dust and manganese dust may explode in air.
- ◆ Mixtures with ammonium nitrate may explode when heated.
- ◆ Powdered metal ignites on contact with fluorine, chlorine + heat, hydrogen peroxide, bromine pentafluoride, and sulfur dioxide + heat.
- ◆ violent reaction with NO_2 and oxidants
- ◆ incandescent reaction with phosphorus, nitryl fluoride, and nitric acid
- ◆ will react with water or steam to produce hydrogen
- ◆ can react with oxidizing materials

EXPOSURE ROUTES
- ◆ naturally occurring substance in many types of rock
- ◆ water
- ◆ soil
- ◆ food
- ◆ air
- ◆ production of manganese ore

♦ use of manganese to make
    other products

REGULATORY STATUS/INFORMATION
♦ CERCLA and Superfund
    • list of chemicals. 40CFR372.65

ADDITIONAL/OTHER COMMENTS
♦ Symptoms
    • inhalation of dust or fumes: languor, sleepiness, weakness,
      emotional disturbances, spastic gait, paralysis
♦ First Aid
    • wash eyes and skin immediately with large amounts
      of water
    • remove to fresh air immediately
♦ fire fighting: use special dry chemical
♦ personal protection: self-contained breathing apparatus, pro-
    tective clothing, and goggles
♦ storage: store in dry area away from acids, oxidizers, and halo-
    gen gases

SOURCES
♦ *Health Effects Notebook for Hazardous Air Pollutants*, U.S.
    Environmental Protection Agency, December 1994
♦ *Material Safety Data Sheets (MSDS)*
♦ *Sax's Dangerous Properties of Industrial Materials*, 9th edi-
    tion, 1996
♦ *Hawley's Condensed Chemical Dictionary*, 12th edition, 1993
♦ *Environmental Law Index to Chemicals*, 1996
♦ *Merck Index*, 12th edition, 1996

MERCURY COMPOUNDS

CAS #: Mercury compounds have variable CAS #s. The CAS # for
mercury is 7439-97-6.
MOLECULAR FORMULA: Mercury compounds have variable molecular formulas. The molecular formula for mercury is Hg.
Note: For all hazardous air pollutants (HAPs) that contain the word
"compounds" in the name, and for glycol ethers, the following applies:
Unless otherwise specified, these HAPs are defined by the 1990
amendments to the Clean Air Act as "including any unique chemical
substance that contains the named chemical (i.e., antimony, arsenic,
etc.) as part of that chemical's infrastructure."
FORMULA WEIGHT: Mercury compounds have variable formula
weights. The formula weight for mercury is 200.59.
SYNONYMS: colloidal mercury, kwik (Dutch), mercure (French),
mercurio (Italian), mercury, metallic (DOT), Quecksilber (German),
quick silver, RTEC (Polish)

USES
♦ Elemental Mercury
    • thermometers   • pressure-sensing devices
    • barometers   • batteries
    • hydrometers   • lamps

- switches
- mercury boilers
- extracting gold and silver from ores
- electric rectifiers
- pharmaceuticals
- agricultural chemicals
- antifouling paints
- industrial processes
- refining
- lubrication oils
- dental amalgams

◆ Inorganic Mercury
  - laxatives
  - skin-lightening creams and soaps
  - latex paint

PHYSICAL PROPERTIES

- ◆ silver-white metal
- ◆ mobile liquid
- ◆ melting point −38.89°C
- ◆ boiling point 356.9°C
- ◆ density 13.534 at 25°C
- ◆ vapor pressure 0.002 mm Hg at 25°C
- ◆ surface tension 484 dynes/cm
- ◆ insoluble in hydrochloric acid
- ◆ soluble in sulfuric acid upon boiling
- ◆ insoluble in water, alcohol, and ether

CHEMICAL PROPERTIES

- ◆ heat capacity 6.687 cal/mol deg at constant pressure and 25°C
- ◆ pure: does not tarnish on exposure to air
- ◆ when heated to boiling point, oxidized to HgO
- ◆ forms alloys with most metals except iron
- ◆ combines with sulfur; reacts with ammonia solutions in air to form Hg_2NOH
- ◆ mercury salts: when heated with Na_2CO_3 yield metallic Hg
- ◆ mercury salts: reduced to metal by H_2O_2 in presence of alkali hydroxide
- ◆ soluble ionized mercuric salts: give a yellow precipitate of HgO with NaOH and a red precipitate of HgI_2 with alkali iodide
- ◆ mercurous salts: give a black precipitate with alkali hydroxides and a white precipitate of calomel with HCl or soluble chlorides; slowly decomposed by sunlight

HEALTH RISK

- ◆ Toxic Exposure Guidelines
  - NIOSH IDLH (mercury vapor) 28 mg/m^3
  - NIOSH IDLH (organo(alkyl) mercury compounds) 10 mg/m^3
  - OSHA PEL (mercury vapor) 0.1 mg/m^3
  - ACGIH TLV (all forms except alkyl vapor) 0.1 mg/m^3
  - OSHA PEL (mercury vapor) 0.05 mg/m^3
  - ACGIH TLV (mercury vapor) 0.05 mg/m^3
  - NIOSH TLV (mercury vapor) 0.05 mg/m^3
  - OSHA STEL (alkyl compounds) 0.03 mg/m^3
  - ACGIH STEL (alkyl compounds) 0.03 mg/m^3
  - ACGIH TLV (alkyl compounds) 0.01 mg/m^3
  - NIOSH TLV (alkyl compounds) 0.01 mg/m^3

- ◆ Toxicity Data (Mercury)
  - inhalation-rat TCLo 1 mg/m³/24H
  - inhalation-rat TCLo 7440 ng/m³/24H
  - intraperitoneal-rat TDLo 400 mg/kg/14D-I:ETA
  - inhalation-man TDLo 44,300 µg/m³/8H:CNS,LIV, MET
  - inhalation-woman TCLo 150 µg/m³/46D:CNS, GIT
  - skin-man TDLo 129 mg/kg/5H-C:EAR, CNS, SKN
  - inhalation-rabbit LCLo 29 mg/m³/30H
- ◆ Acute Risks
  - central nervous system effects
  - hallucinations
  - delirium
  - suicidal tendencies
  - gastrointestinal effects
  - respiratory effects
  - chest pains
  - dyspnea
  - cough
  - pulmonary function impairment
  - interstitial pneumonitis
  - metallic taste in mouth
  - nausea
  - vomiting
  - severe abdominal pain
  - diarrhea
  - kidney damage
  - soluble salts: violent corrosive effects on skin and mucous membranes
- ◆ Chronic Risks
  - erethism (increased excitability)
  - irritability
  - excessive shyness
  - insomnia
  - severe salivation
  - inflammation of mouth and gums
  - gingivitis
  - loosening of the teeth
  - dark line on the gum margins
  - tremor
  - affects the kidney
  - proteinuria
  - jerky gait
  - spasms of extremities
  - personality changes
  - depression
  - nervousness
  - acrodynia in children
  - severe leg cramps
  - paresthesia
  - peeling hands, feet, and nose
  - kidney damage from inorganic mercury
  - increase in rate of spontaneous abortions from elemental mercury
  - elemental mercury: EPA Group D: not classifiable as to human carcinogenicity
- ◆ Chronic Risks (Methyl Mercury)
  - damage to central nervous system
  - paresthesia
  - blurred vision
  - malaise
  - deafness
  - speech difficulties
  - constriction of the visual field
  - significant developmental defects
  - developmental delays and abnormal reflexes in infants
  - inorganic mercury and methyl mercury: EPA Group C: possible human carcinogens

- ◆ Chronic Risks (Mercuric Chloride)
  - • increased incidence of forestomach and thyroid tumors in rats
  - • increased incidence of renal tumors in mice

HAZARD RISK
- ◆ Many mercury compounds are explosively unstable or undergo hazardous reactions.
- ◆ Decomposition emits toxic fumes of Hg.

EXPOSURE ROUTES (Elemental Mercury)
- ◆ inhalation in occupational settings
- ◆ dental amalgam fillings

EXPOSURE ROUTES (Organic: Methyl Mercury)
- ◆ fish and fish products
- ◆ fungicide-treated grains
- ◆ meat from animals fed grains

EXPOSURE ROUTES (Inorganic Mercury)
- ◆ old cans of latex paint

REGULATORY STATUS/INFORMATION
- ◆ CERCLA and Superfund
  - • list of chemicals 40CFR372.65

ADDITIONAL/OTHER COMMENTS
- ◆ Symptoms
  - • exposure: irritability, shyness, tremors, changes in vision or hearing, memory problems, lung damage, nausea, vomiting, diarrhea, increase in blood pressure, skin rashes, eye irritation
- ◆ First Aid
  - • wash eyes immediately with large amounts of water
  - • wash skin immediately with soap and water
  - • remove to fresh air immediately or provide respiratory apparatus if necessary
- ◆ fire fighting: water spray, fog, or foam
- ◆ personal protection: self-contained breathing apparatus and protective clothing
- ◆ storage: in plastic, glass/steel containers in an isolated, cool, and well-ventilated area

SOURCES
- ◆ *Health Effects Notebook for Hazardous Air Pollutants*, U.S. Environmental Protection Agency, December 1994
- ◆ *Material Safety Data Sheets (MSDS)*
- ◆ *Sax's Dangerous Properties of Industrial Materials*, 9th edition, 1996
- ◆ *Hawley's Condensed Chemical Dictionary*, 12th edition, 1993
- ◆ *Environmental Law Index to Chemicals*, 1996
- ◆ *Merck Index*, 12th edition, 1996
- ◆ *Toxicological Profile for Mercury*
- ◆ Agency for Toxic Substances and Disease Registry

METHANOL (mz)

CAS #: 67-56-1
MOLECULAR FORMULA: CH_4O
FORMULA WEIGHT: 32.05
SYNONYMS: alcohol methylique (French), alcool metilico (Italian), carbinol, colonial spirit, columbian spirit, metanolo (Italian), methanol (DOT), methyl alcohol (ACGIH:DOT:OSHA), Methylalkohol (German), methyl hydrate, methyl hydroxide, methylol, metylowy alkohol (Polish), monohydroxymethane, pyroxylic spirit, RCRA waste number U154, UN1230 (DOT), wood alcohol, wood naphtha, wood spirit

USES
◆ Manufacture of Other Chemicals
  • raw material for formaldehyde
  • methyl esters
  • organic and inorganic acids
  • antifreeze
  • gasoline fuel and octane improver
  • extractant for animal and vegetable oils
  • to denature ethanol
  • pharmaceutical products
  • solvent in streptomycin, vitamins, hormones, polymers, plastics
  • dehydrating pipelines, as a deicing agent
  • production of methylamines and chlorine dioxide
  • primarily as a substitute solvent and rubefacient for ethyl alcohol in liniments
  • used on household contents, mortuary instruments, and human clothing
  • used on onions during soil treatment against onion smut
  • used on elms as injection treatment against Dutch elm disease
  • used on timbers, wood fence posts, wood poles/posts, and lumber for soil contact nonfumigation treatment against wood rot and decay fungi
  • effective solvent for the removal of 2,4-dinitrotoluene from spent carbons
  • removal of toxic organic pollutants from soil with supercritical CO_2 and toluene

PHYSICAL PROPERTIES

◆ melting point −97.8°C
◆ boiling point 64.7°C
◆ specific gravity 0.791
◆ flash point 11°C (52°F), 12.2°C for open cup
◆ critical temperature 239.6°C
◆ critical pressure 80.9 atm

◆ density 0.792 at 0°C, 0.791 at 20°C/4°C, and 0.7914 cm³ at 20°C
◆ vapor pressure 92 mm Hg at 20°C
◆ surface tension 22.61 mN/m at 20°C
◆ viscosity 0.614 mPa sec

335

- miscible with water, alcohols, and ethers
- heat of vaporization 39.2 kJ/mol
- clear, colorless, very mobile liquid

CHEMICAL PROPERTIES
- autoignition temperature 454°C
- flammability limits 7.3% lower and 36% upper (by volume)
- heat of combustion 723 kJ/mol

- bulk density 6.59 lb/gal (20°C)
- vapor density 1.11 g/cm³
- specific heat 0.295–0.605 at 20–25°C

- pungent odor when crude
- forms azeotropes with many compounds
- burns with nonluminous bluish flame
- highly polar liquid
- slt alcoholic odor when pure

HEALTH RISK
- Toxic Exposure Guidelines
  - ACGIH TLV TWA 200 ppm, STEL 250 ppm (skin)
  - MSHA STANDARD-AIR TWA 200 ppm (260 mg/m³) (skin)
  - OSHA PEL TWA 200 ppm, STEL 250 ppm (skin)
  - DFG MAK 200 ppm (260 mg/m³; BAT 30 mg/L in urine at end of shift)
  - NIOSH REL TWA 200 ppm; Cl 800 ppm/15M
- Toxicity Data
  - skin-rabbit 20 mg/24H MOD
  - eye-rabbit 100 mg/24H MOD
  - dni-human:lym 300 mmol/L
  - mammal-mouse:lym 7900 mg/L
  - oral-rat TDLo 7500 mg/kg (17–19 D preg)
  - inhalation-rat TDLo 10,000 ppm/7H (7–15 Dpreg)
  - oral-man LDLo 6422 mg/kg:CNS,PUL,GIT
  - oral-man TDLo 3429 mg/kg:EYE
  - oral-human LDLo 428 mg/kg:CNS,PUL
  - oral-human LDLo 143 mg/kg:EYE,PUL,GIT
  - oral-woman TDLo 4 g/kg:EYE,PUL,GIT
  - inhalation-human TDLo 86,000 mg/m³:EYE,PUL
  - inhalation-human TDLo 300 ppm:EYE,CNS,PUL
  - oral-woman TDLo 4 g/kg
  - oral-rat LD$_{50}$ 5628 mg/kg
  - inhalation-rat LC$_{50}$ 64,000 ppm/4H
  - intraperitoneal-rat LD$_{50}$ 7529 mg/kg
  - intravenous-rat LD$_{50}$ 2131 mg/kg
  - oral-mouse LD$_{50}$ 7300 mg/kg
  - intraperitoneal-mouse LD$_{50}$ 10,765 mg/kg
  - subcutaneous-mouse LD$_{50}$ 9800 mg/kg
  - intravenous-mouse LD$_{50}$ 4710 mg/kg
  - oral-monkey LDLo 7000 mg/kg
  - inhalation-monkey LCLo 1000 ppm
  - skin-monkey LDLo 393 mg/kg

◆ Acute Risks
  • harmful if inhaled or swallowed
  • may be harmful if absorbed through skin
  • causes eye and skin irritation
  • Material may be irritating to mucous membranes and upper respiratory tract.
  • gastrointestinal disturbances
  • convulsions
  • targets eyes, kidneys, liver, and heart
◆ Chronic Risks
  • conjunctivitis
  • headaches
  • giddiness
  • insomnia
  • blindness
  • gastric and visual disturbances
  • causes visual impairment
  • moderately toxic via all routes at moderate exposure levels

HAZARD RISK
◆ flammable, dangerous fire risk
◆ toxic by ingestion (causes blindness)
◆ explosive in the form of vapor
◆ reacts vigorously with oxidizing materials
◆ Decomposition causes acrid smoke and irritating fumes.
◆ toxic fumes of carbon monoxide and carbon dioxide
◆ an accumulation poison

EXPOSURE ROUTES
◆ Environmental/Occupational
  • vapor irritation
  • inhalation of fumes such as automobile exhaust
  • dermal contact with consumer products such as paint thinners
  • solvent use
  • consumption of various foods
  • ingestion of liquid
  • absorption through skin
  • hygroscopic

REGULATORY STATUS/INFORMATION
◆ Clean Air Act, Title III: hazardous air pollutants
◆ Safe Drinking Water Act
  • waste specific prohibitions—solvent wastes, 40CFR148.10
  • U waste #: 261.33.f.
◆ Superfund
  • designation of hazardous substances, 40CFR302.4
  • chemicals and chemical categories, 40CFR372.65
◆ OSHA, 29CFR1910.1000 Table Z1

ADDITIONAL/OTHER COMMENTS
◆ First Aid
  • wash out eyes immediately with large amounts of water
  • remove contaminated clothing
  • if inhaled, exit to fresh air
  • if not breathing, artificial respiration

- if breathing difficulty, give oxygen
- if swallowed, wash out mouth and call physician
◆ fire fighting: use dry chemical powder or foam and carbon dioxide
◆ personal protection: wear appropriate NIOSH/MSHA approved respirator, chemical-resistant gloves, safety goggles, and other protective clothing

SOURCES

◆ *Health Effects Notebook for Hazardous Air Pollutants*, U.S. Environmental Protection Agency, December 1994
◆ *Material Safety Data Sheets (MSDS)*
◆ *Sax's Dangerous Properties of Industrial Materials*, 9th edition, volume 3, 1996
◆ *Environmental Law Index to Chemicals*, 1996
◆ *Hawley's Condensed Chemical Dictionary*, 12th edition, 1993
◆ *Environmental Contaminant Reference Databook*, volume 1, 1995
◆ *NIOSH Pocket Guide to Chemical Hazards*, June 1994

METHOXYCHLOR (mz)

CAS #: 72-43-5
MOLECULAR FORMULA: $C_{16}H_{15}Cl_3O_2$
FORMULA WEIGHT: 345.66
SYNONYMS: 1,1'-(2,2,2-trichloroethylidene) bis (4-methoxy-)benzene, 2,2-bis (p-anisyl)-1,1,1-trichlororethane, 1,1,-bis (p-methoxyphenyl)-2,2,2-trichloroethane, 2,2-bis (p-methoxyphenyl)-1,1,1-trichloro ethane, dianisyltrichloroethane, 2,2-di-p-anisyl-1,1,1-trichloroethane, dimethoxy-DDT, p,p'-dimethoxydiphenyltrichloroethanem, 2,2-di(p-methoxyphenyl)-1,1,1-trichloroethane, di(p-methoxyphenyl)-trichloromethyl methane, DMDT, p,p'-DMDT
USES
◆ Manufacture of Other Chemicals
- insecticide (against biting flies and mosquitoes)
- replacement for DDT
- dairy barns
- against the elm bark-beetle vectors of the Dutch elm disease
- ectoparasiticide (a medicine used to kill parasites that live on the exterior of the host)

PHYSICAL PROPERTIES
◆ melting point 78°C
◆ boiling point: decomposes
◆ density 1.41 lb/cu ft at 25°C
◆ vapor density 12 mm
◆ insoluble in water
◆ soluble in alcohol and acetone
◆ white crystalline solid
◆ orange to white powder
◆ vapor pressure very low

CHEMICAL PROPERTIES
◆ not compatible with alkaline materials
◆ resistant to heat and combustion
◆ less readily dehydrochlorinated than DDT by alcoholic alkali

HEALTH RISK
◆ Toxic Exposure Guidelines
- OSHA PEL TWA Total Dust 10 mg/m³, 5 mg/m³

- ACGIH TLV TWA 10 mg/m^3
- DFG MAK 15 mg/m^3
- MSHA STANDARD-AIR TWA 10 mg/m^3
- NIOSH REL AIR CA 0.07 mg/m^3 LOQ
◆ Toxicity Data
  - spm-rat-oral 28 g/kg/10W-C
  - cyt-hamster-intraperitoneal 50 mg/kg
  - oral-rat TDLo 4250 mg/kg (female 24D pre–21D post): REP
  - oral-rat TDLo 2 g/kg (6–15D preg):TER
  - oral-rat TDLo 18,200 mg/kg/2Y-C:CAR,TER
  - oral-mouse TDLo 56,700 mg/kg/90W-C:CAR,TER
  - oral-dog TDLo 383 g/kg/3Y-C:ETA
  - oral-rat TD 80 g/kg/2Y-C:CAR,TER
  - oral-rat TD 72,800 mg/kg/2Y-C:CAR
  - oral-rat TD 87,360 mg/kg/2Y-C:CAR
  - oral-human LDLo 6430 mg/kg
  - skin-human TDLo 2414 mg/kg: CNS
  - oral-rat LD$_{50}$ 5000 mg/kg
  - oral-mouse LD$_{50}$ 1000 mg/kg
  - oral-rabbit LD$_{50}$ >6000 mg/kg
  - LOAEL-rabbit 3505 mg/kg/d
  - NOAEL-rabbit 5.01 mg/kg/d
  - RfD 0.005 mg/kg/d
  - intraperitoneal-hamster LD$_{50}$ 500 mg/kg
◆ Acute Risks
  - harmful if ingested, inhaled, or absorbed through the skin
  - may cause irritation
  - trembling
  - convulsions
  - mild kidney and liver damage
◆ Chronic Risks
  - Overexposure may cause reproductive disorders, based upon tests with laboratory animals.
  - Lab tests have shown mutagenic effects.
  - targets the central nervous system and kidneys
  - suspected carcinogen with experimental data

HAZARD RISK
  ◆ emits highly toxic Cl fumes from heating to decomposition
  ◆ hazardous combustion of carbon dioxide, carbon monoxide, and hydrogen chlorine gas
  ◆ reproductive concern

EXPOSURE ROUTES
  ◆ inhalation by being involved in manufacturing, handling, or application
  ◆ dermal contact
  ◆ ingestion of food
  ◆ drinking contaminated water
  ◆ absorption through the skin
  ◆ vapor fumes that cause irritation

REGULATORY STATUS/INFORMATION
- ◆ Clean Air Act, Title III: hazardous air pollutants
- ◆ Safe Drinking Water Act
  - • 83 contaminants required to be regulated under the SDWA of 1986, 47FR9352
  - • organic chemicals other than total trihalomethanes, sampling and analytical requirements, 40CFR1641.24
  - • public notification, 40CFR141.32
  - • maximum contaminant level goal for organic contaminant, 40CFR141.50(b)
  - • MCL for organic chemicals, 40CFR141.61
  - • variances and exemptions from the MCL's for organic and inorganic chemicals, 40CFR142.62
- ◆ Federal Insecticide, Fungicide, and Rodenticide Act
  - • tolerance and exemptions from tolerances for pesticide chemicals in or on raw agricultural commodities, 40CFR180.102-1147
  - • class of chlorinated organic pesticides, 180.3(4)
- ◆ Resource Conservation Recovery Act
  - • design criteria for municipal and solid waste landfill, 40CFR258.40, 40CFR261
  - • groundwater monitoring chemicals list, 40CFR264
  - • reference air concentrations, 40CFR266
  - • health based limits for exclusion of waste-derived residues, 40CFR266
  - • toxicity characteristics (Table 1: max. conc. of contaminants for the toxicity), 40CFR261.24
  - • designation of hazardous substances, 40CFR302.4
  - • U waste # 261.33.f
- ◆ Superfund
  - • designation of hazardous substances, 40CFR302.4
  - • chemicals and chemical categories, 40CFR372.65
- ◆ Clean Water Act
  - • designation of hazardous substances, 40CFR116.4
- ◆ OSHA 29CFR1910.1000, Table Z1

ADDITIONAL/OTHER COMMENTS
- ◆ First Aid
  - • wash out eyes immediately with large amounts of water
  - • remove contaminated clothing
  - • if inhaled, exit to fresh air
  - • if not breathing, artificial respiration
  - • if breathing difficulty, give oxygen
  - • if swallowed, wash out mouth and call physician
- ◆ fire fighting: use dry chemical powder or foam and carbon dioxide
- ◆ personal protection: wear appropriate NIOSH/MSHA approved respirator, chemical-resistant gloves, safety goggles, and other protective clothing

SOURCES

◆ *Health Effects Notebook for Hazardous Air Pollutants*, U.S. Environmental Protection Agency, December 1994

◆ *Material Safety Data Sheets (MSDS)*

◆ *Sax's Dangerous Properties of Industrial Materials*, 9th edition, volume 3, 1996

◆ *Environmental Law Index to Chemicals*, 1996

◆ *Hawley's Condensed Chemical Dictionary*, 12th edition, 1993

◆ *Environmental Contaminant Reference Databook*, volume 1, 1995

◆ *NIOSH Pocket Guide to Chemical Hazards*, June 1994

METHYL BROMIDE (sf)

CAS #: 74-83-9
MOLECULAR FORMULA: CH_3Br
FORMULA WEIGHT: 94.95
SYNONYMS: bromomethane, brom-o-gas, dawson 100, dowfume, dowfume MC-2 soil fumigant, EDCO, embafume, fumigant-1 (obs.), halon 1001, iscobrome, kayafume, MB, MBX, MEBR, metafume, methogas, monobromomethane, pestmaster (obs.), profume (obs.), R 40B1, RCRA waste number U029, rotox, terabol, terr-o-cide II, terr-o-gas 100, zytox, many others

USES

◆ Used as a Fumigant, Especially in Agricultural Storage to Control
  • rats
  • insects
  • fungus
  • weeds
  • nematodes

◆ Other Uses
  • herbicide
  • wool degreaser
  • methylating agent
  • fire extinguishing agent
  • extracting oils from nuts and seeds
  • solvent for aniline dyes
  • refrigerant

PHYSICAL PROPERTIES

◆ colorless, transparent, easily liquefied gas or highly volatile liquid

◆ density (20°C) 3.3 g/mL

◆ specific gravity 1.732 (0°/0°C) (water = 1)

◆ odor threshold 80 mg/m³

◆ burning taste, chloroform-like odor

◆ melting point −4°C (24.5°F)

◆ boiling point 4°C (38.9°F, 1 atm)

◆ vapor density 3.27 (air = 1)

◆ surface tension 22.36×10^{-3} N/m

◆ viscosity 369.1×10^{-6} Pa s at 25°C

◆ heat capacity 132.3 J/gmol-K (liquid at 25°C)

◆ heat of vaporization 23.91 kJ/gmol

◆ slightly soluble in water

♦ miscible with alcohol, acetone, chloroform, ether, and carbon tetrachloride, and most other common organic solvents
♦ vapor pressure 1250 mm Hg at 20°C

CHEMICAL PROPERTIES

♦ flash point none
♦ autoignition temperature 535°C (996°F)
♦ nonflammable in air but burns in oxygen
♦ explosion limits in air lower 10%, upper 16% but ignition difficult
♦ enthalpy (heat) of formation at 25°C (liquid) −54.9, (gas) −35.5 kJ/gmol
♦ heat of combustion −705.4 kJ/mol
♦ very stable
♦ will not polymerize
♦ incompatible with plastics, rubber, alkali metals, aluminum and its alloys, dimethyl sulfoxide, ethylene oxide, and all strong oxidizers. See the fire fighting section under "Additional/Other Comments" for warning against use of rubber gloves and caution against use with metallic components of zinc, aluminum, and magnesium and their alloys.
♦ Explosive sensitivity in air enhanced by presence of aluminum, magnesium, and zinc and their alloys. See the fire fighting section under "Additional/Other Comments."
♦ hazardous combustion or decomposition products; may produce CO, CO_2, and hydrogen bromide gas

HEALTH RISK

♦ Toxic Exposure Guidelines
  • AGCIH TLV TWA 5 ppm (skin)
  • OSHA PEL TWA 5 ppm (skin)
  • NIOSH REL (monohalomethanes) reduce to lowest feasible level
♦ Toxicity Data
  • inhalation-adult human, LCLo 60,000 ppm/2H, TCLo, 35 ppm: GIT
  • inhalation-child human, LCLo 1 mg/m^3/2H
  • inhalation-rat LC_{50} 302 ppm/8H
  • oral-rat TDLo 3250 mg/kg/13W-I:CAR
  • oral-rat LD_{50} 214 mg/kg
♦ Other Toxicity Indicators
  • EPA Cancer Risk Level: Group D, not classifiable
♦ Acute Risks
  • may be fatal if inhaled: fatal as result of spasm, inflammation, and edema
  • of the larynx and bronchi, chemical pneumonitis, and pulmonary edema
  • extremely destructive to skin, eyes, mucous membranes, and upper respiratory tract
  • harmful if swallowed, inhaled, or absorbed through skin
  • Splashing on skin causes severe itching, dermatitis, and possibly severe skin lesions and appearance of vesicles or blebs.

- metabolized to hydrogen bromide in body, increasing bromide ion concentration in blood
- unconsciousness
- injury to nervous system
- paralysis
- convulsions and shaking
- burning sensation
- coughing and wheezing
- laryngitis
- shortness of breath
- headaches
- dizziness
- numbness and tingling of arms and legs
- blurred vision
- nausea
- vomiting
- slurred speech and other speech defects
- confusion
- Effects may be delayed by 2 to 48 hours.
- ◆ Chronic Risks
  - human mutation data reported
  - possible carcinogen
  - damage to eyes, liver, kidney, lungs
  - Effects are cumulative.

HAZARD RISK

- ◆ NFPA rating Health 3, Flammability 1, Reactivity 0
- ◆ flammable
- ◆ highly toxic
- ◆ inhalation hazard
- ◆ causes burns
- ◆ Incompatible or reacts strongly with plastics, rubber, alkali metals, aluminum and its alloys, dimethyl sulfoxide, ethylene oxide, and all strong oxidizers. See the fire fighting section under "Additional/Other Comments" for warning against use of rubber gloves and caution against use with metallic components of zinc, aluminum, and magnesium and their alloys.
- ◆ Explosive sensitivity in air enhanced by presence of aluminum, magnesium and zinc and their alloys. See the fire fighting section under "Additional/Other Comments."
- ◆ Hazardous combustion or decomposition products. May produce CO, CO_2, and hydrogen bromide gas.

EXPOSURE ROUTES

- ◆ primarily by inhalation
- ◆ absorption through skin and eyes
- ◆ ingestion
- ◆ air
- ◆ occupational
- ◆ drinking water
- ◆ Some is formed naturally by algae and kelp in ocean.

REGULATORY STATUS/INFORMATION

- ◆ Clean Air Act Amendments of 1990 Section 112 Hazardous Air Pollutant
- ◆ Superfund Amendment and Reauthorization Act
  - subject to section 313 reporting requirements
  - health assessment rank 56
- ◆ Federal Insecticide, Fungicide, and Rodenticide Act
  - pesticide classified for restriction, 40CFR152-175
- ◆ Resource Conservation and Recovery Act
  - constituent for detection monitoring, 40CFR258—Appendix 1
  - list of hazardous inorganic and organic constituents, 40CFR258—Appendix 2
  - list of hazardous constituents, 40CFR261—Appendix 8, and also see 40CFR261.11
  - groundwater monitoring chemicals list, 40CFR264—Appendix 9, RCRA designation U029
- ◆ CERCLA and Superfund
  - designation of hazardous substances, 40CFR302.4—Appendices A and B
  - list of extremely hazardous substances and their threshold planning quantities (TPQ), 40CFR355-AB, 40CFR355—Appendix B
  - TPQ 1000 lb
  - Reportable quantity (RQ) is 1000 lb.
- ◆ OSHA Table Z1 and Table Z5
- ◆ Department of Transportation
  - identification number UN1062
  - hazard class/division: 2.3
  - labels: poison gas
- ◆ Community Right-to-Know list

ADDITIONAL/OTHER COMMENTS

- ◆ Symptoms
  - irritation of skin, eyes, and respiratory system
  - unconsciousness
  - injury to nervous system
  - paralysis
  - convulsions and shaking
  - burning sensation
  - coughing and wheezing
  - laryngitis
  - shortness of breath
  - headaches
  - dizziness
  - numbness and tingling of arms and legs
  - blurred vision
  - nausea
  - vomiting
  - slurred speech and other speech defects
  - confusion

- Effects may be delayed by 2 to 48 hours.
◆ First Aid
  - flush eyes and/or skin with large amounts of water for at least 15 minutes while removing contaminated garments and shoes
  - separate eyelids with fingers while flushing
  - seek fresh air
  - provide respiratory support and oxygen
  - if swallowed wash out mouth with water if conscious
  - seek medical help immediately
  - Symptoms may be delayed, so keep under medical supervision.
◆ Fire Fighting
  - "Metallic components of zinc, aluminum, and magnesium (or their alloys) are unsuitable for service with bromomethane because of the **formation of pyrophoric Grignard-Type compounds. A severe explosion due to the ignition of a bromomethane-air mixture by pyrophoric methylaluminum bromides produced by corrosion of an aluminum fitting has been reported.** Reaction with methyl sulfoxide (DMSO) produces trimethylsulfoxonium bromide, a metastable compound which decomposes even in solution after a few hours, leading to an exothermic accelerating reaction with vigorous evolution of vapor." (Aldrich *MSDS*)
  - do not extinguish burning gas if flow cannot be shut off immediately
  - use water spray or fog nozzle to keep cylinder cool
  - move cylinder away from fire if there is no risk
  - self-contained breathing apparatus to avoid breathing toxic fumes
  - Protective clothing to prevent contact with skin and eyes. "WARNING: Methyl bromide readily penetrates rubber gloves. Use gloves constructed from tetrafluoroethylene polymer." (Aldrich *MSDS*)
  - may form explosive mixtures with air
  - emits toxic fumes under fire conditions
◆ Personal Protection
  - wear appropriate NIOSH/MSHA approved respirator
  - **special chemical-resistant gloves,** "WARNING: Methyl bromide readily penetrates rubber gloves. Use gloves constructed from tetrafluoroethylene polymer." [Aldrich *MSDS*], safety goggles, protective clothing
  - exercise extreme caution
  - use only in chemical fume hood
  - wash thoroughly after handling
◆ Storage
  - "Use with equipment rated for cylinder pressure and with compatible materials of construction. Close valve when not in use and when empty. Make sure cylinder is properly se-

cured when not in use or stored. WARNING: Suckback into cylinder may cause rupture. Use flow-back preventative device in piping." (Aldrich *MSDS*)
- outdoor storage preferred
- keep container closed
- cool, dry place away from heat and open flame
- isolate from active metals

SOURCES
- ◆ *AGCIH Threshold Limit Values (TLVs) for 1996–1997*
- ◆ *Aldrich Chemical Company Catalog*, 1996–1997
- ◆ *CRC Handbook of Chemistry and Physics*, 77th edition, 1996–1997
- ◆ *CRC Handbook of Thermophysical and Thermochemical Data*, 1974
- ◆ *Environmental Law Index to Chemicals*, 1996 edition
- ◆ *Fire Protection Guide to Hazardous Materials*, 11th edition, 1994
- ◆ *Hawley's Condensed Chemical Dictionary*, 12th edition, 1993
- ◆ *(MSDS)*, Aldrich Chemical Company, Inc.
- ◆ *Physical and Thermodynamic Properties of Pure Chemicals: Data Compilation*, AiChE and NSRSD, 1996
- ◆ *Sax's Dangerous Properties of Industrial Materials*, 9th edition, 1996
- ◆ *Suspect Chemicals Sourcebook*, 1996 edition
- ◆ *Health Effects Notebook for Hazardous Air Pollutants*, U.S. Environmental Protection Agency, 1994

METHYL CHLORIDE (mz)

CAS #: 74-87-3
MOLECULAR FORMULA: CH_3Cl
FORMULA WEIGHT: 54.49
SYNONYMS: arctic, chloor-methaan (Dutch), Chlor-methan (German), chloromethane, chlorure de methyle (French), clorometano (Italian), cloruro di metile (Italian), Methylchlorid (German), methyl chloride, metylu chlorek (Polish), monochloromethane, R-40, RCRA waste number U045, UN1063 (DOT)
USES
- ◆ Manufacture of Other Chemicals
  - protective colloid in water-based paints to prevent flocculation of pigment
  - production of silicones
  - agricultural chemicals
  - methyl cellulose
  - quaternary amines
  - butyl rubber
  - film and sheeting
  - binder in ceramic glazers
  - leather tanning
  - dispersing agent
  - thickening agent
  - sizing agent
  - adhesive
  - food additive
  - refrigerant
  - cooling media in old equipment only
  - lab reagent
  - local anesthetic
  - catalyst carrier in lower temperatures

PHYSICAL PROPERTIES
- melting point −97°C
- boiling point −24.4°C
- flash point <32°C
- density 0.918 g/mL at 20°C/4°C
- vapor density 1.74 g/m³
- specific gravity 0.915
- vapor pressure 3796 mm at 20°C
- critical temperature 143.2°C
- critical pressure 66.8 atm (970 psia)
- bulk density 7.68 lb/gal
- colorless compressed gas or liquid
- faintly sweet
- ethereal odor
- slightly soluble in water, by which it is decomposed
- soluble in alcohol, chloroform, benzene, carbon tetrachloride, and glacial acetic acid
- very soluble in ethyl alcohol

CHEMICAL PROPERTIES
- explosion limits in air 7% lower and 19% upper (by volume mass)
- autoignition temperature 631°C (1169°F)
- attacks aluminum, magnesium, and zinc

HEALTH RISK
- Toxic Exposure Guidelines
  - OSHA PEL TWA 50 ppm; STEL 100 ppm
  - ACGIH TLV TWA 50 ppm; STEL 100 ppm
  - DFG MAK 50 ppm (105 mg/m³) suspected carcinogen
  - NIOSH REL (monohalomethanes) TWA reduce to lowest level
  - MSHA standard air-Cl 100 ppm (210 mg/m³)
- Toxicity Data
  - oms-human:lym 3 pph
  - sce-human:lym 3 pph
  - inhalation-mouse TDLo 750 ppm/6H (female 6–17D post):TER
  - inhalation-rat TDLo 3000 ppm/6H (male 5D pre):rep
  - inhalation-human LCLo 20,000 ppm/2H:EYE,CNS,GIT
  - oral-rat LD$_{50}$ 1800 mg/kg
  - inhalation-rat LC$_{50}$ 5300 mg/m³/4H
  - inhalation-rat LCLo 2200 ppm/6H
  - inhalation-dog LCLo 14,661 ppm/6H
  - inhalation-guinea pig LCLo 20,000 ppm/2H
  - inhalation-rat TCLo mg/m³/4H/26W-I
  - inhalation-mouse TCLo 150 ppm/22H/11D-C
- Acute Risks
  - harmful if inhaled
  - causes eye and skin irritation
  - Material is irritating to mucous membranes and upper respiratory tract.
  - absorbed through the lungs
  - narcotic effects

- ◆ Chronic Risks
  - • damage to lungs, liver, kidneys, heart
  - • possible carcinogen
  - • may alter genetic material
  - • Overexposure may cause reproductive disorders based on tests with lab animals.
  - • dizziness
  - • drowsiness
  - • incoordination
  - • confusion
  - • nausea
  - • vomiting
  - • abdominal pains
  - • delirium
  - • convulsions
  - • coma

HAZARD RISK
- ◆ highly flammable
- ◆ dangerous fire risk
- ◆ narcotic
- ◆ psychic effects
- ◆ toxic by inhalation
- ◆ possible carcinogen and teratogen
- ◆ formation of corrosive vapors
- ◆ explodes on contact with interhalogens
- ◆ reacts with metals
- ◆ Substance decomposes on heating.
- ◆ Combustion or decomposition products are CO, CO_2, HCl, and phosgene gas.

EXPOSURE ROUTES
- ◆ Environmental
  - • home burning of wood, coal, and certain plastics
  - • use of chlorinated swimming pools
  - • formed in the ocean and found all over the world
  - • found in some lakes and streams and in drinking water at low levels
  - • cigarette smoking
  - • polystyrene insulation
  - • aerosol propellants
  - • absorption through the skin and lungs
- ◆ Occupation
  - • building contractor
  - • metal industries
  - • transportation
  - • car dealers
  - • service station attendants

REGULATORY STATUS/INFORMATION
- ◆ Clean Air Act, Title III: hazardous air pollutants
- ◆ FIFRA, tolerances for pesticides in food, 40CFR185
- ◆ Resource Conservation Recovery Act
  - • constituents for detection monitoring, 40CFR258
  - • list of hazardous organic and inorganic constituents, 40CFR258, 40CFR261
  - • U waste # 261.33.f
- ◆ Superfund
  - • designation of hazardous substances, 40CFR302.4
- ◆ Clean Water Act
  - • total toxic organics (TTOs), 40CFR413.02
  - • 126 priority pollutants, 40CFR423

- ◆ TSCA, hazardous waste constituents subject to testing, 40CFR799.5055
- ◆ Occupational Safety and Health Act
  - 29CFR1910.1000 Table Z1 and Table Z2
  - 29CFR1910.119: list of highly hazardous chemicals, toxics, and reactives

ADDITIONAL/OTHER COMMENTS
- ◆ First Aid
  - wash out eyes immediately with large amounts of water
  - remove contaminated clothing
  - if inhaled, exit to fresh air
  - if not breathing, artificial respiration
  - if breathing difficulty, give oxygen
  - if swallowed, wash out mouth and call physician
- ◆ fire fighting: use dry chemical powder or foam and carbon dioxide
- ◆ personal protection: wear appropriate NIOSH/MSHA approved respirator, chemical-resistant gloves, safety goggles, and other protective clothing

SOURCES
- ◆ *Health Effects Notebook for Hazardous Air Pollutants*, U.S. Environmental Protection Agency, December 1994
- ◆ *Material Safety Data Sheets (MSDS)*
- ◆ *Sax's Dangerous Properties of Industrial Materials*, 9th edition, volume 3, 1996
- ◆ *Environmental Law Index to Chemicals*, 1996 edition
- ◆ *Hawley's Condensed Chemical Dictionary*, 12th edition, 1993
- ◆ *Environmental Contaminant Reference Databook*, volume 1, 1995
- ◆ *NIOSH Pocket Guide to Chemical Hazards*, June 1994

METHYL CHLOROFORM (sc)

CAS #: 71-55-6
MOLECULAR FORMULA: CH_3CCl_3
FORMULA WEIGHT: 133.4
SYNONYMS: aerothene TT, a-T, chloroetene, chloroethene, chlorothane NU, chlorothene, chlorothene (inhibited), chlorothene NU, chlorothene VG, chlorten, 1,1,1-ethane (French), inhibisol, methyltrichloromethane, NC1-Co4626, RCRA waste number U226, solvent III, strobane, 1,1,1- TCE, 1,1,1-trichloroethane, a-trichloroethane, 1,1,1-trichloroorethaan (Dutch), 1,1,1-tricloroetano (Italian), tri-ethane.

USES
- ◆ Manufacture of Other Chemicals
  - resins
  - waxes
  - alkaloids
  - organic compounds
  - tar
  - substitute for carbon tetra-chloride
- ◆ Used as a Component or in the Manufacturing Process
  - dry cleaning
  - inks

- postharvest fumigation
- coatings
- coolant
- plastic cleaning
- drain cleaner
- aerosols
- coolant
- precision instrument cleaner

- pesticides
- extraction solvent
- photoresist polymers
- adhesives
- lubricants
- textile processing
- degreasing

PHYSICAL PROPERTIES

- ◆ clear, colorless liquid
- ◆ melting point $-30.4°C$
- ◆ specific gravity 1.338
- ◆ surface tension 25.4 dynes/cm
- ◆ heat of vaporization -8012.7 cal/g mol
- ◆ viscosity 0.858 cP at 25°C
- ◆ miscible with acetone, benzene, methanol, and carbon disulfide

- ◆ normal boiling point 74.1°C
- ◆ density 1.3376 g/mL at 20°C
- ◆ chloroform-like odor
- ◆ odor threshold 44 ppm
- ◆ vapor pressure: 124 mm Hg at 25°C, 100 mm Hg at 20°C
- ◆ negligible solubility in water
- ◆ log octanol/water partition coefficient 2.49

CHEMICAL PROPERTIES

- ◆ generally stable
- ◆ flash point none
- ◆ will react violently with acetone, N_2O_4, liquid and gaseous oxygen, aluminums, sodium, sodium hydroxide

- ◆ flammability limits 8.0% lower and 7.1% upper by volume in air
- ◆ autoignition temperature 537°C
- ◆ heat of combustion 2600 cal/g (110×10^2 kJ/kg)

HEALTH RISK

- ◆ Toxic Exposure Guidelines
  - ACGIH TLV TWA 350 ppm
  - OSHA PEL TWA 350 ppm
  - NIOSH REL CL 350 ppm
  - MSHA not listed
- ◆ Toxicity Data
  - skin-rabbit 20 mg/24H MOD
  - eye-rabbit 100 mg MLD
  - eye-rabbit 2 mg/24H SEV
  - inhalation-rat TCLo 2100 ppm/24H
  - inhalation-man TCLo 350 ppm:CNS
  - oral-man TDLo 670 mg/kg:GIT
  - inhalation-man TCLo 920 ppm/70M: EYE CNS
  - inhalation-man TCLo 200 ppm/4H: CNS
  - oral-rat LD_{50}: 9600 mg/kg
  - inhalation-rat LC_{50}: 18,000 ppm/4H
  - intraperitoneal-rat LD_{50}: 3595 mg/kg
  - oral-mouse LD_{50}: 6 g/kg
  - inhalation-mouse LC_{50}: 3911 ppm/2H
  - intraperitoneal-mouse LD_{50}: 3636 mg/kg

- oral-dog LD_{50}: 750 mg/kg
- intraperitoneal-dog LD_{50}: 3100 mg/kg
- intrarenal-dog LDLo: 95 mg/kg

◆ Acute Risks
- irritation of skin and eyes
- narcotic in high concentrations
- functional depression of nervous system
- death in high exposures
- respiratory failure
- cardiac arrest if massively inhaled
- diarrhea
- vomiting
- hypotension
- mild hepatic effects

◆ Chronic Risks
- fatty degeneration of liver
- growth depression
- ventricular arrhythmias

HAZARD RISK
◆ NFPA rating Health 2, Flammability 3, Reactivity 0, Special Hazards none
◆ moderate health hazard
◆ combustion by-products: hydrogen chloride, phosgene
◆ Vapors in containers explode when subject to high energy sources.
◆ reacts with oxidizing materials, alkalies, and active metals

EXPOSURE ROUTES
◆ primarily by inhalation
◆ absorption through skin and eyes
◆ ingestion
◆ toxic fumes under fire conditions
◆ use of metal degreasing agents, paints, glues, and cleaning products
◆ consumption of contaminated drinking water

REGULATORY STATUS/INFORMATION
◆ Safe Drinking Water Act
- one of 83 contaminants required to be regulated under the 1986 act 47FR9352
- public notification. 40CFR141.32
- maximum contaminant level goal (MCLG) for organic contaminants 40CFR141.5
- maximum contaminant level (MCL) for organic chemicals. 40CFR141.61
- variances and exemptions from MCL. 40CFR142.62
- waste specific prohibitions—solvent wastes. 40CFR148.10
◆ Federal Insecticide, Fungicide, and Rodenticide Act
- tolerances and exemptions from tolerances for pesticide chemicals in or on raw agricultural commodities. 40CFR180.102-1147
◆ Resource Conservation and Recovery Act
- design criteria for detection monitoring for municipal solid waste landfills. 40CFR258.40

- list of hazardous inorganic and organic constituents. 40CFR258
- groundwater monitoring chemicals list. 40CFR264
- PICs found in stack effluents. 40CFR266
◆ CERCLA and Superfund
  - designation of hazardous substances. 40CFR302.4
  - chemicals and chemical categories list. 40CFR372.65
◆ Clean Water Act
  - total toxic organics (TTOs). 40CFR413.02
  - 126 priority pollutants. 40CFR423
◆ Toxic Substance Control Act
  - list of substances. 40CFR716.120
  - testing consent orders for mixtures without CAS registry #s. 40CFR799.5000
◆ Department of Transportation
  - hazard class/division: 6.1
  - identification number: UN2831
  - labels: keep away from food
◆ Community Right-to-Know Threshold Planning Quantity (unavailable)

ADDITIONAL/OTHER COMMENTS
◆ Symptoms
  - inhalation: irritation of eyes and skin
  - skin absorption: dizziness, nausea
◆ First Aid
  - flush eyes with water
  - wash skin with soap and water
  - if swallowed, wash out mouth with water
◆ fire fighting: water spray, carbon dioxide, dry chemical powder, or appropriate foam
◆ personal protection: wear self-containing breathing apparatus if exposed to high level vapor composition
◆ storage: keep tightly closed and store in cool, dry place

SOURCES
◆ *Environmental Contaminant Reference Databook*, volume 1, 1995
◆ *Environmental Law Index to Chemicals*, 1996 edition
◆ *Fire Protection Guide to Hazardous Materials*, 11th edition, 1994
◆ *Hawley's Condensed Chemical Dictionary*, 12th edition, 1993
◆ *Material Safety Data Sheets (MSDS)*
◆ *Sax's Dangerous Properties of Industrial Materials*, 9th edition, 1996
◆ *Health Effects Notebook for Hazardous Air Pollutants*, U.S. Environmental Protection Agency, December 1994

4,4'-METHYLENE BIS (2-CHLOROANILINE). *See* **p. 373** following methyl tert butyl ether.

METHYLENE CHLORIDE. *See* **p. 375** following 4,4'-methylene bis(2-chloroaniline)

4,4'-METHYLENEDIANILINE. *See* **p. 383** following methylene diphenyl diisocyanate.

METHYLENE DIPHENYL DIISOCYANATE. *See* **p. 379** following methylene chloride.

METHYL ETHYL KETONE (sc)

CAS #: 78-93-3
MOLECULAR FORMULA: C_4H_8O
FORMULA WEIGHT: 72.12
SYNONYMS: Aethylmethylketon (German), butanone 2 (French), 2-butanone, ethyl methyl cetone (French), ethylmethyl keton (Dutch), ethyl methyl ketone (DOT), FEMA no. 2170, MEK, methyle acetone (DOT), metiletilchtone (Italian), metyloetylokketon (Polish), RCRA waste number U159

USES
◆ Manufacture of Other Chemicals
  • resins
  • cleaning fluids
  • vinyl films
  • adhesives
  • paint removers
  • cements
  • acrylic coating
◆ Used as a Component or in the Manufacturing Process
  • organic synthesis
  • solvent in vinyl films coatings
  • manufacture of smokeless powder
  • solvent in nitrocellulose
  • printing
  • catalyst carrier

PHYSICAL PROPERTIES
◆ clear, colorless, volatile liquid
◆ melting point −85.9°C
◆ refractive index 1.379 at 20°C
◆ explosion limits 1.8% upper, 11.5% lower by volume in air
◆ viscosity 0.40 cP at 25°C
◆ vapor density 2.42 g/mL
◆ odor threshold 5.4 ppm
◆ miscible with alcohol, ether, fixed oils, and water
◆ acetone-like odor
◆ normal boiling point 79.57°C
◆ specific heat 0.549 cal/g
◆ bulk density 6.71 lb/gal at 20°C
◆ vapor pressure 71.2 mm Hg at 25°C
◆ specific gravity 0.805
◆ log octanol/water partition coefficient 0.261

CHEMICAL PROPERTIES
◆ flash point 22°F
◆ autoignition temperature 960°F

HEALTH RISK
- ◆ Toxic Exposure Guidelines
  - • ACGIH TLV TWA 200 ppm; STEL 300 ppm
  - • OSHA PEL TWA 200 ppm; STEL 300 ppm
  - • DFG MAK 590 mg/m^3
  - • NIOSH REL (ketones) TWA 590 mg/m^3
  - • MSHA (unavailable)
  - • NOAEL (unavailable)
- ◆ Toxicity Data
  - • eye-hmn 350 ppm
  - • skin-rabbit 500 mg/24H MOD
  - • skin-rabbit 402 mg/24H MLD
  - • eye-rabbit 80 mg
  - • inhalation-rat TCLo 100 ppm
  - • inhalation-human TCLo 100 ppm
  - • oral-rat LD$_{50}$ 2737 mg/kg
  - • inhalation-rat LD$_{50}$ 23,500 mg/m^3/8h
  - • intraperitoneal-rat LD$_{50}$ 607 mg/kg
  - • oral-mouse LD$_{50}$ 4050 mg/kg
  - • inhalation-mouse LC$_{50}$ 40 g/m^3/2H
  - • intraperitoneal-mouse LD$_{50}$ 616 mg/kg
  - • skin-rabbit LD$_{50}$ 6480 mg/kg
  - • intraperitoneal-guinea pig LDLo 2 g/kg
  - • inhalation-rat TCLo 5000 ppm/6h
- ◆ Acute Risks
  - • harmful by inhalation and ingestion
  - • irritation of skin and eyes
  - • central nervous system depression
  - • irritating to mucous membranes and upper respiratory tract
  - • narcotic effects
  - • nausea, dizziness, and headache
  - • gastrointestinal disturbances
  - • dermatitis
- ◆ Chronic Risks
  - • effects on central nervous system
  - • effects on liver and respiratory system

HAZARD RISK
- ◆ NFPA rating (unavailable)
- ◆ protect from moisture
- ◆ incompatible with oxidizing agents, bases, strong reducing agents
- ◆ Mixture with 2-propanol produces explosive peroxides during storage.
- ◆ heat- and shock-sensitive explosive product with hydrogen peroxide and nitric acid reaction
- ◆ Container explosion may occur under fire conditions.

EXPOSURE ROUTES
- ◆ vapor inhalation
- ◆ absorption through skin and eyes
- ◆ ingestion

- ◆ intraperitoneal route
- ◆ photooxidation of butane and other hydrocarbons
- ◆ drinking water and surface water

REGULATORY STATUS/ INFORMATION
- ◆ Resource Conservation and Recovery Act
  - • list of hazardous inorganic and organic constituents. 40CFR258
  - • discarded commercial chemical products, off-specification species, container residues, and spill residues thereof. 40CFR261.33
- ◆ CERCLA and Superfund
  - • designation of hazardous substances. 40CFR302.4
- ◆ Occupational Safety and Health Act
  - • limits for air contaminants under the Occupational Safety and Health Act. 29CFR1910.1000
- ◆ Department of Transportation
  - • hazard class/division: 3
  - • identification number: UN1193
  - • labels: flammable liquid
- ◆ Community Right-to-Know Threshold Planning Quantity (unavailable)

ADDITIONAL/OTHER COMMENTS
- ◆ Symptoms
  - • inhalation: irritation of eyes and skin
  - • absorption: dizziness and nausea
- ◆ First Aid
  - • flush eyes with water
  - • if inhaled, remove to fresh air
  - • if swallowed, wash out mouth with water
- ◆ fire fighting: water spray, carbon dioxide, dry chemical powder, or appropriate foam
- ◆ personal protection: wear respirator, chemical-resistant gloves, safety goggles
- ◆ storage: keep away from heat, sparks, and open flames

SOURCES
- ◆ *Environmental Contaminant Reference Databook*, volume 1, 1995
- ◆ *Environmental Law Index to Chemicals*, 1996 edition
- ◆ *Hawley's Condensed Chemical Dictionary*, 12th edition, 1993
- ◆ *Material Safety Data Sheets (MSDS)*
- ◆ *Sax's Dangerous Properties of Industrial Materials*, 9th edition, 1996
- ◆ *Health Effects Notebook For Hazardous Air Pollutants*, U.S. Environmental Protection Agency, December 1994

METHYL HYDRAZINE(sf)

CAS #: 60-34-4
MOLECULAR FORMULA: CH_3NHNH_2
FORMULA WEIGHT: 46.1
SYNONYMS: hydrazomethane, 1-methyl hydrazine, methyl hydrazine (DOT), MMH, monomethyl hydrazine, RCRA waste number P068

USES
- used as high energy fuel in military applications such as rocket fuel and propellant and as fuel for small electrical power generating units
- Manufacture of Other Chemicals
  - chemical intermediate
  - solvent

PHYSICAL PROPERTIES
- colorless liquid
- ammonia-like fishy smell
- melting point −20.9°C
- density 0.987 g/mL
- odor threshold 2.1 ppm (mg/m³)
- boiling point 87.8°C at 1 atm
- specific gravity 0.874 at 25°C (water = 1)
- vapor density 2 (air = 1)
- heat capacity 134.9 J/gmol-K (liquid at 25°C)
- heat of vaporization 36.12 kJ/gmol
- reacts with oxygen
- vapor pressure 37.5 mm Hg at 20°C, 206 mm Hg at 55°C

CHEMICAL PROPERTIES
- flash point 70°F (21°C)
- autoignition temperature 385°C (195°F)
- flammability limits 2.5% lower and 97% upper
- enthalpy (heat) of formation at 25°C (liquid) 54.0 kJ/gmol
- unstable, may spontaneously combust in air
- reacts with oxygen, oxidizing agents, and peroxides
- Hazardous combustion or decomposition products. Thermal decomposition may produce CO, CO_2, and NO_x.

HEALTH RISK
- Toxic Exposure Guidelines
  - AGCIH TLV CL 0.2 ppm 0.35 mg/m³; suspected human carcinogen (Proposed: CL 0.01 ppm 0.35 mg/m³; suspected human carcinogen)
  - OSHA PEL CL 0.2 ppm (skin)
  - NIOSH REL CL 0.08 mg/m³/2H
- Toxicity Data
  - dnd-human: fbr 116 pmol
  - oral-rat LD_{50} 32 mg/kg
  - inhalation-rat LC_{50} 64 mg/m³ (34 ppm)/4H
  - inhalation-rat TCLo 20 ppb/6H/1Y-I: CAR
  - skin-rabbit LD_{50} 95 mg/kg

◆ Other Toxicity Indicators
  • EPA Cancer Risk Level: nonthreshold category, 1/ED10 is 4.1 mg/kg/d probable human carcinogen, group B2
◆ Acute Risks
  • may be fatal if inhaled, swallowed, or absorbed through skin
  • extremely destructive to skin, eyes, mucous membranes, and upper respiratory tract
  • Inhalation causes spasm, inflammation, and edema of the larynx and bronchi, and chemical pneumonitis and pulmonary edema which may be fatal.
  • may cause allergic respiratory and skin reactions
  • burning sensation
  • diarrhea
  • ataxia
  • anoxia
  • cyanosis
  • tremors
  • convulsions
  • coughing
  • wheezing
  • laryngitis
  • shortness of breath
  • headaches
  • nausea
  • vomiting
  • liver and kidney damage
◆ Chronic Risks
  • Chronic inhalation exposure may impair function of kidneys and liver and affect the blood and spleen.
  • shown to cause convulsions in animals

HAZARD RISK
◆ NFPA rating Health 4, Flammability 3, Reactivity 2
◆ dangerous fire hazard
◆ unstable, may spontaneously combust in air
◆ reacts with oxygen, oxidizing agents, and peroxides
◆ Hazardous combustion or decomposition products. Thermal decomposition may produce CO, CO_2, and NO_x.
◆ carcinogen
◆ mutagen

EXPOSURE ROUTES
◆ primarily by inhalation
◆ absorption through skin and eyes
◆ ingestion
◆ exposure in workplace
◆ may be present in ambient atmosphere during its use as rocket fuel and from spills, leaks, and venting during loading, transfer, and storage
◆ exposure through ingestion of a species of edible mushroom

REGULATORY STATUS/INFORMATION
- ◆ Clean Air Act Title III Hazardous Air Pollutant
  - high-risk pollutant; weighting factor is 10
  - accidental release prevention substance
  - Threshold quantity (TQ) is 15,000 lb.
- ◆ Superfund Amendment and Reauthorization Act
  - subject to section 313 reporting requirements
- ◆ CERCLA and Superfund
  - subject to listing under 40CFR372.65
  - designation of hazardous substances, 40CFR302.4
- ◆ Appendix A, CAS registry of CERCLA hazardous substances
- ◆ Appendix B, radionuclides
  - list of extremely hazardous substances and their threshold planning quantities, 40CFR355-AB
  - Reportable quantity (RQ) is 10 lb.
  - Threshold planning quantity (TPQ) is 500 lb.
- ◆ Resource Conservation and Recovery Act
  - hazardous constituent, 40CFR261, see also 40CFR261.11
  - risk specific doses, 40CFR266 Appendix 5
  - health based limits for exclusion of waste-derived residues, 40CFR266 Appendix 7
  - RCRA designation is P068.
- ◆ OSHA Table Z1 and Table Z5
  - TQ is 100 lb.
- ◆ Department of Transportation
  - identification number UN2480
  - hazard class/division 6.1
  - label: poison material

ADDITIONAL/OTHER COMMENTS
- ◆ Symptoms
  - may be fatal if inhaled, swallowed, or absorbed through skin
  - extremely destructive to skin, eyes, mucous membranes, and upper respiratory tract
  - red, irritated, bloodshot eyes
  - Inhalation causes spasm, inflammation, and edema of the larynx and bronchi, and chemical pneumonitis and pulmonary edema which may be fatal.
  - may cause allergic respiratory and skin reactions
  - burning sensation
  - diarrhea
  - ataxia
  - anoxia
  - cyanosis
  - tremors
  - convulsions
  - coughing
  - wheezing
  - laryngitis
  - shortness of breath
  - headaches
  - malaise

- nausea
- vomiting
- liver and kidney damage

◆ First Aid
  - flush eyes and/or skin with large amounts of water for at least 15 minutes while removing contaminated garments and shoes.
  - separate eyelids with fingers while flushing
  - seek fresh air
  - provide respiratory support and oxygen
  - if swallowed wash out mouth with water if conscious
  - seek medical help immediately

◆ Fire Fighting
  - water spray, carbon dioxide, dry chemical powder, or appropriate foam
  - self-contained breathing apparatus and protective clothing to prevent contact with eyes and skin

◆ Personal Protection
  - wear appropriate NIOSH/MSHA approved respirator
  - safety goggles, face shield (8 in. minimum)
  - use only in chemical fume hood
  - chemical-resistant gloves
  - rubber apron, sleeves, and other protective clothing
  - avoid prolonged or repeated exposure
  - wash thoroughly after handling

◆ Storage
  - stainless steel or glass container only
  - keep container tightly closed
  - handle and store under nitrogen
  - cool, dry place away from heat and open flame

SOURCES
  ◆ *AGCIH Threshold Limit Values (TLV) for 1996*
  ◆ *Aldrich Chemical Company Catalog*, 1996–1997
  ◆ *CRC Handbook of Chemistry and Physics*, 77th edition, 1996–1997
  ◆ *CRC Handbook of Thermophysical and Thermochemical Data*, 1994
  ◆ *Environmental Law Index to Chemicals*, 1996 edition
  ◆ *Hawley's Condensed Chemical Dictionary*, 12th edition, 1993
  ◆ *(MSDS)*, Aldrich Chemical Company, Inc.
  ◆ *Physical and Thermodynamic Properties of Pure Chemicals: Data Compilation*, AiChE and NSRSD, 1996
  ◆ *Sax's Dangerous Properties of Industrial Materials*, 9th edition, 1996
  ◆ *Suspect Chemicals Sourcebook*, 1996 edition
  ◆ *Health Effects Notebook for Hazardous Air Pollutants*, U.S. Environmental Protection Agency, 1994

METHYL IODIDE (sc)

CAS #: 74-88-4
MOLECULAR FORMULA: CH_3I
FORMULA WEIGHT: 141.94
SYNONYMS: halon 10001, iodometano (Italian), iodomethane, iodure de methyle (French), Jod-methan (German), joodmathaan (Dutch), Methyljodid (German), Methyljodide (Dutch), metylu jodek (Polish), monoioduro di metile (Italian), RCRA waste number U138

USES
- ◆ organic synthesis
- ◆ pharmaceuticals and pesticides
- ◆ methylation processes
- ◆ microscopy
- ◆ testing for pyridine
- ◆ fire extinguisher
- ◆ insecticidal fumigant

PHYSICAL PROPERTIES
- ◆ clear, colorless liquid
- ◆ normal boiling point 42.5°C
- ◆ density 2.279 g/mL at 20°C
- ◆ specific gravity 2.28
- ◆ soluble in water at 20°C
- ◆ miscible in alcohol and ether
- ◆ log octanol/water partition coefficient 1.51
- ◆ turns brown on exposure to light
- ◆ melting point −64°C
- ◆ refractive index 1.526 at 25°C
- ◆ pleasant odor
- ◆ vapor pressure 7.89 psi at 20°C and 24.09 psi at 55°C
- ◆ vapor density 4.89 g/mL

CHEMICAL PROPERTIES
- ◆ nonflammable
- ◆ flash point none at 300°C
- ◆ violently reacts with oxygen

HEALTH RISK
- ◆ Toxic Exposure Guidelines
  - • ACGIH TLV TWA 2 ppm; suspected human carcinogen
  - • OSHA PEL TWA 2 ppm
  - • DFG MAK animal carcinogen, suspected human carcinogen
  - • MSHA 28 mg/m³
  - • NIOSH REL 10 mg/m³ (monohalomethanes) reduce to lowest level
- ◆ Toxicity Data
  - • skin-human 1 g/10M MLD
  - • skin-rat 1 g/30M MLD
  - • intraperitoneal-mouse TDLo 44 mg/kg:ETA
  - • oral-rat LDLo 150 mg/kg
  - • inhalation-rat LC_{50} 1300 mg/m³/4H
  - • skin-rat LDLo 800 mg/kg
  - • intraperitoneal-rat 101 mg/kg
  - • inhalation-mouse LCLo 5000 mg/m³/1H
  - • intraperitoneal-mouse LD_{50} 172 mg/kg
  - • intraperitoneal-guinea pig LD_{50} 51 mg/kg

- ◆ Acute Risks
  - fatal if inhaled, swallowed, or absorbed through skin
  - causes severe irritation of skin
  - destructive to tissues of the mucous membranes and upper respiratory tract
  - nausea
  - weakness, drowsiness
  - convulsions
  - narcotic effects
  - pulmonary edema (effects may be delayed)
  - vertigo
  - skin blistering
- ◆ Chronic Risks
  - carcinogen
  - may alter genetic material
  - reproductive disorders

HAZARD RISK
- ◆ NFPA rating (unavailable)
- ◆ emits toxic fumes of I^- when heated to decomposition
- ◆ explosive reaction with trialkylphosphines and silver chlorite

EXPOSURE ROUTES
- ◆ readily absorbed through skin
- ◆ inhalation
- ◆ ingestion
- ◆ intraperitoneal and subcutaneous routes
- ◆ product of marine algae

REGULATORY STATUS/ INFORMATION
- ◆ Resource Conservation and Recovery Act
  - constituents for detection monitoring. 40CFR261.33
  - list of hazardous inorganic and organic constituents. 40CFR258
- ◆ CERCLA and Superfund
  - designation of hazardous substances. 40CFR302.4
- ◆ Department of Transportation
  - hazard class/division: 6.1
  - identification number: UN2644
  - labels: Poison
- ◆ Community Right-to-Know Threshold Planning Quantity (unavailable)

ADDITIONAL/OTHER COMMENTS
- ◆ Symptoms
  - inhalation: burning sensation, coughing, wheezing, shortness of breath
  - absorption: blistering on contact with skin
- ◆ First Aid
  - flush eyes with water
  - if inhaled, remove to fresh air
  - if swallowed, wash out mouth with water

- fire fighting: water spray, carbon dioxide, dry chemical powder, or appropriate foam
- personal protection: wear respirator, chemical-resistant gloves, safety goggles
- storage: keep tightly closed; protect from moisture; store in cool, dry place; light sensitive

SOURCES

- *Environmental Law Index to Chemicals*, 1996 edition
- *Hawley's Condensed Chemical Dictionary*, 12th edition, 1993
- *Material Safety Data Sheets (MSDS)*
- *Sax's Dangerous Properties of Industrial Materials*, 9th edition, 1996
- *Health Effects Notebook for Hazardous Air Pollutants*, U.S. Environmental Protection Agency, December 1994

METHYL ISOBUTYL KETONE

CAS #: 108-10-1
MOLECULAR FORMULA: $(CH_3)_2CHCH_2COCH_3$
FORMULA WEIGHT: 100.18
SYNONYMS: hexanone, hexone, isobutyl-methyl-ketone, isopropylacetone, ketone-isobutyl methyl, methyl-isobutyl-ketone, methyl-2-oxo-pentane(4-), methyl-2-pentanone, methyl-2-pentanone(4-), methyl-4-pentanone(2-), Methyl-propyl-methyl ketone(2-), MIBK, MIK, pentanone(2-), 4-methyl-

USES
- Used as a Solvent
  - gums
  - resins
  - paints
  - varnishes
  - lacquers
  - nitrocellulose
  - alcohol denaturant
  - extraction of rare metals
  - synthetic flavoring adjuvant
  - organic synthesis
  - manufacture of methyl amyl alcohol

PHYSICAL PROPERTIES
- colorless, flammable liquid
- fruity, ethereal odor
- odor threshold 0.10 ppm
- melting point −80.2°C
- boiling point 116.8°C
- specific gravity 0.801
- vapor pressure 15 mm Hg at 20°C
- vapor density 3.5
- moderately soluble in water
- miscible with alcohol and ether
- log K_{ow} 1.09

CHEMICAL PROPERTIES
- flash point 13°C
- autoignition temperature 448°C
- explosion limits in air 1.2% lower and 8% upper at 93°C
- Hazardous polymerization will not occur.
- incompatible with oxidizing agents, reducing agents, and strong bases

HEALTH RISK
- ◆ Toxic Exposure Guidelines
  - OSHA PEL TWA 50 ppm
  - OSHA STEL 75 ppm
  - ACGIH TLV TWA 50 ppm
  - ACGIH TLV STEL 75 ppm
  - DFG MAK 100 ppm
  - NIOSH REL (ketones) TWA 200 mg/m^3
  - MSHA standard 410 mg/m^3
- ◆ Toxicity Data
  - eye-human 200 ppm/15M
  - skin-rabbit 500 mg/24H MLD
  - eye-rabbit 500 mg/24H MLD
  - eye-rabbit 40 mg SEV
  - inhalation-mouse TCLo 3000 ppm/6H
  - oral-rat LD$_{50}$ 2080 mg/kg
  - intraperitoneal-rat LD$_{50}$ 400 mg/kg
  - oral-mouse LD$_{50}$ 2671 mg/kg
  - inhalation-mouse LC$_{50}$ 23,300 mg/m^3
  - intraperitoneal-mouse LD$_{50}$ 268 mg/kg
  - oral-guinea pig LD$_{50}$ 1600 mg/kg
- ◆ Acute Risks
  - irritation of eyes
  - irritation of mucous membranes
  - weakness
  - headache
  - nausea
  - light-headedness
  - vomiting
  - dizziness
  - incoordination
  - narcosis
  - redness of eyes
  - tearing and blurred vision
  - coma
  - death
- ◆ Chronic Risks
  - nausea
  - headache
  - burning in eyes
  - weakness
  - insomnia
  - intestinal pain
  - slight enlargement of liver
  - lethargy in rats and mice
  - increased kidney and liver weights in rats and mice
  - maternal toxicity in rats and mice
  - neurological effects in fetuses of rats and mice
  - EPA Group D: not classifiable as to human carcinogenicity

HAZARD RISK
- ◆ flammable liquid when exposed to heat, flame, or oxidizers
- ◆ ignites on contact with potassium-tert-butoxide
- ◆ moderately explosive in the form of vapor when exposed to heat or flame
- ◆ may form explosive peroxides upon exposure to air
- ◆ can react vigorously with reducing materials
- ◆ incompatible with air, potassium-tert-butoxide, oxidizing agents, reducing agents, and strong bases

EXPOSURE ROUTES
- inhalation of vapors in workplace
- skin and eye contact
- inhalation and dermal contact
- during use of consumer products
- released to environment in effluent and emissions
- exhaust gas from vehicles
- land disposal
- ocean dumping of waste
- inhalation of contaminated air
- ingestion of contaminated water

REGULATORY STATUS/INFORMATION
- Clean Air Act Amendment, Title III: hazardous air pollutants
- Safe Drinking Water Act
  - priority list of drinking water contaminants. 55FR1470
  - waste specific prohibitions-solvent wastes. 40CFR148.10
- Resource Conservation and Recovery Act
  - constituents for detection monitoring (for municipal solid waste landfill). 40CFR258
  - list of hazardous constituents. 40CFR258
  - U waste #: U161
- CERCLA and Superfund
  - designation of hazardous substances. 40CFR302.4
  - list of chemicals. 40CFR372.65
- Toxic Substance Control Act
  - list of substances. 40CFR716.120.a
  - testing consent orders for substances and mixtures with CAS registry #s. 40CFR799.5000
- chemical regulated by the State of California

ADDITIONAL/OTHER COMMENTS
- Symptoms
  - vapor or mist exposure: irritation to eyes, mucous membranes, and upper respiratory tract, skin irritation
  - contact with eyes: redness, tearing, blurred vision
  - contact with skin: defatting and dermatitis
- First Aid
  - wash eyes and skin immediately with large amounts of water
  - remove to fresh air immediately or provide respiratory apparatus if necessary
  - wash out mouth with water
- fire fighting: use carbon dioxide, dry chemical powder, or appropriate foam
- personal protection: self-contained breathing apparatus, protective clothing, and goggles
- storage: keep tightly closed away from heat, sparks, and open flame in a cool, dry place

SOURCES
- *Health Effects Notebook for Hazardous Air Pollutants*, U.S. Environmental Protection Agency, December 1994

◆ *Material Safety Data Sheets (MSDS)*
◆ *Sax's Dangerous Properties of Industrial Materials*, 9th edition, 1996
◆ *Hawley's Condensed Chemical Dictionary*, 12th edition, 1993
◆ *Aldrich Chemical Company Catalog*, 1996–1997
◆ *Environmental Law Index to Chemicals*, 1996 edition

METHYL ISOCYANATE(sf)

CAS #: 624-83-9
MOLECULAR FORMULA: C_2H_3NO
FORMULA WEIGHT: 57.05
SYNONYMS: isocyanate methyl methane, isocyanic acid methyl ester, iso-cyanoto methane MIC, RCRA waste number P064

USES
◆ Manufacture of Other Chemicals
  • chemical intermediate in the production of carbamate insecticides and herbicides

PHYSICAL PROPERTIES
◆ colorless liquid
◆ sharp odor
◆ melting point −17°C
◆ boiling point 37–39°C 1 atm
◆ specific gravity 0.987 (water = 1)
◆ density 0.987 g/mL
◆ odor threshold 2.1 ppm (mg/m^3)
◆ vapor density 2 (air = 1)
◆ viscosity 0.127 cP at 20°C
◆ reacts vigorously with water
◆ vapor pressure 348 mm Hg at 20°C, 1339 mm Hg at 55°C

CHEMICAL PROPERTIES
◆ flash point 20°F (−6°C)
◆ autoignition temperature 533°C (994°F)
◆ flammability limits 5.3% lower and 26% upper
◆ enthalpy (heat) of formation at 25°C (liquid) −92.0 kJ/gmol
◆ unstable
◆ polymerizes easily, and polymer may also be hazardous
◆ reacts with water, alcohols, acids, amines, strong bases, strong oxidizing agents, steel, heat
◆ Hazardous combustion or decomposition products. Thermal decomposition may produce CO, CO_2, NO_x, and HCN.

HEALTH RISK
◆ Toxic Exposure Guidelines
  • AGCIH TLV TWA 0.02 ppm 0.05 mg/m^3 (skin)
  • OSHA PEL TWA 0.02 ppm (skin)
  • NIOSH REL TWA 5 ppm (skin)
◆ Toxicity Data
  • inhalation-human TCLo 2 ppm: NOSE, EYE, PUL
  • oral-rat LD_{50} 51.5 mg/kg
  • inhalation-rat LC_{50} 6100 ppb/6H

- inhalation-rat TCLo 3 ppm/6H/4D-I
- skin-rabbit LD$_{50}$ 213 mg/kg
◆ Other Toxicity Indicators
  - EPA Cancer Risk Level: not classifiable
◆ Acute Risks
  - may be fatal if inhaled, swallowed, or absorbed through skin
  - extremely destructive to skin, eyes, mucous membranes, and upper respiratory tract
  - may cause permanent fibrosis
  - Inhalation causes spasm, inflammation and edema of the larynx and bronchi, and chemical pneumonitis and pulmonary edema which may be fatal.
  - bronchial pneumonia
  - bronchitis
  - burning sensation
  - coughing
  - wheezing
  - laryngitis
  - shortness of breath
  - labored breathing
  - headaches
  - cyanosis
  - dizziness
  - nausea
  - vomiting
  - gastritis
  - sweating
  - chills
  - fever
  - liver and kidney damage
◆ Chronic Risks
  - reproductive risks observed following accident in Bhopal, India
    - leukorrhea
    - pelvic inflammatory disease
    - excessive menstrual bleeding
    - supression of lactation
    - spontaneous abortions
    - increased number of stillbirths
  - no other information available

HAZARD RISK
  ◆ NFPA rating Health 2, Flammability 3, Reactivity 3; avoid use of water
  ◆ dangerous fire hazard
  ◆ unstable
  ◆ polymerizes easily, and polymer may also be hazardous
  ◆ reacts with water, alcohols, acids, amines, strong bases, strong oxidizing agents, steel, heat
  ◆ Hazardous combustion or decomposition products. Thermal decomposition may produce CO, CO_2, NO_x, and HCN.

EXPOSURE ROUTES
- ◆ primarily by inhalation
- ◆ absorption through skin and eyes
- ◆ ingestion
- ◆ exposure to workers who use insecticides or herbicides produced from methyl isocyanate
- ◆ detected in cigarette smoke
- ◆ no information available on levels in ambient air or water

REGULATORY STATUS/INFORMATION
- ◆ Clean Air Act Title III Hazardous Air Pollutant
  - • high-risk pollutant; weighting factor is 10
  - • accidental release prevention substance
  - • Threshold quantity (TQ) is 10,000 lb.
- ◆ Superfund Amendment and Reauthorization Act
  - • subject to section 313 reporting requirements
- ◆ CERCLA and Superfund
  - • subject to listing under 40CFR372.65
  - • designation of hazardous substances, 40CFR302.4
    - • Appendix A, CAS registry of CERCLA hazardous substances
    - • Appendix B, radionuclides
  - • list of extremely hazardous substances and their threshold planning quantities, 40CFR355-AB
  - • Reportable quantity (RQ) is 10 lb.
  - • Threshold planning quantity (TPQ) is 500 lb.
- ◆ Resource Conservation and Recovery Act
  - • hazardous constituent, 40CFR261, see also 40CFR261.11
  - • discarded commercial chemical products, off-specification species, container residues, and spills thereof, 40CFR261.33
  - • RCRA designation is P064.
- ◆ OSHA Table Z1 and Table Z5
  - • TQ is 250 lb.
- ◆ Department of Transportation
  - • identification number UN2480
  - • hazard class/division 3, 6.1
  - • label: poison, flammable liquid

ADDITIONAL/OTHER COMMENTS
- ◆ Symptoms
  - • Inhalation causes spasm, inflammation, and edema of the larynx and bronchi, and chemical pneumonitis and pulmonary edema which may be fatal.
  - • bronchial pneumonia
  - • bronchitis
  - • burning sensation
  - • coughing
  - • wheezing
  - • laryngitis
  - • shortness of breath
  - • labored breathing

- headaches
- cyanosis
- dizziness
- nausea
- vomiting
- gastritis
- sweating
- chills
- fever
◆ First Aid
  - flush eyes and/or skin with large amounts of water for at least 15 minutes while removing contaminated garments and shoes
  - separate eyelids with fingers while flushing
  - seek fresh air
  - provide respiratory support and oxygen
  - if swallowed wash out mouth with water if conscious
  - seek medical help immediately
◆ Fire Fighting
  - carbon dioxide or dry chemical powder only
  - self-contained breathing apparatus and protective clothing to prevent contact with eyes and skin
◆ Personal Protection
  - wear appropriate NIOSH/MSHA approved respirator
  - safety goggles, face shield (8 in. minimum)
  - long chemical-resistant gauntlet gloves
  - rubber apron, sleeves, and other protective clothing
  - avoid prolonged exposure
◆ Storage
  - stainless steel container only
  - keep container tightly closed
  - store under nitrogen
  - cool, dry place away from heat and open flame

SOURCES

◆ *AGCIH Threshold Limit Values (TLV) for 1996*
◆ *Aldrich Chemical Company Catalog*, 1996–1997
◆ *CRC Handbook of Chemistry and Physics*, 77th edition, 1996–1997
◆ *CRC Handbook of Thermophysical and Thermochemical Data*, 1994 edition
◆ *Environmental Law Index to Chemicals*, 1996 edition
◆ *Handbook of Viscosity*, Carl L. Yaws, Gulf Publishing, Houston, TX, 1995
◆ *Hawley's Condensed Chemical Dictionary*, 12th edition, 1993
◆ *MSDS*, Aldrich Chemical Company, Inc.
◆ *Physical and Thermodynamic Properties of Pure Chemicals: Data Compilation*, AiChE and NSRSD, 1996
◆ *Sax's Dangerous Properties of Industrial Materials*, 9th edition, 1996
◆ *Suspect Chemicals Sourcebook*, 1996 edition

METHYL METHACRYLATE (sc)

CAS #: 80-62-6
MOLECULAR FORMULA: $C_5H_8O_2$
FORMULA WEIGHT: 100.13
SYNONYMS: acrylic acid, 2-methyl-methyle ester, diakon, metakrylan metylu (Polish), methacrylate de methyle (French), methacrylic acid methyl ester (AK), methacrylsaeuremethyl ester, Methacrylsaeuremethyl ester (German), methylester kyseliny methakrylove, methylmethacrylaat (Dutch), Methyl-methacrylat (German), methyl methacrylate monomer, inhibited (DOT), methyl-a-methylacrylate, methyl-2-methyl-2-propenoate, 2-methyl-2-propenoic acid methyl ester, metil metacrilato (Italian), MME, "monocite" methacrylate monomer, NCI-C50680, 2-propenoic acid, 2-methyl methyl ester, RCRA waste number U162

USES
♦ Manufacture of Other Chemicals
  • monomer for polymethacrylate resins
♦ Used as a Component or in the Manufacturing Process
  • impregnation of concrete
  • advertising signs
  • bone cement
  • lighting fixtures
  • glazing and skylights
  • building panels and sidings
  • plumbing and bathroom fixtures
  • molding/extrusion powder

PHYSICAL PROPERTIES
♦ clear, colorless, volatile liquid
♦ melting point $-50°C$
♦ sharp, fruity odor
♦ explosion limits 2.12% lower, 12.5% upper by volume in air
♦ vapor pressure 40 mm at 25.5°C
♦ soluble in acetone
♦ normal boiling point 101.0°C
♦ specific gravity 0.936
♦ vapor density 3.45 g/ml
♦ very slightly soluble in water
♦ log octanol/water partition coefficient 0.79

CHEMICAL PROPERTIES
♦ autoignition temperature 815°F
♦ potentially violent reaction with polymerization initiators
♦ flash point 50°F

HEALTH RISK
♦ Toxic Exposure Guidelines
  • ACGIH TLV TWA 100 ppm
  • OSHA PEL TWA 100 ppm
  • DFG MAK 50 ppm (210 mg/m³)
  • MSHA 410 mg/m³
♦ Toxicity Data
  • skin-rabbit 10 g/kg open
  • eye-rabbit 150 mg

369

- inhalation-rat TCLo 109 g/m^3/54M
- inhalation-rat TCLo 4480 mg/m^3/2H
- implant-rat TDLo 1620 mg/kg
- inhalation-human TCLo 125 ppm
- inhalation-human TCLo 60 mg/m^3
- oral-rat LD$_{50}$ 7872 mg/kg
- inhalation-rat LC$_{50}$ 3750 ppm
- intraperitoneal-rat LD$_{50}$ 1328 mg/kg
- subcutaneous-rat LD$_{50}$ 7500 mg/kg
- oral-mouse LD$_{50}$ 5204 mg/kg
- inhalation-mouse TCLo 5000 ppm/6H
- inhalation-mouse LC$_{50}$ 18,500 mg/m^3/2H
- intraperitoneal-mouse LD$_{50}$ 1000 mg/kg
- subcutaneous-mouse LD$_{50}$ 6300 mg/kg
- oral-dog LDLo 5000 mg/kg
- subcutaneous-dog LD$_{50}$ 4500 mg/kg

◆ Acute Risks
  - harmful if swallowed
  - destructive to tissue of the mucous membranes and upper respiratory tract
  - spasm, inflammation, and edema of larynx and bronchi
  - burning sensation
  - coughing
  - wheezing, laryngitis, shortness of breath
  - headache, nausea, and vomiting

◆ Chronic Risks
  - kidney and liver lesions
  - sleeping disturbances
  - cardiovascular disorders
  - effects on the nasal cavity

HAZARD RISK
◆ NFPA rating (not available)
◆ very dangerous fire hazard when exposed to heat or flame
◆ can react with oxidizing materials
◆ Vapor is explosive when exposed to heat or flame.
◆ Monomer may undergo spontaneous, explosive polymerization.
◆ explodes on evaporation at 60°C
◆ may ignite on contact with benzoyl peroxide
◆ emits acrid smoke and irritating fumes when heated to decomposition

EXPOSURE ROUTES
◆ primarily occupational
◆ inhalation
◆ intraperitoneal routes
◆ ingestion
◆ consumption of contaminated water

REGULATORY STATUS/INFORMATION
◆ Resource Conservation and Recovery Act
  - constituents for detection monitoring. 40CFR258

- hazardous constituents. 40CFR261
- groundwater monitoring chemicals list. 40CFR264
- discarded commercial chemical products, off-specification species, container residues, and spill residues thereof. 40CFR261.33
◆ CERCLA and Superfund
- designation of hazardous substances. 40CFR302.4
- chemicals and chemical categories to which this part applies. 40CFR372.65
◆ Clean Water Act
- designation of hazardous substances. 40CFR116.4
- determination of reportable quantities. 40CFR117.3
◆ Toxic Substance Control Act
- subpart E—significant new uses of chemical substances. 40CFR721.225-9975
◆ Department of Transportation
- hazard class/division (unavailable)
- identification number NA1247
- labels (unavailable)
◆ Community Right-to-Know Threshold Planning Quantity (unavailable)

ADDITIONAL/OTHER COMMENTS
◆ Symptoms
- inhalation: burning sensation, coughing, wheezing, shortness of breath
◆ First Aid
- flush eyes with water
- if inhaled, remove to fresh air
- if swallowed, wash out mouth with water
◆ fire fighting: water spray, carbon dioxide, dry chemical powder, or appropriate foam
◆ personal protection: wear respirator, chemical-resistant gloves, safety goggles
◆ storage: keep tightly closed; keep away from heat, sparks, and open flame

SOURCES
◆ *Environmental Law Index to Chemicals*, 1996 edition
◆ *Hawley's Condensed Chemical Dictionary*, 12th edition, 1993
◆ *Material Safety Data Sheets (MSDS)*
◆ *Sax's Dangerous Properties of Industrial Materials*, 9th edition, 1996
◆ *Health Effects Notebook for Hazardous Air Pollutants*, U.S. Environmental Protection Agency, December 1994

METHYL tert BUTYL ETHER (sc)

CAS #: 1634-04-4
MOLECULAR FORMULA: C₅Hr₁₂O
FORMULA WEIGHT: 88.17
SYNONYMS: 2-methoxy-2-methylpropane, methyl 1,1-dimethylethyl ether, methyl tert-butyl ether (DOT), mtbe, propane 2-methoxy-2-methyl- (9CI)

USES
◆ Manufacture of Other Chemicals
  • octane booster in unleaded gasoline
◆ Used as a Component or in the Manufacturing Process
  • isobutene

PHYSICAL PROPERTIES
◆ clear, colorless liquid
◆ specific gravity 0.74
◆ slightly soluble in water

◆ normal boiling point 54°C
◆ vapor pressure: 245 mm Hg at 25°C

CHEMICAL PROPERTIES
◆ flash point 14°F

HEALTH RISK
◆ Toxic Exposure Guidelines
  • ACGIH TLV TWA 40 ppm
◆ Toxicity Data
  • oral-rat LD$_{50}$ 4 g/kg
  • inhalation-rat LC$_{50}$ 23,576 ppm/4H
  • intraperitoneal-rat LD >148 mg/kg
  • intravenous-rat LDLo 148 mg/kg
  • inhalation-mouse LC$_{50}$ 141 g/m³/15M
◆ Acute Risks
  • harmful by inhalation, ingestion, or skin absorption
  • irritation to eyes
  • irritation to mucous membranes and upper respiratory tract
  • skin irritation
  • nausea
  • vomiting
  • diarrhea
◆ Chronic Risks (No Information Available for Humans)
  • damage to kidneys (female rats)
  • decreased brain weight (female rats)
  • increased liver and kidney weights (female rats)
  • swollen periocular tissue (female rats)

HAZARD RISK
◆ NFPA rating (unavailable)
◆ flammable when exposed to heat or flame
◆ emits acrid smoke and irritating fumes when heated to decomposition

EXPOSURE ROUTES
◆ poison by intravenous route
◆ occupationally via ingestion and inhalation
◆ contaminated air from unleaded gasoline

REGULATORY STATUS/INFORMATION
- ◆ Safe Drinking Water Act
  - • priority list of drinking water contaminants. 55FR1470
- ◆ CERCLA and Superfund
  - • chemicals and chemical categories to which this part applies. 40CFR372.65
- ◆ Toxic Substance Control Act
  - • testing consent orders for mixtures with CAS registry #s. 40CFR799.5000
- ◆ Department of Transportation
  - • hazard class/division: 3
  - • identification number: UN2398
  - • labels: flammable liquid
- ◆ Community Right-to-Know Threshold Planning Quantity (unavailable)

ADDITIONAL/OTHER COMMENTS
- ◆ can be fatal if swallowed (presence in the lungs could result in chemical pneumonia which can be fatal)
- ◆ Symptoms
  - • inhalation: burning sensation
  - • absorption: skin irritation
- ◆ First Aid
  - • flush eyes with water
  - • if inhaled, remove to fresh air
  - • if swallowed, wash out mouth with water
- ◆ fire fighting: carbon dioxide, dry chemical powder, or appropriate foam
- ◆ personal protection: appropriate respirator, chemical-resistant gloves, safety goggles
- ◆ storage: keep tightly closed; keep away from heat, sparks, and open flames; store in cool, dry place

SOURCES
- ◆ *Environmental Law Index to Chemicals*, 1996 edition
- ◆ *Material Safety Data Sheets (MSDS)*
- ◆ *Sax's Dangerous Properties of Industrial Materials*, 9th edition, 1996
- ◆ *Health Effects Notebook for Hazardous Air Pollutants*, U.S. Environmental Protection Agency, December 1994

4,4'-METHYLENE BIS(2-CHLOROANILINE)

CAS #: 101-14-4
MOLECULAR FORMULA: $CH_2(C_6H_4ClNH_2)_2$
FORMULA WEIGHT: 267.17
SYNONYMS: benzeneamine, 4,4'-methylenebis(2-chloro-, bis(3-chloro-4-aminophenyl)-methane, bis(4-amino-3-chlorophenyl)-methane, bisamine, cuamine-MT, curalin-M, curene-442, cyanaset, dacpm, di-amino-3-chlorophenyl-methane, diamet-KH, diamino-3,3'-dichlorodiphenylmethane(4,4'-), dichloro-4,4'-diaminodiphenyl-methane (3,3'-), ethyl-acrylate, MBOCA, methylene-4,4'-bis(o-chloroaniline),

methylene-bis(3-chloro-4-aminobenzene), methylene-bis(2-chloro-aniline), methylene-bis(alpha-chloroaniline)(p,p'-), methylene-bis-(2-chloroaniline)(4,4'-), methylene-bis(2-chlorobenzeneamine) (4,4'-), methylene-bis(o-chloroaniline)(4,4'-), methylene-bis (o-chloroani-line)(p,p'-), methylene-bis-(2-chlorobenzamine)(4,4'-), methylene-bis-(2-chlorobenzenamine)(4,4'-), methylene-bis-orthochloroaniline, millionate-M, MOCA, quodorole

USES
- ◆ Used in the Manufacturing Process
  - curing agent for epoxy resins
  - curing agent for liquid polyurethane elastomers
  - shoe soles
  - rolls for postage stamp machines
  - cutting bars in plywood manufacturing
  - rolls and belt drivers in cameras
  - computers
  - reproducing equipment
  - pulleys for escalators and elevators

PHYSICAL PROPERTIES
- ◆ colorless solid
- ◆ used as yellow, tan, or brown pellets
- ◆ nearly odorless
- ◆ melting range 99–107°C
- ◆ density 1.44
- ◆ vapor pressure 1×10^{-5} mm Hg at 25°C
- ◆ log K_{ow} 3.94
- ◆ soluble in hot methyl ethyl ketone, acetone, esters, and aromatic hydrocarbons

CHEMICAL PROPERTIES
- ◆ none available for 4,4'-methylenebis(2-chloroaniline)

HEALTH RISK
- ◆ Toxic Exposure Guidelines
  - OSHA PEL TWA 0.02 ppm
  - ACGIH TLV TWA 0.02 ppm
  - DFG MAK animal carcinogen, suspected human carcinogen
  - NIOSH REL 0.003 mg/m³
- ◆ Toxicity Data
  - otr-mouse:fbr 10 µg/L
  - sce-hamster:ovr 500 µg/L
  - oral-rat TDLo 4050 mg/kg/77W-C:CAR
  - subcutaneous-rat TDLo 25 g/kg/89W-C:CAR
  - oral-rat TD 27 g/kg/79W-C:ETA
  - oral-rat LD$_{50}$ 2100 mg/kg
  - oral-mouse LD$_{50}$ 880 mg/kg
  - intraperitoneal-mouse LD$_{50}$ 64 mg/kg
- ◆ Acute Risks
  - gastrointestinal distress
  - burning face and eyes
  - moderate to high acute toxicity from oral exposure in animals
- ◆ Chronic Risks
  - effects on lung and liver in animals
  - bladder cancer
  - tumors in the bladder

- tumors of liver, lung, and mammary glands in animals
- EPA Group B2: probable human carcinogen

HAZARD RISK
  ◆ Decomposition emits toxic fumes of Cl$^-$ and NO$_x$
  ◆ consider evacuation

EXPOSURE ROUTES
  ◆ factories that manufacture MBOCA
  ◆ factories that produce plastic products
  ◆ general public unlikely to be exposed

REGULATORY STATUS/INFORMATION
  ◆ Clean Air Act Amendment, Title III: hazardous air pollutants
  ◆ Resource Conservation and Recovery Act
    • list of hazardous constituents. 40CFR261, 40CFR261.11
    • risk specific doses. 40CFR266
    • health based limits for exclusion of waste-derived residues. 40CFR266
    • U waste #: U158
  ◆ CERCLA and Superfund
    • designation of hazardous substances. 40CFR302.4
    • list of chemicals. 40CFR372.65
  ◆ TSCA: 704.175
  ◆ OSHA 29CFR1910.1000 Table Z1
  ◆ chemical regulated by the State of California

ADDITIONAL/OTHER COMMENTS
  ◆ symptoms: gastrointestinal distress, burning face and eyes
  ◆ First Aid
    • wash eyes immediately with large amounts of water
    • flush skin with plenty of soap and water
    • remove to fresh air immediately
  ◆ personal protection: compressed air/oxygen apparatus, gas-tight suit
  ◆ storage: keep away from flames or sparks

SOURCES
  ◆ *Health Effects Notebook for Hazardous Air Pollutants*, U.S. Environmental Protection Agency, December 1994
  ◆ *Material Safety Data Sheets (MSDS)*
  ◆ *Sax's Dangerous Properties of Industrial Materials*, 9th edition 1996
  ◆ *Hawley's Condensed Chemical Dictionary*, 12th edition, 1993
  ◆ *Environmental Law Index to Chemicals*, 1996 edition
  ◆ *Merck Index*, 12th edition, 1996

METHYLENE CHLORIDE (sf)

CAS #: 75-09-2
MOLECULAR FORMULA: CH_2Cl_2
FORMULA WEIGHT: 84.93
SYNONYMS: dichloromethane, methane dichloride, methylene bichloride, methylene chloride (ACGIH, DOT, OSHA), methylene dichlo-

ride, NCI-C50102, R-30 (refrigerant), RCRA waste number U080, UN1593 (DOT)

USES
- ◆ Used as Solvent in Manufacture of Other Chemicals and Products
  - drugs
  - paint
  - plastics
  - cellulose acetate
- ◆ Other Uses
  - laboratory solvent
  - propellant
  - refrigerant
  - parts degreaser
  - degreening agent for citrus fruit
  - finishing-cleaning circuit boards after manufacture
  - blowing agent in foams

PHYSICAL PROPERTIES
- ◆ clear liquid
- ◆ density 1.325 g/mL
- ◆ specific gravity 1.325 (water = 1)
- ◆ odor threshold 250 ppm (mg/m³)
- ◆ slightly sweet odor
- ◆ melting point −97°C(−143°F)
- ◆ boiling point 39.8–40°C (103.3–103.7°F) at 1 atm
- ◆ vapor density 2.93 (air = 1)
- ◆ surface tension 26.52 dynes/cm
- ◆ viscosity 0.430 cP at 20°C
- ◆ heat capacity 101.1 J/gmol-K (liquid at 25°C)
- ◆ heat of vaporization 28.06 kJ/gmol
- ◆ very slightly soluble in water
- ◆ miscible with alcohol, acetone, chloroform, ether, and carbon tetrachloride
- ◆ partially miscible with less polar organic solvents
- ◆ vapor pressure 349 mm Hg at 20°C, 1250 mm Hg at 55°C

CHEMICAL PROPERTIES
- ◆ flash point none
- ◆ autoignition temperature 662°C (1224°F)
- ◆ nonflammable
- ◆ explosion limits in air lower 14%, upper 22% by volume but ignition difficult
- ◆ enthalpy (heat) of formation at 25°C (liquid) 124.1 kJ/gmol
- ◆ heat of combustion −513.9 kJ/mol
- ◆ very stable
- ◆ will not polymerize
- ◆ reacts violently with alkali metals, aluminum, and potassium-tert-butoxide
- ◆ Hazardous combustion or decomposition products. May produce CO, CO_2, hydrogen chloride gas, and phosgene gas.

HEALTH RISK
- ◆ Toxic Exposure Guidelines
  - AGCIH TLV TWA 50 ppm
  - OSHA PEL (8H) TWA 500 ppm
  - NIOSH REL reduce to lowest feasible level

- ◆ Toxicity Data
  - oral-human LDLo 357 mg/kg
  - inhalation-human LDLo 500 ppm/8H
  - oral rat LD_{50} 1600 mg/kg
  - inhalation-rat LC_{50} 52 g/m^3, LCLo 88,000 mg/m^3/30M
  - intraperitoneal-rat LD_{50} 916 mg/kg
- ◆ Other Toxicity Indicators
  - EPA Cancer Risk Level (1 in a million excess lifetime risk) 2×10^{-3} mg/m^3
- ◆ Acute Risks
  - irritant to skin, eyes, mucous membranes, and upper respiratory tract
  - harmful if swallowed, inhaled, or absorbed through skin
  - metabolized to carbon monoxide in body, increasing carboxyhemoglobin levels and reducing oxygen carrying capacity of blood.
  - headaches
  - nausea
  - dizziness
- ◆ Chronic Risks
  - paresthesia, somnolence, altered sleep time, convulsions, euphoria, and change in cardiac rate
  - confirmed carcinogen and mutagen
  - experimental teratogen
  - not thoroughly investigated

HAZARD RISK

- ◆ NFPA rating Health 2, Flammability 1, Reactivity 0
- ◆ Mixtures in air with methanol are flammable.
- ◆ slight fire hazard, may explode in confined space
- ◆ incompatible or reacts violently with alkali metals, aluminum, and potassium-tert-butoxide
- ◆ hazardous combustion or decomposition products; may produce CO, CO_2, hydrogen chloride gas, and phosgene gas

EXPOSURE ROUTES

- ◆ primarily by inhalation
- ◆ absorption through skin and eyes
- ◆ ingestion
- ◆ drinking water
- ◆ ambient air
- ◆ occupational or consumer exposure from indoor spray painting or other aerosol uses

REGULATORY STATUS/INFORMATION

- ◆ Clean Air Act Amendments of 1990 Section 112 Hazardous Air Pollutant
- ◆ Safe Drinking Water Act
  - one of 83 contaminants to be regulated under the 1986 act
  - waste specific prohibitions—solvent wastes, 40CFR148.10
  - maximum contaminant level (MCL) 0.005 mg/L
  - maximum contaminant level goal (MCLG) 0 mg/L
- ◆ Superfund Amendment and Reauthorization Act

- subject to section 313 reporting requirements
- health assessment rank 65
◆ Federal Insecticide, Fungicide, and Rodenticide Act
  - tolerances and exemptions from tolerances for pesticide chemicals in or on raw agricultural commodities
◆ Resource Conservation and Recovery Act
  - constituent for detection monitoring, 40CFR258—Appendix 1
  - list of hazardous inorganic and organic constituents, 40CFR258—Appendix 2
  - list of hazardous constituents, 40CFR261—Appendix 8, and also see 40CFR261.11
  - groundwater monitoring chemicals list, 40CFR264—Appendix 9
  - risk specific doses, 40CFR266—Appendix 5
  - health based limits for exclusion of waste-derived residues, 40CFR266-Appendix 7
  - PICs found in stack effluents, 40CFR266—Appendix 8
◆ CERCLA and Superfund
  - designation of hazardous substances, 40CFR302.4—Appendices A and B
  - Reportable quantity is 1000 lb.
◆ Toxic Substances Control Act
  - listed, 40CFR716.120.a
◆ OSHA Table Z1 and Table Z2
◆ Department of Transportation
  - identification number UN1593
  - hazard class/division: 6.1
  - labels: keep away from food
◆ Community Right-to-Know Threshold Planning Quantity: not listed

ADDITIONAL/OTHER COMMENTS
◆ Symptoms
  - irritation of skin, eyes, and respiratory system
◆ First Aid
  - flush eyes and/or skin with large amounts of soap and water for a long time while removing contaminated garments and shoes
  - separate eyelids with fingers while flushing
  - seek fresh air
  - provide respiratory support and oxygen
  - if swallowed wash out mouth with water if conscious
  - seek medical help immediately
◆ Fire Fighting
  - use extinguishing media appropriate to surrounding fire
  - conditions
  - self-contained breathing apparatus to avoid breathing toxic fumes
  - protective clothing
◆ Personal Protection

- wear appropriate NIOSH/MSHA approved respirator
- chemical-resistant gloves, safety goggles, protective clothing
- use only in chemical fume hood
- wash thoroughly after handling
◆ Storage
  - outdoor storage preferred
  - keep container closed
  - cool, dry place away from heat and open flame
  - isolate from active metals

SOURCES
◆ *AGCIH Threshold Limit Values (TLVs) for 1996*
◆ *Aldrich Chemical Company Catalog*, 1996–1997
◆ *CRC Handbook of Chemistry and Physics*, 77th edition, 1996–1997
◆ *CRC Handbook of Thermophysical and Thermochemical Data*, 1994
◆ *Environmental Law Index to Chemicals*, 1996 edition
◆ *Fire Protection Guide to Hazardous Materials*, 11th edition, 1994
◆ *Hawley's Condensed Chemical Dictionary*, 12 edition, 1993
◆ *MSDS*, Aldrich Chemical Company, Inc.
◆ *Physical and Thermodynamic Properties of Pure Chemicals: Data Compilation*, AiChE and NSRSD, 1996
◆ *Sax's Dangerous Properties of Industrial Materials*, 9th edition, 1996
◆ *Suspect Chemicals Sourcebook*, 1996 edition
◆ *Health Effects Notebook for Hazardous Air Pollutants*, U.S. Environmental Protection Agency, 1994

METHYLENE DIPHENYL DIISOCYANATE (jl)

CAS #: 101-68-8
MOLECULAR FORMULA: $C_{15}H_{10}N_2O_2$
FORMULA WEIGHT: 250.26
SYNONYMS: bis(1,4-isocyanatophenyl)methane, bis(p-isocyanatophenyl)methane, caradate 30, desmodur 44, 4,4-diisocyanodiphenylmethane, 4,4'-diisocyanatodiphenylmethane, diphenyl methane diisocyanate, 4,4-diphenyl methane diisocyanate, diphenylmethane diisocyanate, 4,4'-diphenylmethane diisocyanate, hylene m50, isocyanic acid, methylenedi-p-phenylene ester, isonate 125m, isonate 125mf, MDI, 4,4'-methylenediphenyl diisocyanate, mdi, 4,4'-methylenediphenylisocyanate, methylenebis(phenylisocyanate), 1,1-methylenebis(4-isocyanatobenzene), methylenebis(p-phenylene isocyanate), 4,4-methylene bisphenyl isocyanate, 4,4'-methylenebisphenyl isocyanate, methylenebis(p-phenyl isocyanate), methylenedi-p-phenylene diisocyanate, methylenedi-p-phenylene isocyanate, 4,4'-methylenediphenylene isocyanate, methylenediphenyl isocyanate, methylenebis(4-isocyanatobenzene), nocconate 300

USES
- production of polyurethanes
- various plastic production processes

PHYSICAL PROPERTIES
- light-yellow fused solid or crystals
- melting point 37.2°C
- boiling point 172°C at 1 atm
- density 1198 kg/m^3
- specific gravity 1.2 (water = 1)
- vapor density not available
- surface tension not available
- viscosity not available
- heat capacity not available
- heat of vaporization not available
- odorless
- vapor pressure 5×10^{-6} mm Hg at 25°C
- insoluble in water at 20°C

CHEMICAL PROPERTIES
- flash point 218°C
- flammability limits not available
- autoignition temperature not available
- heat of combustion not available
- enthalpy (heat of formation) not available
- Hazardous polymerization will occur.
- incompatible with
  - water
  - amines
  - alcohols
  - heat

HEALTH RISK
- Toxic Exposure Guidelines
  - harmful by inhalation, in contact with skin, and if swallowed
  - possible mutagen
  - ACGIH TLV TWA 0.005 ppm
  - OSHA PEL 0.02 ppm
- Toxicity Data
  - TCLo (inhalation-human) 130 ppb/30M
  - LC$_{50}$ (female rat) 380 mg/m^3
  - LC$_{50}$ (rat) 178 mg/m^3
- Acute Risks
  - Dermal contact has induced dermatitis and eczema in workers.
  - has been observed to irritate the skin and eyes of rabbits
  - Tests involving acute exposure of animals, such as the LC$_{50}$ and LD$_{50}$ tests in rats and mice, have demonstrated MDI to have high to extreme acute toxicity by inhalation and moderate acute toxicity by oral exposure.

- High concentrations are extremely destructive to tissues of the mucous membranes and upper respiratory tract, eyes, and skin.
- Symptoms of exposure
  - wheezing, laryngitis, shortness of breath, burning sensation, coughing, headache, nausea and vomiting
  - Isocyanate sensitization may occur.
- ◆ Chronic Risks
  - Noncancer
    - Chronic (long-term) inhalation exposure to MDI has been shown to cause dyspnea, respiratory distress, asthma, and other respiratory impairments in workers.
    - No information is available on the chronic effects of MDI in animals.
    - The RfC is 0.00002 mg/m^3 based on hyperplasia of the olfactory epithelium in rats.
    - may cause sensitization by inhalation and skin contact
    - possible risk of irreversible effects
  - Cancer Hazard
    - a probable cancer causing agent in humans
    - has been shown to cause liver, bladder, lung, and skin cancer in animals
    - other long-term effects
    - Laboratory experiments have shown a mutagenic effect.
    - lungs
    - immunological including allergic (increased immune response)
    - nutritional and gross metabolic (body temperature increase)
  - Target Organs
    - lungs
  - immunological including allergic (increased immune response)
  - nutritional and gross metabolic (body temperature increase)

HAZARD RISK
- ◆ Unusual Fire and Explosion Hazards
  - Container explosion may occur under fire conditions.
  - emits toxic fumes under fire conditions
- ◆ Hazardous Decomposition Products
  - carbon monoxide
  - carbon dioxide
  - nitrogen oxides
  - hydrogen cyanide

EXPOSURE ROUTES
- ◆ ingestion
- ◆ inhalation
- ◆ absorption
- ◆ eye contact
- ◆ skin contact

REGULATORY STATUS/INFORMATION
- ◆ TSCA Section 4(e) Interagency Testing Committee (ITC) Priority List
- ◆ Canadian Workplace Hazardous Materials Information System (WHMIS) Ingredient Disclosure List—CN2—Ingredient must be disclosed at a concentration of 0.1%
- ◆ Massachusetts Substance List—MA1CE—Carcinogen, extraordinarily hazardous substance
- ◆ OSHA Air Contaminants (29CFR1910)

ADDITIONAL/OTHER COMMENTS
- ◆ First Aid
  - • Eye Contact
    - • first check the victim for contact lenses and remove if present
    - • flush victim's eyes with water or normal saline solution for 20 to 30 minutes while simultaneously calling a hospital or poison control center
    - • do not put any ointments, oils, or medication in the victim's eyes without specific instructions from a physician
    - • immediately transport the victim to a hospital after flushing eyes even if no symptoms (such as redness or irritation) develop
  - • Skin Contact
    - • immediately flood affected skin with water while removing and isolating all contaminated clothing
    - • gently wash all affected skin areas thoroughly with soap and water
    - • immediately call a hospital or poison control center even if no symptoms (such as redness or irritation) develop
    - • immediately transport the victim to a hospital for treatment after washing the affected areas
  - • Inhalation
    - • immediately leave the contaminated area; take deep breaths of fresh air
    - • immediately call a physician and be prepared to transport the victim to a hospital even if no symptoms (such as wheezing, coughing, shortness of breath, or burning in the mouth, throat, or chest) develop
  - • Ingestion
    - • If the victim is conscious and not convulsing, give 1 or 2 glasses of water to dilute the chemical and IMMEDIATELY call a hospital or poison control center.
    - • Generally, the induction of vomiting is NOT recommended outside of a physician's care due to the risk of aspirating the chemical into the victim's lungs.
    - • immediately transport the victim to a hospital.
- ◆ Personal Protection
  - • Respiratory/Eye Protection

- where the neat test chemical is weighed and diluted, wear a NIOSH approved half face respirator equipped with a combination filter cartridge, i.e., organic vapor/acid gas/HEPA (specific for organic vapors, HCl, acid gas, SO_2 and a high efficiency particulate filter)
- Skin Protection
  - if Tyvek-type disposable protective clothing is not worn during handling of this chemical, wear disposable Tyvek-type sleeve taped to your gloves
  - Rubber gloves are recommended.
◆ fire fighting: use water spray, carbon dioxide, dry chemical powder, or appropriate foam
◆ Spills or leak
  - evacuate area
  - wear self-contained breathing apparatus, rubber boots and gloves
  - sweep up, place in a bag, and hold for waste disposal
  - avoid raising dust

SOURCES
◆ *MSDS*, Aldrich Chemical Company, Inc.
◆ *Suspect Chemicals Sourcebook*, 1990 edition, update, p. xxv
◆ *Environmental Contaminant Reference Databook*, volume 1, 1995
◆ *Chemical Status Report*, United States National Toxicology Program. NTP Chemtrack System, Research Triangle Park, NC, November 6, 1990, listed

4,4′-METHYLENEDIANILINE

CAS #: 101-77-9
MOLECULAR FORMULA: $CH_2(C_6H_4NH_2)_2$
FORMULA WEIGHT: 198.29
SYNONYMS: 4-(4-aminobenzyl) aniline, ancamine TL, araldite Hardener 972, benzenamine, 4,4′-methylenebis-, bis-p-aminofenylmethan, bis(p-aminophenyl) methane, bis(4-aminophenyl) methane, curithane, dadpm, dapm, DDM, p,p′-diaminodifenylmethan, p,p′-diaminodiphenylmethane, 4,4′-diaminodiphenylmethane (DOT), Di-(4-aminophenyl) methane, dianilinomethane, 4,4′-diphenylmethanediamine, epicure DDM, epikure DDM, HT 972, jeffamine AP-20, MDA, methylenebis (aniline), 4,4′-methylenebisaniline, p,p′-methylenedianiline, sumicure M, tonox

USES
◆ Used as an Intermediate
  - manufacture of polyurethane foams
  - manufacture of elastomeric fibers
◆ Used in Industry
  - curing agent for epoxy resins
  - curing agent for urethane elastomers
  - corrosion preventative for iron
  - antioxidant for lubricating oils

- rubber processing chemical
- preparation of azo dyes
- production of polyamides
- determination of tungsten and sulfates
- isocyanate resins
- epoxy-resin hardening agent

PHYSICAL PROPERTIES
- ◆ pale-yellow crystals
- ◆ darkens when exposed to air
- ◆ faint amine-like odor
- ◆ melting point 93°C
- ◆ boiling point 232°C at 9 mm
- ◆ very soluble in alcohol, benzene, and ether
- ◆ vapor pressure 10 mm Hg at 25°C
- ◆ log K_{ow} 1.59–2.52
- ◆ slightly soluble in water

CHEMICAL PROPERTIES
- ◆ combustible
- ◆ flash point 440°F
- ◆ incompatible with strong oxidizers
- ◆ Hazardous polymerization will not occur.

HEALTH RISK
- ◆ Toxic Exposure Guidelines
  - ACGIH TLV TWA 0.1 ppm (skin)
  - DFG MAK animal carcinogen, suspected human carcinogen
- ◆ Toxicity Data
  - eye-rabbit 100 mg/24H MOD
  - mmo-sat 250 µg/plate
  - mma-sat 50 µg/plate
  - dnd-rat:intraperitoneal 370 µmol/kg
  - sce-mouse:intraperitoneal 9 mg/kg
  - oral-rat TDLo 320 mg/kg
  - oral-man TDLo 8420 µg/kg
  - oral-rat LD_{50} 347 mg/kg
  - subcutaneous-rat LD_{50} 200 mg/kg
  - oral-mouse LD_{50} 745 mg/kg
  - intraperitoneal-mouse LD_{50} 74 mg/kg
  - oral-dog LDLo 300 mg/kg
  - subcutaneous-dog LDLo 400 mg/kg
  - oral-rabbit LD_{50} 620 mg/kg
  - oral-guinea pig 260 mg/kg
- ◆ Acute Risks
  - liver damage
  - causative agent in "Epping jaundice"
  - jaundice
  - weakness
  - abdominal pain
  - nausea
  - vomiting
  - fever
  - chills
  - anorexia
  - rigidity
  - irritation of skin and eyes
- ◆ Chronic Risks
  - no information available on chronic effects on humans
  - effects on liver in rats and mice
  - lesions of thyroid and liver in rats and mice

- mineralization within the kidneys in rats and mice
- reduced body weight gain in rats and mice
- increases in liver tumors in rats and mice
- increased thyroid tumors in rats and mice
- hepatomas in rats
- IARC Group 2B: possible human carcinogen

HAZARD RISK
◆ combustible when exposed to heat or flame
◆ incompatible with strong oxidizers
◆ hazardous decomposition products: carbon dioxide, carbon monoxide, and oxides of nitrogen
◆ Decomposition emits toxic fumes of aniline and NO_x

EXPOSURE ROUTES
◆ inhalation and dermal contact during its manufacture and use as an intermediate
◆ eating contaminated bread

REGULATORY STATUS/INFORMATION
◆ Clean Air Act Amendment, Title III: hazardous air pollutants
◆ CERCLA and Superfund
- list of chemicals. 40CFR372.65
◆ chemical regulated by the State of California

ADDITIONAL/OTHER COMMENTS
◆ Symptoms
- exposure: jaundice, weakness, abdominal pain, nausea, vomiting, fever, chills, irritation of skin and eyes
◆ First Aid
- wash eyes and skin immediately with large amounts of water
- remove to fresh air immediately and provide respiratory apparatus if necessary
- induce vomiting
◆ fire fighting: use water spray, carbon dioxide, and dry chemical
◆ personal protection: wear self-contained breathing apparatus and protective clothing
◆ storage: keep from contact with oxidizing materials

SOURCES
◆ *Health Effects Notebook for Hazardous Air Pollutants*, U.S. Environmental Protection Agency, December 1994
◆ *Material Safety Data Sheets (MSDS)*
◆ *Sax's Dangerous Properties of Industrial Materials*, 9th edition, 1996
◆ *Hawley's Condensed Chemical Dictionary*, 12th edition, 1993
◆ *Aldrich Chemical Company Catalog*, 1996–1997
◆ *Environmental Law Index to Chemicals*, 1996 edition
◆ *Merck Index*, 12th edition, 1996

NAPHTHALENE (ch)

CAS #: 91-20-3
MOLECULAR FORMULA: $C_{10}H_8$
FORMULA WEIGHT: 128.18
SYNONYMS: albocarbon, camphor tar, mighty 150, mighty RD1, mothballs, moth flakes, naftalen (polish), naphthalene (ACGIH: DOT:OSHA), naphthalene crude or refined (UN1334) (DOT), naphthalene molten (UN2304) (DOT), naphthalin, naphthaline, naphthene, NCI-C52904, RCRA waste number U165, tar camphor, UN1334 (DOT), UN2304 (DOT), white tar

USES
  ◆ Manufacture of Other Chemicals
    • phthalic and anthranilic acids
    • naphthols
    • naphthylamines
    • sulfonic acid
    • synthetic resins
    • celluloid
    • lampblack
    • smokeless powder
    • hydronaphthalenes
  ◆ Used as a Component or in the Manufacturing Process
    • anthraquinone
    • salicyclic acid
    • phthalic anhydride
    • beta-naphthol
    • perylene
    • 1-naphthyl-N-methylcarbamate
    • mono-, di-, tri-, and tetranaphthalenesulfuric acids
    • insecticide
    • externally on livestock and poultry to control lice
    • moth repellents and toilet bowl deodorants
    • surfactants
    • indigo
    • medication
    • synthetic tanning chemicals
    • intestinal vermifuge and wood preservative

PHYSICAL PROPERTIES
  ◆ white crystalline flakes or solid
  ◆ melting point 80°C (176°F) at 760 mm Hg
  ◆ boiling point 218°C (424°F) at 760 mm Hg
  ◆ density 1.175 g/cm³ at 25°C
  ◆ specific gravity 1.14 (water = 1.00)
  ◆ vapor density 4.42 (air = 1.00)
  ◆ surface tension (liquid) 31.8 dynes/cm at 100°C
  ◆ heat capacity 159.28 J/mol-K at 15.5°C and 101.3 kPa
  ◆ heat of vaporization 43.5 kJ/mol
  ◆ aromatic odor, odor of mothballs
  ◆ odor threshold 6.80 ppm (in water)

- vapor pressure 0.01 kPa at 298.15K
- very slightly soluble in water
- miscible with 1,2,3,4-tetrahydronaphthalene, phenols, ethers, carbon disulfide, chloroform, benzene, coal-tar naphtha, carbon tetrachloride, acetone, decadronaphthalene

CHEMICAL PROPERTIES
- combustible
- sublimes at room temperature
- will react vigorously with oxidizing materials and chromic anhydride
- flammability limits 0.9% lower and 5.9% upper (percent by volume in air)
- autoignition temperature 525°C (978°F)
- heat of combustion −9287 cal/g (−5158 kJ/mol) at 15.5°C and 101.3 kPa
- enthalpy (heat) of formation at 298.15K 150.3 kJ/mol (gas)

HEALTH RISK
- Toxic Exposure Guidelines
  - ACGIH TLV TWA 10 ppm
  - OSHA PEL TWA 10 ppm
  - NIOSH REL TWA 10 ppm
  - IDLH 250 ppm
- Toxicity Data
  - skin-rabbit 495 mg open MLD
  - eye-rabbit 100 mg MLD
  - oral-mouse TDLo 2400 mg/kg (7–14D preg): REP
  - intraperitoneal-rat TDLo 5925 mg/kg (1–15D preg): TER
  - subcutaneous-rat TDLo 3500 mg/kg/12W-I: ETA
  - oral-child LDLo 100 mg/kg
  - unreported-human LDLo 29 mg/kg
  - unreported-man LDLo 74 mg/kg
  - oral-rat LD_{50} 490 mg/kg
  - oral-mouse LD_{50} 533 mg/kg
  - intraperitoneal-mouse LD_{50} 150 mg/kg
  - subcutaneous-mouse LD_{50} 969 mg/kg
  - intravenous-mouse LD_{50} 100 mg/kg
  - oral-dog LDLo 400 mg/kg
  - oral-cat LDLo 1000 mg/kg
  - oral-rabbit LDLo 3 g/kg
  - oral-guinea pig LD_{50} 1200 mg/kg
- Acute Risks
  - Onset may be delayed 2 to 4 hours.
  - harmful if swallowed, inhaled, or absorbed through skin
  - eye and skin irritation
  - irritates the mucous membranes and upper respiratory tract
  - Absorption into body leads to the formation of methemoglobin, which in high concentration causes cyanosis.
  - allergic skin reaction

- Systematic absorption of its vapors above 15 ppm may result in cataracts, optical neuritis, injuries to the cornea, and marked eye irritation.
- Ingestion of large quantities may cause severe hemolytic anemia and hemoglobinuria.

◆ Chronic Risks
  - carcinogen
  - target organs: eyes, blood, kidneys, lungs

HAZARD RISK

◆ NFPA rating
◆ Health 2, Flammability 2, Reactivity 0, Special Hazards none
◆ moderate health hazard
◆ combustible solid
◆ volatilizes at room temperature
◆ reacts with oxidizing materials
◆ incompatible or reacts violently with chromic anhydride

EXPOSURE ROUTES

◆ through the use of mothballs
◆ during its manufacture and use (i.e., coal-tar production, wood preserving, tanning, or ink and dye production)
◆ released through the air from the burning of coal and oil and from the use of mothballs. Coal-tar production, wood preserving, and other industries release small amounts.
◆ also detected in tobacco smoke

REGULATORY STATUS/INFORMATION

◆ Clean Air Act hazardous air pollutants. Title III. (The original list was published in 42USC7412.)
◆ Safe Drinking Water Act
  - priority list of drinking water contaminants. 55FR1470
  - monitoring at discretion of the state. 40CRF141.40.j
◆ Resource Conservation and Recovery Act
  - list of hazardous inorganic and organic constituents. 40CFR258—Appendix 2
  - hazardous constituents. 40CFR261—Appendix 8. See also 40CFR261.11.
  - groundwater monitoring chemicals list. 40CFR264—Appendix 9
  - health based limits for exclusion of water-derived residues. 40CFR266—Appendix 7.
  - PICs found in stack effluents. 40CFR266-Appendix 8
  - U Waste #: U165
◆ CERCLA and Superfund
  - designation of hazardous substances. 40CFR302.4. Appendix A: sequential CAS registry # list of CERCLA hazardous substances. Appendix B: radionuclides
  - chemicals and chemical categories to which this part applies (CAS # listing). 40CFR372.65
◆ Clean Water Act

- designation of hazardous substances. 40CFR116.4. Table 116.4A: list of hazardous substances. Table 116.4B: list of hazardous substances by CAS #.
- determination of reportable quantities. 40CFR117.3. Table 117.3: reportable quantities of hazardous substances designated pursuant to sec 311 of CWA
  - toxic pollutants (identical to compounds in 40CFR403— Appendix B) 40CFR401.15
  - total toxic organics (TTOs). 40CFR413.02
  - 126 priority pollutants. 40CFR423—Appendix A
- Toxic Substance Control Act
  - list of substances. 40CFR716.120.a
- OSHA 29CFR1910.1000 Table Z1
- chemical regulated by the State of California

ADDITIONAL/OTHER COMMENTS

- Symptoms
  - burning sensation
  - coughing
  - wheezing
  - laryngitis
  - shortness of breath
  - headache
  - nausea and vomiting
- First Aid
  - wash out eyes immediately with large amount of water
  - remove contaminated clothing
  - if inhaled, exit to fresh air
  - if not breathing, artificial respiration
  - if breathing difficulty, give oxygen
  - if swallowed, wash out mouth and call physician
- fire fighting: use dry chemical, foam, carbon dioxide, or water spray
- personal protection: wear full protective clothing and positive self-contained breathing apparatus
- storage: store in cool, dry, well-ventilated location. Separate from oxidizing materials. May be stored under nitrogen.

SOURCES

- *Environmental Contaminant Reference Databook*, volume 1, 1995
- *Environmental Law Index to Chemicals*, 1996 edition
- *Fire Protection Guide to Hazardous Materials*, 11th edition, 1994
- *Health Effects Notebook for Hazardous Air Pollutants*, U.S. Environmental Protection Agency, December 1994
- *Material Safety Data Sheets (MSDS)*
- *Merck Index*, 12th edition, 1996
- *NIOSH Pocket Guide to Chemical Hazards*, June 1994
- *Sax's Dangerous Properties of Industrial Materials*, 9th edition, volume 2, 1996

NICKEL COMPOUNDS

CAS #: Nickel compounds have variable CAS #s. The CAS # for nickel is 7440-02-0.

MOLECULAR FORMULA: Nickel compounds have variable molecular formulas. The molecular formula for nickel carbonyl is $Ni(CO)_4$. The molecular formula for nickel sulfate is $NiSO_4$.

Note: For all hazardous air pollutants (HAPs) that contain the word "compounds" in the name, and for glycol ethers, the following applies: Unless otherwise specified, these HAPs are defined by the 1990 Amendments to the Clean Air Act as "including any unique chemical substance that contains the named chemical (i.e., antimony, arsenic, etc.) as part of that chemical's infrastructure."

FORMULA WEIGHT: Nickel compounds have variable formula weights. The formula weight for nickel carbonyl is 170.7. The formula weight for nickel sulfate is 154.75.

SYNONYMS: nickel: alnico, C.I. 7775, nickel 200, nickel 201, nickel 205, nickel 270, NP2[b]

nickel sulfate: nickel monosulfate, nickelous sulfate, nickel (II) sulfate, sulfuric acid, nickel (II) salt

USES

- ◆ Nickel and Alloys Used in the Manufacturing Process
  - metal coins
  - jewelry
  - valves
  - heat exchangers
  - batteries
  - textile dyes
  - spark plugs
  - machinery parts
  - stainless steel
  - nickel-chrome resistance wires
  - nickel compounds used in industry
  - nickel plating
  - to color ceramics
  - catalysts

PHYSICAL PROPERTIES (Nickel)

- ◆ silver-white
- ◆ malleable
- ◆ mp 1455°C
- ◆ density 8.90 g/cm³
- ◆ vapor pressure 1 mm Hg at 1810°C
- ◆ boiling point 2730°C
- ◆ insoluble in water

PHYSICAL PROPERTIES (Nickel Sulfate)

- ◆ greenish yellow
- ◆ melting point 840°C
- ◆ density 3.68 g/cm³
- ◆ odorless
- ◆ soluble in water at 20°C
- ◆ insoluble in ether and acetone

PHYSICAL PROPERTIES (Nickel Carbonyl)

- ◆ colorless to brownish
- ◆ sooty/musty odor
- ◆ volatile liquid
- ◆ melting point 13°F (−25°C)
- ◆ boiling point 109°F (43°C)
- ◆ insoluble in water
- ◆ specific gravity 1.32 at 17°C
- ◆ vapor pressure 400 mm Hg at 25.8°C

CHEMICAL PROPERTIES (Nickel Sulfate)

- ◆ not flammable

CHEMICAL PROPERTIES (Nickel Carbonyl)

- ◆ decomposes explosively at 140°F

- ◆ flash point <−4°F (<−20°C)
- ◆ autoignition temperature 140°F
- ◆ flammability limits 2% lower and 34% upper
- ◆ reacts with nitric acid, halogens, oxidizing materials
- ◆ oxidizes in air

HEALTH RISK

- ◆ Toxic Exposure Guidelines
  - ACGIH TLV TWA 1 mg/m³
  - OSHA PEL TWA (soluble compounds) 0.1 mg (Ni)/m³
  - OSHA PEL TWA (insoluble compounds) 1 mg (Ni)/m³
  - DFG MAK human carcinogen
  - NIOSH REL (inorganic nickel) TWA 0.015 mg (Ni)/m³
  - OSHA PEL (nickel carbonyl) 0.007 mg/m³
  - NIOSH REL (nickel carbonyl) 0.007 mg/m³
- ◆ Toxicity Data
  - EPA inhalation unit risk estimates of $2.4 \times 10^{-4} (\mu g/m^3)^{-1}$ for nickel refinery dusts and $4.8 \times 10^{-4} (\mu g/m^3)^{-1}$ for nickel subsulfide
  - nickel disulfide: intramuscular-rat TDLo 117 mg/kg: CAR, and intrarenal-rat TDLo 58,580 ug/kg: NEO
- ◆ Other Toxicity Indicators
  - EPA classifies nickel refinery dust and nickel subsulfide as Group A, human carcinogens; nickel carbonyl as Group B2, probable human carcinogen.
  - Nickel carbonyl produces lung tumors in rats through inhalation.
- ◆ Acute Risks (Nickel Carbonyl)
  - irritation of throat and airways
  - headaches
  - irritability
  - nausea
  - vomiting
  - chest pains
  - gastrointestinal symptoms
  - severe weakness
  - lung damage
  - vertigo
  - insomnia
  - dry coughing
  - sweating
  - dizziness
  - skin allergy
  - cyanosis
  - visual disturbances
- ◆ Chronic Risks (Noncancer)
  - contact dermatitis
  - itching of fingers, wrists, and forearms ('nickel itch')
  - asthma
  - chronic respiratory tract infections
  - pulmonary damage in rodents exposed to nickel dust, $NiCl_2$, or NiO
  - Divalent nickel salts cause hyperglycemia, immune system effects, kidney damage, liver damage, and heart effects.
- ◆ Chronic Risks (Nickel Carbonyl)
  - lung and nasal sinus cancer
  - damage to the heart muscle, liver, and/or kidney
  - permanent lung damage
  - skin allergy with rash
  - lung tumors

◆ Chronic Risks (Nickel Sulfate)
  • mutations in living cells
  • lung allergy with wheezing
  • infertility in males

HAZARD RISK
◆ NFPA rating for nickel carbonyl Flammability 3 Reactivity 3
◆ flammable/reactive
◆ poisonous gases produced in fire
◆ Containers may explode in fire.
◆ Nickel carbonyl is a probable carcinogen.
◆ Vapors may travel to a source of ignition and flash back.
◆ can produce asphyxia by decomposing to form carbon monoxide
◆ Nickel sponge catalyst may ignite spontaneously in air.

EXPOSURE ROUTES
◆ inhalation
◆ ingestion
◆ absorption through skin
◆ discharges from electroplating and smelting industries
◆ weathering of rocks
◆ emissions from burning coal and fossil fuels
◆ welding

REGULATORY STATUS/INFORMATION
◆ Clean Air Act Title III: Hazardous Air Pollutants
◆ CERCLA and Superfund
  • designation of hazardous substances. 40CFR302.4
◆ Clean Water Act
  • toxic pollutant list. 40CFR401.15

ADDITIONAL/OTHER COMMENTS
◆ Symptoms
  • skin absorption: rashes and asthma
◆ First Aid
  • immediately flush eyes with large amounts of water
  • wash contaminated skin with soap and water
  • medical observation for up to 3 days
◆ fire fighting: use dry chemical, carbon dioxide, water spray, or foam
◆ personal protection: enclose operations and use local exhaust ventilation; respirators must be worn
◆ storage: avoid contact with strong oxidizers; tightly closed containers in cool, well-ventilated area away from strong acids

SOURCES
◆ *Hawley's Condensed Chemical Dictionary*, 12th edition, 1993
◆ *Health Effects Notebook for Hazardous Air Pollutants*, U.S. Environmental Protection Agency, December 1994
◆ *Sax's Dangerous Properties of Industrial Materials*, 9th edition, 1996
◆ *Fire Protection Guide to Hazardous Materials*, 11th edition, 1994
◆ *Environmental Law Index to Chemicals*, 1996 edition
◆ *Toxicological Profile for Nickel*

NITROBENZENE (ch)

CAS #: 98-95-3
MOLECULAR FORMULA: $C_6H_5NO_2$
FORMULA WEIGHT: 123.12
SYNONYMS: essence of mirbane, essence of myrbane, mirbane oil, NCI-C60082, nitrobenzeen (Dutch), nitrobenzen (Polish), nitrobenzene liquid (DOT), nitrobenzol (DOT), nitrobenzol liquid (DOT), oil of mirbane (DOT), oil of myrbane, RCRA waste number U169

USES
♦ Manufacture of Other Chemicals
  • aniline
  • cellulose ethers
  • cellulose acetate
♦ Used as a Component or in the Manufacturing Process
  • metal polishes
  • soaps
  • shoe polishes
  • refining lubricating oils
  • pyroxylin compound
  • preservative in spray paints
  • floor polishes
  • substitute for almond essence
  • perfume industry
  • benzidine
  • metanilic acid
  • dinitrobenzene
  • dyes (e.g., nigrosines and magenta)
  • isocyanates
  • pesticides
  • rubber chemicals
  • pharmaceuticals (acetaminophen)

PHYSICAL PROPERTIES
♦ greenish-yellow crystals or yellow, oily liquid
♦ melting point 5–6°C at 760 mm Hg
♦ boiling point 210–211°C at 760 mm Hg
♦ density 1.203 g/cm³ at 20°C
♦ specific gravity 1.2 (water = 1)
♦ vapor density 4.3 (air = 1)
♦ surface tension 43.9 dynes/cm at 20°C
♦ viscosity 1.863 mPa-s at 298.15K
♦ heat capacity 185.8 J/mol-K
♦ heat of vaporization 55.01 kJ/mol at 298.15K
♦ odor of volatile oil almond; nitrobenzene has a pungent, shoe-polish smell
♦ odor threshold 1.46×10^{-2} mg/L vapor, purity not specified

- ◆ vapor pressure 0.04 kPa at 298.15K
- ◆ soluble in about 500 parts water; water solubility 1.780 ppm
  - • soluble in alcohol, benzene, ether, oils, acetate

CHEMICAL PROPERTIES
- ◆ combustible liquid
- ◆ sublimes at room temperature
- ◆ incompatible with strong oxidizing agents, strong reducing agents, strong bases, aluminum chloride and phenol, aniline and glycerine, nitric acid, nitrogen tetroxide, silver perchlorate
- ◆ flammability limits 1.8% lower and 40.0% upper (percent by volume in air)
- ◆ autoignition temperature 899°F (481°C)
- ◆ heat of combustion at 25°C −5.791 cal/g, −242.5×10^5 J/kg, −10.420 Btu/lb
- ◆ enthalpy (heat) of formation at 298.15K 67.5 kJ/mol (gas), 12.5 kJ/mol (liquid)

HEALTH RISK
- ◆ Toxic Exposure Guidelines
  - • ACGIH TLV TWA 1 ppm
  - • OSHA PEL TWA 1 ppm
  - • NIOSH 1 ppm
  - • IDLH 200 ppm
- ◆ Toxicity Data
  - • skin-rabbit 500 mg/24H MOD
  - • eye-rabbit 500 mg/24H MLD
  - • cytogenetic analysis-saccharomyces cerevisiae 10 mmol/tube
  - • inhalation-rat TCLo 1260 µg/m³/4H (female 1–21D post): TER
  - • inhalation-rat TCLo 1260 µg/m³/4H (female 1–21D post): REP
  - • oral-woman TDLo 200 mg/kg: CNS, CVS, PUL
  - • unreported-man LDLo 35 mg/kg
  - • skin-rat LD_{50} 2100 mg/kg
  - • intraperitoneal-rat LD_{50} 640 mg/kg
  - • subcutaneous-rat LDLo 800 mg/kg
  - • oral-mouse LD_{50} 590 mg/kg
  - • subcutaneous-mouse LDLo 286 mg/kg
  - • oral-dog LDLo 740 mg/kg
  - • intravenous-dog LDLo 150 mg/kg
  - • oral-cat LDLo 1 g/kg
  - • skin-cat LDLo 25 g/kg
  - • oral-rabbit LDLo 700 mg/kg
  - • skin-rabbit LDLo 600 mg/kg
- ◆ Acute Risks
  - • Onset may be delayed 2 to 4 hours or longer.
  - • fatal if inhaled, swallowed, or absorbed
  - • irritating to eyes, mucous membranes, upper respiratory tract

- skin irritation
- Skin absorption leads to formation of methemoglobin; if sufficient, cyanosis occurs.
- nervous system disturbances
◆ Chronic Risks
  - target organs: blood, central nervous system, male reproductive system, liver, spleen

HAZARD RISK
◆ NFPA rating Health 3, Flammability 2, Reactivity 1, Special Hazards none
◆ serious health hazard
◆ combustible liquid
◆ incompatible with strong oxidizing agents, strong reducing agents, strong acids, strong bases

EXPOSURE ROUTES
◆ has not been detected in ambient air or in drinking water
◆ possibly in factories that produce nitrobenzene or use nitrobenzene to produce other products
◆ may occur for those persons living near a waste site where nitrobenzene has been found or near a manufacturing or processing plant

REGULATORY STATUS/INFORMATION
◆ Clean Air Act hazardous air pollutants. Title III. (The original list was published in 42USC7412.)
◆ Safe Drinking Water Act
  - priority list of drinking water contaminants. 55FR1470
  - waste specific prohibitions—solvent wastes. 40CFR148.10
◆ Resource Conservation and Recovery Act
  - list of hazardous inorganic and organic constituents. 40CFR258—Appendix 2
  - hazardous constituents. 40CFR261—Appendix 8, see also 40CFR261.11
  - groundwater monitoring chemicals list. 40CFR264—Appendix 9
  - reference air concentrations. 40CFR266—Appendix 4
  - health based limits for exclusion of waste-derived residues. 40CFR266—Appendix 7
  - D waste #: D036
  - U waste #: U169
◆ CERCLA and Superfund
  - designation of hazardous substances. 40CFR302.4. Appendix A: sequential CAS registry # list of CERCLA hazardous substances. Appendix B: radionuclides
  - list of extremely hazardous substances and their threshold planning quantities (CAS # order). 40CFR355-AB: 40CFR355—Appendix B
  - chemicals and chemical categories to which this part applies (CAS # listing). 40CFR372.65

- ◆ Clean Water Act
  - • designation of hazardous substances. 40CFR116.4. Table 116.4A: list of hazardous substances. Table 116.4B: list of hazardous substances by CAS #
  - • determination of reportable quantities. 40CFR117.3. Table 117.3: reportable quantities of hazardous substances designated pursuant to sec 311 of CWA
  - • toxic pollutants (identical to compounds in 40CFR403—Appendix B). 40CFR401.15
  - • total toxic organics (TTOs). 40CFR413.02
  - • 126 priority pollutants. 40CFR423—Appendix A.
- ◆ Toxic Substance Control Act
  - • list of substances. 40CFR716.120.a
- ◆ OSHA 29CFR1910.1000 Table Z1
- ◆ chemical regulated by the State of California

ADDITIONAL/OTHER COMMENTS

- ◆ Chemical, physical, and toxicological properties have not been thoroughly investigated.
- ◆ Symptoms
  - • may be delayed
  - • irritation to eyes, skin, respiratory system
  - • headache
  - • nausea
  - • drowsiness
  - • vomiting
  - • cyanosis
- ◆ First Aid
  - • flush eyes and skin with large quantity of water for at least 15 min
  - • remove contaminated clothing and shoes
  - • remove to fresh air
  - • artificial respiration and/or oxygen
  - • if swallowed, wash out mouth with water *if conscious* and call physician immediately
- ◆ fire fighting: use dry chemical, foam, carbon dioxide, or water spray
- ◆ personal protection: wear special protective clothing and positive pressure self-contained breathing apparatus
- ◆ storage: store in cool, dry, well-ventilated, dark location. Separate from acids, bases, oxidizing materials, and metals.

SOURCES

- ◆ *Environmental Contaminant Reference Databook*, volume 1, 1995
- ◆ *Environmental Law Index to Chemicals*, 1996 edition
- ◆ *Fire Protection Guide to Hazardous Materials*, 11th edition, 1994
- ◆ *Health Effects Notebook for Hazardous Air Pollutants*, U.S. Environmental Protection Agency, December 1994
- ◆ *Material Safety Data Sheets (MSDS)*
- ◆ *Merck Index*, 12th edition, 1996
- ◆ *NIOSH Pocket Guide to Chemical Hazards*, June 1994

4-NITROBIPHENYL (rs)

CAS #: 92-93-3
MOLECULAR FORMULA: $C_{12}H_9NO_2$ or $C_6H_5C_6H_4NO_2$
FORMULA WEIGHT: 199.22
SYNONYMS: o-nitrobiphenyl, p-nitrobiphenyl, 4-nitro-1,1-biphenyl, p-nitrodiphenyl, 4-nitrodiphenyl, p-phenyl-nitrobenzene, 4-phenyl-nitrobenzene, ONB, o-nitrodiphenyl, PNB

USES
◆ Manufacture of Other Chemicals
  • formerly in preparation of p-biphenyl amine
◆ Used as a Component or in the Manufacturing Process of
  • fungicides
  • plasticizer for cellulosic materials
  • wood preservative
  • dye intermediate
  • not manufactured, imported, used, or sold in the United States at the present time

PHYSICAL PROPERTIES
◆ white to light-yellow to reddish crystalline solid or liquid
◆ melting point 112–114°C (233.6–237.2°F)
◆ boiling point 340°C (644°F) at 760 mm Hg, 223.7–224.1°C at 30 mm Hg
◆ flash point 143°C (290°F)
◆ density 1.203 g/cm^3 (10 lb/gal)
◆ soluble in carbon tetrachloride, mineral spirits, pine oil, turpentine, benzene, acetone, glacial acetic acid, perchloroethylene; insoluble in water
◆ sweet odor

CHEMICAL PROPERTIES
◆ lower explosive limit (LEL) not determined
◆ upper explosive limit (UEL) not determined
◆ autoignition temperature 180°C (356°F)
◆ combustible
◆ incompatible with strong reducers, oxidizing agents

HEALTH RISK
◆ Toxic Exposure Guidelines
  • OSHA cancer suspect agent
  • NIOSH REL carcinogen
  • IDLH carcinogen, no level detected
  • ACGIH TLV confirmed human carcinogen
◆ Toxicity Data
  • mutation in microorganisms, bacillus subtilis, 50 mmol/L
  • unscheduled DNA synthesis, rat 100 µmol/L
  • oral-dog TDLo 7000 mg/kg/2Y-intermittently, neoplastic effects

- oral-dog TD 7000 mg/kg/3Y-intermittently, equivocal tumorigenic agent
- oral-rat LD$_{50}$ 2230 mg/kg
- intraperitoneal-mouse LD$_{50}$ 347 mg/kg
- oral-rabbit, LD$_{50}$ 1970 mg/kg
- ◆ Acute Risks
  - toxic by ingestion and skin contact
  - poison by intraperitoneal route
- ◆ Chronic Risks
  - highly toxic suspected human carcinogen
  - mutation data reported
- ◆ Target Organs
  - bladder
  - blood

HAZARD RISK
- ◆ Decomposition emits toxic fumes of NO$_x$.

EXPOSURE ROUTES
- ◆ dermal
- ◆ inhalation
- ◆ ingestion

REGULATORY STATUS/INFORMATION
- ◆ Clean Air Act
  - Title III: regulated hazardous air pollutant
- ◆ CERCLA and Superfund
  - 40CFR372.65, designation as a specific toxic chemical
- ◆ Occupational Safety and Health Act
  - specially regulated substances (carcinogen), 29CFR1910.1001-1910.1048
  - limits for air contaminants, 29CFR1910.1000 table Z1

ADDITIONAL/OTHER COMMENTS
- ◆ Symptoms
  - dizziness
  - headache
  - lethargy
  - dyspnea
  - ataxia
  - weakness
  - methemoglobinemia
  - urinary burning
  - acute hemorrhagic cystitis
- ◆ First Aid
  - eyes: irrigate immediately
  - inhalation: artificial respiration
  - skin: remove all contaminated clothing; wash with soap and water immediately
  - ingestion: seek immediate medical attention
- ◆ Fire Fighting
  - use dry chemical, foam, or carbon dioxide
  - emits toxic fumes

- ◆ personal protection: full protective clothing and self-contained breathing apparatus
- ◆ Spill or Leak
  - cover spill with dry lime or soda ash; keep in closed container and dispose of properly
  - ventilate area and wash spill site after complete material pickup

SOURCES
- ◆ *Aldrich Chemical Company Catalog*, 1996–1997
- ◆ *Environmental Law Index to Chemicals*, 1994 edition
- ◆ *MSDS*, Aldrich Chemical Company, Inc., 1996
- ◆ *Merck Index*, 12th edition, 1996
- ◆ *NIOSH Pocket Guide to Chemical Hazards*, June 1994
- ◆ *Sax's Dangerous Properties of Industrial Materials*, 9th edition, Volumes 1 and 2, 1996
- ◆ *Hawley's Condensed Chemical Dictionary*, 12th edition, 1993
- ◆ *EPA Health Effects Notebook for Hazardous Air Pollutants*, U.S. Environmental Protection Agency, December 1994

4-NITROPHENOL (rs)

CAS #: 100-02-7
MOLECULAR FORMULA: $C_6H_5NO_3$ or $NO_2C_6H_4OH$
FORMULA WEIGHT: 139.11
SYNONYMS: 4-hydroxynitrobenzene, niphen, nitrophenol, p-nitrophenol, PNP, RCRA waste number U170

USES
- ◆ Manufacture of Other Chemicals
  - intermediate in organic synthesis
- ◆ Used as a Component or in the Manufacturing Process
  - parathion
  - dyes
  - drugs
  - fungicide for leather
  - indicator in 0.1% alcohol solution

PHYSICAL PROPERTIES
- ◆ colorless to slightly yellow crystals
- ◆ sweetish, then burning taste
- ◆ odorless
- ◆ sublimes
- ◆ melting point 113–114°C (235.4–237.2°F)
- ◆ boiling point 279°C (534.2°F), decomposes
- ◆ specific gravity 1.48
- ◆ vapor pressure 7 mm Hg at 165°C
- ◆ density 1.479–1.495 g/cm^3 at 20°C
- ◆ moderately soluble in cold water
- ◆ soluble in hot water, alcohol, ether, chloroform, solutions of fixed alkali solutions, hydroxides, and carbonates
- ◆ slightly volatile with steam

CHEMICAL PROPERTIES
- ◆ autoignition temperature 283°C (541°F)
- ◆ decomposes at elevated temperatures (release oxides of nitrogen)
- ◆ incompatible with oxidizing materials, strong bases
- ◆ combustible solid

HEALTH RISK
- ◆ Toxic Exposure Guidelines
  - none available
- ◆ Toxicity Data
  - DNA damage, e-coli 50 μmol/L
  - DNA inhibition, human fibroblast 1 mmol/L
  - subcutaneous-rat LDLo 200 mg/kg
  - oral-mouse LD_{50}, 380 mg/kg
  - intraperitoneal-mouse LD_{50}: 75 mg/kg
  - intravenous-dog LDLo 10 mg/kg
  - subcutaneous-guinea pig LDLo 200 mg/kg
  - intramuscular-pigeon LDLo 65 mg/kg
- ◆ Acute Risks
  - poison by ingestion, subcutaneous, intraperitoneal, intravenous, and intramuscular routes
  - serious health hazard
  - harmful if absorbed through skin or inhaled
  - irritating to skin, eyes, and respiratory system
  - human mutation data reported
- ◆ Chronic Risks
  - possible risk of irreversible effects
  - possible formation of cyanosis
  - mutagenic effects
- ◆ Target Organs
  - central nervous system
  - blood

HAZARD RISK
- ◆ Exothermic decomposition causes a dangerous fast pressure increase.
- ◆ Mixtures with diethyl phosphite may explode when heated.
- ◆ Decomposition emits toxic fumes of NO_x.

EXPOSURE ROUTES
- inhalation
- ingestion
- dermal

REGULATORY STATUS/INFORMATION
- ◆ Clean Air Act
  - Title III: regulated under hazardous pollutant and accident prevention provisions
- ◆ CERCLA and Superfund
  - 40CFR302.4, designation as hazardous substance
  - table 302.4, list of hazardous substances and reportable quantities

- 40CFR372.65, list of substances subject to Community Right-to-Know
◆ Clean Water Act
  - 40CFR413.02, general definitions, as total toxic organic, see also §433.11, §464.02, §464.11, §464.21, §464.31, §464.41, §465.02, §467.02, §469.12, §469.22, §469.31
  - 40CFR423, appendix A, 126 priority pollutants
◆ Chemical regulated by the State of California
◆ DOT Class 6.1, poisonous material

ADDITIONAL/OTHER COMMENTS

◆ Symptoms
  - headache
  - drowsiness
  - nausea
  - cyanosis
◆ First Aid
  - eyes: rinse immediately with plenty of water for at least 15 minutes, assure adequate flushing by separating eyelids
  - inhalation: fresh air (artificial respiration if needed)
  - respiratory distress: oxygen inhalation
  - skin: remove all contaminated clothing; rinse with water for at least 15 minutes
  - ingestion: wash out mouth with water (if person is conscious)
◆ Fire Fighting
  - use dry chemical, foam, or carbon dioxide, water spray; use water spray to keep containers cool
  - approach fire from upwind
  - fight fire from protected location or maximum possible distance
  - Closed containers may rupture violently when heated.
  - when heated to decomposition, emits toxic fumes of nitrous oxides
◆ Personal Protection
  - full protective clothing and positive pressure self-contained breathing apparatus
  - rubber boots and heavy rubber gloves
◆ storage: store in cool, dry location
◆ spill or leak: shovel into suitable dry container
◆ preventative measures: avoid breathing vapors; keep upwind
◆ NFPA diamond: Health 3, Flammability 1, Reactivity 2
◆ DOT: UN 1663

SOURCES

◆ *Environmental Law Index to Chemicals*, 1994 edition
◆ *Fire Protection Guide to Hazardous Materials*, 11th edition, 1994
◆ *MSDS*, Aldrich Chemical Company, Inc., 1996
◆ *Merck Index*, 12th edition, 1996
◆ *Sax's Dangerous Properties of Industrial Materials*, 9th edition, volumes 1 and 2, 1996

♦ *Hawley's Condensed Chemical Dictionary*, 12th edition, 1993
♦ *Health Effects Notebook for Hazardous Air Pollutants*, U.S. Environmental Protection Agency, December 1994

2-NITROPROPANE

CAS #: 79-46-9
MOLECULAR FORMULA: $C_3H_7NO_2$
FORMULA WEIGHT: 89.11
SYNONYMS: carbamic-chloride, dimethyl-, carbamoyl-chloride, dimethyl-, carbamyl-chloride, n,n-dimethyl-, chloro-formic-acid, dimethylamide, dimethyl-carbamic chloride, dimethyl-carbamidoyl-chloride, dimethyl-carbamidoyl-chloride(n,n-), dimethyl-nitromethane, isonitropropane, nipar-S-20, nitroisopropane, propane, 2-nitro-

USES
♦ Used as a Solvent
  • organic compounds
  • vinyl and epoxy coatings
  • nitrocellulose and chlorinated rubbers
  • inks
  • dyes
  • adhesives
  • vinyl resins
  • to improve drying time
♦ Used in Manufacture of Products
  • explosives
  • propellants, rocket propellant
  • fuels (in racing cars)
  • gasoline additive
  • smoke depressant in diesel fuel
  • paint and varnish remover
♦ Used as an Intermediate
  • synthesis of pharmaceuticals
  • dyes
  • insecticides
  • textile chemicals

PHYSICAL PROPERTIES
  • colorless liquid
  • mild odor
  • melting point −93°C
  • boiling point 120.3°C at 760 mm Hg
  • density 0.9821 at 25°C
  • odor threshold 2.49×10^4 ppb
  • surface tension 30 dynes/cm
  • vapor pressure 20 mm Hg at 25°C
  • heat of vaporization 9.88 kcal/mol at 25°C
  • heat of formation −43.78 kcal/mol at 25°C
  • vapor density 3.06
  • specific gravity 0.98
  • solubility in water 1.7 mL/100 mL
  • soluble in chloroform
  • miscible with most aromatic hydrocarbons, ketones, esters, ethers, and lower carboxilic acids

CHEMICAL PROPERTIES
  • autoignition temperature 428°C
  • flash point 82°F
  • flammability limits 2.6% lower and 11.0% upper
  • heat of combustion 477.60 kcal/mol at 25°C
  • violent decomposition above 220°C

- incompatible with alkalies, combustible materials, lead, copper
- will not polymerize

HEALTH RISK
◆ Toxic Exposure Guidelines
  - ACGIH TLV TWA 10 ppm
  - OSHA PEL TWA 10 ppm
  - NIOSH REL TWA reduce to lowest feasible level
  - DFG TRK 5 ppm
◆ Toxicity Data
  - mmo-sat 1 mg/plate
  - mma-sat 1 mg/plate
  - intraperitoneal-rat TDLo 2550 mg/kg
  - oral-rat TDLo 4277 mg/kg/16W-I:CAR
  - inhalation-rat TCLo 207 ppm/7H/26W-I:CAR
  - inhalation-rat TC 207 ppm/26W-I:ETA
  - inhalation-man TCLo 20 ppm
  - oral-rat LD_{50} 720 mg/kg
  - inhalation-rat LC_{50} 400 ppm/6H
  - inhalation-mouse LC_{50} 10 g/m^3/2H
  - intraperitoneal-mouse LD_{50} 75 mg/kg
  - inhalation-cat LCLo 714 ppm/5H
  - oral-rabbit LDLo 500 mg/kg
  - inhalation-rabbit LCLo 2381 ppm/5H
  - inhalation-guinea pig LCLo 4622 ppm/5H
◆ Acute Risks
  - severe liver damage
  - kidney damage
  - vapor irritating to eyes and respiratory tract
  - mild irritation to skin
  - nausea
  - vomiting
  - headache
  - diarrhea
◆ Chronic Risks
  - methemoglobinemia
  - cyanosis
  - possible cancer hazard
  - anorexia
  - nausea
  - vomiting
  - diarrhea
  - severe headaches
  - pulmonary irritation
  - liver effects
  - hypermotility
  - hepatocellular carcinomas in rats
  - EPA Group 2B: probable human carcinogen

HAZARD RISK
- very dangerous fire hazard
- may explode on heating
- violent reactions with chlorosulfonic acid, oleum
- may react with amines and heavy metal oxides to form explosive salts
- ignites on contact with mixtures of carbon + hopcalite
- Flashback along vapor trail may occur.
- incompatibility with alkalies, combustible materials, lead, copper

EXPOSURE ROUTES
- industrial construction and maintenance industries
- printing industries
- highway maintenance (traffic markings)
- shipbuilding and maintenance (marine coatings)
- furniture and plastic products
- tobacco smoke

REGULATORY STATUS/INFORMATION
- ◆ Clean Air Act Amendment, Title III: hazardous air pollutants
- ◆ Safe Drinking Water Act: waste specific prohibitions–solvent wastes. 40CFR148.10
- ◆ Resource Conservation and Recovery Act
  - list of hazardous constituents. 40CFR261, 40CFR261.11
  - risk specific doses. 40CFR266
  - U waste #: U171
- ◆ CERCLA and Superfund
  - designation of hazardous substances. 40CFR302.4
  - list of chemicals. 40CFR372.65
- ◆ Toxic Substance Control Act
  - OHSA chemicals in need of dermal absorption testing. 40CFR712.30.e10, 40CFR716.120.d10
- ◆ OSHA 29CFR1910.1000 Table Z1: limits for air contaminants
- ◆ chemical regulated by the State of California

ADDITIONAL/OTHER COMMENTS
- ◆ Symptoms
  - inhalation: headache, dizziness, coughing, difficult breathing, irritates eyes and skin
  - ingestion: nausea, vomiting
- ◆ First Aid
  - wash eyes immediately with large amounts of water
  - wash skin with large amounts of water
  - induce vomiting
  - provide respiratory support
- ◆ fire fighting: use dry chemical, foam, or carbon dioxide
- ◆ personal protection: protective clothing, goggles, and respiratory apparatus if necessary
- ◆ storage: keep away from heat, sparks, and flame; closed container with adequate ventilation

SOURCES
- ◆ *Health Effects Notebook for Hazardous Air Pollutants*, U.S. Environmental Protection Agency, December 1994
- ◆ *Material Safety Data Sheets (MSDS)*
- ◆ *Sax's Dangerous Properties of Industrial Materials*, 9th edition, 1996
- ◆ *Aldrich Chemical Company Catalog*, 1996–1997
- ◆ *Environmental Law Index to Chemicals*, 1996 edition
- ◆ *Hawley's Condensed Chemical Dictionary*, 12th edition, 1993
- ◆ *Merck Index*, 12th edition, 1996

N-NITROSO-N-METHYLUREA

CAS #: 684-93-5
MOLECULAR FORMULA: $C_2H_5N_3O_2$
FORMULA WEIGHT: 103.10
SYNONYMS: Methylnitroso-harnstoff (German), N-Methyl-n-nitroso-harnstoff (German), methylnitrosourea, n-methyl-N-nitrosourea, 1-methyl-1-nitrosourea, methylnitrosouree (French), MNU, n-nitroso-n-methylcarbamide, n-nitroso-n-methyl-harnstoff (German), nitroso-methylurea, n-nitroso-n-methylurea, 1-nitroso-1-methylurea, NMH, NMU, NSC 23909, SKI 24464, SRI 859

USES
◆ Used Mostly in the Laboratory
   • synthesis of diazomethane
   • cancer chemotherapy agent
   • mutagenic effects on plants
   • antitumoral drug

PHYSICAL PROPERTIES
◆ pale-yellow crystals
◆ melting point 124°C

CHEMICAL PROPERTIES
◆ decomposes at the melting point

HEALTH RISK
◆ Toxic Exposure Guidelines
   • No data is available for toxic exposure guidelines.
◆ Toxicity Data
   • sln-dmg-par 1 mmol/L
   • oms-mam:lym 1 mmol/L
   • intraperitoneal-rat TDLo 5 mg/kg
   • oral-rat TDLo 6 mg/kg:CAR
   • skin-rat TDLo 576 mg/kg/24W-I:NEO,TER
   • intraperitoneal-rat TDLo 20 mg/kg
   • intravenous-rat TDLo 10 mg/kg
   • ocu-rat TDLo 800 µg/kg:ETA
   • par-rat TDLo 5 mg/kg
   • unr-rat TDLo 10 mg/kg:NEO,TER
   • oral-mouse TDLo 300 mg/kg:8W-I:NEO
   • subcutaneous-mouse TDLo 75 mg/kg
   • intravenous-mouse TDLo 50 mg/kg
   • intravenous-hamster LDLo 42 mg/kg
   • oral-uns LD_{50} 110 mg/kg
   • intravenous-uns LD_{50} 110 mg/kg
◆ Acute Risks
   • dermatitis
   • high acute toxicity orally in animals

- ◆ Chronic Risks
  - No information is available on the chronic effects in humans.
  - tumors of the nervous system in offspring of rats
  - tumors of the kidneys in offspring of rats
  - tumors of the kidneys, stomach, pancreas, brain, and mammary glands in animals
  - EPA Group B2: probable human carcinogen

HAZARD RISK
- ◆ explodes at room temperature
- ◆ can detonate with $KOH + CH_2Cl_2$
- ◆ Decomposition emits toxic fumes of NO_x.

EXPOSURE ROUTES
- ◆ primarily in chemical research laboratories

REGULATORY STATUS/INFORMATION
- ◆ Clean Air Act Amendment, Title III: hazardous air pollutants
- ◆ Resource Conservation and Recovery Act
  - list of hazardous constituents. 40CFR261, 40CFR261.11
  - risk specific doses. 40CFR266
  - health based limits for exclusion of waste derived-residues. 40CFR266
  - U waste # 177
- ◆ CERCLA and Superfund
  - designation of hazardous substances. 40CFR302.4
  - list of chemicals. 40CFR372.65
- ◆ chemical regulated by the State of California

ADDITIONAL/OTHER COMMENTS
- ◆ symptoms: dermatitis

SOURCES
- ◆ *Health Effects Notebook for Hazardous Air Pollutants*, U.S. Environmental Protection Agency, December 1994
- ◆ *Material Safety Data Sheets (MSDS)*
- ◆ *Sax's Dangerous Properties of Industrial Materials*, 9th edition, 1996
- ◆ *Environmental Law Index to Chemicals*, 1996 edition

N-NITROSODIMETHYLAMINE

CAS #: 62-75-9
MOLECULAR FORMULA: $(CH_3)_2NNO$
FORMULA WEIGHT: 74.10
SYNONYMS: Dimethylnitrosamin (German), dimethylnitrosamine, N,N-dimethylnitrosamine, dimethylnitrosoamine, DMN, DMNA, N-methyl-n-nitrosomethanamine, NDMA, nitrosodimethylamine
USES
- ◆ Used as a Solvent
  - industrial solvent
  - in fiber
  - in plastics industry
  - in lubricants
- ◆ Used in Industry
  - condensers to increase dielectric constant
  - nematocide
  - softener for copolymers

- formerly in production of rocket fuels
- research chemical
- chemical intermediate for 1,1-dimethylhydrazine
- rubber accelerator
- inhibition of nitrification in soil
- plasticizer for acrylonitrile polymers
- preparation of thiocarbonyl fluoride polymers
- antioxidant

PHYSICAL PROPERTIES

- ◆ yellow, oily liquid
- ◆ faint, characteristic odor
- ◆ boiling point 153°C
- ◆ density 1.0048 at 20°C
- ◆ viscosity low
- ◆ vapor pressure 2.7 mm Hg at 20°C
- ◆ very soluble in water, alcohol, and ether
- ◆ miscible with methylene chloride, vegetable oils
- ◆ log K_{ow} −0.57

CHEMICAL PROPERTIES

- ◆ stable at room temperature
- ◆ slightly less stable in acidic solution
- ◆ sensitive to UV light
- ◆ oxidized to nitramine
- ◆ reduced to hydrazine or amine
- ◆ relatively resistant to hydrolysis
- ◆ photochemically reactive

HEALTH RISK

- ◆ Toxic Exposure Guidelines
  - OSHA PEL cancer suspect agent
  - ACGIH TLV suspected human carcinogen
- ◆ Toxicity Data
  - mma-sat 100 μg/plate
  - msc-human:lym 14 mmol/L
  - intraperitoneal-mouse TDLo 10 mg/kg
  - oral-rat TDLo 30 mg/kg
  - oral-rat TDLo 23 mg/kg/2Y-I:CAR
  - inhalation-rat TCLo 200 μg/m³/45W-C:CAR
  - unr-rat TDLo 7 mg/kg
  - oral-woman LDLo 20 mg/kg/2.5Y:GIT,MET
  - oral-rat LD_{50} 37 mg/kg
  - inhalation-rat LC_{50} 78 ppm/4H
  - intraperitoneal-rat LD_{50} 26,000 μg/kg
  - subcutaneous-rat LD_{50} 45 mg/kg
  - intravenous-rat LDLo 40 mg/kg
  - inhalation-mouse LC_{50} 57 ppm/4H
  - intraperitoneal-mouse LD_{50} 19 mg/kg
  - oral-dog LDLo 20 mg/kg
  - inhalation-dog LCLo 16 ppm/4H
- ◆ Acute Risks
  - liver damage
  - nausea
  - vomiting
  - headaches
  - malaise
  - fever
  - ulceration or bleeding from small intestine
  - hepatomegaly
  - jaundice

- ascites
- diarrhea
- abdominal cramps
◆ Chronic Risks
  - liver damage
  - jaundice
  - swelling
  - low platelet counts
  - increased fetal mortality in rats
  - causes cancer in offspring of rats and mice
  - fatal liver disease in humans
- enlarged liver
- reduced function of liver, kidneys, and lungs

- liver tumors in rats and mice
- increased kidney tumors in rats and mice
- increased lung tumors in rats and mice
- EPA Group B2: probable human carcinogen

HAZARD RISK
◆ very toxic
◆ decomposes on heating
◆ Decomposition emits toxic fumes of NO_x.

EXPOSURE ROUTES
◆ rubber, tanning, fish processing, dye, and surfactant industries
◆ ingestion of food
◆ cured meats and smoked fish
◆ drinking contaminated water
◆ breathing cigarette smoke
◆ contaminated ambient air

REGULATORY STATUS/INFORMATION
◆ Clean Air Act Amendment, Title III: hazardous air pollutants
◆ Resource Conservation and Recovery Act
  - list of hazardous constituents. 40CFR258, 40CFR261, and 40CFR261.11
  - groundwater monitoring chemicals list. 40CFR264
  - P waste # 082
◆ CERCLA and Superfund
  - designation of hazardous substances. 40CFR302.4
  - list of extremely hazardous substances and their threshold planning quantities. 40CFR355-AB
  - list of chemicals. 40CFR372.65
◆ Clean Water Act
  - total toxic organics. 40CFR413.02
  - 126 priority pollutants. 40CFR423
◆ OSHA, 29CFR1910.1000 Table Z1 and specially regulated substances
◆ chemical regulated by the State of California

ADDITIONAL/OTHER COMMENTS
◆ symptoms: headache, fever, weakness, nausea, vomiting, dizziness, diarrhea, gastrointestinal hemorrhage, hepatomegaly, jaundice, ascites
◆ First Aid
  - wash eyes immediately with large amounts of water
  - wash skin immediately with soap and water

◆ personal protection: compressed air/oxygen apparatus, gas-tight suit

SOURCES

- *Health Effects Notebook for Hazardous Air Pollutants*, U.S. Environmental Protection Agency, December 1994
- *Material Safety Data Sheets (MSDS)*
- *Sax's Dangerous Properties of Industrial Materials*, 9th edition, 1996
- *Environmental Law Index to Chemicals*, 1996 edition
- *Merck Index*, 12th edition, 1996
- *Hawley's Condensed Chemical Dictionary*, 12th edition, 1993
- *Environmental Contaminant Reference Databook*, volume 1, 1995
- *Aldrich Chemical Company Catalog*, 1996–1997

N-NITROSO-N-METHYLUREA. *See* p. 405 following 2-nitropropane.

N-NITROSOMORPHOLINE

CAS #: 59-89-2
MOLECULAR FORMULA: $C_4H_8N_2O_2$
FORMULA WEIGHT: 116.14
SYNONYMS: N-Nitrosomorpholin (German), nitrosomorpholine, 4-nitrosomorpholine, NMOR

USES
- used as a solvent for polyacrylonitrile
- used as a microbial agent
- intermediate for n-aminomorpholine
- not used in the United States

PHYSICAL PROPERTIES
- exists as yellow crystals
- melting point 29°C
- boiling point 140°C
- soluble in water

CHEMICAL PROPERTIES
- Chemical properties are not available.

HEALTH RISK
- Toxic Exposure Guidelines
  - DFG MAK animal carcinogen, suspected human carcinogen
- Toxicity Data
  - slt-dmg-oral 210 μmol/kg
  - dns-human:fbr 160 μg/L
  - oral-rat TDLo 320 mg/kg:CAR
  - intravenous-rat TDLo 286 mg/kg/57W-I:ETA
  - oral-mouse TDLo 3140 mg/kg/28W-C:NEO
  - oral-hamster TDLo 365 mg/kg/61W-C:CAR
  - oral-rat TD 180 mg/kg/I:ETA,REP
  - oral-rat LD_{50} 282 mg/kg
  - intraperitoneal-rat LD_{50} 282 mg/kg
  - subcutaneous-rat LD_{50} 170 mg/kg

- intravenous-rat LD$_{50}$ 98 mg/kg
- inhalation-mouse LCLo 1000 mg/m^3/10M
- intraperitoneal-mouse LD$_{50}$ 54 mg/kg
- oral-hamster LD$_{50}$ 956 mg/kg
- subcutaneous-hamster LD$_{50}$ 491 mg/kg
- ◆ Acute Risks
  - No information is available on the acute risks in humans.
  - moderate to high acute toxicity from oral exposure in animals
- ◆ Chronic Risks
  - No information is available on the chronic risks in humans.
  - effects on the liver in animals
  - tumors of the lung in animals
  - tumors of the liver and kidneys in animals
  - tumors of the blood vessels in animals
  - IARC Group 2B: possible human carcinogen

HAZARD RISK
- ◆ Decomposition emits toxic fumes of NO$_x$.

EXPOSURE ROUTES
- ◆ airborne contaminant inside some cars
- ◆ rubber-stoppered blood collection tubes

REGULATORY STATUS/INFORMATION
- ◆ Clean Air Act Amendment, Title III: hazardous air pollutants
- ◆ Resource Conservation and Recovery Act
  - list of hazardous constituents. 40CFR261, 40CFR261.11
  - groundwater monitoring chemicals list. 40CFR264
- ◆ CERCLA and Superfund
  - list of chemicals. 40CFR372.65
- ◆ chemical regulated by the State of California

ADDITIONAL/OTHER COMMENTS
- ◆ symptoms: moderately toxic by inhalation

SOURCES
- ◆ *Health Effects Notebook for Hazardous Air Pollutants*, U.S. Environmental Protection Agency, December 1994
- ◆ *Material Safety Data Sheets (MSDS)*
- ◆ *Sax's Dangerous Properties of Industrial Materials*, 9th edition, 1996
- ◆ *Environmental Law Index to Chemicals*, 1996 edition
- ◆ *Merck Index*, 12th edition, 1996

PARATHION

CAS #: 56-38-2
MOLECULAR FORMULA: C$_{10}$H$_{14}$NO$_5$PS
FORMULA WEIGHT: 291.28
SYNONYMS: AAT, AATP, ACC-3422, alkron, alleron, American-cyanamid 3422, aphamite, aralo, bayer-E 605, bladan-F, diethyl-O-4-nitrophenyl-phosphorothioate(o,o-), diethyl-O-p-nitro-phenyl-phosphorothioate (O-O-), diethyl-4-nitrophenyl-phosphorothionate, diethyl-p-nitrophenyl-phosphorothionate, diethyl-p-nitro-phenyl-thio-

phosphate, diethyl parathion, E-605, E-605 F, ethyl-parathion, etilon, folidol, folidol-E, folidol-E-605, folidol-oil, fosferno, fostex, gearphos, lirothion, niran, nitrostigmine, niuif-100, nourithion, oleofos-20, oleoparathene, oleoparathion, OMS-19, pac01, paramar-50, paraphos, parathene, parathion-ethyl, penncap-E, phosphoro-thioic-acid O,O-diethyl-o-(4-nitrophenyl)-ester, RB, rhodiasol, rhodiatox, selephos, super-rodiatox, thiomex-thiophos 3422, thiophos, vitrex

USES
♦ Used as an Insecticide
- fruit
- cotton
- wheat
- vegetables
- nut crops
♦ used as an acaricide

PHYSICAL PROPERTIES
♦ pure: pale-yellow liquid
♦ technical grades: brown
♦ garlic-like odor
♦ odor threshold 0.47 mg/m^3
♦ melting point 6°C
♦ boiling point 275°C
♦ density 1.26 at 25°/4°
♦ vapor pressure 3.78×10^{-5} mm Hg at 20°C
♦ viscosity 15.30 cP at 25°C
♦ surface tension 39.2 dynes/cm at 25°C
♦ practically insoluble in water
♦ insoluble in petr ether, kerosene, and the usual spray oils
♦ soluble in alcohols, esters, ethers, ketones, aromatic hydrocarbons
♦ log K$_{ow}$ 3.83

CHEMICAL PROPERTIES
♦ flammable
♦ flash point 120°C open cup
♦ reacts with bases
♦ incompatible with substances having a pH higher than 7.5

HEALTH RISK
♦ Toxic Exposure Guidelines
- OSHA PEL TWA 0.1 mg/m^3 (skin)
- ACGIH TLV TWA 0.1 mg/m^3 (skin)
- DFG MAK 0.1 mg/m^3
- NIOSH REL (parathion) TWA 0.005 mg/m^3
- MSHA standard 0.1 mg/m^3
♦ Toxicity Data
- mma-sat 500 µg/plate
- dns-human:fbr 10 µmol/L
- oral-rat TDLo 360 µg/kg
- subcutaneous-rat TDLo 9800 µg/kg
- oral-rat TDLo 1260 mg/kg/80W-C:ETA
- oral-man TDLo 429 µg/kg/4D-I:SYS
- oral-human TDLo LD$_{50}$ 171 µg/kg
- oral-woman TDLo 5670 µg/kg
- oral-human LD$_{50}$ 3 mg/kg
- skin-human LDLo 7143 µg/kg
- intratracheal-human LDLo 714 µg/kg
- oral-rat·LD$_{50}$ 2 mg/kg
- inhalation-rat LC$_{50}$ 84 mg/m^3/4H

- skin-rat LD_{50} 6800 µg/kg
- intramuscular-rat LD_{50} 6 mg/kg
- oral-mouse LD_{50} 5 mg/kg
- inhalation-mouse LCLo 15 mg/kg
- skin-mouse LD_{50} 19 mg/kg

◆ Acute Risks
- nausea
- vomiting
- abdominal cramps
- diarrhea
- excessive salivation
- headache
- weakness
- difficult breathing
- blurring of vision
- dimness of vision
- convulsions
- central nervous system depression
- paralysis
- coma
- respiratory failure
- death
- affects the blood
- affects the eyes and skin
- anorexia
- pupillary constriction
- bronchoconstriction
- muscle twitching
- Cholinesterase inhibitor effects are cumulative.

◆ Chronic Risks
- depressed red blood cell cholinesterase activity
- nausea
- headaches
- depressed plasma
- degenerative changes in the liver
- Methyl parathion is linked with human birth defects.
- reduced number of offspring and decreased fetal weight in rats
- high postnatal mortality in animals
- increased tumors in the adrenal gland in rats
- thyroid follicular adenomas in rats
- pancreatic islet-cell carcinomas in rats

HAZARD RISK
◆ combustible when exposed to heat or flame
◆ violent reaction with endrin
◆ highly dangerous
◆ Shock can shatter the container, releasing the contents.
◆ Decomposition emits highly toxic fumes of NO_x, PO_x, and SO_x.

EXPOSURE ROUTES
◆ field application and formulation
◆ dermal and inhalation exposure from spray drift
◆ workplace during manufacture of parathion
◆ ingesting food containing parathion residues

REGULATORY STATUS/INFORMATION
◆ Clean Air Act Amendment, Title III: hazardous air pollutants
◆ Federal Insecticide, Fungicide, and Rodenticide Act
- class of cholinesterase-inhibiting pesticides. 180.3(5)
◆ Resource Conservation and Recovery Act
- list of hazardous constituents. 40CFR258, 40CFR261, 40CFR261.11

- groundwater monitoring list. 40CFR264
- P waste #: U089
◆ CERCLA and Superfund
  - designation of hazardous substances. 40CFR302.4
  - list of extremely hazardous substances and their threshold planning quantities. 40CFR255-AB
  - list of chemicals. 40CFR372.65
◆ Clean Water Act
  - designation of hazardous substances. 40CFR4116.4
  - determination of reportable quantities. 40CFR4117.3
◆ OSHA 29CFR1910.1000 Table Z1
◆ chemical regulated by the State of California

ADDITIONAL/OTHER COMMENTS
◆ Symptoms
  - skin: prickling
  - eye: bad sight, narrowing pupil, redness, pain, irritating, flood of tears
  - oral: vomiting, diarrhea, weakness, headache, dizziness, labored breath, abdominal gripes
◆ First Aid
  - wash eyes immediately with large amounts of water
  - flush skin with plenty of soap and water
  - remove to fresh air immediately or provide oxygen if necessary
  - drink water and induce vomiting
◆ fire fighting: use dry chemical, water spray, foam, or carbon dioxide
◆ personal protection: compressed air/oxygen apparatus, gastight suit
◆ storage: keep away from flames or sparks

SOURCES
◆ *Health Effects Notebook for Hazardous Air Pollutants*, U.S. Environmental Protection Agency, December 1994
◆ *Material Safety Data Sheets (MSDS) for Arsine*
◆ *Sax's Dangerous Properties of Industrial Materials*, 9th edition, 1996
◆ *Environmental Law Index to Chemicals*, 1996 edition
◆ *Merck Index*, 12th edition, 1996

PENTACHLORONITROBENZENE (ahs)

CAS #: 82-68-8
MOLECULAR FORMULA: $C_6Cl_5NO_2$
FORMULA WEIGHT: 295.32
SYNONYMS: benzene, nitropentachloro-, brassicol, PCNB, quintozene, terrachlor

USES
◆ soil fungicide
◆ seed disinfectant for peanuts

PHYSICAL PROPERTIES
- light-yellow crystalline powder
- boiling point 328°C at 760 mm Hg
- specific gravity 1.723 (water = 1)
- surface tension not available
- musty odor
- slightly soluble in ethyl alcohol
- insoluble in water
- melting point 140–146°C (284–295°F)
- density 1.718 g/mL at 25°C
- vapor density 10.2 (air = 1)
- heat capacity not available
- vapor pressure 0.013 mm Hg at 25°C
- soluble in benzene and chloroform

CHEMICAL PROPERTIES
- flash point not available
- flammability limits not available
- autoignition temperature not available
- heat of combustion not available
- heat of formation not available
- will react vigorously with strong oxidizing agents, strong acids, and strong bases
- generally very stable
- heat of vaporization not available
- will not polymerize

HEALTH RISK
- Toxic Exposure Guidelines
  - ACGIH TLV TWA 0.5 mg/m^3
- Toxicity Data
  - oral-rat LD$_{50}$ 1100 mg/kg
  - inhalation-rat LC$_{50}$ 1400 mg/m^3
  - intraperitoneal-rat LD$_{50}$ 5 g/kg
  - oral-mouse LD$_{50}$ 1400 mg/kg
  - inhalation-mouse LC$_{50}$ 2 g/m^3
  - intraperitoneal-mouse LD$_{50}$ 4500 mg/kg
  - oral-rabbit LD$_{50}$ 800 mg/kg
- Acute Risks
  - irritation of skin and eyes
  - irritation to upper respiratory tract
  - irritation to mucous membranes
  - allergic respiratory and skin reactions
- Chronic Risks
  - possible carcinogen
  - target organ: thyroid
  - possible mutagenic effects

HAZARD RISK
- NFPA ratings not available
- emits highly toxic fumes of NO$_x$ and chlorine when heated to decomposition
- emits hydrogen chloride gas and phosgene gas when heated to decomposition

EXPOSURE ROUTES
- primarily to workers engaged in the manufacture, formulation, and application of pentachloronitrobenzene exposed via inhalation or skin contact

- general public is most likely to be exposed by the ingestion of contaminated food

REGULATORY STATUS/INFORMATION
- Clean Air Act Title III, hazardous air pollutants
- Resource Conservation and Recovery Act
  - list of hazardous inorganic and organic constituents, 40CFR258—Appendix 2
  - hazardous constituents, 40 CFR261—Appendix 8
  - groundwater monitoring chemicals list, 40CFR264—Appendix 9
  - health based limits for exclusion of waste-derived residues, 40CFR266—Appendix 2
  - U waste #: U185
- CERCLA and Superfund
  - designation of hazardous substances, 40CFR302.4
  - sequential CAS registry # list of CERCLA hazardous substances, 40CFR302—Appendix A
  - radionuclides, 40CFR302—Appendix B
  - reportable quantity (RQ) 100 lbs (45.4 kg)
- Federal Insecticide, Fungicide, and Rodenticide Act
  - tolerances and exemptions from tolerances for pesticide chemicals in or on raw agricultural commodities, 40CFR180.102-1147
- chemical regulated by the State of California
- Department of Transportation
  - hazard class/division not available
  - identification number not available
  - labels not available

ADDITIONAL/OTHER COMMENTS
- First Aid
  - flush eyes or skin with large amounts of water for a minimum of 15 minutes
  - remove contaminated clothing
  - provide respiratory support
  - if swallowed, wash out mouth with water provided person is conscious
- fire fighting: water spray; carbon dioxide, dry chemical powder, or appropriate foam. Wear self-contained breathing apparatus and protective clothing to prevent accidental contact with skin and eyes.
- personal protection: wear appropriate NIOSH/MSHA approved respirator, chemical-resistant gloves, safety goggles, and other protective clothing

SOURCES
- *CRC Handbook of Chemistry and Physics*, 77th edition, 1996–1997
- *Environmental Contaminant Reference Databook*, volume 1, 1995
- *Fire Protection Guide to Hazardous Materials*, 11th edition, 1994

♦ *Health Effects Notebook for Hazardous Air Pollutants*, U.S. Environmental Protection Agency, December 1994
♦ *MSDS*, Aldrich Chemical Company, Inc., catalog number 24232-2
♦ *NIOSH Pocket Guide to Chemical Hazards*, June 1994
♦ *Sax's Dangerous Properties of Industrial Materials*, 9th edition, 1996
♦ *Merck Index*, 12th edition, 1996
♦ *ACGIH Threshold Limit Values (TLVs) for 1996*
♦ *Suspect Chemicals Sourcebook*, 1996 edition
♦ *Handbook of Toxic and Hazardous Chemicals and Carcinogens*, 2nd edition, 1985

PENTACHLOROPHENOL (ahs)

CAS #: 87-86-5
MOLECULAR FORMULA: C_6HCl_5O
FORMULA WEIGHT: 266.32
SYNONYMS: PCP, penta, santophen 20

USES
♦ Manufacture of Other Products
  • insecticides
  • fungicides
  • algicides
  • herbicides
♦ preservation of wood and wood products
♦ manufacture of sodium pentachlorophenate

PHYSICAL PROPERTIES

♦ off-white powder
♦ melting point 188–191°C (370–376°F)
♦ density 1.978 g/mL at 22°C
♦ vapor density 9.2 (air = 1)
♦ heat capacity not available
♦ vapor pressure 0.00011 mm at 25°C, 40 mm at 211.2°C
♦ soluble in benzene
♦ Impure pentachlorophenol is dark gray to brown.

♦ boiling point 310°C (decomposes)
♦ specific gravity 1.982 (water = 1)
♦ surface tension not available
♦ very pungent, phenolic oder only when hot
♦ highly soluble in ethyl ether and ethyl alcohol
♦ almost insoluble in water

CHEMICAL PROPERTIES

♦ flash point not available
♦ autoignition temperature not available
♦ heat of combustion not available
♦ hazardous decomposition products of carbon monoxide, carbon dioxide, and hydrogen

♦ flammability limits not available
♦ reacts with strong oxidizing agents, strong bases, acid chlorides, and acid anhydrides
♦ heat of vaporization 70.05 kJ/mole at 25°C

HEALTH RISK

- ◆ Toxic Exposure Guidelines
  - ACGIH TLV TWA 0.5 mg/m³
  - OSHA PEL TWA 0.5 mg/m³
  - NIOSH REL TWA 0.5 mg/m³
- ◆ Toxicity Data
  - oral-human LDLo 401 mg/kg
  - oral-rate LD_{50} 27 mg/kg
  - inhalation-rat LC_{50} 335 mg/m³
  - skin-rat LD_{50} 96 mg/kg
  - intraperitoneal-rat LD_{50} mg/kg
  - oral-mouse LD_{50} 36 mg/kg
  - inhalation-mouse LC_{50} 225 mg/m³
  - intraperitoneal-mouse LD_{50} 32 mg/kg
  - oral-hamster LD_{50} 168 mg/kg
  - oral-gerbil LD_{50} 294 mg/kg
  - oral-duck LD_{50} 380 mg/kg
- ◆ Acute Risks
  - may be fatal if inhaled, swallowed, or absorbed through skin
  - causes eye and skin irritation
  - irritating to mucous membranes and upper respiratory tract
  - convulsions
  - dermatitis
- ◆ Chronic Risks
  - damage to liver and kidneys
  - carcinogen
  - target organs: liver and kidney
  - may cause congenital malformation in the fetus
  - may alter genetic material

HAZARD RISK

- ◆ NFPA rating Health 3, Flammability 0, Reactivity 0
- ◆ emits highly toxic fumes of chlorine when heated to decomposition

EXPOSURE ROUTES

- ◆ contaminated air in pressure treated log homes
- ◆ detected at low levels in drinking water and food
- ◆ contaminated air at wood treatment facilities and lumber mills
- ◆ dermal contact with pressure treated lumber

REGULATORY STATUS/INFORMATION

- ◆ Clean Air Act Title III, hazardous air pollutants
- ◆ Safe Drinking Water Act
  - 83 contaminants to be regulated under the SDWA of 1986, 49FR9352
- ◆ Resource Conservation and Recovery Act
  - list of hazardous inorganic and organic constituents, 40CFR258—Appendix 2
  - hazardous constituents, 40CFR261—Appendix 8
  - reference air concentrations, 40CFR264—Appendix 4

- groundwater monitoring chemicals list, 40CFR264—Appendix 9
- health based limits for exclusion of waste-derived residues, 40CFR266—Appendix 2
- PICs found in stack effluents, 40CFR266—Appendix 6
- D waste #: D037
- F waste #: F027

◆ CERCLA and Superfund
  - designation of hazardous substances, 40CFR302.4
  - chemicals and chemical categories to which this part applies (CAS # listing), 40CFR372.65
  - reportable quantity (RQ) 10 lb (4.54 kg)
◆ Clean Water Act
  - designation of hazardous substances, 40CFR116.4
  - list of hazardous substances, 40CFR116.4A
  - list of hazardous substances by CAS #, 40CFR116.4B
  - determination of reportable quantities, 40CFR117.3
  - reportable quantities of hazardous substances designated pursuant to section 311 of the Clean Water Act, Table 117.3
  - toxic pollutants, 40CFR401.15
  - total toxic organics, 40CFR413.02
  - 126 priority pollutants, 40CFR432—Appendix A
◆ chemical regulated by the State of California
◆ Occupational Safety and Health Act
  - limits for air concentration under the Occupational Safety and Health Act, 29CFR1910.1000 Table Z1
◆ SARA Health Assessment Rank: 51
◆ Department of Transportation
  - hazard class/division 6.1
  - identification number UN 3155
  - labels: poison
  - severe marine pollutant

ADDITIONAL/OTHER COMMENTS
◆ First Aid
  - flush eyes or skin with large amounts of water for a minimum of 15 minutes
  - remove contaminated clothing
  - provide respiratory support
  - if swallowed, wash out mouth with water provided person is conscious
◆ fire fighting: water spray; carbon dioxide, dry chemical powder, or appropriate foam. Wear self-contained breathing apparatus and protective clothing to prevent contact with eyes and skin.
◆ personal protection: wear appropriate NIOSH/MSHA approved respirator, chemical-resistant gloves, safety goggles, and other protective clothing. Use only in a chemical fume hood.

SOURCES
- *CRC Handbook of Chemistry and Physics*, 77th edition, 1996–1997
- *Environmental Contaminant Reference Databook*, volume 1, 1995
- *Fire Protection Guide to Hazardous Materials*, 11th edition, 1994
- *Health Effects Notebook for Hazardous Air Pollutants*, U.S. Environmental Protection Agency, December 1994
- *MSDS*, Aldrich Chemical Company, Inc., catalog number P260-4
- *NIOSH Pocket Guide to Chemical Hazards*, June 1994
- *Sax's Dangerous Properties of Industrial Materials*, 9th edition, 1996
- *Merck Index*, 12th edition, 1996
- *ACGIH Threshold Limit Values (TLVs) for 1996*
- *Suspect Chemicals Sourcebook*, 1996 edition
- *Handbook of Toxic and Hazardous Chemicals and Carcinogens*, 2nd edition, 1985

PHENOL (ahs)

CAS #: 108-95-2
MOLECULAR FORMULA: C_6H_6O
FORMULA WEIGHT: 94.12
SYNONYMS: carbolic acid, hydroxybenzene, oxybenzene, phenic acid, phenylic acid, phenyl hydroxide

USES
- Manufacture of Colorless or Light-Colored Artificial Resins
  - plywood
  - appliance
  - construction
  - automotive
- Medical Uses
  - ear and nose drops
  - general disinfectant
  - throat lozenges
  - pharmaceutic aid (preservative)
  - mouthwashes
- industrial organic compounds and dyes
- production of caprolactam and bisphenol A (intermediates in the production of nylon and epoxy resins)
- reagent in chemical analysis

PHYSICAL PROPERTIES
- colorless or white solid when pure
- melting point (ultrapure material) 40.85°C (105.5°F)
- density 1.0545 g/mL at 45°C
- most often sold as a liquid
- boiling point 182°C at 760 mm Hg
- specific gravity 1.065 (water = 1)
- vapor density 3.24 (air = 1)
- heat capacity 127.4 J/mole-K (liquid at 25°C)
- vapor pressure 1 mm at 40.1°C

- soluble in water and ethyl alcohol
- miscible with acetone
- surface tension 40.9 dynes/cm at 20°C

CHEMICAL PROPERTIES
- flash point 79°C
- autoignition temperature 715°C
- heat of combustion 3053.5 kJ/mole at 25°C
- heat of formation −165.1 kJ/mole (crystal at 25°C)
- entropy 144.0 J/mole-K (crystal)

- characteristic odor, very strong and sweet
- very soluble in ethyl ether
- almost insoluble in petroleum ether

- flammability limits 1.8% lower and 8.6% upper
- reacts with aluminum chloride in the presence of nitromethane or nitrobenzene
- heat of vaporization 45.69 kJ/mole at 181.8°C
- heat of formation −96.4 kJ/mole (gas at 25°C)
- light sensitive

HEALTH RISK
- Toxic Exposure Guidelines
  - ACGIH TLV TWA 5 ppm (19 mg/m³)
  - OSHA PEL TWA 5 ppm (19 mg/m³)
  - NIOSH REL TWA 5 ppm (19 mg/m³)
- Toxicity Data
  - oral-infant LDLo 10 mg/kg
  - oral-human LDLo 14 mg/kg
  - oral-rat LD_{50} 317 mg/kg
  - inhalation-rat LC_{50} 316 mg/m³
  - skin-rat LD_{50} 669 mg/kg
  - intraperitoneal-rat LD_{50} 127 mg/kg
  - subcutaneous-rat LD_{50} 460 mg/kg
  - oral-mouse LD_{50} 270 mg/kg
  - inhalation-mouse LC_{50} 177 mg/m³
  - intraperitoneal-mouse LD_{50} 180 mg/kg
  - subcutaneous-mouse LD_{50} 344 mg/kg
- Acute Risks
  - irritation of skin, eyes, and mucous membranes
  - muscle weakness and tremors
  - paralysis
  - coma
  - weak pulse
  - reduced blood pressure
  - very toxic to humans
  - loss of coordination
  - convulsions
  - respiratory arrest
  - cardiac depression
- Chronic Risks
  - muscle pain
  - elevated levels of liver enzymes
  - enlarged liver
  - dermal inflammation and necrosis
  - affects liver, kidney, respiratory, cardiovascular, and central nervous systems

HAZARD RISK
- NFPA rating Health 3, Flammability 2, Reactivity 0

- potentially explosive reaction with aluminum chloride and nitromethane at 110°C and 100 bar
- potentially explosive reactions with formaldehyde, peroxysulfuric acid, peroxymonosulfuric acid
- emits acrid smoke and irritating fumes when heated to decomposition

EXPOSURE ROUTES
- primarily through breathing contaminated air
- dermal exposure
- exposure through the use of phenol-containing medicinal products
- eating certain foods, such as fried chicken, mountain cheese, and some types of fish
- smoking tobacco

REGULATORY STATUS/INFORMATION
- Clean Air Act Title III, hazardous air pollutants
- Resource Conservation and Recovery Act
  - list of hazardous inorganic and organic constituents, 40CFR258—Appendix 2
  - hazardous constituents, 40CFR261—Appendix 8
  - reference air concentrations, 40CFR264—Appendix 4
  - groundwater monitoring chemicals list, 40CFR264—Appendix 9
  - health based limits for exclusion of waste-derived residues, 40CFR266—Appendix 2
  - PICs found in stack effluents, 40CFR266—Appendix 6
  - U waste #: U188
- CERCLA and Superfund
  - designation of hazardous substances, 40CFR302.4
  - list of extremely hazardous substances and their threshold planning quantities, 40CFR355—Appendix B
  - reportable quantity (RQ) 1000 lb (454 kg)
  - threshold planning quantity (TPQ) 500/10,000 lbs
- Clean Water Act
  - designation of hazardous substances, 40CFR116.4
  - list of hazardous substances, 40CFR116.4A
  - list of hazardous substances by CAS #, 40CFR116.4B
  - determination of reportable quantities, 40CFR117.3
  - reportable quantities of hazardous substances designated pursuant to section 311 of the Clean Water Act, Table 117.3
  - toxic pollutants, 40CFR401.15
  - total toxic organics, 40CFR413.02
  - 126 priority pollutants, 40CFR432—Appendix A
- chemical regulated by the State of California
- Occupational Safety and Health Act
  - limits for air concentration under the Occupational Safety and Health Act, 29CFR1910.1000 Table Z1
- Toxic Substance Control Act
  - list of substances, 40CFR717.120.a
- SARA Health Assessment Rank: 129

- ◆ Department of Transportation
  - hazard class/division 6.1
  - identification number UN1671 (solid); UN 2821 (solutions)
  - labels: poison

ADDITIONAL/OTHER COMMENTS
- ◆ First Aid
  - flush eyes or skin with large amounts of water for a minimum of 15 minutes
  - remove contaminated clothing
  - provide respiratory support
  - if swallowed, wash out mouth with water provided person is conscious
- ◆ fire fighting: water spray; carbon dioxide, dry chemical powder, or appropriate foam
- ◆ personal protection: wear appropriate NIOSH/MSHA approved respirator, chemical-resistant gloves, safety goggles, and other protective clothing

SOURCES
- ◆ *CRC Handbook of Chemistry and Physics*, 77th edition, 1996–1997
- ◆ *Environmental Contaminant Reference Databook*, volume 1, 1995
- ◆ *Fire Protection Guide to Hazardous Materials*, 11th edition, 1994
- ◆ *Health Effects Notebook for Hazardous Air Pollutants*, U.S. Environmental Protection Agency, December 1994
- ◆ *MSDS*, Aldrich Chemical Company, Inc., catalog number 24232-2
- ◆ *NIOSH Pocket Guide to Chemical Hazards*, June 1994
- ◆ *Sax's Dangerous Properties of Industrial Materials*, 9th edition, 1996
- ◆ *Merck Index*, 12th edition, 1996
- ◆ *ACGIH Threshold Limit Values (TLV) for 1996*
- ◆ *Suspect Chemicals Sourcebook*, 1996 edition
- ◆ *Handbook of Toxic and Hazardous Chemicals and Carcinogens*, 2nd edition, 1985

p-PHENYLENEDIAMINE

CAS #: 106-50-3
MOLECULAR FORMULA: $C_6H_4(NH_2)_2$
FORMULA WEIGHT: 108.16
SYNONYMS: p-aminoaniline, 4-aminoaniline, basf ursol D, p-benzenediamine, 1,4-benzenediamine, benzofur D, C.I. 76060, C.I. developer 13, C.I. oxidation base 10, developer 13, developer PF, p-diaminobenzene, 1,4-diaminobenzene, durafur black R, fenylenodwuamina (Polish), fouramine D, fourinne D, fourrine I, fur black 41867, fur brown 41866, furrod D, fur yellow, futramine D, nako H, orsin, para, paraphenylen-diamine, pelagol D, pelagol DR, pelagol grey

D, peltol D, 1,4-phenylenediamine, phenylenediamine, para, solid (DOT), PPD, renal PF, santoflex IC, tertral D, ursol D, USAF EK-394, vulkanox 4020, zoba black D

USES

- ◆ Used in Industry
  - dye intermediate
  - dye
  - photographic developing agent
  - chemical intermediate
  - vulcanization accelerator
  - antioxidant in rubber compounds
  - photochemical measurements
  - accelerator for synthetic fibers
  - laboratory reagent
  - dyeing hair and fur

PHYSICAL PROPERTIES

- ◆ white or slightly red crystals
- ◆ darkens on exposure to air
- ◆ melting point 146°C
- ◆ boiling point 267°C
- ◆ slightly soluble in water
- ◆ soluble in alcohol, chloroform, and ether
- ◆ vapor pressure 1 mm Hg at 21°C
- ◆ vapor density 3.72
- ◆ log K_{ow} −0.25

CHEMICAL PROPERTIES

- ◆ combustible
- ◆ flash point 155°C
- ◆ affected by light
- ◆ incompatible with strong oxidizers
- ◆ Hazardous polymerization will not occur.

HEALTH RISK

- ◆ Toxic Exposure Guidelines
  - OSHA PEL TWA 0.1 mg/m³ (skin)
  - ACGIH TLV TWA 0.1 mg/m³
  - DFG MAK 0.1 mg/m³
  - NIOSH REL 0.1 mg/m³
  - MSHA standard 0.1 mg/m³
- ◆ Toxicity Data
  - skin-human 250 mg/24H MLD
  - skin-mouse 250 mg/24H MLD
  - skin-dog 250 mg/24H MLD
  - skin-rabbit 12,500 µg/24H MLD
  - skin-rabbit 250 mg/24H MOD
  - skin-pig 250 mg/24H MLD
  - skin-guinea pig 250 mg/24H MLD
  - mma-sat 10 µg/plate
  - otr-rat:emb 1850 ng/plate
  - sln-dmg-oral 15,500 µmol/L/3D
  - subcutaneous-rat TDLo 2625 mg/kg/30W-C:ETA
  - oral-man TDLo 71 mg/kg
  - oral-rat LD₅₀ 80 mg/kg
  - intravenous-rat LDLo 50 mg/kg
  - subcutaneous-mouse LDLo 140 mg/kg
  - intraperitoneal-rabbit LDLo 150 mg/kg
  - intravenous-rabbit LDLo 300 mg/kg

- intraperitoneal-rat LD$_{50}$ 37 mg/kg
- subcutaneous-rat LDLo 170 mg/kg
- intraperitoneal-mouse LD$_{50}$ 50 mg/kg
- subcutaneous-dog LDLo 100 mg/kg
- intravenous-dog LDLo 17 mg/kg
- oral-cat LDLo 100 mg/kg
- oral-rabbit LDLo 250 mg/kg
- skin-rabbit LDLo 5000 mg/kg
- subcutaneous-rabbit LDLo 200 mg/kg

◆ Acute Risks
  - severe dermatitis
  - eye irritation
  - tearing
  - asthma
  - gastritis
  - renal failure
  - vertigo
  - tremors
  - convulsions
  - coma

◆ Chronic Risks
  - eczematoid contact der-matitis
  - fatal liver damage
  - depressed body weights in rats and mice
  - elevated liver and kidney weights in rats
  - lung tumors in rats
  - thyroid tumors in rats
  - tumors of the lungs, hematopoietic system, and liver in mice

HAZARD RISK
◆ combustible when exposed to heat or flame
◆ can react vigorously with oxidizing materials
◆ Decombustion emits acrid smoke and irritating fumes.

EXPOSURE ROUTES
◆ manufacture and use of p-phenylenediamine
◆ contact with hair dye

REGULATORY STATUS/INFORMATION
◆ Clean Air Act Amendment, Title III: hazardous air pollutants
◆ Resource Conservation and Recovery Act
  - list of hazardous constituents. 40CFR258
  - groundwater monitoring list. 40CFR264
◆ CERCLA and Superfund
  - designation of hazardous substances. 40CFR302.4
  - list of chemicals. 40CFR372.65
◆ Toxic Substance Control Act
  - list of substances. 40CFR716.120.a
  - unsubstituted phenylene-diamines. 40CFR799.3300
◆ OHSA 29CFR1910.1000 Table Z1
◆ chemical regulated by the State of California

ADDITIONAL/OTHER COMMENTS
◆ Symptoms
  - exposure: dermatitis, eye irritation and tearing, asthma, gastritis, renal failure, vertigo, tremors, convulsions, coma
◆ First Aid
  - wash eyes and skin immediately with large amounts of water

- remove to fresh air immediately and provide respiratory apparatus if necessary
- wash out mouth with milk or water and induce vomiting
- ◆ fire fighting: use water spray, carbon dioxide, and dry chemical
- ◆ personal protection: wear self-contained breathing apparatus and protective clothing
- ◆ storage: keep away from contact with oxidizing materials

SOURCES
- ◆ *Health Effects Notebook for Hazardous Air Pollutants*, U.S. Environmental Protection Agency, December 1994
- ◆ *Material Safety Data Sheets (MSDS)*
- ◆ *Sax's Dangerous Properties of Industrial Materials*, 9th edition, 1996
- ◆ *Hawley's Condensed Chemical Dictionary*, 12th edition, 1993
- ◆ *Aldrich Chemical Company Catalog*, 1996–1997
- ◆ *Environmental Law Index to Chemicals*, 1996 edition
- ◆ *Merck Index*, 12th edition, 1996

PHOSGENE (ahs)

CAS #: 75-44-5
MOLECULAR FORMULA: $COCl_2$
FORMULA WEIGHT: 98.91
SYNONYMS: carbonic dichloride, carbonyl chloride, chloroformyl chloride

USES
- ◆ Widely Used Chemical Intermediate
  - isocyanate based polymers
  - carbonic acid esters
  - acid chlorides
- ◆ Manufacture of Other Products
  - dyestuffs
  - insecticides
  - pharmaceuticals
- ◆ metallurgy
- ◆ used as a chemical warfare agent in World Wars I and II

PHYSICAL PROPERTIES
- ◆ colorless, highly toxic gas
- ◆ boiling point 8.2°C at 760 mm Hg
- ◆ specific gravity 1.432 (water = 1)
- ◆ surface tension 34.6 mN/m at 0°C
- ◆ suffocating odor, reminiscent of moldy hay
- ◆ melting point −128°C
- ◆ density 1.3719 g/mL at 25°C
- ◆ vapor density 3.41 (air = 1)
- ◆ heat capacity 57.7 J/mole-K (gas at 25°C)
- ◆ vapor pressure 568 mm at 0°C, 1215 mm Hg at 20°C
- ◆ soluble in benzene, toluene, chloroform, and carbon tetrachloride
- ◆ slightly soluble in water

CHEMICAL PROPERTIES
- ◆ flash point not available
- ◆ autoignition temperature not available

- heat of vaporization 6224.2 gcal/gmol
- Gibbs free energy −204.9 kJ/mole
- flammability limits not available
- heat of combustion not available
- heat of formation −219.1 kJ/mole (gas at 25°C)
- entropy 283.5 J/mole-K (gas)

HEALTH RISK
- Toxic Exposure Guidelines
  - ACGIH TLV TWA 0.1 ppm (0.40 mg/m^3)
  - OSHA PEL TWA 0.1 ppm (0.40 mg/m^3)
  - NIOSH REL TWA 0.1 ppm (0.40 mg/m^3)
- Toxicity Data
  - inhalation-human LC$_{50}$ 3200 mg/m^3
  - inhalation-human TCLo 25 ppm/30M:PUL
  - inhalation-man LCLo 360 mg/m^3/30M
  - inhalation-mammal LCLo 50 ppm/5M
  - inhalation-cat LCLo 190 mg/m^3/15M
  - inhalation-mice LC$_{50}$ 1800 mg/m^3
  - inhalation-rat LC$_{50}$ 1400 mg/m^3
- Acute Risks
  - severe respiratory effects
  - possible death
  - strong eye and skin irritant
  - pulmonary edema
  - pulmonary emphysema
  - inhalation causes choking, chest constriction, coughing, painful breathing, and bloody sputum
- Chronic Risks

HAZARD RISK
- NFPA rating Health 4, Flammability 0, Reactivity 1
- undergoes hazardous reactions with aluminum, sodium, lithium, and potassium under appropriate conditions
- produces toxic and corrosive fumes of chlorine when heated to decomposition or when exposed to water or steam

EXPOSURE ROUTES
- inhalation in the workplace during its manufacture, handling, or use
- in ambient air from direct industrial emissions
- in ambient air from the decomposition of chlorinated hydrocarbons
- in ambient air from the photooxidation of chloroethylenes

REGULATORY STATUS/INFORMATION
- Clean Air Act Title III, hazardous air pollutants
- Resource Conservation and Recovery Act
  - hazardous constituents, 40 CFR261—Appendix 8
  - P waste #: P095
- CERCLA and Superfund
  - designation of hazardous substances, 40CFR302.4
  - list of extremely hazardous substances and their threshold planning quantities, 40CFR355—Appendix B

- chemicals and chemical categories to which this part applies (CAS # listing), 40CFR372.65
- reportable quantity (RQ) 10 lb (4.54 kg)
- threshold planning quantity (TPQ) 10 lb (4.54 kg)
◆ Clean Water Act
  - designation of hazardous substances, 40CFR116.4
  - list of hazardous substances, 40CFR116.4A
  - list of hazardous substances by CAS #, 40CFR116.4B
  - determination of reportable quantities, 40CFR117.3
  - reportable quantities of hazardous substances designated pursuant to section 311 of the Clean Water Act, Table 117.3
◆ chemical regulated by the State of California
◆ Occupational Safety and Health Act
  - limits for air concentration under the Occupational Safety and Health Act, 29CFR1910.1000 Table Z1
  - list of highly hazardous chemicals, toxics, and reactives, 29CFR1910.1000 Table Z5
◆ EPCRA Section 313 de minimis concentration 1.0%
◆ Department of Transportation
  - hazard class/division 2.3
  - identification number UN 2199
  - labels: poison gas; corrosive

ADDITIONAL/OTHER COMMENTS
◆ First Aid
  - flush eyes or skin with large amounts of water
  - provide respiratory support
  - if swallowed, do not induce vomiting
◆ personal protection: wear full face gas masks with phosgene canisters or self-contained breathing apparatus. Automatic continuous monitors with alarm indicators should be used whenever exposure to phosgene exists. Where the potential of exposure exists, wear safety goggles and other protective clothing.

SOURCES
◆ *CRC Handbook of Chemistry and Physics*, 77th edition, 1996–1997
◆ *Environmental Contaminant Reference Databook*, volume 1, 1995
◆ *Fire Protection Guide to Hazardous Materials*, 11th edition, 1994
◆ *Health Effects Notebook for Hazardous Air Pollutants*, U.S. Environmental Protection Agency, December 1994
◆ *NIOSH Pocket Guide to Chemical Hazards*, June 1994
◆ *Sax's Dangerous Properties of Industrial Materials*, 9th edition, 1996
◆ *Merck Index*, 12th edition, 1996
◆ *ACGIH Threshold Limit Values (TLV) for 1996*
◆ *Suspect Chemicals Sourcebook*, 1996 edition
◆ *Handbook of Toxic and Hazardous Chemicals and Carcinogens*, 2nd edition, 1985

PHOSPHINE (ahs)

CAS #: 7803-51-2
MOLECULAR FORMULA: PH_3
FORMULA WEIGHT: 34.00
SYNONYMS: H_3P, phosphorus hydride

USES
- ◆ Insecticide for
  - fumigation of grains
  - animal feeds
  - leaf stored tobacco
- ◆ Chemical Intermediate
  - doping agent for n-type semiconductors
  - polymerization inhibitor
  - synthesis of flame retardants for cotton fabrics
  - condensation catalyst
- ◆ Rodenticide

PHYSICAL PROPERTIES
- ◆ colorless, flammable gas
- ◆ boiling point −87.5 to −87.7°C at 760 mm Hg
- ◆ specific gravity 0.76 (water = 1), 0.57 at 20 atm
- ◆ surface tension not available
- ◆ odor of decaying fish
- ◆ slightly soluble in ethyl ether and ethyl alcohol
- ◆ melting point −132.5 to −133°C
- ◆ density 1.492 g/mL
- ◆ vapor density 1.17 (air = 1)
- ◆ heat capacity 37.1 J/mole-K (at 25°C)
- ◆ vapor pressure greater than 760 mm Hg at 20°C
- ◆ insoluble in water

CHEMICAL PROPERTIES
- ◆ flash point not available
- ◆ autoignition temperature 100°C
- ◆ heat of combustion not available
- ◆ heat of formation 5.4 kJ/mole (gas at 25°C)
- ◆ flammability limits 1.6% lower and estimated at 98% upper
- ◆ reacts violently with oxygen and the halogens
- ◆ heat of vaporization 14.6 kJ/mole at −87°C
- ◆ spontaneously flammable in air if there is a trace of P_2H_4 present

HEALTH RISK
- ◆ Toxic Exposure Guidelines
  - ACGIH TLV TWA 0.3 ppm (0.42 mg/m³)
  - OSHA PEL TWA 0.3 ppm (0.4 mg/m³)
  - NIOSH REL TWA 0.3 ppm (0.4 mg/m³)
- ◆ Toxicity Data
  - inhalation-human LCLo 1000 ppm
  - inhalation-rat LC_{50} 11 ppm/4H
  - inhalation-mouse LCLo 380 mg/m³/2H
  - inhalation-cat LCLo 70 mg/m³/2H
  - inhalation-rabbit LCLo 2500 ppm/20M
  - inhalation-guinea pig LCLo 140 mg/m³/4H

- Acute Risks
  - headaches
  - fatigue
  - dizziness
  - burning substernal pain
  - nausea
  - cough with fluorescent green sputum
  - labored breathing
  - pulmonary edema
  - vomiting
  - gastrointestinal distress
  - pulmonary irritation
  - tremors
- Chronic Risks
  - inflammation of the nasal cavity and throat
  - liver effects
  - dizziness
  - weakness
  - increased blood density
  - jaundice
  - nausea, gastrointestinal, cardiorespiratory, and central nervous system symptomology

HAZARD RISK

- NFPA rating Health 4, Flammability 4, Reactivity 2
- extreme fire hazard by spontaneous chemical reaction
- explosive reaction with dichlorine oxide, silver nitrate, concentrated nitric acid, nitrogen trichloride, and oxygen
- ignition or violent reaction with air, bromine, chlorine, iodine, and other oxidants
- emits highly toxic fumes of PO_x when heated to decomposition

EXPOSURE ROUTES

- Occupational exposure to phosphine may occur during its manufacture or use.
- inhalation of contaminated air
- ingestion of foods contaminated with residue

REGULATORY STATUS/INFORMATION

- Clean Air Act Title III, hazardous air pollutants
- Resource Conservation and Recovery Act
  - list of hazardous inorganic and organic constituents, 40CFR258—Appendix 2
  - hazardous constituents, 40CFR261—Appendix 8
  - reference air concentrations, 40CFR264—Appendix 4
  - health based limits for exclusion of waste-derived residues, 40CFR266—Appendix 2
  - P waste #: P096
- CERCLA and Superfund
  - designation of hazardous substances, 40CFR302.4
  - list of extremely hazardous substances and their threshold planning quantities, 40CFR355—Appendix B
  - chemicals and chemical categories to which this part applies (CAS # listing), 40CFR372.65
  - reportable quantity (RQ) 100 lb (45.4 kg)
  - threshold planning quantity (TPQ) 500 lb (227 kg)
- chemical regulated by the State of California
- Occupational Safety and Health Act
  - limits for air concentration under the Occupational Safety and Health Act, 29CFR1910.1000 Table Z1

- list of highly hazardous chemicals, toxics, and reactives, 29CFR1910.1000 Table Z5
◆ EPCRA Section 313 de minimis concentration 1.0%
◆ Department of Transportation
  - hazard class/division 2.3
  - identification number UN 10766
  - labels: poison gas, flammable gas

ADDITIONAL/OTHER COMMENTS
◆ first aid: provide respiratory support
◆ fire fighting: water spray; carbon dioxide, dry chemical powder, or appropriate foam
◆ personal protection: wear appropriate NIOSH/MSHA approved respirator or self-contained breathing apparatus

SOURCES
◆ *CRC Handbook of Chemistry and Physics*, 77th edition, 1996-1997
◆ *Environmental Contaminant Reference Databook*, volume 1, 1995
◆ *Fire Protection Guide to Hazardous Materials*, 11th edition, 1994
◆ *Health Effects Notebook for Hazardous Air Pollutants*, U.S. Environmental Protection Agency, December 1994
◆ *Material Safety Data Sheet (MSDS)* for air products
◆ *NIOSH Pocket Guide to Chemical Hazards*, June 1994
◆ *Sax's Dangerous Properties of Industrial Materials*, 9th edition, 1996
◆ *Merck Index*, 12th edition, 1996
◆ *ACGIH Threshold Limit Values (TLVs) for 1996–1997*
◆ *Suspect Chemicals Sourcebook*, 1996 edition
◆ *Handbook of Toxic and Hazardous Chemicals and Carcinogens*, 2nd edition, 1985

PHOSPHORUS

CAS #: 7723-14-0
MOLECULAR FORMULA: The molecular formula for white/yellow phosphorus is P_4. The molecular formula for red phosphorus is P.
FORMULA WEIGHT: The molecular weight for white/yellow phosphorus is 123.88. The molecular weight for red phosphorus is 30.97.
SYNONYMS: phosphorus, amorphous (DOT)

USES
◆ White Phosphorus
  - manufacture of rat poisons
  - smoke screens
  - gas analysis
◆ Red Phosphorus
  - pyrotechnics
  - manufacture of safety matches
  - organic synthesis
  - manufacture of phosphoric acid
  - manufacture of phosphine

- manufacture
  of phosphoric anhydride
- manufacture
  of phosphorus pentachloride and phosphorus trichloride

- manufacture of fertilizers
- manufacture of pesticides
- incendiary shells
- smoke bombs
- tracer bullets

PHYSICAL PROPERTIES (White/Yellow Phosphorus)

- cubic, colorless crystals or
  yellow leaflets
- wax-like solid
- melting point 44.1°C
- boiling point 280°C
- density 1.88
- moderately soluble in water

- vapor pressure 1 mm Hg at
  76.6°C
- vapor density 4.42
- Mohs hardness 0.5
- soluble in carbon disulfide
- exhibits phosphorescence at
  room temperature

PHYSICAL PROPERTIES (Red Phosphorus)

- reddish-brown powder
- melting point 590°C at 43
  atm
- boiling point 280°C
- density 2.34

- vapor density 4.77
- insoluble in organic solvents
- soluble in phosphorus tribromide

CHEMICAL PROPERTIES (White/Yellow Phosphorus)

- spontaneously flammable in
  air
- autoignition temperature
  86°F
- ignites at 30°C in moist air
- combines directly with the
  halogens to form tri- or pen-
  tahalides
- combines with sulfur to form
  sulfides

- reacts with several metals to
  form phosphides
- yields orthophosphoric acid
  when treated with nitric acid
- reacts with alkali hydroxides
- incompatible with sulfur, iodine, oil of turpentine, and
  potassium chlorate

CHEMICAL PROPERTIES (Red Phosphorus)

- autoignition temperature
  500°F
- ignites at the boiling point
- sublimes at 416°C
- less reactive than white phosphorus

- reacts only at high temperature
- yields the white modifica-
  tion when distilled at 290°C

HEALTH RISK

- Toxic Exposure Guidelines (White/Yellow Phosphorus)
  - OSHA PEL TWA 0.1 mg/m^3
  - ACGIH TLV TWA 0.1 mg/m^3
- Toxic Exposure Guidelines (Red Phosphorus)
  - DFG MAK 0.1 mg/m^3
- Toxicity Data (White/Yellow Phosphorus)
  - oral-rat TDLo 11 µg/kg
  - oral-woman LDLo 22 mg/kg:CVS
  - oral-woman TDLo 11 mg/kg
  - oral-human LDLo 1400 µg/kg
  - oral-woman LDLo 4600 µg/kg:PUL,GIT,SKN

- oral-woman TDLo 2600 µg/kg
- oral-rat LD_{50} 3030 µg/kg
- oral-mouse LD_{50} 4820 µg/kg
- oral-cat LDLo 4 mg/kg
- oral-dog LDLo 10 mg/kg
- subcutaneous-dog LDLo 2 mg/kg
- subcutaneous-rabbit LDLo 10 mg/kg
- oral-duck LDLo 3 mg/kg
- oral-uns LDLo 200 mg/kg

◆ Acute Risks (White/Yellow Phosphorus)
- severe gastrointestinal irritation
- bloody diarrhea
- liver damage
- skin eruptions
- oligurua
- circulatory collapse
- coma
- convulsions
- death
- severe burns
- severe eye damage
- photophobia with myosis
- dilation of pupils
- retinal hemorrhage
- congestion of blood vessels

◆ Acute Risks (Red Phosphorus)
- relatively nontoxic
- mild to severe destruction of tissue
- spontaneous fractures
- anemia
- weight loss
- cardiomyopathy
- cyanosis
- nausea
- vomiting
- sweating

◆ Chronic Risks (White/Yellow Phosphorus)
- boney necrosis
- necrosis of the jaw (phossy-jaw)

◆ Chronic Risks (Red Phosphorus)
- stomach pains
- vomiting
- diarrhea

HAZARD RISK (White/Yellow Phosphorus)
◆ dangerous fire hazard when exposed to heat or flame or by chemical reaction with oxidizers
◆ ignites spontaneously in air
◆ very reactive
◆ Combustion in a confined space causes asphyxiation.
◆ dangerous explosion hazard by chemical reactions with various compounds
◆ reacts vigorously with oxidizing materials
◆ Decomposition emits toxic fumes of PO_x.

HAZARD RISK (Red Phosphorus)
◆ dangerous fire hazard when exposed to heat or by chemical reaction with oxidizers
◆ can react with reducing materials
◆ moderate explosion hazard by chemical reaction or on contact with organic materials
◆ may explode on impact

- ◆ catches fire when heated in air and burns with formation of the pentoxide
- ◆ burns when heated in an atmosphere of chlorine
- ◆ Decomposition emits toxic fumes of PO_x.

EXPOSURE ROUTES
- ◆ No information is available on the exposure routes for either form of phosphorus.

REGULATORY STATUS/INFORMATION (Phosphorus)
- ◆ Clean Air Act Amendment, Title III: hazardous air pollutants
- ◆ CERCLA and Superfund
  - • designation of hazardous substances. 40CFR302.4
  - • list of extremely hazardous substances and their threshold planning quantities. 40CFR355-AB
  - • list of chemicals. 40CFR372.65
- ◆ Clean Water Act
  - • designation of hazardous substances. 40CFR116.4
  - • determination of reportable quantities. 40CFR117.3
- ◆ chemical regulated by the State of California

REGULATORY STATUS/INFORMATION (White/Yellow Phosphorus)
- ◆ TSCA: list of substances. 40CFR716.120.a
- ◆ OSHA 29CFR1910.1000 Table Z1

ADDITIONAL/OTHER COMMENTS
- ◆ First Aid
  - • wash eyes and skin immediately with large amounts of water
  - • remove to fresh air immediately or provide respiratory apparatus if necessary
  - • drink water
- ◆ fire fighting: use water
- ◆ personal protection: self-contained breathing apparatus, gloves, goggles, and protective clothing
- ◆ storage: store red phosphorus in a cool, dry place and white phosphorus under water and away from heat

SOURCES
- ◆ *Material Safety Data Sheets (MSDS)*
- ◆ *Sax's Dangerous Properties of Industrial Materials*, 9th edition, 1996
- ◆ *Environmental Law Index to Chemicals*, 1996 edition
- ◆ *Merck Index*, 12th edition 1996
- ◆ *Hawley's Condensed Chemical Dictionary*, 12th edition, 1993
- ◆ *Aldrich Chemical Company Catalog*, 1996–1997

PHTHALIC ANHYDRIDE

CAS #: 85-44-9
MOLECULAR FORMULA: $C_6H_4(CO)_2O$
FORMULA WEIGHT: 148.12
SYNONYMS: anhydride phthalique (French), anidride ftalica (Italian), 1,2-benzenedicarboxylic acid anhydride, 1,3-dioxophthalan, ESEN, ftaalzuuranhydride (Dutch), ftalowy bezwodnik (Polish), 1,3-isobenzo-

furandione, 1,3-phthalandione, phthalic acid anhydride, Phthalsaeu-reanhydrid (German), retarder AK, retarder Esen, retarder PD

USES
- ◆ Used in the Manufacturing Process
  - phthaleins
  - phthalates
  - benzoic acid
  - synthetic indigo
  - artificial resins (glyptal)
  - primary amines
  - agricultural fungicide phthaltan
  - thalidomide
  - rubber retarder
  - alizarine
  - xanthene
  - anthraquinone
  - rhodamine
- ◆ Used as a Chemical Intermediate
  - many phthalate esters
  - plasticizers in synthetic resins
  - dioctyl phthalates
- ◆ Used as a Precursor
  - anthraquinone
  - phthalein
  - rhodamine
  - phthalocyanine
  - fluorescein
  - xanthene dyes

PHYSICAL PROPERTIES
- white, lustrous needles
- mild characteristic choking odor
- odor threshold air 0.053 μl/L
- melting point 130.8°C
- boiling point 295°C
- density 1.20 at 135°C
- surface tension 35.5 dynes/cm
- slightly soluble in water and ether
- soluble in alcohol
- vapor pressure 5.14×10^{-4} mm Hg at 25°C

CHEMICAL PROPERTIES
- flash point 305°F
- autoignition temperature 570°C
- flammability limits 1.7% lower and 10.4% upper
- heat of combustion 783.4 kcal
- heat of vaporization 13,919.0 gcal/gmol

HEALTH RISK
- ◆ Toxic Exposure Guidelines
  - OSHA PEL TWA 1 ppm
  - ACGIH TLV TWA 1 ppm
  - DFG MAK 1 mg/m^3 as total dust
  - MSHA standard 12 mg/m^3
- ◆ Toxicity Data
  - skin-rabbit 500 mg/24H MLD
  - eye-rabbit 100 mg SEV
  - intraperitoneal-mouse TDLo 203 mg/kg
  - oral-rat TDLo 309 g/kg/13W-I
  - oral-rat TDLo 63,700 mg/kg/1Y-C
  - oral-rat LD$_{50}$ 4020 mg/kg
  - oral-mouse LD$_{50}$ 1500 mg/kg
  - oral-cat LD$_{50}$ 800 mg/kg
  - oral-guinea pig LDLo 100 mg/kg

- ◆ Acute Risks
  - • eye irritation
  - • respiratory tract irritation
  - • skin irritation
  - • sneezing
  - • acute nose pain
  - • bloody nasal discharge
  - • coughing
  - • abdominal pain
  - • diarrhea
  - • vomiting
  - • dermatitis
- ◆ Chronic Risks
  - • conjunctivitis
  - • irritation of skin
  - • irritation of mucous membranes of respiratory tract
  - • bloody sputum
  - • emphysema
  - • bronchitis
  - • bronchial asthma
  - • lowering of blood pressure
  - • central nervous system excitation
  - • nasal congestion in animals
  - • injury of lung cells in animals

HAZARD RISK
- ◆ combustible when exposed to heat or flame
- ◆ can react with oxidizing materials
- ◆ moderate explosion hazard in the form of dust when exposed to flame
- ◆ Production has caused many industrial explosions.
- ◆ Mixtures with copper oxide or sodium nitrite explode when heated.
- ◆ violent reaction with nitric acid + sulfuric acid above 80°C
- ◆ may generate electrostatic charges
- ◆ can ignite by electric sparks

EXPOSURE ROUTES
- ◆ manufacture of phthalate-derived products
- ◆ plastics from which phthalate plasticizers are leached
- ◆ blood bags
- ◆ plastic syringes
- ◆ plastic tubing
- ◆ Phthalate esters are environmental pollutants.

REGULATORY STATUS/INFORMATION
- ◆ Clean Air Act Amendment, Title III: hazardous air pollutants
- ◆ Resource Conservation and Recovery Act
  - • list of hazardous constituents. 40CFR261, 40CFR261.11
  - • reference air concentrations. 40CFR266
  - • U waste # 190
- ◆ CERCLA and Superfund
  - • designation of hazardous substances. 40CFR302.4
  - • list of chemicals. 40CFR372.65
- ◆ OSHA 29CFR1910.1000 Table Z1
- ◆ chemical regulated by the State of California

ADDITIONAL/OTHER COMMENTS
- ◆ Symptoms
  - • labored breath, coughing, sore throat, shortness of breath, irritation

- skin contact: pain, redness, caustic, burns
- ingestion: vomiting, diarrhea, sore throat, cramps, abdominal pain
◆ First Aid
  - wash eyes immediately with large amounts of water
  - wash skin immediately with soap and water
  - remove to fresh air immediately or provide respiratory apparatus if necessary
  - drink water
◆ fire fighting: use carbon dioxide or dry chemical
◆ personal protection: compressed air/oxygen apparatus, gastight suit, fireproof suit

SOURCES

◆ *Health Effects Notebook for Hazardous Air Pollutants*, U.S. Environmental Protection Agency, December 1994
◆ *Material Safety Data Sheets (MSDS)*
◆ *Sax's Dangerous Properties of Industrial Materials*, 9th edition, 1996
◆ *Environmental Law Index to Chemicals*, 1996 edition
◆ *Merck Index*, 12th edition, 1996
◆ *Hawley's Condensed Chemical Dictionary*, 12th edition, 1993
◆ *Environmental Contaminant Reference Databook*, volume 1, 1995
◆ *Aldrich Chemical Company Catalog*, 1996–1997

POLYCHLORINATED BIPHENYLS

CAS #: 1336-36-3
MOLECULAR FORMULA: Polychlorinated biphenyls are a class of industrial chemical that contains 209 individual compounds.
FORMULA WEIGHT: Polychlorinated biphenyls have many molecular weights. The molecular weight for one particular PCB (Aroclor 1260) is 375.7.
SYNONYMS: aroclor, aroclor 1016, aroclor 1221, aroclor 1232, aroclor 1242, aroclor 1248, aroclor 1254, aroclor 1260, aroclor 1262, aroclor 1268, aroclor 2565, aroclor 4465, aroclor 5442, biphenyl, polychloro-, chlophen, chlorextol, chlorinated biphenyl, chlorinated diphenyl, chlorinated diphenylene, chloro biphenyl, chloro 1,1-biphenyl, clophen, dykanol, fenclor, fenclor 42, intereen, kanechlor, kanechlor 300, kanechlor 400, montar, noflamol, PCB, PCBs, phenochlor, phenoclor, polychlorobiphenyl, pyralene, pyranol, santotherm, santotherm FR, solvol, therminol FR-1

USES

◆ Used in Industry
  - hydraulic systems
  - heat transfer systems
  - microscopy mounting medium
  - research and development
  - electrical capacitors
  - electrical transformers
  - vacuum pumps
  - gas transmission turbines
  - enzyme inducers

- improving water-
  proof characteristics
  of surface coatings
- carbonless copy paper
- printing inks
- plasticizers
- special adhesives

- lubricating additives
- vacuum pump fluids
- wax extenders
- dedusting agents
- pesticide extenders
- cutting oils
- fire retardants

PHYSICAL PROPERTIES

- ◆ Lower chlorinated aroclors are colorless mobile oils.
- ◆ Increasing chlorine content results in viscous liquids or sticky resins.
- ◆ Aroclors 1268 and 1270 are white powders.
- ◆ melting point 340–375°C
- ◆ density 1.44 at 30°C

- ◆ extremely low solubility in water
- ◆ soluble in oils and organic solvents
- ◆ Vapor pressure of aroclor 1260 is 4.05×10^{-5} mm Hg at 25°C.
- ◆ log K_{ow} for aroclor 1260 is 6.8.

CHEMICAL PROPERTIES

- ◆ very inert
- ◆ stable to conditions of hydrolysis and oxidation in industrial use

HEALTH RISK

- ◆ Toxic Exposure Guidelines
  - OSHA PEL (chlorodiphenyl, 42% chlorine) 1 mg/m^3
  - ACGIH TLV (chlorodiphenyl, 42% chlorine) 1 mg/m^3
  - OSHA PEL (chlorodiphenyl, 54% chlorine) 0.5 mg/m^3
  - ACGIH TLV (chlorodiphenyl, 54% chlorine) 0.5 mg/m^3
  - NIOSH REL TWA 0.001 mg/m^3
  - DFG MAK suspected carcinogen
- ◆ Toxicity Data
  - intraperitoneal-rat TDLo 700 mg/kg
  - oral-rat TDLo 16,800 mg/kg/2Y-C:ETA
  - oral-mouse TDLo 1250 mg/kg/25W-I:CAR
  - oral-rat TD 1250 mg/kg/25W-I-CAR
  - oral-mouse LD$_{50}$ 1900 mg/kg
- ◆ Acute Risks
  - strong irritant
  - the higher the chlorine content, the more toxic
  - acne from skin contact

  - effects on liver, kidney, and central nervous system in animals
- ◆ Chronic Risks
  - respiratory tract symptoms
  - cough
  - tightness of chest
  - gastrointestinal effects
  - anorexia
  - weight loss
  - nausea
  - vomiting

  - abdominal pain
  - mild liver effects
  - effects on skin and eyes
  - chloracne
  - skin rashes
  - eye irritation
  - cardiovascular effects
  - hypertension
  - pigmentation of skin

- acne
- lower birth weights
- shortened gestational age
- developmental effects
- motor deficits at birth
- impaired psychomotor index in infants
- impaired visual recognition in infants
- deficits in short-term memory in infants
- learning deficits in offspring of animals
- impaired immune functions in offspring of animals
- focal liver necrosis, cellular alterations of the thyroid in the offspring of animals
- decreased fertility and conception and prolonged menstruation in animals
- suggestive evidence of liver cancer
- increase in liver tumors in rats and mice
- EPA Group B2: probable human carcinogens

HAZARD RISK
- ◆ combustible when exposed to heat or flame
- ◆ Decomposition emits highly toxic fumes of Cl^-.

EXPOSURE ROUTES
- ◆ redistribution of PCBs in soil and water
- ◆ disposal sites containing transformers, capacitors, and other PCB wastes
- ◆ incineration of PCB-containing wastes
- ◆ improper disposal of the compounds to open areas
- ◆ indoor air
- ◆ fluorescent lighting ballasts
- ◆ wastewater
- ◆ drinking water
- ◆ food, primarily in fish

REGULATORY STATUS/INFORMATION
- ◆ Clean Air Act Amendment, Title III: hazardous air pollutants
- ◆ Safe Drinking Water Act
  - 83 contaminants required to be regulated. 47FR9352
  - organic chemicals other than total trihalomethanes, sampling and analytical requirements. 40CFR141.24
  - public notification. 40CFR141.32
  - maximum contaminant level goal for organic constituents. 40CFR141.50(a)
  - maximum contaminant level for organic chemicals. 40CFR141.61
  - variances and exemptions from the maximum contaminant levels for organic and inorganic chemicals. 40CFR142.62
- ◆ Resource Conservation and Recovery Act
  - list of hazardous constituents. 40CFR258, 40CFR261, 40CFR261.11
  - groundwater monitoring chemicals list. 40CFR264
  - risk specific doses. 40CFR266
  - health based limits for exclusion of waste-derived residues. 40CFR266
- ◆ CERCLA and Superfund
  - designation of hazardous substances. 40CFR302.4
  - list of chemicals. 40CFR372.65

438

- ◆ Clean Water Act
  - • designation of hazardous substances. 40CFR116.4
  - • determination of reportable quantities. 40CFR117.3
  - • toxic pollutants. 40CFR401.15
  - • total toxic organics (TTOs). 40CFR413.02
- ◆ chemical regulated by the State of California

ADDITIONAL/OTHER COMMENTS
- ◆ symptoms: liver damage and skin changes
- ◆ First Aid
  - • wash eyes immediately with large amounts of water
  - • wash skin immediately with soap and water
  - • remove to fresh air immediately or provide respiratory apparatus if necessary
  - • drink water and induce vomiting
- ◆ fire fighting: use agent suitable for surrounding fire
- ◆ personal protection: compressed air/oxygen apparatus, gas-tight suit

SOURCES
- ◆ *Health Effects Notebook for Hazardous Air Pollutants*, U.S. Environmental Protection Agency, December 1994
- ◆ *Material Safety Data Sheets (MSDS)*
- ◆ *Sax's Dangerous Properties of Industrial Materials*, 9th edition, 1996
- ◆ *Environmental Law Index to Chemicals*, 1996 edition
- ◆ *Hawley's Condensed Chemical Dictionary*, 12th edition, 1993
- ◆ *Environmental Contaminant Reference Databook*, volume 1, 1995

POLYCYCLIC ORGANIC MATTER

CAS #: There are various CAS #s for polycyclic organic matter. The CAS # for benzo(a)pyrene is 50-32-8.

MOLECULAR FORMULA: Polycyclic organic matter is made up of chemicals formed during the incomplete burning of coal, gas, wood, or other organic substances. The molecular formula for benzo(a)pyrene is $C_{20}H_{12}$.

Note: The 1990 Amendments to the Clean Air Act state that the hazardous air pollutant "polycyclic organic matter" "includes organic compounds with more than one benzene ring, and which have a boiling point greater than or equal to 100°C."

FORMULA WEIGHT: There are various molecular weights for polycyclic organic matter. The molecular weight for benzo(a)pyrene is 252.3.

SYNONYMS: benzo(a)pyrene: benzo[def]chrysene, 3,4-benzopirene (Italian), 3,4-benzopyrene, 6,7-benzopyrene, 3,4-Benzpyren (German), 3,4-benzpyrene, benz[a]pyrene, 3,4-benzypyrene, BP, B[a]P

USES
- ◆ The majority of the polycyclic organic compounds have no commercial uses.
- ◆ A Few Are Used in Industry
  - • medicines
  - • dyes

- plastics
- pesticides
- asphalt used in road construction

PHYSICAL PROPERTIES (Polycyclic Aromatic Hydrocarbons)
- ◆ colorless, white, or pale yellow-green solids
- ◆ faint, pleasant odor

PHYSICAL PROPERTIES (Benzo(a)pyrene)
- melting point 179°C
- boiling point 310–312°C
- density 1.351 g/cm³
- vapor pressure 5.6×10^{-9} mm Hg at 25°C
- solubility in water 3.8×10^{-3} g/L
- sparingly soluble in ethanol and methanol
- soluble in benzene, toluene, xylene, and ether
- log K_{ow} 6.06
- log K_{oc} 6.74
- Henry's Law constant 4.9×10^{-7} atm-m³/mol

CHEMICAL PROPERTIES
- ◆ There is no information available on the chemical properties of polycyclic organic matter.

HEALTH RISK
- ◆ Toxic Exposure Guidelines
  - ACGIH TLV (phenanthrene, cyclohexane extractable fraction) 1030 mg/m³
  - OSHA PEL (coal tar pitch volatiles-benzene soluble) 0.2 mg/m³
  - ACGIH TLV (coal tar pitch volatiles-benzene soluble) 0.2 mg/m³
  - NIOSH REL (coal tar pitch volatiles, benzo(a)pyrene) 0.1 mg/m³
- ◆ Toxicity Data (Benzo(a)pyrene)
  - skin-mouse 14 µg MLD
  - dnd-sal:spr 3 g/L
  - dnd-human:oth 1500 nmol/L
  - msc-human:oth 100 nmol/L
  - intraperitoneal-rat TDLo 60 mg/kg
  - oral-mouse TDLo 1280 mg/kg
  - oral-rat TDLo 15 mg/kg:CAR
  - intraperitoneal-rat TDLo 16 mg/kg:ETA
  - subcutaneous-rat TDLo 455 µg/kg/60D-I:NEO
  - intramuscular-rat TDLo 2400 µg/kg:CAR
  - subcutaneous-rat LD_{50} 50 mg/kg
  - intraperitoneal-mouse LDLo 500 mg/kg
  - irn-frog LDLo 11 mg/kg
- ◆ Acute Risks
  - There is no information on the acute risks in humans.
  - enzyme alterations in the mucosa of the gastrointestinal tract in animals
  - increases in liver weights in animals
  - high acute toxicity in rats from oral exposure
- ◆ Chronic Risks (Benzo(a)pyrene)
  - dermatitis
  - photosensitization in sunlight

- irritation of the eyes
- cataracts
- effects on the blood in animals
- effects on the liver in animals
- effects on the immune system in animals
- induces reproductive toxicity in animals
- reduced incidence of pregnancy and decreased fertility in animals
- reduced viability of litters in animals
- reduced mean pup weight in animals
- increase in lung cancer in humans exposed to coke oven emissions, roofing tar emissions, and cigarette smoke
- EPA Group B2: probable human carcinogen
- respiratory tract tumors in animals
- forestomach tumors, leukemia, and lung tumors in animals

HAZARD RISK (Benzo(a)pyrene)
- ◆ Decomposition emits acrid smoke and fumes.

EXPOSURE ROUTES (Polycyclic Organic Matter)
- ◆ ambient air
- ◆ cigarette smoke
- ◆ vehicle exhaust
- ◆ asphalt roads
- ◆ coal
- ◆ coal tar
- ◆ agricultural burning
- ◆ residential wood burning
- ◆ hazardous waste sites
- ◆ drinking water supplies
- ◆ coal tar production plants
- ◆ coking plants
- ◆ coal-gasification sites
- ◆ smokehouses
- ◆ municipal trash incinerators

REGULATORY STATUS/INFORMATION
- ◆ Clean Air Act Amendment, Title III: hazardous air pollutants
- ◆ CERCLA and Superfund
  - designation of hazardous substances. 40CFR302.4

ADDITIONAL/OTHER COMMENTS
- ◆ symptoms: irritation

SOURCES
- ◆ *Health Effects Notebook for Hazardous Air Pollutants*, U.S. Environmental Protection Agency, December 1994
- ◆ *Material Safety Data Sheets (MSDS)*
- ◆ *Sax's Dangerous Properties of Industrial Materials*, 9th edition, 1996
- ◆ *Environmental Law Index to Chemicals*, 1996 edition
- ◆ *Merck Index*, 12th edition, 1996
- ◆ *Toxicological Profile for Polycyclic Aromatic Hydrocarbons (PAHs)*

1,3-PROPANE SULTONE (kjc)

CAS #: 1120-71-4
MOLECULAR FORMULA: $C_3H_6O_3S$
FORMULA WEIGHT: 122.15
SYNONYMS: 3-hydroxy-1-propanesulfonic acid gamma-sultone, 3-hydroxy-1-propanesulphonic acid sulfone, 3-hydroxy-1-propanesulphon-

ic acid sultone, 1,2-oxathrolane 2,2-dioxide, 1-propanesulfonic acid-3-hydroxy-gamma-sultone, propane sultone (ACGIH), 1,3-propane sultone, propanesultone

USES

- ◆ fungicides
- ◆ insecticides
- ◆ cation-exchange resins
- ◆ dyes
- ◆ vulcanization accelerator

- ◆ used as an intermediate to introduce sulfopropyl groups into molecules
- ◆ used to confer water solubility and ionic character

PHYSICAL PROPERTIES

- ◆ white to light-brown liquid
- ◆ melting point 31.11°C (88°F)
- ◆ boiling point 180°C (356°F) at 30 mm Hg
- ◆ density 1.392 g/mL at 25°C
- ◆ specific gravity 1.392 (water = 1.00)
- ◆ vapor density not available
- ◆ surface tension not available
- ◆ viscosity not available
- ◆ heat capacity not available

- ◆ heat of vaporization not available
- ◆ foul odor
- ◆ odor threshold not established
- ◆ vapor pressure not available
- ◆ solubility not available
- ◆ soluble in water and organic solvents
- ◆ log octanol/water partition coefficient not available

CHEMICAL PROPERTIES

- ◆ volatile (sublimes readily)
- ◆ combustible
- ◆ nonstaining
- ◆ noncorrosive
- ◆ reacts vigorously with strong oxidizing agents, strong bases, and strong acids
- ◆ flash point 109°C (228°F)

- ◆ flammability limits not available
- ◆ autoignition temperature not available
- ◆ heat of combustion not available
- ◆ enthalpy of formation not available

HEALTH RISK

- ◆ Toxic Exposure Guidelines
  - • not available
- ◆ Toxicity Data
  - • skin-rabbit 500 mg MLD
  - • intravenous-rat LDLo 80 mg/kg (15D preg):TER
  - • oral-rat LDLo 7840 mg/kg/60W-I:CAR
  - • subcutaneous-rat LDLo 10 mg/kg:ETA
  - • intravenous-rat LDLo 20 mg/kg (15D preg):CAR,TER
  - • skin-mouse LDLo 1000 mg/kg:CAR
  - • subcutaneous-mouse LDLo 756 mg/kg/63W-I:CAR
  - • subcutaneous-rat LD_{50} 135 mg/kg
  - • skin-mouse LDLo 1000 mg/kg
  - • intraperitoneal-mouse LD_{50} 467 mg/kg
- ◆ Other Toxicity Indicators
  - • California no-significant risk level: 0.3 mcg/d
- ◆ Acute Risks
  - • allergic skin reaction

- extremely destructive to tissue of the mucous membranes, upper respiratory tract, eyes, and skin in high concentrations
◆ Chronic Risks
  - alters genetic material
  - damage to the peripheral and central nervous systems
  - bone marrow disorders
  - EPA Group B2: probable human carcinogen
  - target organs: ears, small intestine, large intestine

HAZARD RISK
◆ NFPA rating Health 2, Flammability 2, Reactivity 3, Special Hazards none

EXPOSURE ROUTES
◆ primarily by inhalation
◆ Absorption through skin and eyes is possible.
◆ ingestion
◆ during manufacture and use as a fungicide and insecticide
◆ in the production of vulcanized rubber materials; car tires

REGULATORY STATUS/INFORMATION
◆ Clean Air Act Amendments of 1990 Section 112 Statutory Air Pollutants
◆ Resource Conservation and Recovery Act
  - RCRA Hazardous Substances
  - RCRA Land Disposal Prohibitions: Scheduled Second Third Wastes
  - discarded commercial chemical products, off-specification species, container residues, and spill residues thereof
  - U waste #: U193
◆ CERCLA and Superfund
  - designation of hazardous substances. 40CFR302.4
  - list of chemicals. 40CFR372.65
  - Reportable Quantity (RQ): 10 lb
  - SARA Title III Section 313 Toxic Chemicals
  - CERCLA Hazardous Substances
◆ National Toxicology Program (NTP) Annual Report on Carcinogens
◆ Department of Transportation
  - hazard class/division: 3
  - identification number: not available
  - labels: not available
  - emergency response guide #: not available
◆ Emergency Planning and Community Right-to-Know Act
  - section 313 de minimis concentration: 0.1%

ADDITIONAL/OTHER COMMENTS
◆ Symptoms
  - inhalation: eye irritation
  - skin absorption: skin burns, mucous membrane irritation
◆ First Aid
  - wash eyes immediately with large amounts of water
  - wash skin immediately with water
  - provide respiratory support

- ◆ fire fighting: use dry chemical, foam, carbon dioxide, or water spray
- ◆ personal protection
  - wear self-contained breathing apparatus
  - do not wear contact lenses when working with this chemical
  - wear protective clothing
  - use only in chemical fume hood
  - avoid all contact
- ◆ Storage
  - keep tightly closed
  - moisture sensitive
  - store in cool, dry place

SOURCES

- ◆ *Environmental Containment Reference Databook*, volume 1, 1995
- ◆ *Environmental Law Index to Chemicals*, 1996 edition
- ◆ *Fire Protection Guide to Hazardous Materials*, 11th edition, 1994
- ◆ *Material Safety Data Sheets (MSDS)*
- ◆ *NIOSH Pocket Guide to Chemical Hazards*, June 1996
- ◆ *Sax's Dangerous Properties of Industrial Materials*, 9th edition, volume 2, 1996

beta-PROPIOLACTONE

CAS #: 57-57-8
MOLECULAR FORMULA: OCH_2CH_2CO
FORMULA WEIGHT: 72.07
SYNONYMS: betaprone, BPL, hydracrylic acid beta-lactone, 3-hydroxypropionic acid lactone, propanolide, propiolactone, 1,3-propiolactone, 3-propiolactone, beta-propionolactone, beta-propriolactone, beta-prolactone

USES

- ◆ Used in Medical Field
  - vaccines
  - tissue grafts
  - surgical instruments
  - enzymes
- ◆ Used as a Sterilant
  - blood plasma
  - water
  - milk
  - nutrient broth
  - vapor phase disinfectant in enclosed spaces
- ◆ Used against Viruses
  - vegetative bacteria
  - pathologic fungi
- ◆ Used as a Chemical Intermediate
  - organic synthesis

PHYSICAL PROPERTIES

- ◆ clear, colorless liquid
- ◆ pungent, slightly sweetish smell
- ◆ melting point −33°C
- ◆ boiling point 155°C
- ◆ specific gravity 1.1460

◆ vapor pressure 3.4 mm Hg at 25°C

◆ very soluble in water: 67 g/100 mL
◆ log K_{ow} 0.462

CHEMICAL PROPERTIES
◆ decomposition temperature 155°C
◆ reacts with alcohol
◆ flash point 75°C
◆ upper explosion limit 2.90
◆ stable when stored in glass at refrigeration temperature
◆ incompatible with strong oxidizing agents and strong bases
◆ combustible

HEALTH RISK
◆ Toxic Exposure Guidelines
  • ACGIH TLV TWA 0.5 ppm
  • OSHA carcinogen
  • DFG MAK animal carcinogen, suspected human carcinogen
◆ Toxicity Data
  • slt-dmg-oral 200 mmol/L
  • otr-human:fbr 28 µmol/L
  • oral-rat TDLo 2868 mg/kg/1Y-I:CAR
  • inhalation-rat TCLo 5 ppm/6H/30D-I:CAR
  • Intratracheal-rat TDLo 72 mg/kg/30W-I:ETA
  • subcutaneous-mouse TDLo 69 mg/kg/43W-I:NEO
  • intraperitoneal-mouse LD_{50} 405 mg/kg
◆ Acute Risks
  • severe irritation of eyes, nose, throat
  • irritation of respiratory tract
  • irritation of skin
  • blistering
  • burns of skin
  • permanent corneal opacification
  • burns of mouth and stomach
  • muscular spasms
  • respiratory difficulty
  • convulsions
  • death
  • liver and kidney tubular damage
◆ Chronic Risks
  • skin irritation
  • scarring in mice
  • hair loss in mice
  • squamous cell carcinomas of the forestomach in mice
  • lymphomas in mice
  • hepatomas in mice
  • skin tumors in animals
  • EPA Group B2: probable human carcinogen

HAZARD RISK
◆ combustible
◆ incompatible with strong oxidizing agents and strong bases
◆ hazardous decomposition products: carbon monoxide, carbon dioxide
◆ Decomposition emits acrid smoke and irritating fumes.

EXPOSURE ROUTES
◆ inhalation during manufacture and use
◆ dermal contact during manufacture and use

REGULATORY STATUS/INFORMATION
- ◆ Clean Air Act Amendment, Title III: hazardous air pollutants
- ◆ CERCLA and Superfund
  - • list of chemicals. 40CFR372.65
- ◆ OSHA 29CFR1910.1000 Table Z1 and specially regulated substances
- ◆ chemical regulated by the State of California

ADDITIONAL/OTHER COMMENTS
- ◆ Symptoms
  - • contact: irritation to skin, burns, blistering
  - • inhalation: irritation of eyes, nose, throat, and respiratory tract
- ◆ First Aid
  - • wash eyes immediately with large amounts of water
  - • flush skin with plenty of soap and water
  - • remove to fresh air immediately
- ◆ fire fighting: data on extinguishing media not available
- ◆ personal protection: wear self-contained breathing apparatus and protective clothing

SOURCES
- ◆ *Health Effects Notebook for Hazardous Air Pollutants*, U.S. Environmental Protection Agency, December 1994
- ◆ *Material Safety Data Sheets (MSDS)*
- ◆ *Sax's Dangerous Properties of Industrial Materials*, 9th edition, 1996
- ◆ *Hawley's Condensed Chemical Dictionary*, 12th edition, 1993
- ◆ *Aldrich Chemical Company Catalog*, 1996-1997
- ◆ *Environmental Law Index to Chemicals*, 1996 edition
- ◆ *Merck Index*, 12th edition, 1996

PROPIONALDEHYDE

CAS #: 123-38-6
MOLECULAR FORMULA: C_2H_5CHO
FORMULA WEIGHT: 58.09
SYNONYMS: aldehyde propionique (French), FEMA no. 2923, methylacetaldehyde, NCI-C61029, propanal, propionic aldehyde, propyl aldehyde, propylic aldehyde

USES
- ◆ Used in Industry
  - • manufacture of polyvinyl and other plastics
  - • synthesis of rubber chemicals
  - • disinfectant
  - • preservative
  - • medicinal and agricultural chemical preparations
- ◆ Used as a Chemical Intermediate
  - • propionic acid
  - • n-propyl alcohol
  - • trimethylol ethane
  - • polyvinyl acetals

PHYSICAL PROPERTIES
- ◆ colorless liquid
- ◆ suffocating, fruity odor
- ◆ odor threshold 1 ppm
- ◆ melting point −81°C

- melting point −81°C
- boiling point 48°C
- density 0.8071 at 20°C
- taste of cocoa, coffee
- surface tension 23.4 dynes/cm
- viscosity 0.47 cP at 760 mm Hg
- miscible with alcohol, ether, and water
- vapor pressure 317 mm Hg at 25°C
- vapor density 2.0

CHEMICAL PROPERTIES
- flash point 15–19°F
- autoignition temperature 405°F
- heat of combustion 434.1 kcal at 25°C
- heat of vaporization 211 Btu/lb
- flammability limits 2.6% lower and 16.1% upper in air

HEALTH RISK
- Toxic Exposure Guidelines
  - No information is available for toxic exposure guidelines
- Toxicity Data
  - skin-rabbit 500 mg open MLD
  - eye-rabbit 41 mg SEV
  - oral-rat LD_{50} 1410 mg/kg
  - inhalation-rat LCLo 8000 ppm/4H
  - subcutaneous-rat LD_{50} 820 mg/kg
  - oral-mouse LDLo 800 mg/kg
  - inhalation-mouse LC_{50} 21,800 mg/m^3/2H
  - subcutaneous-mouse LD_{50} 680 mg/kg
  - skin-rabbit LD_{50} 5040 mg/kg
  - inhalation-mammal LC_{50} 21,800 mg/m^3
  - oral-uns LD_{50} 1400 mg/kg
  - inhalation-uns LC_{50} 21,800 mg/m^3
- Acute Risks
  - irritation of eyes
  - irritation of nose
  - irritation of throat
  - nausea
  - vomiting
  - anesthesia in animals
  - liver damage in animals
  - increased blood pressure in animals
- Chronic Risks
  - No information is available on the chronic effects in humans or animals.

HAZARD RISK
- flammable liquid
- dangerous fire hazard when exposed to heat or flame
- reacts vigorously with oxidizers
- explosive in the form of vapor when exposed to heat or flame
- vigorous polymerization reaction with methyl methacrylate
- Decomposition emits acrid smoke and irritating fumes.

EXPOSURE ROUTES
- ambient air
- manufacturing facilities
- municipal waste incinerators
- combustion of wood, gasoline, diesel fuel, and polyethylenes

- ◆ tobacco smoke
- ◆ drinking water
- ◆ coffee

REGULATORY STATUS/INFORMATION
- ◆ Clean Air Act Amendment, Title III: hazardous air pollutants
- ◆ CERCLA and Superfund
  - • list of chemicals. 40CFR372.65
- ◆ chemical regulated by the State of California

ADDITIONAL/OTHER COMMENTS
- ◆ symptoms: irritation of eyes, nose, throat, nausea, vomiting
- ◆ First Aid
  - • wash eyes immediately with large amounts of water
  - • wash skin immediately with soap and water
  - • remove to fresh air immediately or provide respiratory apparatus if necessary
  - • drink water
- ◆ fire fighting: use alcohol foam, carbon dioxide, or dry chemical
- ◆ personal protection: compressed air/oxygen apparatus, gastight suit, fireproof suit

SOURCES
- ◆ *Health Effects Notebook for Hazardous Air Pollutants*, U.S. Environmental Protection Agency, December 1994
- ◆ *Material Safety Data Sheets (MSDS)*
- ◆ *Sax's Dangerous Properties of Industrial Materials*, 9th edition, 1996
- ◆ *Environmental Law Index to Chemicals*, 1996 edition
- ◆ *Merck Index*, 12th edition, 1996
- ◆ *Hawley's Condensed Chemical Dictionary*, 12th edition, 1993
- ◆ *Environmental Contaminant Reference Databook*, volume 1, 1995
- ◆ *Aldrich Chemical Company Catalog*, 1996–1997

PROPOXUR

CAS #: 114-26-1
MOLECULAR FORMULA: $(CH_3)_2CHOC_6H_4OOCNHCH_3$
FORMULA WEIGHT: 209.27
SYNONYMS: aprocarb, bay 9010, bayer 39007, baygon, bifex, blattanex, boygon, ENT 25,671, o-IMPC, invisigard, isocarb, o-isopropoxyphenyl methylcarbamate, o-isopropoxyphenyl-n-methylcarbamate, 2-isopropoxyphenyl-n-methylcarbamate, 2-(1-methylethoxy) phenol methylcarbamate, n-methyl-2-isopropoxyphenylcarbamate, OMS-33, PHC, propoksuru (Polish), propyon, sendran, suncide, tugon Fliegenkugel, unden

USES
- ◆ Used as a Nonfood Insecticide
  - • control cockroaches
  - • control flies
  - • control lawn and turf insects
  - • malaria control activities
  - • flea collars for pets

PHYSICAL PROPERTIES

- white to tan crystalline powder
- faint characteristic odor
- melting point 84–87°C
- vapor pressure 0.1 mm Hg at 120°C
- relatively insoluble in cold water
- soluble in hot water and in all polar organic solvents
- log K_{ow} 0.14

CHEMICAL PROPERTIES

- decomposes at high temperature forming methyl isocyanate
- unstable in high alkaline media
- stable under normal use conditions

HEALTH RISK

- Toxic Exposure Guidelines
  - OSHA PEL TWA 0.5 mg/m³
  - ACGIH TLV TWA 0.5 mg/m³
  - DFG MAK 2 mg/m³
  - NIOSH REL 0.5 mg/m³
- Toxicity Data
  - mmo-sat 250 ng/plate
  - cyt-hamster-intraperitoneal 250 mg/kg
  - oral-rat TLDo 1600 mg/kg
  - oral-mouse TDLo 600 mg/kg
  - oral-woman LDLo 24 mg/kg
  - oral-rat LD_{50} 70 mg/kg
  - inhalation-rat LC_{50} 1440 mg/m³/1H
  - skin-rat LD_{50} 800 mg/kg
  - intraperitoneal-rat LD_{50} 30 mg/kg
  - subcutaneous-rat LD_{50} 56 mg/kg
  - intravenous-rat LD_{50} 11 mg/kg
  - ims-rat LD_{50} 53 mg/kg
  - oral-mouse LD_{50} 23,500 μg/kg
- Acute Risks
  - cholinesterase inhibition of red blood cells
  - blurred vision
  - nausea
  - vomiting
  - sweating
  - tachycardia
  - cholinesterase depression in rats
- Chronic Risks
  - depressed cholinesterase levels
  - headaches
  - vomiting
  - nausea
  - depressed body weight in animals
  - increased liver weight in animals
  - effects to the bladder in animals
  - slight increase in neuropathy in animals
  - tumors of bladder in rats
  - tumors of uterus in rats
  - EPA Group B2: probable human carcinogen

HAZARD RISK

- decomposes on heating
- Decomposition emits toxic fumes of NO$_x$.

EXPOSURE ROUTES
♦ primarily dermal exposure
♦ inhalation
♦ manufacture, formation, and application of insecticide
♦ surface water

REGULATORY STATUS/INFORMATION
♦ Clean Air Act Amendment, Title III: hazardous air pollutants
♦ Resource Conservation and Recovery Act
  • list of hazardous constituents. 40CFR261, 40CFR261.11
  • U waste # 411
♦ CERCLA and Superfund
  • list of chemicals. 40CFR372.65
♦ OSHA 29CFR1910.1000 Table Z1
♦ chemical regulated by the State of California

ADDITIONAL/OTHER COMMENTS
♦ Symptoms
  • skin: prickling
  • eye: prickling
  • oral: narrowing pupil, prickling, drowsiness, vomiting, diarrhea, cramps, headache
♦ First Aid
  • wash eyes and skin immediately with soap and water
  • remove to fresh air immediately or provide respiratory apparatus if necessary
  • drink water and induce vomiting
♦ personal protection: compressed air/oxygen apparatus, gastight suit

SOURCES
♦ *Health Effects Notebook for Hazardous Air Pollutants*, U.S. Environmental Protection Agency, December 1994
♦ *Material Safety Data Sheets (MSDS)*
♦ *Sax's Dangerous Properties of Industrial Materials*, 9th edition, 1996
♦ *Environmental Law Index to Chemicals*, 1996 edition
♦ *Merck Index*, 12th edition, 1996
♦ *Hawley's Condensed Chemical Dictionary*, 12th edition, 1993

PROPYLENE DICHLORIDE

CAS #: 78-87-5
MOLECULAR FORMULA: $CH_3CHClCH_2Cl$
FORMULA WEIGHT: 112.99
SYNONYMS: bichlorure de propylene (French), alpha,beta-dichloropropane, 1,2-dichloropropane, dwuchloropropan (Polish), ENT 15,406, NCI-C55141, propylene chloride, alpha,beta-propylene dichloride

USES
♦ Used as a Chemical Intermediate
  • rubber processing   • perchloroethylene

- carbon tetrachloride

◆ Used in Industry
  - livestock
  - dry cleaning fluids
  - degreasing
  - insecticidal fumigant
  - mixtures
  - lead scavenger for anti-knock fluids
  - spotting agent
◆ Used as a Solvent
  - plastics
  - resins
  - metal industries
  - oils and fats

- chlorinated organic chemicals
- ion exchange manufacture
- toluene diisocyanate production
- photographic film manufacture
- paper coating
- petroleum catalyst regeneration

- waxes
- gums
- cellulose esters
- ethers

PHYSICAL PROPERTIES
◆ colorless liquid
◆ sweet, chloroform-like odor
◆ odor threshold 0.25 ppm
◆ melting point $-100.4°C$
◆ boiling point 96.4°C
◆ density 1.159 at 25°C
◆ surface tension 29 dynes/cm
◆ vapor pressure 49.67 mm Hg at 25°C

◆ vapor density 3.9
◆ critical temperature 304°C
◆ critical pressure 42.73 atm
◆ slightly soluble in water
◆ miscible with organic solvents
◆ $\log K_{ow}$ 1.99

CHEMICAL PROPERTIES
◆ flammable
◆ sensitive to heat
◆ flash point 60°F
◆ autoignition temperature 1035°F

◆ flammability limits 3.4% lower and 14.5% upper
◆ heat of combustion -7300 Btu/lb
◆ heat of vaporization 8428.5 gcal/gmole

HEALTH RISK
◆ Toxic Exposure Guidelines
  - OSHA PEL TWA 75 ppm
  - OSHA PEL STEL 110 ppm
  - ACGIH TLV TWA 75 ppm
  - ACGIH TLV STEL 110 ppm
  - DFG MAK suspected carcinogen
◆ Toxicity Data
  - eye-rabbit 500 mg MLD
  - mmo-sat 100 µg/plate
  - mma-sat 333 µg/plate
  - oral-mouse TDLo 130 g/kg/2Y-I:CAR
  - oral-rat LD_{50} 1947 mg/kg
  - inhalation-rat LC_{50} 14 g/m^3/8H
  - oral-mouse LD_{50} 860 mg/kg
  - oral-rat LD_{50} 2196 mg/kg

- inhalation-mouse LCLo 1000 ppm/2H
- oral-dog LDLo 5000 mg/kg
- skin-rabbit LD_{50} 8750 mg/kg
- oral-guinea pig LD_{50} 2000 mg/kg

◆ Acute Risks
  - effects on gastrointestinal system
  - effects on blood
  - effects on liver
  - effects on kidneys
  - effects on central nervous system
  - effects on the lung
  - chest discomfort
  - dyspnea
  - cough
  - conjunctival hemorrhages
  - dermatitis
  - drowsiness
  - light-headedness

◆ Chronic Risks
  - No information is available on the chronic effects in humans.
  - effects on the respiratory system in animals
  - effects on the blood in animals
  - effects on the liver in animals
  - effects on the central nervous system in animals
  - increased incidence of mammary gland tumors in rats
  - liver tumors in rats
  - EPA Group B2: probable human carcinogen

HAZARD RISK
◆ flammable liquid
◆ very dangerous fire hazard when exposed to heat or flame
◆ reacts with aluminum to form aluminum chloride which can lead to explosions
◆ can react vigorously with oxidizing materials
◆ Decomposition emits toxic fumes of Cl^-.

EXPOSURE ROUTES
◆ ambient air
◆ drinking water
◆ groundwater
◆ production of propylene dichloride
◆ use in chemical reactions or as a solvent
◆ evaporation from wastewater containing the chemical

REGULATORY STATUS/INFORMATION
◆ Clean Air Act Amendment, Title III: hazardous air pollutants
◆ Resource Conservation and Recovery Act
  - constituents for detection monitoring. 40CFR258
  - list of hazardous constituents. 40CFR258, 40CFR261, 40CFR261.11
  - U waste # 083
◆ CERCLA and Superfund
  - designation of hazardous substances. 40CFR302.4
◆ chemical regulated by the State of California

ADDITIONAL/OTHER COMMENTS
◆ symptoms: contact with eyes or skin causes irritation

- ◆ First Aid
  - wash eyes and skin immediately with large amounts of water
  - remove to fresh air immediately or provide respiratory apparatus if necessary
  - drink water or milk
- ◆ fire fighting: use water, foam, carbon dioxide, or dry chemical
- ◆ personal protection: compressed air/oxygen apparatus, gastight suit, fireproof suit

SOURCES
- ◆ *Health Effects Notebook for Hazardous Air Pollutants*, U.S. Environmental Protection Agency, December 1994
- ◆ *Material Safety Data Sheets (MSDS)*
- ◆ *Sax's Dangerous Properties of Industrial Materials*, 9th edition, 1996
- ◆ *Environmental Law Index to Chemicals*, 1996 edition
- ◆ *Merck Index*, 12th edition, 1996
- ◆ *Hawley's Condensed Chemical Dictionary*, 12th edition, 1993
- ◆ *Environmental Contaminant Reference Databook*, volume 1, 1995
- ◆ *Aldrich Chemical Company Catalog*, 1996–1997

PROPYLENE OXIDE

CAS #: 75-56-9
MOLECULAR FORMULA: OCH_2CHCH_3
FORMULA WEIGHT: 58.09
SYNONYMS: epoxypropane, 1,2-epoxypropane, 2,3-epoxypropane, methyl ethylene oxide, methyl oxirane, NCI-C50099, oxyde de propylene (French), propene oxide, propylene epoxide, 1,2-propylene oxide

USES
- ◆ Used in Industry
  - sterilization of packaged food products in fumigation chambers
  - herbicide
  - preparation of lubricants
  - surfactants
  - oil demulsifiers
  - solvent
  - fumigant
  - soil sterilant
  - stabilizer for dichloromethane
  - detergents
  - component in brake fluids
  - microbicide
  - insecticide
  - miticide
  - bacteriostat
  - fungicide
- ◆ Used as a Chemical Intermediate
  - polyurethane polyols
  - propylene glycol
  - dipropylene glycol
  - glycol ethers
  - nonpolyurethane polyols
  - isopropanolamines
  - propoxylated surfactants
  - propylene carbonate
  - hydroxypropylated cellulose
  - glycerol via allyl alcohol

PHYSICAL PROPERTIES

- colorless, ethereal liquid
- sweet, alcoholic, ether-like odor
- odor threshold 44 ppm
- melting point −112.13°C
- boiling point 34.23°C
- density 0.8304 at 20°C
- surface tension 24.5 dynes/cm
- viscosity 0.28 cP at 25°C
- vapor pressure 445 mm Hg at 20°C
- vapor density 2.0
- critical temperature 209.1°C
- critical pressure 48.6 atm
- very soluble in water
- soluble in alcohol and ether
- log K_{ow} −.13

CHEMICAL PROPERTIES

- extremely flammable
- flash point −35°F
- autoignition temperature 449°C
- flammability limits 2.3% lower and 36% upper
- heat of combustion −3,000 Btu/lb
- heat of vaporization 205 Btu/lb
- polymerizes exothermically

HEALTH RISK

- Toxic Exposure Guidelines
  - OSHA PEL TWA 20 ppm
  - ACGIH TLV TWA 20 ppm
  - MSHA standard 240 mg/m³
- Toxicity Data
  - skin-rabbit 415 mg open MOD
  - skin-rabbit 50 mg/6M SEV
  - eye-rabbit 20 mg SEV
  - eye-rabbit 20 mg/24H MOD
  - mmo-sat 350 µg/plate
  - sce-human:lym 25,000 ppm
  - inhalation-rat TCLo 500 ppm/7H
  - oral-rat TDLo 10,798 mg/kg/2Y-I:CAR
  - inhalation-mouse TCLo 400 ppm/6H/2Y-I:CAR
  - inhalation-rat TCLo 100 ppm/7H/2Y-I:NEO
  - subcutaneous-rat TDLo 1500 mg/kg/46W-I:ETA
  - oral-rat LD_{50} 380 mg/kg
  - inhalation-man TCLo 1400 g/m³/10M:CNS,GIT
  - inhalation-rat LCLo 4000 ppm/4H
  - intraperitoneal-rat LD_{50} 150 mg/kg
  - oral-mouse LD_{50} 440 mg/kg
  - inhalation-mouse LC_{50} 1740 ppm/4H
  - intraperitoneal-mouse LD_{50} 175 mg/kg
  - inhalation-dog LCLo 2005 ppm/4H
  - skin-rabbit LD_{50} 1245 mg/kg
  - oral-guinea pig LD_{50} 660 mg/kg
  - inhalation-guinea pig LCLo 4000 ppm/4H
- Acute Risks
  - eye irritation
  - respiratory tract irritation
  - coughing
  - dyspnea
  - pulmonary edema
  - pneumonia
  - skin irritation
  - necrosis

- mild central nervous system depressant
- headache
- motor weakness
- incoordination
- ataxia
- coma

◆ Chronic Risks
  - No information is available on the chronic effects in humans.
  - decrease in body weight gain in rodents
  - increased mortality in rodents
  - inflammatory lesions of the nasal cavity, trachea, and lungs in rodents
  - neuropathological changes in monkeys
  - neurotoxicity in rats
  - tumors at or near site of administration in rats
  - forestomach tumors in rodents
  - nasal tumors in rodents
  - EPA Group B2: probable human carcinogen

HAZARD RISK

◆ flammable liquid
◆ very dangerous fire and explosion hazard when exposed to heat or flame
◆ explosive reaction with epoxy resin and sodium hydroxide
◆ forms explosive mixtures with oxygen
◆ reacts with ethylene oxide + polyhydric alcohol to form the thermally unstable polyether alcohol
◆ incompatible with NH_4OH, chlorosulfonic acid, HCl, HF, HNO_3, oleum, and H_2SO_4
◆ can react vigorously with oxidizing materials
◆ Decomposition emits acrid smoke and fumes.

EXPOSURE ROUTES

◆ inhalation and dermal
◆ production of propylene oxide
◆ storage and transport of propylene oxide
◆ use of propylene oxide
◆ use and production as an intermediate or fumigant and soil sterilant
◆ contaminated air
◆ fumigated food products
◆ consumption of contaminated food

REGULATORY STATUS/INFORMATION

◆ Clean Air Act Amendment, Title III: hazardous air pollutants
◆ Federal Insecticide, Fungicide, and Rodenticide Act
  - tolerances for pesticides in food. 40CFR185
◆ CERCLA and Superfund
  - designation of hazardous substances. 40CFR302.4
  - list of extremely hazardous substances and their threshold planning quantities. 40CFR355-AB
  - list of chemicals. 40CFR372.65
◆ Clean Water Act
  - designation of hazardous substances. 40CFR116.4
  - determination of reportable quantities. 40CFR117.3
◆ TSCA: chemicals subject to test rules or consent orders for which the testing reimbursement period has passed. 40CFR799.18

- ◆ OSHA 29CFR1910.1000 Table Z1
- ◆ chemical regulated by the State of California

ADDITIONAL/OTHER COMMENTS

- ◆ Symptoms
  - inhalation: headache, nausea, vomiting, unconsciousness, mild depression of central nervous system, lung irritation
  - contact: very irritating to skin and eyes
- ◆ First Aid
  - wash eyes and skin immediately with large amounts of water
  - remove to fresh air immediately or provide respiratory apparatus if necessary
  - drink water
- ◆ fire fighting: use alcohol foam, carbon dioxide, or dry chemical
- ◆ personal protection: compressed air/oxygen apparatus, gastight suit, fireproof suit
- ◆ storage: keep away from heat and open flame

SOURCES

- ◆ *Health Effects Notebook for Hazardous Air Pollutants*, U.S. Environmental Protection Agency, December 1994
- ◆ *Material Safety Data Sheets (MSDS)*
- ◆ *Sax's Dangerous Properties of Industrial Materials*, 9th edition, 1996
- ◆ *Environmental Law Index to Chemicals*, 1996 edition
- ◆ *Merck Index*, 12th edition, 1996
- ◆ *Hawley's Condensed Chemical Dictionary*, 12th edition, 1993
- ◆ *Environmental Contaminant Reference Databook*, volume 1, 1995
- ◆ *Aldrich Chemical Company Catalog*, 1996–1997

1,2-PROPYLENIMINE (pf)

CAS #: 75-55-8
MOLECULAR FORMULA: C_3H_7N
FORMULA WEIGHT: 57.11
SYNONYMS: 2-methylazacyclopropane, 2-methylaziridine, 2-methyl aziridine, methylethylenimine, 2-methylethylenimine, propylene imine, 1,2-propylene imine, 1,2-propyleneimine, 1,2-propyleneimine (ACGIH:OSHA), propyleneimine (inhibited (DOT)), propylenimine, RCRA waste number PO67, UN1921 (DOT)

USES

- ◆ Chemical Intermediate
  - latex surface coating resins
  - textile industry
  - paper industry
  - dyes
  - photography
  - gelatins
  - oil additives, for viscosity control, high pressure performance, and oxidation resistance

- rubber industry
- pharmaceutical industry for medicines
- adhesive industry
- petroleum refining
- rocket propellant fuels
- agricultural chemicals

PHYSICAL PROPERTIES

- ◆ fuming, colorless, oily liquid
- ◆ melting point $-65°C$ ($-85°F$)
- ◆ boiling point $66-67°$ C ($150.8-152.6°$ F) at 760 mm Hg
- ◆ density $0.8039-0.8070$ g/cm^3 at 25°C
- ◆ specific gravity 0.808 (water =1.00)
- ◆ vapor density 2.0 (air = 1)
- ◆ surface tension not available
- ◆ viscosity not available
- ◆ heat capacity not available
- ◆ heat of vaporization 250 Btu/lb
- ◆ strong ammonia-like odor
- ◆ odor threshold intolerable at 14 ppm
- ◆ vapor pressure 112 mm at 20°C, 140 mm at 25°C
- ◆ soluble in water
- ◆ miscible with ethanol, chloroform, ether

CHEMICAL PROPERTIES

- ◆ flammable, corrosive
- ◆ polymerizes with exposure to acids or acid fumes
- ◆ hygroscopic liquid
- ◆ reacts vigorously with oxidizing materials
- ◆ incompatible with acid, acid chlorides, acid anhydrides, and sensitive to moisture
- ◆ flash point $-10°C$ ($14°F$)
- ◆ explosion limits in air not available
- ◆ autoignition temperature not available
- ◆ heat of combustion $-15,500$ Btu/lb
- ◆ enthalpy (heat) of formation not available

HEALTH RISK

- ◆ Toxic Exposure Guidelines
  - ACGIH TLV TWA 2 ppm (5 mg/m^3) (skin)
  - OSHA PEL TWA 2 ppm (5 mg/m^3) (skin), suspected human carcinogen
  - DFG MAK animal carcinogen, suspected human carcinogen
- ◆ Toxicity Data
  - eye-rabbit 250 µg open SEV
  - oral-rat TDLo 1120 mg/kg/28W-I:CAR
  - oral-rat TD 3920 mg/kg/27W-C:CAR
  - oral-rat LD$_{50}$ 19 mg/kg
  - inhalation-rat LCLo 500 ppm/4H
  - inhalation-guinea pig LCLo 500 ppm/1H
  - skin-guinea pig LD$_{50}$ 43 mg/kg

◆ Acute Risks
  • Inhalation causes severe irritation to the eyes and upper res-
    piratory tract, as well as headaches, dizziness, nausea, bron-
    chitis, shortness of breath, and edema of the lungs.
  • extremely destructive to tissue of the mucous membranes
  • Dermal exposure causes irritation and burns on human
    skin.
  • extreme acute toxicity from oral and dermal exposure in
    rats
  • Ingestion causes burns of the mouth and stomach.
  • may be fatal if inhaled, swallowed, or absorbed through the
    skin
◆ Chronic Risks
  • Inhalation and oral exposure affect the kidneys, blood, and
    gastrointestinal system of animals.
  • EPA has stated that there is inadequate data to establish a
    reference concentration.
  • reference dose not established by EPA
  • no information on reproductive or developmental effects
    on humans or animals
  • tumors reported in the mammary glands and intestine as
    well as leukemias in animals due to oral exposure
  • probable human carcinogen
  • implicated as brain carcinogen
  • mutation data reported and may cause heritable genetic
    damage
  • targets the blood, nerves, and ears
HAZARD RISK
  ◆ highly toxic and corrosive
  ◆ dangerous fire hazard
  ◆ Flashback along vapor trail may occur.
  ◆ Container explosion may occur under fire conditions.
  ◆ emits toxic fumes under fire conditions
  ◆ polymerizes explosively on exposure to acids or acid fumes
  ◆ Heating to decomposition also results in toxic substances
    of carbon monoxide, carbon dioxide, and nitrogen oxides.
  ◆ can react vigorously with oxidizing materials
  ◆ incompatible with acid, acid chlorides, acid anhydrides; sensi-
    tive to moisture
EXPOSURE ROUTES
  ◆ inhalation
  ◆ contact through skin or eyes
  ◆ ingestion
  ◆ due to spills, leaks, and venting during loading, transfer, and
    storage
  ◆ occupational exposure for workers in the paint and chemical
    industries
REGULATORY STATUS/INFORMATION
  ◆ NTP 7th Annual Report on Carcinogens
  ◆ IARC Cancer Review: Group 2B IMEMDT 7,56,87
  ◆ Animal Limited Evidence IMEMDT 9,61,75

- ◆ EPA Genetic Toxicology Program
- ◆ Community Right-to-Know list
- ◆ EPA Extremely Hazardous Substances list
- ◆ Reported in EPA TSCA Inventory
- ◆ EPA Office of Air Quality Planning and Standards, for a hazard ranking under section 112(g) of the Clean Air Act Amendments, has ranked the contaminant in the nonthreshold category.
- ◆ Resource Conservation and Recovery Act
  - discarded commercial chemical products, off-specification species, container residues, and spill residues thereof 40CFR261.33, 40CFR261.5(e)
- ◆ CERCLA and Superfund
  - list of extremely hazardous substances and their threshold planning quantities 40CFR355—Appendix B
  - chemicals and chemical categories to which this part applies 40CFR372.65
- ◆ OSHA, 29CFR1910.1000 Table Z1: limits for air contaminants under the Occupational Safety and Health Act
- ◆ chemical regulated by the State of California
- ◆ Department of Transportation
  - hazard class division: 3
  - identification number: UN 1921
  - labels: flammable, liquid

ADDITIONAL/OTHER COMMENTS

- ◆ Symptoms
  - inhalation: nose, eyes, and throat irritation, burning sensation, coughing, wheezing, laryngitis, shortness of breath, headaches, nausea, and vomiting caused in humans
  - Contact with skin causes burns that are slow to heal.
  - Ingestion will cause burns in the mouth and stomach.
  - may be fatal if inhaled, swallowed, or absorbed through the skin
- ◆ First Aid
  - if inhaled remove to fresh air, give artificial respiration if not breathing, or oxygen if breathing is difficult, and get medical attention
  - if contacted with the eyes, flush with water for at least 30 minutes and get medical attention
  - if contacted with skin flush with water and rinse with vinegar and water, while removing clothing; get medical attention if symptoms persist after washing
  - if swallowed drink large amounts of milk or water and seek immediate medical attention
- ◆ fire fighting: carbon dioxide or dry chemical
- ◆ personal protection: use in a hood with chemical-resistant gloves, boots, apron, goggles or a face shield, and a NIOSH approved vapor respirator if needed due to poor ventilation
- ◆ storage: store away from combustible materials and ignition sources, since container may explode under fire conditions

SOURCES
♦ *Material Safety Data Sheets (MSDS)*
♦ *Sax's Dangerous Properties of Industrial Materials*, 9th edition, 1996
♦ *Merck Index*, 12th edition, 1996
♦ *Hawley's Condensed Chemical Dictionary*, 12th edition, 1993
♦ *Environmental Law Index to Chemicals*, 1996 edition
♦ *Health Effects Notebook for Hazardous Air Pollutants*, U.S. Environmental Protection Agency, December 1994
♦ *Environmental Containment Reference Databook*, volume 1, 1995

QUINOLINE

CAS #: 91-22-5
MOLECULAR FORMULA: $C_6H_4N{=}CHCH{=}CH$
FORMULA WEIGHT: 129.17
SYNONYMS: 1-azanaphthalene, B-500, 1-benzazine, 1-benzine, benzo(b)pyridine, chinoleine, chinolin, chinoline, leucol, leucoline, leukol, USAF EK-218

USES
♦ Used in Industry
  • preparation of hydroxyquinoline sulfate
  • preparation of niacin
  • dyes
  • anatomical specimen
  • preservative
  • solvent for resins and terpenes
  • decarboxylation reagent
  • paints
  • antimalarial medication
  • flavoring
  • metallurgical processes
  • polymers and agricultural chemicals
  • corrosion inhibitor
♦ Used as a Chemical Intermediate
  • 8-hydroxyquinoline
  • fungistat
  • pharmaceuticals

PHYSICAL PROPERTIES
♦ colorless hygroscopic liquid
♦ darkens with age
♦ penetrating, pungent odor
♦ odor threshold 71 ppm
♦ melting point $-15°C$
♦ boiling point 237.7°C at 760 mm Hg
♦ density 1.0900 at 25°C
♦ surface tension 45 dynes/cm at 20°C
♦ viscosity 2.997 cP at 30°C
♦ vapor pressure 0.0091 mm Hg at 25°C
♦ vapor density 4.45
♦ critical temperature 509.15°C
♦ critical pressure 37.5 atm
♦ sparingly soluble in cold water
♦ easily soluble in hot water
♦ miscible with alcohol, ether, and carbon disulfide
♦ log K_{ow} 2.03

CHEMICAL PROPERTIES

- autoignition temperature 896°F
- heat of combustion −8710 cal/g
- heat of vaporization 86 cal/g
- may attack some forms of plastic
- volatile with steam
- forms water-soluble salts with strong acids

HEALTH RISK

- Toxic Exposure Guidelines
  - No information is available on the toxic exposure guidelines for Quinoline.
- Toxicity Data
  - skin-rabbit 10 mg/24H open MLD
  - eye-rabbit 250 μg open SEV
  - mma-sat 1 μmol/plate
  - dnd-esc 30 μmol/L
  - mma-hamster:ovr 80 μmol/L
  - sce-hamster:ovr 110 μg/L
  - oral-rat TDLo 7770 mg/kg/37W-C:NEO
  - oral-mouse TDLo 50 g/kg/30W-C:ETA
  - intraperitoneal-mouse TDLo 9042 μg/kg/15D-I:CAR
  - oral-rat LD_{50} 331 mg/kg
  - intraperitoneal-mouse LDLo 64 mg/kg
  - skin-rabbit LD_{50} 540 mg/kg
  - subcutaneous-rabbit LDLo 200 mg/kg
  - subcutaneous-frog LDLo 150 mg/kg
- Acute Risks
  - eye irritation
  - nose irritation
  - throat irritation
  - headaches
  - dizziness
  - nausea
  - coma
  - shortness of breath
- Chronic Risks
  - effects on liver
  - effects on kidneys
  - liver damage in rats
  - liver hemangioendotheliomas in rats
  - EPA Group C: possible human carcinogen

HAZARD RISK

- combustible when exposed to heat or flame
- Preparation has caused many industrial explosions.
- potentially explosive reaction with hydrogen peroxide
- violent reaction with dinitrogen tetraoxide and perchromates
- incompatible with seed oil + thionyl chloride and maleic anhydride
- unpredictably violent
- Decomposition emits toxic fumes of NO_x.

EXPOSURE ROUTES

- inhalation and ingestion of particulates
- dermal contact
- inhalation of ambient air
- petroleum refining
- coal mining
- quenching and coking
- release in shale oil
- synthetic coal conversion wastewaters

- ◆ wood preservative wastewaters
- ◆ underground coal gasification
- ◆ consumption
  of contaminated water
- ◆ cigarette smoke

REGULATORY STATUS/INFORMATION
- ◆ Clean Air Act Amendment, Title III: hazardous air pollutants
- ◆ CERCLA and Superfund
  - • designation of hazardous substances. 40CFR302.4
  - • list of chemicals. 40CFR372.65
- ◆ Clean Water Act
  - • designation of hazardous substances. 40CFR116.4
  - • determination of reportable quantities. 40CFR117.3
- ◆ chemical regulated by the State of California

ADDITIONAL/OTHER COMMENTS
- ◆ Symptoms
  - • inhalation: irritation to nose and throat, headaches, diz-
    ziness, nausea
  - • ingestion: irritation of mouth and stomach, vomiting
- ◆ First Aid
  - • wash eyes and skin immediately with large amounts
    of water
  - • remove to fresh air immediately or provide respiratory ap-
    paratus if necessary
  - • drink water
- ◆ fire fighting: use water, dry chemical, foam, or carbon dioxide
- ◆ personal protection: compressed air/oxygen apparatus, gas-
  tight suit, fireproof suit

SOURCES
- ◆ *Health Effects Notebook for Hazardous Air Pollutants*, U.S.
  Environmental Protection Agency, December 1994
- ◆ *Material Safety Data Sheets (MSDS)*
- ◆ *Sax's Dangerous Properties of Industrial Materials*, 9th edi-
  tion, 1996
- ◆ *Environmental Law Index to Chemicals*, 1996 edition
- ◆ *Merck Index*, 12th edition, 1996
- ◆ *Hawley's Condensed Chemical Dictionary*, 12th edition, 1993
- ◆ *Environmental Contaminant Reference Databook*, volume 1,
  1995
- ◆ *Aldrich Chemical Company Catalog*, 1996–1997

QUINONE

CAS #: 106-51-4
MOLECULAR FORMULA: $C_6H_4 O_2$
FORMULA WEIGHT: 108.10
SYNONYMS: Benzo-chinon (German), 1,4-benzoquine, benzoquinone (DOT), p-benzoquinone, 1,4-benzoquinone, chinon (Dutch, German), p-Chinon (German), chinone, cyclohexadienedione, 1,4-cyclohexa-dienedione, 2,5-cyclohexadiene-1,4-dione, 1,4-cyclohexadiene diox-

ide, 1,4-Dioxy-benzol (German), NCI-C55845, p-quinone, USAF P-220

USES

◆ Used in Industry
- tanning hides
- making gelatin insoluble
- strengthening animal fibers
- rubber accelerator
- quinhydrone used in pH electrodes
- cosmetic industry
- polymerization inhibitor
- raw material for hydroquinone
- manufacture of fungicides

- analytical reagent
- oxidizing agent
- photography
- adhesive mixtures
- coal analysis
- pharmaceutical industry
- production of cortisone
- polymer and resin industry
- leather industry
- chemical intermediate

PHYSICAL PROPERTIES

◆ yellow prisms
◆ penetrating odor like chlorine
◆ odor threshold 0.400 mg/m^3
◆ melting point 115.7°C
◆ boiling point sublimes

◆ density 1.318 at 20°C
◆ vapor pressure 0.1 mm Hg at 25°C
◆ soluble in EtOH and Et$_2$O
◆ sparingly soluble in water
◆ log K$_{ow}$ 0.20

CHEMICAL PROPERTIES

◆ flash point 38°C
◆ autoignition temperature 560°C

◆ may darken on standing
◆ heat of combustion 656.6 kcal/gmol at 25°C

HEALTH RISK

◆ Toxic Exposure Guidelines
- OSHA PEL TWA 0.1 ppm
- ACGIH TLV TWA 0.1 ppm
- DFG MAK 0.1 ppm
- ACGIH IDLH 331 mg/m^3
- ACGIH STEL 2 mg/m^3
◆ Toxicity Data
- oms-human:lym 5 µmol/L
- sce-human:lym 5 µmol/L
- skin-mouse TDLo 800 mg/kg/29W-C:ETA
- oral-rat LD$_{50}$ 130 mg/kg
- intravenous-rat LD$_{50}$ 25 mg/kg
- unr-rat LD$_{50}$ 5600 µg/kg
- intraperitoneal-mouse LD$_{50}$ 8500 µg/kg
- subcutaneous-uns LD$_{50}$ 296 mg/kg
◆ Acute Risks
- highly irritating to eyes
- discoloration of the conjunctiva and cornea
- dermatitis
- skin discoloration
- erythema

- effects on kidneys in animals
- generalized swelling
- formation of papules and vesicles

- ◆ Chronic Risks
  - • skin ulceration
  - • visual disturbances
  - • necrosis
  - • IARC Group 3: not classifiable as to carcinogenicity to humans

HAZARD RISK
- ◆ Moist material self-heats and decomposes exothermically above 60°C.
- ◆ Decomposition emits acrid smoke and fumes.

EXPOSURE ROUTES
- ◆ dye industries
- ◆ textile industries
- ◆ chemical industries
- ◆ tanning industries
- ◆ cosmetic industries
- ◆ tobacco smoke

REGULATORY STATUS/INFORMATION
- ◆ Clean Air Act Amendment, Title III: hazardous air pollutants
- ◆ CERCLA and Superfund
  - • designation of hazardous substances. 40CFR302.4
  - • list of chemicals. 40CFR372.65
- ◆ TSCA: list of substances. 40CFR716.120.a
- ◆ OSHA, 29CFR1910.1000 Table Z1
- ◆ chemical regulated by the State of California

ADDITIONAL/OTHER COMMENTS
- ◆ Symptoms
  - • contact: severe damage to skin and mucous membranes, discoloration, irritation, erythema, generalized swelling, papules, vesicles, conjunctivitis, photophobia, eye lacrimation, cornea ulceration and scarring
- ◆ First Aid
  - • wash eyes and skin immediately with large amounts of water
  - • remove to fresh air immediately or provide respiratory apparatus if necessary
  - • drink water
- ◆ fire fighting: use water spray, dry chemical, or carbon dioxide
- ◆ personal protection: compressed air/oxygen apparatus, gas-tight suit

SOURCES
- ◆ *Health Effects Notebook for Hazardous Air Pollutants*, U.S. Environmental Protection Agency, December 1994
- ◆ *Material Safety Data Sheets (MSDS)*
- ◆ *Sax's Dangerous Properties of Industrial Materials*, 9th edition, 1996
- ◆ *Environmental Law Index to Chemicals*, 1996 edition
- ◆ *Merck Index*, 12th edition, 1996
- ◆ *Hawley's Condensed Chemical Dictionary*, 12th edition, 1993
- ◆ *Environmental Contaminant Reference Databook*, volume 1, 1995
- ◆ *Aldrich Chemical Company Catalog*, 1996–1997

RADIONUCLIDES (including radon)

CAS #: Radionuclides have variable CAS's.
MOLECULAR FORMULA: Radionuclides have variable molecular formulas. The molecular formula for uranium is U. The molecular formula for radium is Ra. The molecular formula for radon is Rn.
Note: The 1990 Amendments to the Clean Air Act define the hazardous air pollutant "radionuclides" as "a type of atom which spontaneously undergoes radioactive decay."
FORMULA WEIGHT: Radionuclides have variable formula weights. The formula weight for uranium is 238.03. The formula weight for radium is 226.03. The formula weight for radon is 222.
SYNONYMS: uranium: uranium I (238U), uranium-238
radium: not available
radon: not available

USES
- ◆ Uranium
  - • nuclear power plants
  - • ceramics
  - • photographic chemicals
  - • nuclear weapons
  - • light bulbs
  - • household products
- ◆ Radium
  - • radiation source for treating neoplastic diseases
  - • radon source
  - • radiography of metals
  - • neutron source for research
- ◆ Radon
  - • treating malignant tumors
  - • experimental studies

PHYSICAL PROPERTIES (Uranium)
- ◆ silver
- ◆ contains 3 isotopes: uranium-239, uranium-235, uranium-238
- ◆ melting point 1132.3°C
- ◆ density 19.05 g/cm³
- ◆ vapor pressure 1 mm Hg at 2450°C
- ◆ boiling point 3818°C
- ◆ insoluble in water
- ◆ soluble in acids

PHYSICAL PROPERTIES (Radium)
- ◆ silvery-white
- ◆ melting point 700°C
- ◆ boiling point <1140°C
- ◆ density 5 g/cm³ at 20°C
- ◆ decays in water at 20°C
- ◆ decays in acids

PHYSICAL PROPERTIES (Radon)
- ◆ colorless
- ◆ odorless
- ◆ tasteless
- ◆ radioactive gas
- ◆ melting point 71°C
- ◆ boiling point 61.8°C
- ◆ soluble in water 230 cm³/L
- ◆ slightly soluble in alcohol
- ◆ density 9.96×10^{-3} g/cm³ at 20°C
- ◆ vapor pressure 400 mm Hg at 25.8°C

CHEMICAL PROPERTIES
- ◆ flash point no data available
- ◆ Autoignition temperature for uranium is 1472°F.
- ◆ flammability limits no data available

HEALTH RISK
- ◆ Toxic Exposure Guidelines
  - • ACGIH TLV (insoluble compounds) 0.2 mg/m³
  - • OSHA PEL TWA (soluble compounds) 0.05 mg/m³
  - • OSHA PEL TWA (insoluble compounds) 0.2 mg/m³
  - • NIOSH REL (insoluble compounds) 0.25 mg/m³
  - • NIOSH REL (soluble compounds) 0.05 mg/m³
  - • NCRP (indoor air guideline) 8 pCi/L
  - • EPA (indoor air action level and guideline for schools) 4 pCi/L
- ◆ Toxicity Data
  - • rat LC_{50} (uranium hexafluoride) 120,000 mg/m³
  - • guinea pig LC_{50} (uranium hexafluoride) 62,000 mg/m³
- ◆ Acute Risks
  - • no information available in humans
  - • animals: inflammatory reactions in nasal passages and kidney damage
  - • low to moderate toxicity from inhalation
  - • high toxicity from ingestion
- ◆ Chronic Risks (Uranium)
  - • death from nonmalignant respiratory disease
  - • kidney effects
  - • body weight loss in rabbits
  - • moderate nephrotoxicity in rabbits
  - • tumors of lymphatic and hematopoietic tissues
  - • lung cancer
- ◆ Chronic Risks (Radium)
  - • acute leukopenia
  - • terminal bronchopneumonia
  - • anemia
  - • lung, bone, brain, and nasal passage tumors
  - • necrosis of jaw
- ◆ Chronic Risks (Radon)
  - • chronic lung disease
  - • effects on blood
  - • lung cancer
  - • pneumonia
  - • fibrosis of lung
  - • decrease in body weight

HAZARD RISK
- ◆ Uranium is pyrophoric in finely divided state.
- ◆ Uranium oxide powder results from burning uranium.

EXPOSURE ROUTES
- ◆ occupational exposure in industries that process uranium or phosphate fertilizers
- ◆ Uranium-238 is present in rocks, in soil, and throughout the environment.
- ◆ uranium: through air, food consumption, and drinking water
- ◆ radium: soil, water, plants, and food
- ◆ radon: major source is by inhalation, homes, schools, or office buildings

REGULATORY STATUS/INFORMATION
- ◆ Clean Air Act Title III: Hazardous Air Pollutants
- ◆ CERCLA and Superfund

- list of extremely hazardous substances and their threshold planning quantities. 40CFR355-AB

ADDITIONAL/OTHER COMMENTS
◆ Symptoms
  - oral exposure: lung, bone, brain, and nasal passage tumors
  - inhalation: lung cancer

SOURCES
◆ *Health Effects Notebook for Hazardous Air Pollutants*, U.S. Environmental Protection Agency, December 1994
◆ *Environmental Law Index to Chemicals*, 1996 edition
◆ *Toxicological Profile for Uranium*
◆ *Toxicological Profile for Radium*
◆ *Toxicological Profile for Radon*
◆ *Material Safety Data Sheets (MSDS)* for uranium

SELENIUM COMPOUNDS

CAS #: Selenium compounds have variable CAS #'s.
MOLECULAR FORMULA: Selenium compounds have variable molecular formulas. The molecular formula for hydrogen selenide is H_2Se.
Note: For all hazardous air pollutants (HAPs) that contain the word "compounds" in the name, and for glycol ethers, the following applies: Unless otherwise specified, these HAPs are defined by the 1990 Amendments to the Clean Air Act as "including any unique chemical substance that contains the named chemical (i.e., antimony, arsenic, etc.) as part of that chemical's infrastructure."
FORMULA WEIGHT: Selium compounds have variable formula weights. The formula weight for hydrogen selenide is 80.98.
SYNONYMS: Hydrogen selenide: dihydrogen selenide, hydrogen selenide, selane, selenium anhydride, selenium dihydride, selenium hydride.
Selenious acid: monohydrated selenium dioxide, selenous acid
Sodium selenide: disodium monoselenide
Sodium selenite: disodium selenite, disodium selenium trioxide, selenious acid, disodium salt, sodium selenium oxide
Selenium dioxide: selenious anhydride, selenium oxide, selenous acid anhydride
Selenium sulfide: selenium monosulfide, Selensulfid (German), sulfur selenide (SSe)
Selenium disulfide: selenium disulphide, selenium sulfide, sulfur selenide

USES
◆ Used as a Component or in the Manufacturing Process
  - photographic devices
  - enamels
  - glass
  - paints
  - rubber
  - pesticides
  - inks
  - dandruff shampoos
  - plastics
  - fungicides
  - lubricants
◆ Used as a Catalyst in Preparation of Pharmaceuticals
  - niacincortisone

PHYSICAL PROPERTIES (Hydrogen Selenide)
- colorless
- disagreeable odor
- melting point −67.73°C
- density 2.12 at 42°C
- vapor pressure 1330 at 30°C, 3420 at 0.2°C, 9120 at 30.8°C
- soluble in water, carbon disulfide, carbonyl chloride
- gas at room temperature
- boiling point 41.3°C

PHYSICAL PROPERTIES (Sodium Selenite and Sodium Selenate)
- soluble in water
- white or colorless crystals

PHYSICAL PROPERTIES (Selenium Sulfide)
- bright red-yellow powder
- insoluble
- melting point 118–119°C
- density 3.056 at 0°C

CHEMICAL PROPERTIES (Sodium Selenide and Sodium Selenite)
- not flammable

CHEMICAL PROPERTIES (Selenium)
- reacts with metal amides to form explosive products
- violently reacts with barium carbide, bromine pentafluoride, chromic oxide, fluorine, lithium carbide, lithium silicon, metals, nickel, nitric acid, sodium, nitrogen trichloride, oxygen, potassium, potassium bromate, rubidium carbide, zinc, silver bromate, strontium carbide, thorium carbide, uranium

HEALTH RISK
- Toxic Exposure Guidelines
  - ACGIH TLV TWA (selenium hexafluoride) 0.05 ppm (0.4 mg/m^3)
  - OSHA PEL TWA (selenium hexafluoride) 0.4 mg/m^3
  - DFG MAK 0.1 mg(Se)/m^3
  - NIOSH IDLH (selenium compounds) 100 mg/m^3
  - ACGIH TLV TWA (selenium compound) 0.2 mg/m^3
  - OSHA PEL TWA (selenium compound) 0.2 mg/m^3
  - EPA LC$_{50}$ (hydrogen selenide) (guinea pigs) 12.7 mg/m^3
  - NIOSH TWA (selenium compounds as selenium) 0.2 mg/m^3
- Toxicity Data (Selenium)
  - oral-rat LD$_{50}$ 6700 mg/kg
  - Inhalation-rat LDLo 33 mg/kg/8H
  - intravenous-rabbit LD$_{50}$ 6 mg/kg
  - Intravenous-rabbit LDLo 2500 μg/kg
  - oral-mouse TDLo 134 mg/kg
- Other Toxicity Indicators (Selenium)
  - IRAC cancer review: Group 3 IMEMDT 7,56,87
  - EPA: selenium sulfide is Group B2, probable human carcinogen
- Acute Risks
  - irritation, burning, tearing of eyes
  - irritation of mucous membranes
  - bronchial spasms
  - persistent bronchitis
  - loss of olfaction
  - chemical pneumonia
  - headaches
  - indigestion

- cardiovascular effects
- abdominal pain
- effects on liver
- dizziness
- fatigue
- irritation of skin
- nosebleed
- dyspnea
- ◆ Chronic Risks
  - nose irritation
  - sputum production
  - burns
  - numbness
  - brittle hair
  - excessive tooth decay and discoloration
  - lack of mental alertness
  - listlessness
  - discoloration of skin
- nausea
- malaise
- lesions of lungs
- diarrhea
- vomiting
- chills and tremors
- blind "staggers" disease

- death
- selenosis causing hemiplegia
- skin rashes
- contact dermatitis
- paralysis
- deformed nails or loss of nails
- "Keshan disease"
- "Kashin-Beck disease"

HAZARD RISK

◆ Selenium reacts with metal amides to form explosive products.

◆ When heated to decomposition, selenium compounds emit toxic fumes.

EXPOSURE ROUTES

◆ principally by burning coal and fly ash

◆ inhalation

◆ Drinking water causes absorption into digestive tract.

◆ occupational exposure in metal industries, selenium-recovery processes, paint manufacturing, and special trades

◆ eating locally grown food

◆ hazardous waste sites: in food, water, soil, and air

◆ weathering of rocks

◆ volcanic eruptions

◆ incineration of rubber tires, paper, and municipal waste

REGULATORY STATUS/INFORMATION

◆ Clean Air Act Title III: Hazardous Air Pollutants

◆ CERCLA and Superfund
  - designation of hazardous substances. 40CFR302.4

◆ Clean Water Act
  - toxic pollutant list. 40CFR401.15

ADDITIONAL/OTHER COMMENTS

◆ Symptoms
  - excessive salivation
  - garlic odor of breath
  - shallow breathing
  - diarrhea

SOURCES

◆ *Hawley's Condensed Chemical Dictionary*, 12th ed., 1993

- *Health Effects Notebook for Hazardous Air Pollutants*, U.S. Environmental Protection Agency, December 1994
- *Sax's Dangerous Properties of Industrial Materials*, 9th edition, 1996
- *Fire Protection Guide to Hazardous Materials*, 11th edition, 1994
- *Environmental Law Index to Chemicals*, 1996 edition
- *Handbook of Chemistry and Physics*, 77th edition, 1996–1997
- *Toxicological Profile for Selenium*

STYRENE

CAS #: 100-42-5
MOLECULAR FORMULA: $C_6H_5CH=CH_2$
FORMULA WEIGHT: 104.16
SYNONYMS: cinnamene, cinnamenol, diarex HF 77, ethenylbenzene, NCI-C02200, phenethylene, phenylethene, phenylethylene, stirolo (Italian), styreen (Dutch), styren (Czech), styrene monomer (ACGIH), styrene monomer, inhibited (DOT), Styrol·(German), styrole, styrolene, styron, styropor, vinylbenzen (Czech), vinylbenzene, vinylbenzol

USES
- Used in the Production Process
  - polystyrene
  - resins
  - plastics
  - synthetic rubber
  - insulator
  - acrylonitrile-butadiene-styrene
  - styrene-acrylonitrile polymer resins
  - protective coatings
  - styrenated polyesters
  - copolymer resins
  - construction materials
  - boats
  - synthetic flavoring substance and adjuvant
  - ice cream and candy
  - paints
  - cross-linking agent in polyester resins
- Used as a Chemical Intermediate
  - materials used for ion exchange resins
  - copolymers
  - styrenated phenols
  - styrene oxide
  - styrenated oils

PHYSICAL PROPERTIES
- colorless to yellow, oily liquid
- sweet smell
- odor threshold 0.32 ppm
- melting point −30.63°C
- boiling point 145.2°C
- density 0.9059 at 20°C
- surface tension 32.14 dynes/cm
- viscosity 0.751 mPa
- vapor pressure 5 mm Hg at 20°C
- vapor density 3.6
- slightly soluble in water
- miscible in alcohol and ether
- log K_{ow} 2.95

CHEMICAL PROPERTIES
- exposure to light and air: slow polymerization
- corrodes copper and copper alloys

- flash point 88°F
- autoignition temperature 914°F
- flammability limits 1.1% lower and 6.1% upper
- heat of combustion 4381 kJ/mol at 20°C
- heat of vaporization 86.8 cal/g

HEALTH RISK
- Toxic Exposure Guidelines
  - OSHA PEL TWA 50 ppm
  - OSHA PEL STEL 100 ppm
  - ACGIH TLV TWA 50 ppm
  - ACGIH TLV STEL 100 ppm (skin)
  - DFG MAK 20 ppm
  - NIOSH REL TWA 50 ppm
- Toxicity Data
  - skin-human 500 mg nse
  - skin-rabbit 500 mg open MLD
  - skin-rabbit 100% MOD
  - eye-rabbit 18 mg
  - oral-rat TDLo 8600 mg/kg
  - inhalation-rat TCLo 1500 $\mu g/m^3$/24H
  - inhalation-rat TCLo 100 ppm/4H/5D/1Y-I:CAR
  - inhalation-human LCLo 10,000 ppm/30M
  - inhalation-human TCLo 600 ppm:NOSE,EYE
  - inhalation-human TCLo 20 $\mu g/m^3$:EYE
  - oral-rat LD_{50} 5000 mg/kg
  - inhalation-rat LC_{50} 24 g/m^3/4H
  - intraperitoneal-rat LD_{50} 898 mg/kg
  - oral-mouse LD_{50} 316 mg/kg
  - inhalation-mouse LC_{50} 9500 mg/m^3/4H
  - intraperitoneal-mouse LD_{50} 660 mg/kg
  - intravenous-mouse LD_{50} 90 mg/kg
  - inhalation-guinea pig LCLo 12 mg/m^3/14H
- Acute Risks
  - respiratory effects
  - mucous membrane irritation
  - eye irritation
  - gastrointestinal effects
  - nausea
  - headache
  - vomiting
  - central nervous system depression
  - dermatitis
  - drowsiness
- Chronic Risks
  - effects on central nervous system
  - headache
  - fatigue
  - weakness
  - depression
  - peripheral neuropathy
  - minor effects on kidney enzyme functions
  - effects on blood
  - effects on central nervous system, liver, and kidney in animals
  - irritation of eye and nose in animals
  - increased risk of leukemia
  - increased risk of lymphoma
  - IARC Group 2B: possible human carcinogen
  - EPA Group C: possible human carcinogen

HAZARD RISK
◆ involved in several industrial explosions
◆ storage hazard above 32°C
◆ very dangerous fire hazard when exposed to heat, flame, or oxidants
◆ explosive in the form of vapor when exposed to heat or flame
◆ reacts with oxygen above 40°C to form a heat-sensitive explosive peroxide
◆ Violent or explosive polymerization may be initiated by alkali-metal-graphite composites, butyllithium, dibenzoyl peroxide, and other initiators.
◆ reacts violently with chlorosulfonic acid, oleum, sulfuric acid, chlorine + iron (III) chloride
◆ may ignite when heated with air + polymerizing polystyrene
◆ can react vigorously with oxidizing materials
◆ Decomposition emits acrid smoke and irritating fumes.

EXPOSURE ROUTES
◆ primarily by indoor air
◆ emissions from building materials
◆ emissions from consumer products
◆ tobacco smoke
◆ ambient air
◆ reinforced plastic industry
◆ polystyrene factories

REGULATORY STATUS/INFORMATION
◆ Clean Air Act Amendment, Title III: hazardous air pollutants
◆ Safe Drinking Water Act
  • list of contaminants removed and substituted. 53FR1896
  • public notification. 40CFR141.32
  • maximum contaminant level goal for organic contaminants. 40CFR141.50(b)
  • maximum contaminant level for organic chemicals. 40CFR141.61
  • variances and exemptions from the maximum contaminant levels for organic and inorganic chemicals. 40CFR141.62
◆ Resource Conservation and Recovery Act
  • constituents for detection monitoring. 40CFR258
  • list of hazardous constituents. 40CFR258
  • groundwater monitoring chemicals list. 40CFR264
◆ CERCLA and Superfund
  • designation of hazardous substances. 40CFR302.4
  • list of chemicals. 40CFR372.65
◆ Clean Water Act
  • designation of hazardous substances. 40CFR116.4
  • determination of reportable quantities. 40CFR117.3
◆ OSHA, 29CFR1910.1000 Table Z1 and Table Z2
◆ chemical regulated by the State of California

ADDITIONAL/OTHER COMMENTS
◆ symptoms: irritation of eyes and skin, dizziness, drunkenness, anesthesia

- ◆ First Aid
  - • wash eyes and skin immediately with large amounts of water
  - • remove to fresh air immediately or provide respiratory apparatus if necessary
  - • rinse out mouth with water
- ◆ fire fighting: use foam, carbon dioxide, or dry chemical
- ◆ personal protection: artificial respiration, gloves, goggles, and protective clothing
- ◆ storage: keep tightly closed away from heat, sparks, and open flame; light sensitive

SOURCES

- ◆ *Health Effects Notebook for Hazardous Air Pollutants*, U.S. Environmental Protection Agency, December 1994
- ◆ *Material Safety Data Sheets (MSDS)*
- ◆ *Sax's Dangerous Properties of Industrial Materials*, 9th edition, 1996
- ◆ *Environmental Law Index to Chemicals*, 1996 edition
- ◆ *Merck Index*, 12th edition, 1996
- ◆ *Hawley's Condensed Chemical Dictionary*, 12th edition, 1993
- ◆ *Environmental Contaminant Reference Databook*, volume 1, 1995
- ◆ *Aldrich Chemical Company Catalog*, 1996–1997

STYRENE OXIDE

CAS #: 96-09-3
MOLECULAR FORMULA: C_8H_8O
FORMULA WEIGHT: 120.16
SYNONYMS: epoxyethylbenzene (8CI), 1,2-epoxyethylbenzene, epoxystyrene, alpha,beta-epoxystyrene, NCI-C54977, phenethylene oxide, 1-phenyl-1,2-epoxyethane, phenylethylene oxide, phenyloxirane, 1-phenyloxirane, 2-phenyloxirane, styrene epoxide, styrene-7,8-oxide, styryl oxide

USES

- ◆ Used in Industry
  - • production of styrene glycol and its derivatives
  - • reactive diluent for epoxy resins
  - • raw material for production of phenylethyl alcohol for perfumes
  - • treatment of fibers and textiles
- ◆ Used as a Chemical Intermediate
  - • cosmetics
  - • surface coatings
  - • agricultural and biological chemicals

PHYSICAL PROPERTIES

- ◆ colorless to pale straw-colored liquid
- ◆ sweet, pleasant odor
- ◆ odor threshold 0.4 ppm
- ◆ boiling point 194.2°C
- ◆ melting point −36.7°C
- ◆ density 1.0469 at 25°/4° C
- ◆ vapor pressure 0.3 mm Hg at 20°C
- ◆ vapor density 4.14

- ◆ slightly soluble in water
- ◆ miscible with benzene, acetone, ether, and methanol

CHEMICAL PROPERTIES
- • flash point 74°C
- • combustible

HEALTH RISK
- ◆ Toxic Exposure Guidelines
  - • No information is available on the toxic exposure guidelines for styrene oxide.
- ◆ Toxicity Data
  - • skin-rabbit 10 mg/24H open MLD
  - • skin-rabbit 500 mg open MOD
  - • eye-rabbit 500 mg open
  - • dni-human:hla 4400 μmol/L
  - • oms-mouse:fbr 1 μmol/L
  - • inhalation-rat TCLo 100 ppm/7H
  - • inhalation-rabbit TCLo 15 ppm/7H
  - • oral-rat TDLo 10 g/kg/52W-I:CAR
  - • oral-mouse TDLo 273 g/kg/2Y-C:CAR
  - • skin-mouse TDLo 74 g/kg/62W-I:ETA
  - • unr-mouse TDLo 96 mg/kg:ETA
  - • oral-rat TD 52 g/kg/52W-I:CAR
  - • oral-rat LD_{50} 2000 mg/kg
  - • inhalation-rat LCLo 500 ppm/4H
  - • intraperitoneal-rat LD_{50} 460 mg/kg
  - • oral-mouse LD_{50} 1500 mg/kg
  - • skin-rabbit LD_{50} 1060 mg/kg
- ◆ Acute Risks
  - • skin irritation
  - • eye irritation
  - • corneal injury in rabbits
  - • changes in liver enzymes in rats
  - • central nervous system depression
  - • generation of liver lesions in animals
- ◆ Chronic Risks
  - • increased fetal mortality in rats and rabbits
  - • increased preimplantation loss of fetuses in rats
  - • reduced fetal weights in rats
  - • ossification defects in rats
  - • maternal toxicity in rabbits
  - • increased frequency of resorptions in rabbits
  - • squamous-cell carcinomas in rats and mice
  - • papillomas of the forestomach in rats and mice
  - • hepatocellular neoplasms in mice
  - • IARC Group 2A: probable human carcinogen

HAZARD RISK
- ◆ flammable when exposed to heat, flame, or oxidizers
- ◆ can react with oxidizing materials
- ◆ Substance may polymerize.
- ◆ Decomposition emits acrid smoke and fumes.

EXPOSURE ROUTES
- released to the environment in wastewater
- emissions during production and use
- occupational exposure in workplace

REGULATORY STATUS/INFORMATION
- Clean Air Act Amendment, Title III: hazardous air pollutants
- CERCLA and Superfund
  - list of chemicals. 40CFR372.65
- chemical regulated by the State of California

ADDITIONAL/OTHER COMMENTS
- symptoms: headache, coughing, sore throat, shortness of breath, abdominal pain, nausea
- First Aid
  - wash eyes immediately with large amounts of water
  - wash skin immediately with soap and water
  - remove to fresh air immediately or provide respiratory apparatus if necessary
  - drink water and induce vomiting
- fire fighting: use foam, carbon dioxide, and dry chemical
- personal protection: compressed air/oxygen apparatus, gastight suit, fireproof suit

SOURCES
- *Health Effects Notebook for Hazardous Air Pollutants*, U.S. Environmental Protection Agency, December 1994
- *Material Safety Data Sheets (MSDS)*
- *Sax's Dangerous Properties of Industrial Materials*, 9th edition, 1996
- *Environmental Law Index to Chemicals*, 1996 edition
- *Hawley's Condensed Chemical Dictionary*, 12th edition, 1993
- *Aldrich Chemical Company Catalog*, 1996–1997

2,3,7,8-TETRACHLORODIBENZO-p-DIOXIN

CAS #: 1746-01-6
MOLECULAR FORMULA: $C_{12}H_4Cl_4O_2$
FORMULA WEIGHT: 321.96
SYNONYMS: 2,3,7,8-czterochlorodwubenzo-p-dwuoksyny (Polish), dioksyny (Polish), dioxin, dioxine, TCBDB, TCDD, 2,3,7,8-TCDD, 2,3,7,8-tetrachlorodibenzo(b,e)(1,4)dioxan, 2,3,6,7-tetrachlorodibenzo-p-dioxin, 2,3,7,8-tetrachlorodibenzo-1,4-dioxin, tetradioxin

USES
- Variety of Uses
  - contaminant for herbicides and defoliants
  - contaminant created in the manufacture of Agent Orange
  - used as a defoliant in Vietnam War
  - research chemical
  - wood preservatives (not commercially)

PHYSICAL PROPERTIES
- white, crystalline solid or colorless needles
- melting point 305°C
- boiling point 412.2°C

- density 1.827 g/mL
- vapor pressure 1.52×10^{-9} at 25°C

CHEMICAL PROPERTIES

- very toxic
- caustic

HEALTH RISK

- Toxic Exposure Guidelines
  - NIOSH REL reduce to lowest feasible level
  - DFG MAK animal carcinogen, suspected human carcinogen
- Toxicity Data
  - eye-rabbit 2 mg MOD
  - mmo-smc 10 mg/L
  - oms-mouse:oth 1 nmol/L
  - intraperitoneal-rat TDLo 6 µg/L
  - oral-mouse TDLo 12 µg/kg
  - oral-rat TDLo 52 µg/kg/2Y-I:CAR
  - oral-mouse TDLo 52 µg/kg/2Y-I:CAR
  - oral-rat TD 328 µg/kg/78W-C:ETA
  - oral-mouse TD 36 µg/kg/52W-I:NEO
  - skin-human TDLo 107 µg/kg
  - oral-rat LD_{50} 20 µg/kg
  - intraperitoneal-rat LD_{50} 60 µg/kg
  - oral-mouse LD_{50} 114 µg/kg
  - skin-mouse LDLo 80 µg/kg
  - intraperitoneal-mouse LD_{50} 120 µg/kg
  - oral-dog LD_{50} 1 µg/kg
  - oral-monkey LD_{50} 2 µg/kg
  - skin-rabbit LD_{50} 275 µg/kg
- Acute Risks
  - tightness in chest
  - dizziness
  - headache
  - nausea
  - drowsiness
  - prickling of skin and eyes
  - allergic dermatitis
  - irritation of eyes
  - wasting
  - hepatic necrosis
  - thymic atrophy
  - hemorrhage
  - lymphoid depletion
  - chloracne
- Chronic Risks
  - burnlike skin lesions
  - pancreatic carcinoma
  - bronchogenic carcinoma
  - chloracne
  - porphyrinuria
  - porphyria cutanea tarda
  - anorexia
  - severe weight loss
  - hepatoxicity
  - hepatoporphyria
  - vascular lesions
  - gastric ulcers
  - liver damage, cancer, tumors
  - teratogenicity
  - delayed death
  - EPA Group 2B: probable human carcinogen

- solubility in water 7.91 ng/L at 22°C
- $\log K_{ow}$ 6.15–7.28
- $\log K_{oc}$ 6.0–7.39

HAZARD RISK
- ◆ most toxic member of the 75 dioxins
- ◆ one of the most toxic synthetic chemicals
- ◆ caustic and corrosive
- ◆ Decomposition emits toxic fumes of Cl^-.

EXPOSURE ROUTES
- use of herbicides containing 2,4,5-trichlorophenoxy acids
- production and use of 2,4,5-trichlorophenol in wood preservatives
- production and use of hexachlorophene as a germicide
- pulp and paper manufacturing plants, paper products
- incineration/combustion of wastes
- burning of wood in presence of chlorine
- transformer/capacitor fires involving chlorinated benzenes and biphenyls
- flue gases
- fly ash, soot particles
- exhaust from automobiles
- chlorinated chemical wastes
- low levels in air
- trace levels in urban soils and vegetation
- inhalation of wood dusts
- fish and cow's milk
- by-product of polychlorinated phenols

REGULATORY STATUS/INFORMATION
- ◆ Clean Air Act Amendment, Title III: hazardous air pollutants
- ◆ Safe Drinking Water Act
  - one of 83 contaminants required to be regulated under the SWDA of 1986. 47FR9352
  - sampling and analytical requirements. 40CFR141.24
  - public notification 40CFR141.32
  - maximum contaminant level goal (MCLG) for organic contaminants. 40CFR141.50(a)
  - maximum contaminant level for organic chemicals. 40CFR141.61
  - variances and exemptions from the maximum contaminant levels for chemicals
- ◆ Resource Conservation and Recovery Act
  - list of hazardous constituents. 40CFR261, 40CFR261.11
  - groundwater monitoring list. 40CFR264
  - risk specific doses. 40CFR266
- ◆ CERCLA and Superfund
  - designation of hazardous substances. 40CFR302.4
- ◆ Clean Water Act
  - toxic pollutants. 40CFR401.15
  - total toxic organics (TTOs). 40CFR413.02
  - 126 priority pollutants. 40CFR423
- ◆ chemical regulated by the State of California

ADDITIONAL/OTHER COMMENTS
- ◆ Symptoms
  - skin absorption: prickling

477

- ingestion: tightness of chest, dizziness, headache, nausea, drowsiness
♦ First Aid
  • wash eyes immediately with large amounts of water
  • flush skin with plenty of soap and water
  • provide respiratory apparatus
♦ personal protection: gastight suit and respiratory apparatus; consider evacuation

SOURCES
♦ *Material Safety Data Sheets (MSDS)*
♦ *Sax's Dangerous Properties of Industrial Materials*, 9th edition, 1996
♦ *Environmental Law Index to Chemicals*, 1996 edition
♦ *Merck Index*, 12th edition, 1996
♦ *Hawley's Condensed Chemical Dictionary*, 12th edition, 1996
♦ *Toxicological Profile for 2,3,7,8-Tetrachlorodibenzo-p-dioxin*

1,1,2,2-TETRACHLOROETHANE (pf)

CAS #: 79-34-5
MOLECULAR FORMULA: $C_2H_2Cl_4$
FORMULA WEIGHT: 167.84
SYNONYMS: acetylene tetrachloride, bonoform, cellon, 1,1,2,2-cztero-chlorietan (Polish), 1,1-dichloro-2,2-dichloroethane, NCI-C03554, RCRA waste number U209, TCE, 1,1,2,2-tetrachloorethaan (Dutch), 1,1,2,2-Tetrachloraethan (German), tetrachlorethane, 1,1,2,2-tetrachlorethane (French), tetrachloroethane, 1,1,2,2-tetrachloroethane, s-tetrachloroethane, sym-tetrachloroethane, 1,1,2,2-tetrachloroethane (ACGIH:OSHA), 1,1,2,2-tetracloroetano (Italian), tetrachlorure d'acetylene (French), westron, many others

USES
♦ Used in the Production of Other Chemicals
  • trichloroethylene
  • tetrachloroethylene
  • 1,2-dichloroethylene
♦ Used as a Component or in the Manufacturing Process
  • cleaning and degreasing metals
  • solvent
  • paint removers
  • varnishes and lacquers
  • photographic films
  • an extract for oils and fats
  • resins and waxes
  • alcohol denaturant
  • organic synthesis
  • insecticides
  • weed killer
  • fumigant
PHYSICAL PROPERTIES
♦ dense, heavy, colorless, corrosive liquid

- melting point −43.8°C (−46.84°F)
- boiling point 146.4°C (295.52°F) at 760 mm Hg
- density 1.593 g/cm³ at 25°C
- specific gravity 1.586 (water = 1.00)
- vapor density 5.8 (air = 1)
- surface tension not available
- viscosity not available
- heat capacity not available
- heat of vaporization not available
- sweet, chloroform-like odor
- odor threshold of 1.5 ppm
- vapor pressure 8 mm at 20°C, 5.95 mm at 25°C
- very sparingly soluble in water
- miscible with methanol, ethanol, benzene, ether, carbon tetra-chloride, chloroform, carbon disulfide, dimethylformamide, oils
- not an inert solvent
- highest solvent power of the chlorinated hydrocarbons

CHEMICAL PROPERTIES

- nonflammable, corrosive
- stable
- Hazardous polymerization will not occur.
- reacts violently with N_2O_4, 2,4-dinitrophenyl disulfide, and on contact with sodium or potassium
- Heating with solid potassium hydroxide, spontaneously flammable chloro- or dichloroacetylene gas is produced.
- Interaction with water causes hydrolysis.
- Above 110°C hydrolysis and oxidation become rapid.
- Heating to decomposition yields toxic hydrogen chloride gas fumes.
- incompatible with strong oxidizing agents, strong bases, strong reducing agents
- flash point none, since noncombustible
- explosion limits in air none
- autoignition temperature none
- heat of combustion not available
- enthalpy (heat) of formation not available

HEALTH RISK

- Toxic Exposure Guidelines
  - ACGIH TLV TWA 1 ppm (7 mg/m³) (skin)
  - OSHA PEL TWA 1 ppm (7 mg/m³) (skin)
  - NIOSH REL (7 mg/m³) reduce to lowest level
- Toxicity Data
  - oral-rat TDLo 42 g/kg/78W-I:ETA
  - oral-mouse TDLo 55 g/kg/78W-I:CAR
  - oral-mouse TD 110 g/kg/78W-I:CAR
  - oral-human TDLo 30 mg/kg:CNS
  - inhalation-human TCLo 1000 mg/m³/30M:CNS
  - oral-rat LD_{50} 250 mg/kg
  - inhalation-rat LCLo 1000 ppm/4H

- inhalation-mouse TC_{50} 4500 mg/m³/2H
- intraperitoneal-mouse LDLo 30 mg/kg
- subcutaneous-mouse LD_{50} 1108 mg/kg
- oral-dog LDLo 300 mg/kg
- intravenous-dog LDLo 50 mg/kg
- inhalation-cat LCLo 19 g/m³/45M
- subcutaneous-rabbit LCLo 500 mg/kg

◆ Acute Risks
- at very high levels has caused death in humans due to severe liver destruction
- harmful if swallowed or absorbed
- Inhalation has caused respiratory, neurological, and gastrointestinal effects in humans.
- Vapor or mist is irritating to the eyes, mucous membranes, and upper respiratory tract.
- Exposure may include burning sensation, coughing, wheezing, laryngitis, shortness of breath, headache, abdominal pain, nausea and vomiting, tremors, central nervous system depression, and dermatitis.
- Inhalation has caused effects on the liver, eyes, and central nervous system.
- Tests on rats and mice have shown moderate acute toxicity.

◆ Chronic Risks
- Long-term exposure in humans and animals results in effects on the liver (jaundice and an enlarged liver) and central and peripheral nervous systems (headaches, tremors, dizziness, and drowsiness).
- Overexposure may cause reproductive disorders based on tests in animals.
- may alter genetic material
- blood effects
- damage to the heart
- possible human carcinogen
- incidence of death from genital cancers, leukemia, and other cancers
- Reference concentration not established by EPA.
- reference dose is under review by EPA

HAZARD RISK
◆ hazard toxic by ingestion, inhalation, and absorption
◆ carcinogen
◆ mutagen
◆ corrosive
◆ reproductive hazard
◆ Heating to decomposition also results in toxic substances of carbon monoxide, carbon dioxide, and hydrogen chloride gas.
◆ incompatible with strong oxidizing agents, strong bases, strong reducing agents
◆ reacts violently with N_2O_4, 2,4-dinitrophenyl disulfide, and on contact with sodium or potassium

- Heating with solid potassium hydroxide, spontaneously flammable chloro- or dichloroacetylene gas is produced.

EXPOSURE ROUTES
- inhalation
- absorption through skin or eyes
- ingestion
- Low levels of the contaminant are found in outdoor air (0.005 ppb) and indoor (1.8 ppb) air.
- detected in surface water and groundwater but not in water supplies
- through inhalation or skin contact in industrial settings

REGULATORY STATUS/INFORMATION
- IARC Cancer Review: Group 3 IMEMDT 7,354,87
- Animal Limited Evidence IMEMDT 20,477,79
- NCI Carcinogenesis Bioassay (gavage)
- Clear Evidence: mouse NCITR* NCI-CG-TR-27,78
- Some Evidence: rat NCITR* NCI-CG-TR-27,78
- Reported in EPA TSCA Inventory
- EPA Genetic Toxicology Program
- Community Right-to-Know list
- EPA Extremely Hazardous Substances list
- Clean Air Act of 1970
  - Chapter I, Subchapter C—Air Programs 40CFR50-99
  - Criteria pollutant
- Solid Waste Disposal Act of 1965
  - Chapter I, Subchapter I—Solid Wastes 40CFR240-299
- Safe Drinking Water Act
  - priority list of drinking water contaminants, 55FR1470
  - community water systems and nontransient, noncommunity water systems, 40CFR141.40.e
- Resource Conservation and Recovery Act of 1976
  - Chapter I, Subchapter I—Solid Wastes 40CFR240-299
  - constituents for detection monitoring (for MSWLF) 40CFR258—Appendix 1
  - list of hazardous inorganic and organic constituents 40CFR258—Appendix 2
  - hazardous constituents 40CFR261 Appendix 8, 40CFR261.11
  - groundwater monitoring chemicals list 40CFR264 Appendix 9
  - risk specific doses 40CFR266 Appendix 5
  - health based limits for exclusion of waste-derived residues 40CFR266 Appendix 7
  - discarded commercial chemical products, off-specification species, container
  - residues, and spill residues thereof 40CFR261.33, 261.33.f. (U waste)
- CERCLA and Superfund
  - designation of hazardous substances 40CFR302.4

- Appendix A: sequential CAS registry # list of CERCLA hazardous substances
- Appendix B: radionuclides
- chemicals and chemical categories to which this part applies 40CFR372.65
◆ Clean Water Act of 1977
  - total toxic organics (TTOs) 40CFR413.02
  - 126 priority pollutants 40CFR423—Appendix A
◆ EPA Office of Air Quality Planning and Standards, for a hazard ranking under section 112(g) of the Clean Air Act Amendments, has ranked the contaminant in the nonthreshold category.
◆ OSHA, 29CFR1910.1000 Table Z1: limits for air contaminants
◆ chemical regulated by the State of California

ADDITIONAL/OTHER COMMENTS
◆ Symptoms
  - Inhalation has caused respiratory, neurological, and gastrointestinal effects in humans.
  - Vapor or mist is irritating to the eyes, mucous membranes, and upper respiratory tract.
  - Exposure may cause burning sensation, coughing, wheezing, laryngitis, shortness of breath, headache, abdominal pain, nausea and vomiting, tremors, central nervous system depression, and dermatitis.
◆ First Aid
  - if inhaled remove to fresh air; give artificial respiration if not breathing, or oxygen if breathing is difficult, and get medical attention
  - if contacted with the eyes, flush with water for 15 minutes and get medical attention
  - if contacted with skin wash with water for at least 15 minutes, while removing clothing; get medical attention
  - if swallowed induce vomiting immediately as directed by medical personnel and call poison control center immediately
◆ fire fighting: use water spray, carbon dioxide, dry chemical powder, or appropriate foam
◆ personal protection: use chemical-resistant gloves, boots, apron, goggles or a face shield, and a NIOSH approved vapor respirator if needed due to poor ventilation
◆ storage: keep the storage container tightly closed and away from metals, strong bases, or strong reducing agents

SOURCES
◆ *Material Safety Data Sheets (MSDS)*
◆ *Sax's Dangerous Properties of Industrial Materials*, 9th edition, 1996
◆ *Merck Index*, 12th edition, 1996
◆ *Hawley's Condensed Chemical Dictionary*, 12th edition, 1993
◆ *Environmental Law Index to Chemicals*, 1996 edition

◆ *Health Effects Notebook for Hazardous Air Pollutants,* U.S. Environmental Protection Agency, December 1994
◆ http://hazard.com:80/fish/acros/11/1434.html

TETRACHLOROETHYLENE

CAS #: 127-18-4
MOLECULAR FORMULA: $Cl_2C=CCl_2$
FORMULA WEIGHT: 165.82
SYNONYMS: ankilostin, antisol 1, carbon bichloride, carbon dichloride, czterochloroetylen (Polish), didakene, dow-per, ENT 1,860, ethylene tetrachloride, fedal-UN, NCI-C04580, nema, perawin, perchloorethyleen, per (Dutch), perchlor, perchlorethylene, per (French), perchloroethylene, perclene, percloroetilene (Italian), percosolve, perk, perklone, persec, tetlen, tetracap, tetrachlooretheen (Dutch), Tetrachlorathen (German), tetrachloroethene, 1,1,2,2-tetrachloroethylene, tetracloroetene (Italian), tetraleno, tetralex, tetravec, tetroguer, tetropil

USES
◆ Used in Industry
  • dry cleaning
  • textile processing
  • chemical intermediate
  • degreasing agent
  • rubber coatings
  • solvent soaps
  • printing inks
  • adhesives
  • glues
  • sealants
  • polishes
  • lubricants
  • silicones
  • insulating fluid and cooling gas in electrical transformers

PHYSICAL PROPERTIES
◆ colorless liquid
◆ sharp, sweet odor
◆ odor threshold 4.68 ppm
◆ melting point −23.35°C
◆ boiling point 121.20°C
◆ density 1.6311 at 15°/4°
◆ critical temperature 347.1°C
◆ critical pressure 47 atm
◆ vapor pressure 18.47 mm Hg at 25°C
◆ vapor density 5.83
◆ insoluble in water
◆ miscible with alcohol, ether, chloroform, and benzene
◆ log K_{ow} 3.40

CHEMICAL PROPERTIES
◆ extremely stable
◆ resists hydrolysis
◆ nonflammable

HEALTH RISK
◆ Toxic Exposure Guidelines
  • OSHA PEL TWA 25 ppm
  • ACGIH TLV TWA 50 ppm
  • ACGIH TLV STEL 200 ppm
  • DFG MAK 50 ppm
  • NIOSH REL minimize workplace exposure
◆ Toxicity Data
  • skin-rabbit 819 mg/24H SEV
  • skin-rabbit 500 mg/24H MLD

- eye-rabbit 162 mg MLD
- eye-rabbit 500 mg/24H MLD
- dns-human:lng 100 mg/L
- otr-rat:emb 97 μmol/L
- inhalation-rat TCLo 900 ppm/7H
- inhalation-rat TCLo 1000 ppm/24H
- inhalation-rat TCLo 200 ppm/6H/2Y-I:CAR
- oral-mouse TDLo 195 g/kg/50W-I:CAR
- inhalation-rat TC 200 ppm/6H/2Y-I:NEO
- inhalation-human TCLo 96 ppm/7H:PNS,EYE,CNS
- oral-cld TDLo 545 mg/kg:CNS
- inhalation-man TCLo 600 ppm/10M:EYE,CNS
- inhalation-man TDLo 2857 mg/kg:CNS,PUL
- oral-rat LD_{50} 2629 mg/kg
- inhalation-rat LC_{50} 34,200 mg/m^3/8H
- intraperitoneal-rat LD_{50} 4678 mg/kg
- oral-mouse LD_{50} 8100 mg/kg
- inhalation-mouse LC_{50} 5200 ppm/4H
- subcutaneous-mouse LD_{50} 65 g/kg
- oral-dog LDLo 4000 mg/kg
- intraperitoneal-dog LD_{50} 2100 mg/kg
- intravenous-dog LDLo 85 mg/kg
- oral-cat LDLo 4000 mg/kg

◆ Acute Risks
  - death
  - intense irritation of the upper respiratory tract
  - irritation of eyes
  - kidney dysfunction
  - neurological effects
  - reversible mood changes
  - behavioral changes
  - impairment of coordination
  - anesthetic effects
  - dermatitis
  - nausea
  - flushing of face and neck
  - vertigo
  - dizziness
  - headache
  - somnolence
  - effects on the liver, kidney, and central nervous system system in animals

◆ Chronic Risks
  - neurological effects
  - headache
  - impairment of memory and concentration
  - impairment of intellectual function
  - cardiac arrhythmia
  - liver damage
  - possible kidney effects
  - liver, kidney, and central nervous system effects in animals
  - menstrual disorders
  - spontaneous abortions
  - increased incidence of a variety of tumors
  - increased risk of childhood leukemia
  - liver tumors in mice
  - kidney and mononuclear cell leukemias in rats
  - EPA Group B2/C: probable human carcinogen

HAZARD RISK
◆ extremely stable and resists hydrolysis

♦ reacts violently under the proper conditions with Ba, Be, Li, N_2O_4, metals, and NaOH
♦ Decomposition emits highly toxic fumes of Cl^-.

EXPOSURE ROUTES
- ambient air
- drinking water
- auto brake cleaners
- suede protectors
- water repellents
- silicone lubricants
- dry cleaning establishments
- industries manufacturing or using the chemical

REGULATORY STATUS/INFORMATION
♦ Clean Air Act Amendment, Title III: hazardous air pollutants
♦ Safe Drinking Water Act
- 83 contaminants required to be regulated. 47FR9352
- public notification. 40CFR141.32
- maximum contaminant level goal for organic contaminants. 40CFR141.50(a)
- maximum contaminant level for organic chemicals. 40CFR141.61
- variances and exemptions from the maximum contaminant levels for organic and inorganic chemicals. 40CFR141.62
- waste specific prohibitions—solvent wastes. 40CFR148.10
♦ Resource Conservation and Recovery Act
- constituents for detection monitoring. 40CFR258
- list of hazardous constituents. 40CFR258, 40CFR261, 40CFR262.11
- groundwater monitoring chemicals list. 40CFR264
- risk specific doses. 40CFR266
- health based limits for exclusion of waste-derived residues. 40CFR266
- PICs found in stack effluents. 40CFR266
- D waste # 039
- U waste # 210
♦ CERCLA and Superfund
- designation of hazardous substances. 40CFR302.4
- list of chemicals. 40CFR372.65
♦ Clean Water Act
- toxic pollutants. 40CFR401.15
- total toxic organics (TTOs). 40CFR413.02
- 126 priority pollutants. 40CFR423
♦ OSHA, 29CFR1910.1000 Table Z2
♦ chemical regulated by the State of California

ADDITIONAL/OTHER COMMENTS
♦ symptoms: unconsciousness, dizziness, intoxicating, headache, irritation, drowsiness, stupefying, vomiting, nausea, abdominal pain, diarrhea
♦ First Aid
- wash eyes immediately with large amounts of water
- wash skin immediately with soap and water
- remove to fresh air immediately or provide respiratory apparatus if necessary
- drink water and induce vomiting

♦ personal protection: compressed air/oxygen apparatus, gas-tight suit

SOURCES
♦ *Health Effects Notebook for Hazardous Air Pollutants*, U.S. Environmental Protection Agency, December 1994
♦ *Material Safety Data Sheets (MSDS)*
♦ *Sax's Dangerous Properties of Industrial Materials*, 9th edition, 1996
♦ *Environmental Law Index to Chemicals*, 1996 edition
♦ *Merck Index*, 12th edition, 1996
♦ *Hawley's Condensed Chemical Dictionary*, 12th edition, 1993
♦ *Aldrich Chemical Company Catalog*, 1996–1997

TITANIUM TETRACHLORIDE

CAS #: 7550-45-0
MOLECULAR FORMULA: $TiCl_4$
FORMULA WEIGHT: 189.70
SYNONYMS: tetrachlorure de titane (French), titaantetrachlorid (Dutch), titane (tetrachlorure de) (French), titanio tetracloruro di (Italian), titanium chloride, Titanterachlorid (German)

USES
♦ Used as an Intermediate
  • titanium metal
  • titanium pigments
♦ Used in the Manufacturing Process
  • iridescent glass
  • artificial pearls
  • polymerization catalyst
  • smoke screens
♦ used as an implant material in orthopedics, oral surgery, and neurosurgery
♦ Formerly Used with Potassium Bitartrate
  • mordant in the textile industry
  • dyewoods in dyeing leather

PHYSICAL PROPERTIES
♦ colorless to light-yellow liquid
♦ penetrating acid odor
♦ melting point −24.1°C
♦ boiling point 136.4°C
♦ density 1.726
♦ vapor pressure 9.6 mm Hg at 20°C
♦ soluble in cold water and alcohol

CHEMICAL PROPERTIES
♦ corrosive
♦ evolves dense white fumes
♦ absorbs moisture from air
♦ decomposes by hot water
♦ heat of vaporization 79.7 Btu/lb

HEALTH RISK
♦ Toxic Exposure Guidelines
  • No information is available on the toxic exposure guidelines of Titanium tetrachloride.

- ◆ Toxicity Data
  - inhalation-rat LC_{50} 400 mg/m^3
  - inhalation-mouse LC_{50} 100 mg/m^3/2H
- ◆ Acute Risks
  - irritation to eyes
  - irritation to skin
  - irritation to mucous membranes
  - surface skin burns
  - marked congestion of mucous membranes of the pharynx, vocal cords, and trachea
  - stenosis of the larynx, trachea, and upper bronchi
  - cornea damage
  - endobronchial polyps
  - white corneas in rats
- ◆ Chronic Risks
  - pleural diseases
  - upper respiratory tract irritation
  - chronic bronchitis
  - cough
  - bronchoconstriction
  - wheezing
  - chemical pneumonitis
  - pulmonary edema

HAZARD RISK

- ◆ sensitive to air
- ◆ incompatible with strong oxidizing agents
- ◆ reacts violently with water
- ◆ Decomposition emits toxic fumes of Cl^-.

EXPOSURE ROUTES

- ◆ inhalation during manufacture or use
- ◆ dermal contact during manufacture or use

REGULATORY STATUS/INFORMATION

- ◆ Clean Air Act Amendments, Title III: hazardous air pollutants
- ◆ CERCLA and Superfund
  - list of extremely hazardous substances and their threshold planning quantities. 40CFR355-AB
  - list of chemicals. 40CFR372.65
- ◆ chemical regulated by the State of California

ADDITIONAL/OTHER COMMENTS

- ◆ Symptoms
  - vapors: severe irritation to eyes, coughing, headache, dizziness, lung damage, bronchial pneumonia
  - liquid: thermal and acid burns of eyes, skin, throat, and stomach
  - ingestion: nausea, vomiting, cramps, diarrhea, possible tissue ulceration
- ◆ First Aid
  - wash eyes and skin immediately with large amounts of water
  - remove to fresh air immediately or provide respiratory apparatus if necessary
  - drink water and induce vomiting

- ◆ fire fighting: carbon dioxide, dry chemical powder, or appropriate foam; no water
- ◆ personal protection: self-contained breathing apparatus and protective clothing
- ◆ storage: keep tightly closed in a cool dry place; no contact with water

SOURCES
- ◆ *Health Effects Notebook for Hazardous Air Pollutants*, U.S. Environmental Protection Agency, December 1994
- ◆ *Material Safety Data Sheets (MSDS)*
- ◆ *Sax's Dangerous Properties of Industrial Materials*, 9th edition, 1996
- ◆ *Environmental Law Index to Chemicals*, 1996 edition
- ◆ *Merck Index*, 12th edition, 1996
- ◆ *Hawley's Condensed Chemical Dictionary*, 12th edition, 1993
- ◆ *Environmental Contaminant Reference Databook*, volume 1, 1995
- ◆ *Aldrich Chemical Company Catalog*, 1996–1997

TOLUENE

CAS #: 108-88-3
MOLECULAR FORMULA: $C_6H_5CH_3$
FORMULA WEIGHT: 92.15
SYNONYMS: antisal-1A, benzene, methyl-, casewell no. 859, CP-25, methacide, methacide, toluol, methane, phenyl-, methyl-benzene, methyl-benzol, NCI-CO7272, phenylmethane, RCRA waste number U220, tolueen (Dutch), toluen (Czech), toluol (DOT), toluol, methyl-benzene, toluolo (Italian), tolusol

USES
- ◆ Manufacture of Other Chemicals
  - benzoic acid
  - dyes
  - caprolactam
  - perfumes
  - benzaldehyde
  - artificial leather
  - saccharin
  - benzene and urethane via hydrodealkylation
  - explosives
  - detergent
  - medicines
- ◆ Used as a Solvent
  - paints
  - resins
  - lacquers
  - plastic toys
  - gums
  - model airplanes

PHYSICAL PROPERTIES
- colorless liquid
- density 0.8661 g/mL at 20°C
- melting point −95°C
- odor threshold 2.14 ppm (8 mg/m³)
- vapor pressure 36.7 mm Hg at 30°C
- boiling point 111°C at 760 mm Hg
- soluble in petroleum ether, alcohol, ether, acetate, benzene, ligand
- sweet, pungent odor
- viscosity 0.590 cP at 20°C

- surface tension 29.0 dynes/cm
- specific gravity 0.87
- insoluble in water

CHEMICAL PROPERTIES

- flammable liquid
- autoignition temperature 480°C
- heat of combustion −17,430 Btu/lb
- TLV 100 ppm in air
- lel 1.27%

- miscible with alcohol, chloroform, ether, acetone, glacial acetic acid, carbon disulfide

- flammability limits: 1.2% lower and 7.1% upper by volume in air.
- flash point 40°F
- will react with oxidizing materials

HEALTH RISK

- Toxic Exposure Guidelines
  - ACGIH TLV TWA 50 ppm (188 mg/m³)
  - OSHA PEL TWA 100 ppm (326 mg/m³)
  - DFG MAK 50 ppm (190 mg/m³)
  - NIOSH REL TWA 100 ppm
- Toxicity Data
  - eye-human 300 ppm
  - skin-rabbit 435 mg MLD, 500 MOD
  - eye-rabbit 870 μg MLD, 2 mg/24H SEV, 100 mg/30S rns MLD
  - inhalation-oms-grh 562 mg/L
  - subcutaneous-rat-cyt 12 g/kg/12D-I
  - inhalation-mouse TCLo 400 ppm/7H
  - oral-mouse TDLo 9 g/kg
  - oral-human LDLo 50 mg/kg
  - inhalation-human TCLo 200 ppm: BRN, CNS, BLD
  - inhalation-man TCLo 100 ppm: CNS
  - oral-rat LD_{50} 5000 mg/kg
  - inhalation-rat LCLo 4000 ppm/4H
  - intraperitoneal-rat LD_{50} 1332 mg/kg
  - intravenous-rat LD_{50} 1960 mg/kg
  - unr-rat LD_{50} 6900 mg/kg
  - inhalation-mouse LC_{50} 400 ppm/24H
  - intraperitoneal-mouse LD_{50} 59 mg/kg
  - subcutaneous-mouse LD_{50} 2250 mg/kg
  - unr-mouse LD_{50} 2 g/kg
  - intraperitoneal-mouse LD_{50} 640 mg/kg
  - inhalation-rabbit LCLo 55,000 ppm/40M
  - skin-rabbit LD_{50} 12,124 mg/kg
  - intraperitoneal-guinea pig LD_{50} 500 mg/kg
  - oral-mammal LD_{50} 4 g/kg
  - inhalation-mammal LC_{50} 30 g/m³
- Acute Risks
  - harmful if swallowed, inhaled, or absorbed through skin
  - causes severe irritation
  - destructive to tissues of the mucous membranes, upper respiratory tract, eyes, and skin
  - burning sensation
  - coughing
  - laryngitis
  - shortness of breath
  - headache
  - nausea and vomiting
  - lung irritation

- chest pain and edema
  which may cause death
- wheezing
- ◆ Chronic Risks
  - nervous system disturbances
  - inflammatory and ulcerous lesions of the penis, prepuce, and scrotum in animals

HAZARD RISK
- ◆ a very dangerous fire hazard
- ◆ explodes in the form of vapor when exposed to heat or flames
- ◆ explosive reaction with 1,3-dichloro-5,5-dimethyl-2,4-imidazolididione, dinitrogen tetraoxide, concentrated nitric acid, $H_2SO_4 + HNO_3$, N_2O_4, $AgClO_4$, BrF_3, UF_6, sulfur dichloride
- ◆ forms an explosive mixture with tetranitromethane
- ◆ reacts vigorously with oxidizing materials

EXPOSURE ROUTES
- ◆ derived from coal tar
- ◆ inhalation
- ◆ ingestion

REGULATORY STATUS/INFORMATION
- ◆ Clean Air Act Title III: Hazardous Air Pollutants
- ◆ Safe Drinking Water Act
  - one of 83 contaminants required to be regulated under the 1986 act
  - public notification
  - maximum contaminant level goal (MCLG) for organic contaminants. 40CFR141.50(b)
  - maximum contaminant level (MCL) for organic chemicals. 40CFR141.61
  - variances and exemptions from the maximum contaminant level (MCL). 40CFR142.62
  - waste specific prohibitions. 40CFR148.10
- ◆ Resource Conservation and Recovery Act
  - constituents for detection monitoring for municipal solid waste landfills. 40CFR258
  - list of hazardous constituents. 40CFR258, 40CFR261
  - groundwater monitoring chemicals list. 40CFR264
  - reference air concentrations. 40CFR266
  - health based limits for exclusion of waste-derived residues. 40CFR266
  - PICs found in stack effluents. CFR266
  - discarded commercial chemical products, off-specification species, container residues, and spill residues thereof.
  - U waste # 261.33.f.
- ◆ CERCLA and Superfund
  - designation of hazardous substances. 40CFR302.4
  - list of chemicals. 40CFR372.65
- ◆ Clean Water Act
  - designation of hazardous substances. 40CFR116.4

- reportable quantity list. 40CFR117.3
- toxic pollutants list. 40CFR401.15
- total toxic organics (TTOs). 40CFR413.02
- 126 priority pollutants. 40CFR423
◆ Toxic Substance Control Act
  - list of substances. 40CFR716.120.a
◆ OSHA 29CFR1910.1000 Table Z2
◆ chemical regulated by the State of California

ADDITIONAL/OTHER COMMENTS
◆ Symptoms
  - inhalation: central nervous system recording changes, hallucinations or distorted perceptions, motor activity changes, antipsychotic, psychological test changes, bone marrow changes, irritates eyes and skin
◆ First Aid
  - wash eyes immediately with large amounts of water
  - if inhaled, remove to fresh air
  - provide respiratory support
  - wash clothing before reuse
◆ fire fighting: use carbon dioxide, dry chemical powder, or foam
◆ personal protection: wear self-contained breathing apparatus and protective clothes; use water spray to cool fire-exposed containers
◆ storage: cool, dry area; keep away from sparks, heat, and open flame

SOURCES
◆ *Environmental Contaminant Reference Databook*, volume 1, 1995
◆ *Material Safety Data Sheets (MSDS)*
◆ *Hawleys Condensed Chemical Dictionary*, 12th edition, 1993
◆ *Health Effects Notebook for Hazardous Air Pollutants*, U.S. Environmental Protection Agency, December 1994
◆ *SAX's Dangerous Properties of Industrial Materials*, 9th edition, 1996
◆ *NIOSH Pocket Guide to Chemical Hazards*, 1994 edition
◆ *Handbook of Chemistry and Physics*, 77th edition, 1996-1997
◆ *Fire Protection Guide to Hazardous Materials*, 11th ed., 1994
◆ *ACGIH Threshold Limit Values (TLVs)* for 1996
◆ *Environmental Law Index to Chemicals*, 1996 edition
◆ *Merck Index*, 12th edition 1996
◆ *Aldrich Chemical Company Catalog* 1996–1997

2,4-TOLUENE DIAMINE

CAS #: 95-80-7
MOLECULAR FORMULA: $CH_3C_6H_3(NH_2)_2$
FORMULA WEIGHT: 122.19
SYNONYMS: 3-amino-p-toluidine, 5-amino-o-toluidine, azogen developer H, benzofur MT, C.I. oxidation base, developer H, 1,3-diamino-

4-methylbenzene, 2,4-diamino-1-methylbenzene, 2,4-diaminotoluen (Czech), diaminotoluene, 2,4-diaminotoluene, 2,4-diamino-1-toluene, 2,4-diaminotoluol, eucanine GB, fouramine, fourrine, meta toluylene diamine, 4-methyl-1,3-benzenediamine, 4-methyl-m-phenylenediamine, MTD, nako TMT, pelagol grey J, pontamine developer TN, renal MD, TDA, 2,4-tolamine, m-toluenediamine, m-toluylendiamin (Czech), m-toluylenediamine, 2,4-toluylenediamine (DOT), m-tolyenediamine, m-tolylenediamine, tolylene-2,4-diamine, 2,4-tolylenediamine, 4-m-tolylenediamine, zoba gke, zogen developer H

USES
♦ Used as a Component or in the Manufacturing Process
- toluene diisocyanate
- polyurethane
- direct oxidation black
- dye for hair, furs, and leather
- polyimides with superior wire-coating properties
- impact resistant resins
- benzimidazolethiols (antioxidants)
- hydraulic fluids
- urethane foams
- fungicide stabilizers
- sensitizers for explosives
♦ Used as an Intermediate
- dyes
- heterocyclic compounds
- enhancement of thermal stability in polyimides
- fatigue resistance and dyeability in fibers
♦ used as a chain extender and cross-linker

PHYSICAL PROPERTIES
♦ colorless crystals
♦ melting point 99°C
♦ boiling point 283°C
♦ vapor pressure 1 hPa at 106°C
♦ log K_{ow} 0.337
♦ very soluble in hot water

CHEMICAL PROPERTIES
♦ stable under normal temperatures and pressures
♦ autoignition temperature 365°C
♦ incompatible with acids, acid chlorides, acid anhydrides, chloroformates, and strong oxidizing agents

HEALTH RISK
♦ Toxic Exposure Guidelines
- DFG MAK animal carcinogen, suspected human carcinogen
♦ Toxicity Data
- skin-rabbit 500 mg/24H MLD
- eye-rabbit 100 mg/24H MOD
- mma-sat 50 µg/plate
- dnd-human:fbr 100 µmol/L
- oral-mouse TDLo 1200 mg/kg
- oral-rat TDLo 2100 mg/kg/90W-C:CAR
- oral-mouse TDLo 8050 mg/kg/101W-C:CAR
- intraperitoneal-rat LD_{50} 325 mg/kg
- subcutaneous-rat LDLo 50 mg/kg
- intraperitoneal-mouse LD_{50} 480 mg/kg
- subcutaneous-dog LDLo 200 mg/kg
- subcutaneous-rabbit LDLo 400 mg/kg

- ◆ Acute Risks
  - severe skin and eye irritation
  - permanent blindness
  - asthma
  - stomach gas
- ◆ Chronic Risks
  - no information available for humans
  - liver injury to rats and mice
  - focal degenerative liver changes in rats and mice
  - decrease in births in animals
  - rise in blood pressure
  - dizziness
  - convulsions
  - fainting
  - coma
  - increase in maternal deaths, stillbirths, and resorptions in animals
  - tumors of the liver, mammary gland, subcutaneous fibromas, lung lymphomas
  - leukemia
  - EPA Group B2: probable human carcinogen

HAZARD RISK
- ◆ incompatible with acids, acid chlorides, acid anhydrides, chloroformates, and strong oxidizing agents
- ◆ hazardous decomposition products: nitrogen oxides, carbon monoxide, carbon dioxide, nitrogen
- ◆ Decomposition emits toxic fumes of NO_x.

EXPOSURE ROUTES
- ◆ wastewater samples
- ◆ leaches from boil-in-bags
- ◆ leaches from pouches into water
- ◆ inhalation of air in polyurethane plants

REGULATORY STATUS/INFORMATION
- ◆ Clean Air Act Amendments, Title III: hazardous air pollutants
- ◆ Resource Conservation and Recovery Act
  - list of hazardous constituents. 40CFR261, 40CFR261.11

ADDITIONAL/OTHER COMMENTS
- ◆ Symptoms
  - inhalation: asthma, stomach gas, dizziness, convulsions, fainting
  - irritation of skin and eyes
- ◆ First Aid
  - wash eyes immediately with large amounts of water
  - wash skin immediately with soap and water
  - remove to fresh air
- ◆ fire fighting: use water, dry chemical, chemical foam, or alcohol-resistant foam
- ◆ personal protection: wear protective clothing and respiratory apparatus if necessary

SOURCES
- ◆ *Health Effects Notebook for Hazardous Air Pollutants*, U.S. Environmental Protection Agency, December 1994
- ◆ *Material Safety Data Sheets (MSDS)*

◆ *Sax's Dangerous Properties of Industrial Materials*, 9th edition, 1996
◆ *Hawley's Condensed Chemical Dictionary*, 12th edition, 1993
◆ *Aldrich Chemical Company Catalog*, 1996–1997
◆ *Environmental Law Index to Chemicals*, 1996 edition

2,4-TOLUENE DIISOCYANATE

CAS #: 584-84-9
MOLECULAR FORMULA: $CH_3C_6H_3(NCO)_2$
FORMULA WEIGHT: 174.16
SYNONYMS: cresorcinol diisocyanate, desmodur T80, diisocyanate de toluylene, di-iso-cyanatoluene, 2,4-diisocyanato-1-methylbenzene (9CI), 2,4-diisocyanatotoluene, diisocyanat-toluol, hylene T, hylene TCPA, hylene TLC, hylene TM, hylene TM-65, hylene TRF, isocyanic acid, methylphenylene ester, isocyanic acid, 4-methyl-m-phenylene ester, 4-methyl-phenylene diisocyanate, 4-methyl-phenylene isocyanate, mondur TD, mondur TD-80, mondur TDS, nacconate 100, NCI-C50533, niax TDI, niax TDI-P, rubinate TDI 80/20, TDI (OSHA), 2,4-TDI, TDI-80, tolueen-diisocianato, tolueen-diisocyanaat, toluene diisocyanate, 2,4-toluenediisocyanate, toluene-2,4-diisocyanate, toluilenodwiuzocijjanian, toluylendiisocyanate, toluylene-2,4-diisocyanate, m-tolyene diisocyanate, tolyene-2,4-diisocyante, 2,4-tolyenediisocyanate

USES
◆ Used as a Component or in the Manufacturing Process
- polyurethane foams
- coatings
- cross-linking agent for nylon 6
- elastomers

PHYSICAL PROPERTIES
- clear, faint yellow liquid
- sharp, pungent odor
- melting point 19.5–12.5°C
- odor threshold 0.17 ppm
- vapor pressure 1 mm Hg at 80°C
- boiling point 251°C
- log octanol/water partition coefficient (log K_{ow}) 0–1
- specific gravity 1.22

CHEMICAL PROPERTIES
◆ autoignition temperature >1148°F (619°C)
◆ reacts with water
◆ flammability limits 0.9% lower and 9.5% upper by volume in air
◆ reacts violently with bases, amines, organometallic compounds, alcohols, organic acids
◆ hazardous polymerization
◆ hazardous decomposition with water
◆ self-reacts at elevated temperatures to form dimers, trimers, and polymers
◆ flash point 260°F (121°C)

HEALTH RISK
◆ Toxic Exposure Guidelines
- ACGIH TLV TWA 0.005 ppm (0.036 mg/m³)

- OSHA PEL TWA 0.02 ppm (0.14 mg/m^3)
- DFG MAK 0.01 ppm (0.07 mg/m^3)
- NIOSH REL TWA 0.005 ppm, CI 0.02 ppm/10M
- ACGIH TLV STEL/CEILING 0.002 ppm (0.14 mg/m^3)
- Toxicity Data
  - skin-rabbit LD_{50} >16 mg/kg
  - inhalation-mouse LC_{50} 10 ppm/4H
  - oral-rat LD_{50} 5800 mg/kg
  - inhalation-rat LC_{50} 14 ppm/4H
  - inhalation-rabbit LC_{50} 11 ppm/4H
  - inhalation-guinea pig LC_{50} 13 ppm/4H
  - intravenous-mouse LD_{50} 56 mg/kg
  - oral-bird LD_{50} 100 mg/kg
- Other Toxicity Indicators
  - IRAC classifies toluene 2,4-diisocyanate as a Group 2B, possible human carcinogen.
- Acute Risks
  - irritation of skin, eyes, and nose
  - affects gastrointestinal and central nervous systems
  - burns
  - chest pain and edema
  - allergic reaction
  - asthma
  - fatal if inhaled, swallowed, or absorbed through skin
  - destructive to tissue of mucous membranes and upper respiratory tract
  - lung irritation
- Chronic Risks
  - carcinogen
  - effects on liver, blood, and kidneys
  - decreases in lung function
  - bronchopneumonia
  - weight loss

HAZARD RISK
- combustible when exposed to heat or flame
- explosive in the form of vapor when exposed to heat or flame
- violent polymerization reaction with bases or acyl chlorides
- incompatible with alcohols, strong bases, amines, acids, strong oxidizing agents, heat

EXPOSURE ROUTES
- occupational exposure in industries that use toluene-2,4-diisocyanate in the manufacture and use of polyurethane foam kits
- indoor air
- inhalation
- absorption through skin and eyes

REGULATORY STATUS/INFORMATION
- Clean Air Act Title III: Hazardous Air Pollutants
- CERCLA and Superfund
  - list of extremely hazardous substances and their threshold planning quantities. 40CFR355-AB
- OSHA 29CFR1910.1000 Table Z1

- ◆ chemical regulated by the state of California

ADDITIONAL/OTHER COMMENTS
- ◆ Symptoms
  - • inhalation: irritates eyes, skin, nose, respiratory obstruction, cough, sputum, pulmonary and gastrointestinal
- ◆ First Aid
  - • wash eyes immediately with large amounts of water
  - • provide respiratory support
  - • remove and wash contaminated clothing
- ◆ fire fighting: use carbon dioxide or dry chemical powder
- ◆ personal protection: wear self-contained breathing apparatus
- ◆ storage: store under nitrogen, refrigerate, tightly closed; light and moisture sensitive

SOURCES
- ◆ *Material Safety Data Sheets (MSDS)*
- ◆ *Hawleys Condensed Chemical Dictionary, 12th edition, 1993*
- ◆ *Health Effects Notebook for Hazardous Air Pollutants*, U.S. Environmental Protection Agency, December 1994
- ◆ *Sax's Dangerous Properties of Industrial Materials*, 9th edition, 1996
- ◆ *Fire Protection Guide to Hazardous Materials*, 11th ed., 1004
- ◆ *ACGIH Threshold Limit Values (TLVs)* for 1996
- ◆ *Environmental Law Index to Chemicals*, 1996 edition
- ◆ *Merck Index*, 12th edition, 1996
- ◆ *Aldrich Chemical Company Catalog* 1996–1997
- ◆ *Environmental Contaminant Reference Databook*, volume 1, 1995

o-TOLUIDINE (ch)

CAS #: 95-53-4
MOLECULAR FORMULA: C_7H_9N
FORMULA WEIGHT: 107.17
SYNONYMS: 1-amino-2-methylbenzene, 2-amino-1-methylbenzene, o-aminotoluene, 2-aminotoluene, aniline 2-methyl-, benzenamine, 2-methyl- (9CI), C.I. 37077, 1-methyl-2-aminobenzene, 2-methyl-1-aminobenzene, o-methylbenzenamine, RCRA waste number U328, o-toluidin (Czech), 2-toluidine, o-toluidine (ACGIH:DOT:OSHA), o-toluidyna (Polish), o-tolylamine, UN1708 (DOT)

USES
- ◆ Used as a Component or in the Manufacturing Process
  - • printing textiles blue-black
  - • making various colors fast to acids
  - • preparation of ion exchange resins
  - • synthesis of dyestuffs and other intermediates
  - • chemical intermediate for dyes
  - • antioxidant in rubber manufacturing
  - • laboratory reagent in glucose analysis
  - • rodine products

PHYSICAL PROPERTIES
- light-yellow or colorless liquid
- melting point $-14.7°C$ (5°F) at 760 mm Hg
- boiling point 200.2°C (392°F) at 760 mm Hg
- density 0.9947 g/cm^3 at 298.15K
- specific gravity 1.00 (water = 1.0)
- vapor density 3.70 (air = 1.0)
- surface tension 43.55 dynes/cm, 0.04355 N/m^2 at 20°C
- heat capacity at 298.15K 130.2 J/K-mol
- heat of vaporization 99.5 cal/g, $4.16×10^5$ J/kg, 179.1 Btu/lb
- aromatic, aniline-like odor
- odor threshold 0.25 ppm
- vapor pressure <1 mm Hg at 20°C
- slightly soluble in water
- soluble in alcohol, ether, dilute acids; water solubility 16.6 g/L

CHEMICAL PROPERTIES
- combustible liquid
- will react vigorously with strong oxidizers, nitric acid, bases
- flammability limits 1.5% lower, upper not determined (percent by volume in air)
- autoignition temperature 900°F (482°C)
- heat of combustion -8990 cal/g, $-376×10^6$ J/kg, $-16,180$ Btu/lb
- enthalpy (heat) of formation at 298.15K 56.4 kJ/mol (gas)

HEALTH RISK
- Toxic Exposure Guidelines
  - ACGIH TLV TWA 2 ppm
  - OSHA PEL TWA 5 ppm
  - IDLH 50 ppm
- Toxicity Data
  - skin-rabbit 10 mg/24H open MLD
  - skin-rabbit 500 mg/24H MLD
  - eye-rabbit 750 µg/24H SEV
  - specific locus test-Drosophilia melanogaster-oral 1 mmol/L
  - unscheduled DNA synthesis-human HeLa cell 50 µL/L
  - mutation in mammalian somatic cells-mouse lymphocyte 10 mg/L
  - oral-rat TDLo 109 g/kg/2Y-C: NEO
  - subcutaneous-mouse TDLo 320 mg/kg (15–21D preg): NEO, TER
  - subcutaneous-rabbit TDLo 840 mg/kg/14W-I: ETA
  - inhalation-man TCLo 25 mg/m^3: KID, BLD
  - oral-rat LD_{50} 670 mg/kg
  - oral-mouse LD_{50} 520 mg/kg
  - intraperitoneal-mouse LD_{50} 150 mg/kg
  - oral-cat LDLo 150 mg/kg
  - oral-rabbit LD_{50} 840 mg/kg
  - skin-rabbit LD_{50} 3250 mg/kg
  - oral-frog LDLo 151 mg/kg
- Acute Risks

- Onset may be delayed 2 to 4 hours or longer.
- possibly fatal if inhaled, swallowed, or absorbed through skin
- vapor irritating to eyes, mucous membranes, and upper respiratory tract
- skin irritant
- formation of methemoglobin and possibly cyanosis
◆ Chronic Risks
  - carcinogen; target organs: bladder, kidneys, blood
  - may alter genetic material

HAZARD RISK
◆ NFPA rating
◆ Health 3, Flammability 2, Reactivity 0, Special Hazards none
◆ serious health hazard
◆ combustible liquid
◆ reacts with oxidizing materials, acids, bases

EXPOSURE ROUTES
◆ detected in tobacco smoke and in steam volatiles from the distillation of one type of tobacco leaf
◆ detected in a variety of foods including fresh kale, celery, and carrots and in shelled peas, red cabbage, and black tea aroma
◆ Inhalation or skin contact is possible in work environments utilizing 2-methylaniline.

REGULATORY STATUS/INFORMATION
◆ Clean Air Act hazardous air pollutants. Title III. (The original list was published in 42USC7412.)
◆ Resource Conservation and Recovery Act
  - list of hazardous inorganic and organic constituents. 40CFR258—Appendix 2
  - hazardous constituents. 40CFR261—Appendix 8. see also 40CFR261.11
  - ground water monitoring chemicals list. 40CFR264—Appendix 9.
  - U Waste #: U328
◆ CERCLA and Superfund
  - designation of hazardous substances. 40CFR302.4. Appendix A: sequential CAS registry # list of CERCLA hazardous substances. Appendix B: radionuclides
  - chemicals and chemical categories to which this part applies (CAS # listing). 40CFR372.65
◆ OSHA 29CFR1910.1000 Table Z1
◆ chemical regulated by the State of California

ADDITIONAL/OTHER COMMENTS
◆ Symptoms
  - weakness
  - headache
  - difficulty in breathing
  - air hunger
  - psychic disturbances
  - marked irritation of the kidneys and bladder

♦ First Aid
  • wash out eyes immediately with large amount of water
  • remove contaminated clothing and discard
  • if inhaled, exit to fresh air
  • if not breathing, artificial respiration
  • if breathing difficulty, give oxygen
  • if swallowed, wash out mouth and call physician
♦ fire fighting: use dry chemical, foam, carbon dioxide, or water spray
♦ personal protection: wear full protective clothing and positive pressure self-contained breathing apparatus
♦ storage: store in cool, dry, well-ventilated location. Separate from oxidizing materials. Store away from heat and sunlight.

SOURCES
♦ *Environmental Contaminant Reference Databook*, volume 1, 1995
♦ *Environmental Law Index to Chemicals*, 1996 edition
♦ *Fire Protection Guide to Hazardous Materials*, 11th edition, 1994
♦ *Health Effects Notebook for Hazardous Air Pollutants*, U.S. Environmental Protection Agency, December 1994
♦ *Material Safety Data Sheets (MSDS)*
♦ *Merck Index*, 12th edition, 1996
♦ *NIOSH Pocket Guide to Chemical Hazards*, June 1994
♦ *Sax's Dangerous Properties of Industrial Materials*, 9th edition, volume 2, 1996

TOXAPHENE

CAS #: 8001-35-2
MOLECULAR FORMULA: $C_{10}H_{10}Cl_8$
FORMULA WEIGHT: 413.80
SYNONYMS: alltox, anatox, camphechlor, camphene (C10H16), octachlor-, camphochlor, canfeclor, chlorinated-camphene, esotonox, geniphene, hercules 3956, kamfochlor, M-5055, melipax, motox, PCC, PCHK, phenacide, phenatox, PKHF, pnephene, polychlorocamphene, strobane-T, synthetic-3956, technical-chlorinated-camphene, toxakil, toxyphen

USES
♦ Used for Pest Control
  • cotton crops
  • peas
  • soybeans
  • corn
  • wheat
  • in livestock and poultry
  • field crops
  • banned as a pesticide in 1982 by EPA except for controlling insects on banana and pineapple crops in Puerto Rico and the Virgin Islands and in emergency situations

PHYSICAL PROPERTIES
- yellow-to-amber, waxy solid
- smells like turpentine
- odor threshold 0.14 ppm
- melting point 65–90°C
- density 1.65 at 25°C
- vapor pressure 0.2 to 0.4 mm Hg at 20°C
- almost insoluble in water
- very soluble in aromatic hydrocarbons
- log K_{ow} 2000

CHEMICAL PROPERTIES
- ◆ dehydrochlorinates in presence of alkali, in prolonged exposure to sunlight, and at temperatures above 155°C
- ◆ reacts with bases

HEALTH RISK
- ◆ Toxic Exposure Guidelines
  - OSHA PEL TWA 0.5 mg/m³
  - OSHA PEL STEL 1 mg/m³ (skin)
  - ACGIH TLV TWA 0.5 mg/m³
  - ACGIH TLV TWA 0.5 mg/m³
  - ACGIH TLV STEL 1 mg/m³ (skin)
  - NIOSH REL 0.5 mg/m³
  - DFG MAK 0.5 mg/m³
- ◆ Toxicity Data
  - skin-mammal 500 mg MOD
  - oral-rat TDLo 900 µg/kg (female 5–22D, post): REP
  - oral-mouse TDLo 375 mg/kg (female 7–12D post): REP
  - oral-rat TDLo 250 mg/kg (female 7–16D post): TER
  - oral-rat TDLo 280 mg/kg (10W male):REP
  - oral-rat TDLo 150 mg/kg (female–16D post):TER
  - oral-rat TDLo 30 g/kg/80W-C:ETA
  - oral-mouse TDLo 6600 mg/kg/80W-C:CAR
  - oral-mouse TD 13 g/kg/80W-C:CAR
  - oral-human LDLo 28 mg/kg:CNS
  - oral-man LDLo 29 mg/kg
  - skin-human TDLo 657 mg/kg:SKN
  - oral-rat LD_{50} 50 mg/kg
  - skin-rat LD_{50} 600 mg/kg
  - intraperitoneal-rat LDLo 70 mg/kg
  - oral-mouse LD_{50} 112 mg/kg
  - inhalation-mouse LCLo 2000 mg/m³/2H
  - intraperitoneal-mouse LD_{50} 47 mg/kg
  - oral-dog LD_{50} 15 mg/kg
  - oral-rabbit LD_{50} 75 mg/kg
  - skin-rabbit LD_{50} 1025 mg/kg
  - oral-guinea pig LD_{50} 250 mg/kg
  - oral-hamster LD_{50} 200 mg/kg
  - oral-duck LD_{50} 31 mg/kg
- ◆ Other Toxicity Indicators
  - EPA Cancer Risk Level (1 in a million excess lifetime risk) 3×10^{-6} mg/m³
- ◆ Acute Risks
  - intermittent convulsions
  - vomiting

- diarrhea
- respiratory failure
- seizures
- somnolence
- coma
- allergic skin dermatitis
- skin irritation
◆ Chronic Risks
  - reversible respiratory toxicity
  - loss of weight
  - loss of appetite
  - temporary deafness and disorientation
  - effects on liver, kidney, adrenal and thyroid glands, central nervous system, and immune system in animals
- death
- central nervous system stimulation
- effects on liver, kidneys, and central nervous system in animals

- developmental effects in animals
- behavioral effects and immunosuppression in offspring of rats
- increased incidence of cancer of the thyroid gland in rats
- liver cancer in mice
- EPA Group B2: probable human carcinogen

HAZARD RISK
◆ nonflammable
◆ decomposes on heating
◆ Decomposition emits toxic fumes of Cl⁻.

EXPOSURE ROUTES
◆ outdoor air
◆ soil
◆ food
◆ fish and other seafood
◆ from contaminated water
◆ farmers and pesticide applicators exposed
◆ in past, when toxaphene was manufactured and used as an insecticide, dermal and inhalation

REGULATORY STATUS/INFORMATION
◆ Clean Air Act Amendment, Title III: hazardous air pollutants
◆ Safe Drinking Water Act
  - 83 contaminants required to be regulated under the SDWA of 1986. 47FR9352
  - organic chemicals other than total trihalomethanes, sampling and analytical requirements. 40CFR141.24
  - public notification. 40CFR141.32
  - maximum contaminant level goal for organic contaminants. 40CFR141.50(b)
  - maximum contaminant level for organic chemicals. 40CFR141.61
  - variances and exemptions from the maximum contaminant levels for organic and inorganic chemicals. 40CFR142.62
◆ Resource Conservation and Recovery Act
  - design criteria (for municipal solid waste landfill under RCRA). 40CFR258.40
  - list of hazardous constituents. 40CFR258, 40CFR261, 40CFR261.11
  - groundwater monitoring list. 40CFR264

- risk specific doses. 40CFR266
- health based limits for exclusion of waste-derived residues. 40CFR266
- D waste #: DO15
- P waste #: U123
◆ CERCLA and Superfund
  - designation of hazardous substances. 40CFR302.4
  - list of chemicals. 40CFR372.65
◆ Clean Water Act
  - designation of hazardous substances. 40CFR4116.4
  - determination of reportable quantities. 40CFR4117.3
  - toxic pollutants. 40CFR401.15
  - total toxic organics (TTOs). 40CFR413.02
  - 126 priority pollutants. 40CFR423
◆ Chemical regulated by the State of California

ADDITIONAL/OTHER COMMENTS
◆ Symptoms
  - exposure: salivation, leg and back muscle spasms, nausea, vomiting, hyperexcitability, tremors, shivering, clonic convulsions
◆ First Aid
  - wash eyes immediately with large amounts of water
  - flush skin with plenty of soap and water
  - remove to fresh air immediately or provide oxygen if necessary
  - drink water and induce vomiting
◆ fire fighting: use agent suitable for surrounding fire
◆ personal protection: compressed air/oxygen apparatus, gas-tight suit

SOURCES
◆ *Health Effects Notebook for Hazardous Air Pollutants*, U.S. Environmental Protection Agency, December 1994
◆ *Material Safety Data Sheets (MSDS)*
◆ *Sax's Dangerous Properties of Industrial Materials*, 9th edition, 1996
◆ *Hawley's Condensed Chemical Dictionary*, 12th edition 1993
◆ *Environmental Law Index to Chemicals*, 1996 edition
◆ *Environmental Contaminant Reference Databook*, volume 1, 1995

1,2,4-TRICHLOROBENZENE

CAS #: 120-82-1
MOLECULAR FORMULA: $C_6H_3Cl_3$
FORMULA WEIGHT: 181.44
SYNONYMS: hostetex L-PEC, unsym-trichlorobenzene, 1,2,5-trichlorobenzene, 1,3,4-trichlorobenzene, 1,2,4-trichlorobenzene (ACGIH-OSHA), 1,2,4-trichlorobenzol, trojchlorobenzen (Polish)

USES
◆ Used as a Component or in the Manufacturing Process

- dye carrier
- herbicide intermediates
- dielectric fluid in transformers
- a degreaser
- a lubricant
- synthetic transformer oils
- insecticide against termites
◆ used as a solvent in chemical manufacturing

PHYSICAL PROPERTIES
◆ colorless liquid
◆ aromatic odor
◆ melting point 16°C (60.8°F)
◆ boiling point 214°C (417.2°F) at 1 atm
◆ specific gravity 1.454
◆ slightly soluble in water
◆ sparingly soluble in alcohol
◆ miscible with ether, benzene, petroleum ether, and carbon disulfide
◆ odor threshold 3 ppm (22.3 mg/m^3)
◆ vapor density >6 kg/m^3
◆ vapor pressure 1 mm Hg at 38.4°C
◆ octanol/water partition coefficient (log K_{ow}) 4.02
◆ volatile with steam

CHEMICAL PROPERTIES
◆ flash point 110°C (230°F)
◆ autoignition temperature 570°C (1060°F)
◆ explosion limits 2.4% lower and 6.6% upper by volume in air at 150°C (302°F)

HEALTH RISK
◆ Toxic Exposure Guidelines
  - OSHA PEL CL 5 ppm (37.1 mg/m^3)
  - ACGIH TLV CL 5 ppm (37.1 mg/m^3)
  - DFG MAK 5 ppm (37.1 mg/m^3)
◆ Toxicity Data
  - skin-rabbit 1950 mg/13W-I maximum oral dosage (MOD)
  - oral-rat TDLo 1800 mg/kg/ (9–13D preg):TER
  - oral-rat LD$_{50}$ 756 mg/kg
  - oral-mouse LD$_{50}$ 300 mg/kg
  - oral-mouse LD$_{50}$ 1223 mg/kg
◆ Acute Risks
  - harmful if swallowed, inhaled, or absorbed through skin
  - Vapor may cause irritation to eyes, mucous membranes, and upper respiratory tract.
  - causes skin irritation
◆ Chronic Risks
  - damaging to liver and kidneys
  - dizziness
  - gastrointestinal disturbances
  - headaches
  - nausea
  - EPA Group D: not classifiable as to human carcinogenicity

HAZARD RISK
- ◆ incompatible with strong oxidizers
- ◆ forms carbon monoxide, carbon dioxide, or hydrogen chloride gas upon combustion
- ◆ emits toxic fumes under fire conditions

EXPOSURE ROUTES
- ◆ primarily occupational through inhalation during its manufacture and use
- ◆ ingestion: found in drinking water and food, especially contaminated fish

REGULATORY STATUS/INFORMATION
- ◆ Clean Air Act
  - • Title III: hazardous air pollutant
- ◆ Safe Drinking Water Act
  - • public notification. 40CFR141.32
  - • MCLG (maximum contaminant level goal) for organic contaminants. 40CFR141.50(b)
  - • MCL (maximum contaminant level) for organic chemicals. 40CFR141.61
  - • MCL (maximum contaminant level) for inorganic chemicals. 40CFR141.62
- ◆ Resource Conservation Recovery Act
  - • Appendix 2: list of hazardous inorganic and organic constituents. 40CFR258
  - • Appendix 8: hazardous constituents. 40CFR261, also 40CFR261.11
  - • Appendix 9: groundwater monitoring chemicals list. 40CFR264
  - • Appendix 4: reference air concentrations. 40CFR266
  - • Appendix 8: PICs found in stack effluents. 40CFR266
- ◆ CERCLA and Superfund
  - • designation of hazardous substances. 40CFR302.4
  - • Appendix A: sequential CAS registry # list of CERCLA hazardous substances
  - • Appendix B: radionuclides
  - • chemicals and chemical categories to which this part applies (CAS # listing) 40CFR372.65
- ◆ Clean Water Act
  - • total toxic organics (TTOs). 40CFR413.02
  - • Appendix A: 126 priority pollutants. 40CFR423
- ◆ Toxic Substance Control Act of 1976
  - • reporting on precursor chemical substances. 40CFR766.38
  - • identification of specific chemical substance and mixture testing requirements. 40CFR799
- ◆ chemicals regulated by the State of California

ADDITIONAL/OTHER COMMENTS
- ◆ First Aid
  - • flush eyes or skin with copious amounts of water for at least 15 minutes

- remove contaminated clothing
- if inhaled, remove to fresh air
- if not breathing, give artificial respiration
- if breathing is difficult, give oxygen
- if ingested, wash out mouth with water and call physician
◆ disposal considerations: dissolve or mix the material with a combustible solvent and burn in a chemical incinerator equipped with an afterburner and scrubber
◆ Spill or Leak
- evacuate area and keep personnel upwind
- wear self-contained breathing apparatus, rubber boots, and heavy rubber gloves
- absorb on sand or vermiculite and place in closed containers
- ventilate area and wash spill site after material pickup is complete
- sweep up, place in a bag, and hold for waste disposal

SOURCES
◆ *Encyclopedia of Chemical Technology*, 4th edition, 1991
◆ *Environmental Law Index to Chemicals*, 1996 edition
◆ *Health Effects Notebook for Hazardous Air Pollutants*, U.S. Environmental Protection Agency, 1994
◆ *Handbook of Environmental Data on Organic Chemicals*, 2nd Edition, 1983
◆ *Material Safety Data Sheets (MSDS)*
◆ *Merck Index*, 12th edition, 1996
◆ *Sax's Dangerous Properties of Industrial Materials*, 9th edition, 1996

1,1,2-TRICHLOROETHANE (pf)

CAS #: 79-00-5
MOLECULAR FORMULA: $C_2H_3Cl_3$
FORMULA WEIGHT: 133.40
SYNONYMS: ethane trichloride, NCI-CO4579, RCRA waste number U277, RCRA waste number U359, beta-T, 1,1,2-trichlorethane, beta-trichloroethane, 1,1,2-trichloroethane, 1,2,2-trichloroethane, 1,1,2-trichloroethane (ACGIH:OSHA), trojchloroetan (1,1,2) (Polish), vinyl trichloride, many others

USES
◆ Used as a Chemical Intermediate in the Production of Other Chemicals
- 1,1-dichloroethene
◆ Used as a Solvent for
- chlorinated rubbers
- fats
- oils
- waxes
- resins
- alkaloids
◆ used in organic synthesis

PHYSICAL PROPERTIES
- ◆ clear, colorless liquid
- ◆ melting point −36.5°C (−33.7°F)
- ◆ boiling point 113.7°C (236.66°F) at 760 mm Hg
- ◆ density 1.435 g/cm³ at 20°C
- ◆ specific gravity 1.435 (water = 1)
- ◆ vapor density 4.55 (air = 1)
- ◆ surface tension not available
- ◆ viscosity not available
- ◆ heat capacity not available
- ◆ heat of vaporization not available
- ◆ sweet, slightly irritating odor
- ◆ odor threshold not available
- ◆ vapor pressure 18.8 mm at 20°C, 22.49 mm at 25°C, 40 mm at 35.2°C
- ◆ almost insoluble in water
- ◆ miscible with alcohols, ether, esters, ketones, and many other organic liquids
- ◆ not an inert solvent
- ◆ highest solvent power of the chlorinated hydrocarbons

CHEMICAL PROPERTIES
- ◆ nonflammable liquid
- ◆ stable under normal temperatures and pressures
- ◆ Hazardous polymerization will not occur.
- ◆ Heating to decomposition yields toxic hydrogen chloride gas, phosgene, carbon monoxide, and carbon dioxide fumes.
- ◆ reacts violently with sodium, potassium, magnesium, and aluminum and is heat sensitive
- ◆ incompatible with strong bases, strong oxidizing agents
- ◆ flash point not available
- ◆ explosion limits in air 8.4% lower and 13.3% upper
- ◆ autoignition temperature 460°C (860°F)
- ◆ heat of combustion not available
- ◆ enthalpy (heat) of formation not available

HEALTH RISK
- ◆ Toxic Exposure Guidelines
  - • ACGIH TLV TWA 10 ppm (55 mg/m³) (skin)
  - • OSHA PEL TWA 10 ppm (45 mg/m³) (skin)
  - • NIOSH REL TWA 10 ppm (45 mg/m³) reduce to lowest level
- ◆ Toxicity Data
  - • skin-rabbit 500 mg open MLD
  - • skin-rabbit 810 mg/24H SEV
  - • eye-rabbit 162 mg MLD
  - • skin-guinea pig 1440 mg/15M
  - • oral-mouse TDLo 76/g/kg/78W-I:CAR
  - • oral-mouse TD 152 g/kg/78W-I:CAR
  - • oral-rat LD_{50} 836 mg/kg
  - • inhalation-rat LCLo 2000 ppm/4H
  - • oral-mouse LD_{50} 378 mg/kg

- intraperitoneal-mouse LD$_{50}$ 494 mg/kg
- subcutaneous-mouse LD$_{50}$ 227 mg/kg
- oral-dog LDLo 500 mg/kg
- intraperitoneal-dog LC$_{50}$ 450 mg/kg
- intravenous-dog LDLo 95 mg/kg
- skin-rabbit LD$_{50}$ 5377 mg/kg

◆ Acute Risks
  - Dermal exposure in humans causes stinging and burning sensations with a subsequent whitening of the skin.
  - harmful if swallowed, inhaled, or absorbed
  - Exposure may lead to narcotic effects and dermatitis.
  - Ingestion may cause gastrointestinal irritation with nausea, vomiting, and diarrhea.
  - Vapor or mist is irritating to the eyes, mucous membranes, and upper respiratory tract.
  - Inhalation has caused effects on the liver, kidneys, and central nervous system of animals.
  - Tests on rats and mice have shown moderate to high acute toxicity from inhalation and oral exposures.

◆ Chronic Risks
  - Effects on the liver, eyes, kidneys, cardiovascular system, and immune system have been recorded.
  - mutagen
  - may alter genetic material
  - possible human carcinogen
  - Reference concentration is under review by EPA.
  - The reference dose (with medium confidence) is 0.004 mg/kg/d based on clinical serum chemistry in mice.

HAZARD RISK
◆ hazard toxic by ingestion, inhalation, and absorption
◆ possible human carcinogen
◆ mutagen
◆ harmful liquid and fumes
◆ may alter genetic material
◆ Heating to decomposition yields toxic hydrogen chloride gas, phosgene, carbon monoxide, and carbon dioxide fumes.
◆ reacts violently with sodium, potassium, magnesium, and aluminum and is heat sensitive
◆ incompatible with strong bases, strong oxidizing agents

EXPOSURE ROUTES
◆ inhalation
◆ absorption through skin or eyes
◆ ingestion
◆ Low levels of the contaminant are found in ambient air (0.01–0.05 ppb).
◆ Exposure from contaminated drinking water is rare.
◆ when used as a solvent in industry

REGULATORY STATUS/INFORMATION
◆ IARC Cancer Review: Group 3 IMEMDT 7,56,87
◆ Animal Limited Evidence IMEMDT 20,533,79

- ◆ NCI Carcinogenesis Bioassay (gavage)
- ◆ No Evidence: rat NCITR* NCI-CG-TR-74,78
- ◆ Clear Evidence: mouse NCITR* NCI-CG-TR-74,78
- ◆ reported in EPA TSCA Inventory
- ◆ Community Right-to-Know list
- ◆ Clean Air Act of 1970
  - • Chapter I, Subchapter C—Air Programs 40CFR50-99
  - • Title III: hazardous air pollutants, originally in 42USC7412
- ◆ Solid Waste Disposal Act of 1965
  - • Chapter I, Subchapter I—Solid Wastes 40CFR240-299
- ◆ Safe Drinking Water Act
  - • 83 contaminants required to be regulated under the SDWA of 1986 47FR9352
  - • public notification 40CFR141.32
  - • MCLG (maximum contaminant level goal) for organic contaminants 40CFR141.50(b)
  - • MCL for organic chemicals 40CFR141.61
  - • variances and exemptions from the maximum contaminant levels for organic and inorganic chemicals 40CFR142.62
  - • waste specific prohibitions—solvent wastes 40CFR148.10
- ◆ Resource Conservation Recovery Act of 1976
  - • Chapter I, Subchapter I—Solid Wastes 40CFR240-299
  - • constituents for detection monitoring (for MSWLF) 40CFR258—Appendix 1
  - • list of hazardous inorganic and organic constituents 40CFR258—Appendix 2
  - • hazardous constituents 40CFR261 Appendix 8, 40CFR261.11
  - • groundwater monitoring chemicals list 40CFR264 Appendix 9
  - • risk specific doses 40CFR266 Appendix 5
  - • health based limits for exclusion of waste-derived residues 40CFR266 Appendix 7
  - • discarded commercial chemical products, off-specification species, container
  - • residues, and spill residues thereof 40CFR261.33, 261.33.f. (U waste)
- ◆ CERCLA and Superfund
  - • designation of hazardous substances 40CFR302.4
  - • Appendix A: sequential CAS registry # list of CERCLA hazardous substances
  - • Appendix B: radionuclides
  - • chemicals and chemical categories to which this part applies 40CFR372.65
- ◆ Clean Water Act of 1977
  - • total toxic organics (TTOs) 40CFR413.02
  - • 126 priority pollutants 40CFR423—Appendix A

- EPA Office of Air Quality Planning and Standards, for a hazard ranking under section 112(g) of the Clean Air Act Amendments, has ranked the contaminant in the nonthreshold category.
- OSHA, 29CFR1910.1000 Table Z1: limits for air contaminants
- chemical regulated by the State of California

ADDITIONAL/OTHER COMMENTS

- Symptoms
  - Dermal exposure in humans causes stinging and burning sensations with a subsequent whitening of the skin.
  - Exposure may lead to narcotic effects and dermatitis.
  - Ingestion may cause gastrointestinal irritation with nausea, vomiting, and diarrhea.
  - Vapor or mist is irritating to the eyes, mucous membranes, and upper respiratory tract.
- First Aid
  - if inhaled remove to fresh air; give artificial respiration if not breathing, or oxygen if breathing is difficult, and get medical attention
  - if contacted with the eyes, flush with water for 15 minutes and get medical attention
  - if contacted with skin wash with soap and water for at least 15 minutes, while removing clothing; get medical attention
  - if swallowed give 2–4 cupfuls of milk or water and get medical attention
- fire fighting: use carbon dioxide, dry chemical powder, or appropriate foam
- personal protection: use chemical-resistant gloves, boots, apron, goggles or a face shield, and a NIOSH approved vapor respirator if needed due to poor ventilation
- storage: keep the storage container tightly closed and store in a cool, dry place away from heat, sparks and flames

SOURCES

- *Material Safety Data Sheets (MSDS)*
- *Sax's Dangerous Properties of Industrial Materials*, 9th edition, 1996
- *Merck Index*, 12th edition, 1996
- *Hawley's Condensed Chemical Dictionary*, 12th edition, 1993
- *Environmental Law Index to Chemicals*, 1996 edition
- *Health Effects Notebook for Hazardous Air Pollutants*, U.S. Environmental Protection Agency, December 1994
- http://hazard.com:80/fish/acros/11/1898.html

TRICHLOROETHYLENE

CAS #: 79-01-6
MOLECULAR FORMULA: $CHCl:CCl_2$
FORMULA WEIGHT: 131.38
SYNONYMS: acetylene-trichloride, algylen, anamenth, chlorilen, chloro-2,2-dichloroethylene(1-), chlorylen, densinfluat, dichloro-2-chloroethylene(1,1-), ethene, trichloro-(C_2HCl_3), ethinyl-trichloride, ethylene-trichloride, ethylene, trichloro-, fluate, germalgene, narcogen, narkosoid, threthylen, threthylene, trethylene, tri, trichloran, trichloren, trichlorethylene, trichloro-ethene, triclene, tri-clene, trielene, trielin, trieline, trilen, trilene, trimar, uestrosol

USES
◆ Used as an Extraction Solvent
  • greases
  • oils
  • fats
  • waxes
  • tars
◆ Used in Industry
  • degreaser of metal parts
  • chemical intermediate in the production of other chemicals
  • refrigerant
  • typewriter correction fluids
  • paint removers/strippers
  • adhesives
  • spot removers
  • rug-cleaning fluids
  • in the past as a general anesthetic

PHYSICAL PROPERTIES
◆ clear, colorless liquid
◆ mobile
◆ sweet odor of chloroform
◆ odor threshold 28 ppm
◆ melting point $-84°C$
◆ boiling point 86.7°C
◆ density 1.4649 at 20°/4°
◆ vapor pressure 60 mm Hg at 20°C
◆ vapor density 4.53
◆ viscosity 0.0055 poise
◆ immiscible with water
◆ miscible with alcohol, ether, acetone, carbon tetrachloride
◆ insoluble in water
◆ soluble in most organic solvents
◆ $\log K_{ow}$ 2.42

CHEMICAL PROPERTIES
◆ flash point 90°F
◆ autoignition temperature 788°F
◆ explosion limits 8% lower and 10.5% upper at 77°F (25°C)
◆ slowly decomposes by light in the presence of moisture
◆ nonflammable
◆ incompatible with light, ignition sources, oxidizers, alkalies, chemically active metals, epoxies, and oxidants

HEALTH RISK
◆ Toxic Exposure Guidelines
  • OSHA PEL TWA 50 ppm
  • OSHA PEL STEL 200 ppm
  • ACGIH TLV TWA 50 ppm
  • ACGIH TLV STEL 200 ppm

- DFG MAK 50 ppm
- NIOSH REL TWA 250 ppm
◆ Toxicity Data
  - oral-mouse TDLo 455 g/kg/78W-I:CAR
  - inhalation-hamster TCLo 100 ppm/6H/77W-I:ETA
  - oral-man TDLo 2143 mg/kg:GIT
  - inhalation-human TCLo 6900 mg/m^3/10M:CNS
  - inhalation-human TCLo 160 ppm/83M:CNS
  - inhalation-human TDLo 812 mg/kg
  - inhalation-man TCLo 110 ppm/8H:EYE, CNS
  - oral-rat LD$_{50}$ 5650 mg/kg
  - intraperitoneal-rat LD$_{50}$ 1282 mg/kg
  - oral-human LDLo 7 g/kg
  - inhalation-rat LC$_{50}$ 25,700 ppm/1H
  - oral-mouse LD$_{50}$ 2402 mg/kg
  - inhalation-mouse LC$_{50}$ 8450 ppm/4H
  - intravenous-mouse LD$_{50}$ 33,900 μg/kg
  - intraperitoneal-dog LD$_{50}$ 1900 mg/kg
  - subcutaneous-dog LDLo 150 mg/kg
  - intravenous-dog LDLo 150 mg/kg
  - oral-cat LDLo 5864 mg/kg
  - oral-rabbit LDLo 7330 mg/kg
  - subcutaneous-rabbit LDLo 1800 mg/kg
  - inhalation-guinea pig LCLo 37,200 ppm/40M
◆ Acute Risks
  - death
  - cardiac arrhythmias
  - massive liver damage
  - central nervous system effects
  - sleepiness
  - confusion
  - feeling of euphoria
  - effects on gastrointestinal system
  - effects on liver and kidneys
  - effects on skin
  - effects similar to alcohol inebriation
  - ventricular fibrillation
  - eye effects
  - somnolence
  - jaundice
◆ Chronic Risks
  - central nervous system effects
  - dizziness
  - headache
  - sleepiness
  - nausea
  - confusion
  - blurred vision
  - facial numbness
  - weakness
  - ventricular fibrillation
  - cardiac failure
  - increases in miscarriages in nurses exposed
  - association between elevated levels of chlorinated hydrocarbons and congenital heart disease in children of exposed parents
  - reproductive effects from drinking contaminated water
  - increase in abnormal sperm morphology in mice
  - increase in childhood leukemia
  - increase in lung, liver, and testicular tumors in animals
  - EPA Group B2/C: probable human carcinogen

HAZARD RISK
- ◆ nonflammable
- ◆ High concentrations of vapor in high temperature air can be made to burn mildly if plied with a strong flame.
- ◆ reacts with alkali and epoxides to form the spontaneously flammable gas dichloroacetylene
- ◆ can react violently with Al, Ba, N_2O_4, Li, Mg, liquid O_2, O_3, KOH, KNO_3, Na, NaOH, and Ti
- ◆ reacts with water under heat and pressure to form HCl gas
- ◆ Decomposition emits toxic fumes of Cl^-.

EXPOSURE ROUTES
- ◆ ambient air
- ◆ drinking water
- ◆ groundwater
- ◆ air around factories where trichloroethylene is manufactured
- ◆ use of products containing the chemical
- ◆ evaporation and leaching from waste disposal sites

REGULATORY STATUS/INFORMATION
- ◆ Clean Air Act Amendment, Title III: hazardous air pollutants
- ◆ Safe Drinking Water Act
  - • 83 contaminants required to be regulated under the SDWA of 1986. 47FR9352
  - • public notification. 40CFR141.32
  - • maximum contaminant level goal for organic contaminants. 40CFR141.50(a)
  - • maximum contaminant level for organic chemicals. 40CFR141.61
  - • variances and exemptions from the maximum contaminant levels for organic and inorganic chemicals. 40CFR142.62
  - • waste specific prohibitions—solvent wastes. 40CFR148.10
- ◆ Resource Conservation and Recovery Act
  - • design criteria (for municipal solid waste landfill under RCRA). 40CFR258.40
  - • constituents for monitoring detection for municipal solid waste landfill. 40CFR258
  - • list of hazardous constituents. 40CFR258, 40CFR261, 40CFR261.11
  - • groundwater monitoring list. 40CFR264
  - • risk specific doses. 40CFR266
  - • health based limits for exclusion of waste-derived residues. 40CFR266
  - • PICs found in stack effluents. 40CFR266
  - • D waste #: DO40
  - • U waste #: U228
- ◆ CERCLA and Superfund
  - • designation of hazardous substances. 40CFR302.4
  - • list of chemicals. 40CFR372.65

- ◆ Clean Water Act
  - designation of hazardous substances. 40CFR4116.4
  - determination of reportable quantities. 40CFR4117.3
  - toxic pollutants. 40CFR401.15
  - total toxic organics (TTOs). 40CFR413.02
  - 126 priority pollutants. 40CFR423
- ◆ OSHA 29CFR1910.100 Table Z1 and Table Z2
- ◆ chemical regulated by the State of California

ADDITIONAL/OTHER COMMENTS
- ◆ Symptoms
  - inhalation and ingestion: dizziness, headache, nausea, drowsiness, unconsciousness, death
- ◆ First Aid
  - wash eyes immediately with large amounts of water
  - wash skin with soap and water immediately
  - remove to fresh air immediately or provide respiratory apparatus if necessary
- ◆ fire fighting: use water spray, dry chemical, carbon dioxide, or foam
- ◆ personal protection: protective clothing, goggles, and respiratory apparatus
- ◆ storage
  - tightly closed container away from sources of ignition in a cool, dry, well-ventilated area away from oxidizing materials
  - sealed, light resistant ampules or in frangible, light resistant glass tubes

SOURCES
- ◆ *Health Effects Notebook for Hazardous Air Pollutants*, U.S. Environmental Protection Agency, December 1994
- ◆ *Material Safety Data Sheets (MSDS)*
- ◆ *Sax's Dangerous Properties of Industrial Materials*, 9th edition, 1996
- ◆ *Hawley's Condensed Chemical Dictionary*, 12th edition, 1993
- ◆ *Aldrich Chemical Company Catalog*, 1996–1997
- ◆ *Environmental Law Index to Chemicals*, 1996 edition
- ◆ *Merck Index*, 12th edition, 1996

2,4,5-TRICHLOROPHENOL

CAS #: 95-95-4
MOLECULAR FORMULA: $Cl_3C_6H_2OH$
FORMULA WEIGHT: 197.44
SYNONYMS: collunosol, dowicide 2, dowicide B, nurelle, preventol-I, TCP

USES
- ◆ Used as a Preservative
  - adhesives
  - synthetic textiles
  - rubber
  - wood
  - paints
  - paper manufacture

◆ Used in the Manufacturing Process
- fungicides
- bactericide
- adsorbent in microbial production of vitamin B_{12}
- cooling towers
- pulp mill systems
- hide and leather processing and disinfection
- swimming pool related surfaces
- household sickroom equipment
- food contact surfaces
- hospital rooms

◆ Used as an Intermediate
- for 2,4,5-trichlorophenyl chloroformate
- for the herbicide 2,4,5-trichlorophenoxyacetic acid
- for chemicals used as pesticides

◆ Used as an Antifungal Agent
- in adhesives as a preservative in polyvinyl acetate emulsions
- in auto industry to preserve rubber gaskets
- in textiles to preserve emulsions used for rayon
- medication

PHYSICAL PROPERTIES
◆ gray flakes or needles
◆ strong unpleasant phenolic odor
◆ melting point 67°C
◆ boiling point 253°C at 760 mm Hg
◆ density 1.678 at 25°C
◆ odor threshold 100 μg/L at 30°C in water
◆ heat of vaporization 13,237 gcal/gmol
◆ vapor pressure 0.022 mm Hg at 25°C
◆ log K_{ow} 3.96
◆ very soluble in ethyl ethers
◆ sparingly soluble in water

CHEMICAL PROPERTIES
◆ flash point 0°C
◆ sublimes
◆ nonflammable

HEALTH RISK
◆ Toxic Exposure Guidelines
- meets criteria for OSHA medical records rule
◆ Toxicity Data
- mmo-sat 10 μg/plate
- mma-sat 10 μg/plate
- oral-mouse TDLo 4 g/kg
- skin-mouse TDLo 6700 mg/kg/16W-I:NEO
- oral-rat LD_{50} 820 mg/kg
- intraperitoneal-rat LD_{50} 355 mg/kg
- subcutaneous-rat LD_{50} 2260 mg/kg
- oral-mouse LD_{50} 600 mg/kg
- intravenous-mouse LD_{50} 56 mg/kg
◆ Acute Risks
- skin burns
- redness of skin
- edema
- irritation of eyes, nose, pharynx, and lungs
- swelling of eyes

- asthenia
- anorexia
- loss of weight

◆ Chronic Risks
  - no information available for humans
  - degenerative changes in liver and kidneys of rats
  - chloracne
  - skin ulcerations

- headache
- conjunctivitis

- porphyria cutanea tarda
- EPA Group D: not classifiable as to human carcinogenicity due to inadequate data

HAZARD RISK
◆ When heated to decomposition it explodes.
◆ emits toxic fumes of Cl^- upon decomposition
◆ nonflammable

EXPOSURE ROUTES
◆ primarily by inhalation and dermal contact
◆ workers involved in manufacture, formulation, or application of pesticides
◆ production and use as a pesticide and intermediate
◆ emissions from incinerators to air
◆ drinking water
◆ food

REGULATORY STATUS/INFORMATION
◆ F waste #: F027

ADDITIONAL/OTHER COMMENTS
◆ Symptoms
  - inhalation of dust: swelling of eyes and eye injury, irritation of nose and throat
  - contact: irritates skin on prolonged contact
◆ First Aid
  - wash eyes and skin immediately with large amounts of water
  - remove to fresh air immediately
◆ fire fighting: use dry chemical, foam, or carbon dioxide
◆ personal protection: wear self-contained breathing apparatus and protective clothing

SOURCES
◆ *Health Effects Notebook for Hazardous Air Pollutants*, U.S. Environmental Protection Agency, December 1994
◆ *Material Safety Data Sheets (MSDS)*
◆ *Sax's Dangerous Properties of Industrial Materials*, 9th edition, 1996
◆ *Hawley's Condensed Chemical Dictionary*, 12th edition, 1993
◆ *Aldrich Chemical Company Catalog*, 1996–1997
◆ *Environmental Law Index to Chemicals*, 1996 edition
◆ *Merck Index*, 12th edition, 1996
◆ *Environmental Contaminant Reference Databook*, volume 1, 1995

2,4,6-TRICHLOROPHENOL

CAS #: 88-06-2
MOLECULAR FORMULA: $C_6H_2CL_3OH$
FORMULA WEIGHT: 197.44
SYNONYMS: dowicide-25, omal, phenachlor, 2,4,6-trichlorofenol (Czech), trichloro-2-hydroxybenzene(1,3,5-)

USES
- ◆ no longer used in the United States
- ◆ Previously Used in the Manufacture of Other Chemicals
  - leather
  - pesticide for wood
  - glue preservation
  - antimildew treatment
  - fungicide
  - herbicide
  - defoliant

PHYSICAL PROPERTIES
- ◆ yellow flakes
- ◆ strong phenolic odor (sweet smell)
- ◆ melting point 69°C
- ◆ boiling point 245°C
- ◆ specific gravity 1.7
- ◆ density 1.4901 g/cm³
- ◆ vapor pressure 1 mm Hg at 76.5°C
- ◆ odor threshold 0.0026 ppm
- ◆ log K_{ow} 3.38
- ◆ K_{oc} 2.0×10^3
- ◆ soluble in water, acetone, alcohol, and ether

CHEMICAL PROPERTIES
- ◆ nonflammable
- ◆ stable
- ◆ incompatible with strong oxidizers
- ◆ will not polymerize
- ◆ no flash point

HEALTH RISK
- ◆ Toxic Exposure Guidelines
  - none available
- ◆ Toxicity Data
  - skin-rabbit 500 mg/24H MOD
  - eye-rabbit 50 µg/24H SEV
  - mma-sat 10 µg/plate
  - msc-mouse:lyms 80 mg/L
  - oral-rat TDLo 12,500 mg/kg
  - oral-rat TDLo 185 g/kg/2Y-C:CAR
  - oral-mouse TDLo 441 g/kg/2Y-C:CAR
  - oral-rat LD_{50} 820 mg/kg
  - intraperitoneal-rat LD_{50} 276 mg/kg
  - oral-mammal LD_{50} 454 mg/kg
  - skin-mammal LD_{50} 700 mg/kg
- ◆ Other Toxicity Indicators
  - EPA Cancer Risk Level (1 in a million excess lifetime risk) 0.0003 mg/m³
- ◆ Acute Risks
  - irritation of the upper respiratory tract
  - irritation to skin and eyes
  - burns on skin
  - corneal injury
  - irritation of the gastrointestinal tract

516

- ◆ Chronic Risks
  - respiratory effects: cough, wheezing
  - chronic bronchitis
  - altered pulmonary function
  - pulmonary lesions
  - changes in liver and spleen cells
  - lowers body weight
  - death
  - leukemia and liver cancer in rats
  - hepatic hyperplasia
  - EPA Group 2B: probable human carcinogen

HAZARD RISK
- ◆ Combustion will produce carbon dioxide, carbon monoxide, and hydrogen chloride.
- ◆ capable of creating dust explosions
- ◆ incompatible with strong oxidizers
- ◆ Decomposition emits toxic fumes of Cl^-.

EXPOSURE ROUTES
- ◆ primarily in water
- ◆ detected in air
- ◆ ingestion of food
- ◆ pesticides
- ◆ soil
- ◆ wood
- ◆ leather
- ◆ glue preservatives
- ◆ production sources of chlorinated phenols or waste burners
- ◆ wastewater
- ◆ near hazardous waste sites

REGULATORY STATUS/INFORMATION
- ◆ Clean Air Act Amendment, Title III: hazardous air pollutants
- ◆ Resource Conservation and Recovery Act
  - list of hazardous constituents. 40CFR258, 40CFR261, 40CFR261.11
  - groundwater monitoring list. 40CFR264
  - risk specific doses. 40CFR266
  - health based limits for exclusion of waste-derived residues. 40CFR266
  - PICs found in stack effluents. 40CFR266
  - D waste #: DO42
  - F waste #: F027
- ◆ CERCLA and Superfund
  - designation of hazardous substances. 40CFR302.4
  - list of chemicals. 40CFR372.65
- ◆ Clean Water Act
  - total toxic organics (TTOs). 40CFR413.02
  - 126 priority pollutants. 40CFR423
- ◆ chemical regulated by the State of California

ADDITIONAL/OTHER COMMENTS
- ◆ Symptoms
  - inhalation: irritation of the upper respiratory tract
  - skin contact: irritation and burns
  - ingestion: irritation of the gastrointestinal tract
- ◆ First Aid
  - wash eyes and skin immediately with large amounts of water
  - remove to fresh air immediately

- fire fighting: use dry chemical, water spray or mist, or carbon dioxide
- personal protection: wear self-contained breathing apparatus and protective clothing

SOURCES

- *Health Effects Notebook for Hazardous Air Pollutants*, U.S. Environmental Protection Agency, December 1994
- *Material Safety Data Sheets (MSDS)*
- *Sax's Dangerous Properties of Industrial Materials*, 9th edition, 1996
- *Hawley's Condensed Chemical Dictionary*, 12th edition, 1993
- *Aldrich Chemical Company Catalog*, 1996–1997
- *Environmental Law Index to Chemicals*, 1996 edition
- *Merck Index*, 12th edition, 1996
- *Toxicological Profile for 2,4,6-Trichlorophenol*

TRIETHYLAMINE

CAS #: 121-44-8
MOLECULAR FORMULA: $(C_2H_5)_3N$
FORMULA WEIGHT: 101.22
SYNONYMS: diethyl-ethanamine(n,n-), ethanamine, n,n-diethyl-, triethylamine-anhydrous

USES

- Used in Industry
  - catalytic solvent in chemical synthesis
  - accelerator for rubber
  - corrosion inhibitor
  - curing and hardening agent for polymers
  - propellant
  - manufacture of wetting, penetrating, and water-proofing agents of quaternary ammonium compounds
  - desalination of seawater
  - antilivering agent for urea and melamine based enamels
  - recovery of gelled paint vehicles
  - herbicides
  - pesticides
  - preparation of emulsifiers for pesticides
  - nonnutritive sweeteners, ketenes, and salts
  - drying of printing inks
  - carpet cleaners

PHYSICAL PROPERTIES

- colorless liquid
- strong, fishy, ammonia-like odor
- odor threshold 0.48 ppm
- melting point −115°C
- boiling point 89.3°C
- density 0.7255 at 25°C
- surface tension 20.7 dynes/cm
- vapor pressure 400 mm Hg at 31.5°C
- vapor density 3.48
- soluble in 0.7 parts water
- soluble in alcohol and chloroform
- slightly soluble in benzene
- practically insoluble in ether
- log K_{ow} 1.45

CHEMICAL PROPERTIES

- flammable
- flash point 0°C
- autoignition temperature 842°F
- heat of combustion −1036.8 kcal/gmol wt at 20°C
- heat of vaporization 140 Btu/lb
- explosion limits in air 1.2% lower and 8% upper
- incompatible with acids and oxidizing agents

HEALTH RISK

- Toxic Exposure Guidelines
  - OSHA PEL TWA 10 ppm
  - OSHA PEL STEL 15 ppm
  - ACGIH TLV TWA 10 ppm
  - ACGIH TLV STEL 15 ppm
  - DFG MAK 10 ppm
  - MSHA standard 100 mg/m^3
- Toxicity Data
  - skin-rabbit 10 mg/24H open MLD
  - skin-rabbit 365 mg open MLD
  - eye-rabbit 250 mg open SEV
  - eye-rabbit 50 ppm/30D-I SEV
  - oral-rabbit TDLo 6900 μg/kg
  - inhalation-human TCLo 12 mg/m^3/11W-C:EYE
  - oral-rat LD$_{50}$ 460 mg/kg
  - inhalation-rat LCLo 1000 ppm/4H
  - oral-mouse LD$_{50}$ 546 mg/kg
  - skin-rabbit LD$_{50}$ 570 mg/kg
  - inhalation-mouse LD$_{50}$ 6 g/m^3/2H
  - intraperitoneal-mouse LD$_{50}$ 405 mg/kg
  - inhalation-guinea pig LCLo 1000 ppm/4H
  - inhalation-mammal LC$_{50}$ 6 g/m^3
- Acute Risks
  - eye irritation
  - corneal swelling
  - halo vision
  - seeing "blue haze"
  - "smoky vision"
  - irritation of skin and mucous membranes
- Chronic Risks
  - reversible corneal edema
  - inflammation of the nasal passage in rats
  - thickening of the interalveolar walls of the lungs in rats
  - mucous accumulation in the alveolar spaces in the lungs in rats
  - hematological effects in rats
  - irritation of lungs and edema in rabbits
  - moderate peribronchitis in rabbits
  - vascular thickening in rabbits
  - eye lesions in rabbits

HAZARD RISK

- very dangerous fire hazard when exposed to heat, flame, or oxidizers
- explosive in the form of vapor when exposed to heat or flame

- ◆ Complex with dinitrogen tetraoxide explodes below 0°C when undiluted with solvent.
- ◆ exothermic reaction with maleic anhydride above 150°C
- ◆ reacts with oxidizing materials
- ◆ incompatible with $N_2 O_2$
- ◆ Decomposition emits toxic fumes of NO_x.

EXPOSURE ROUTES
- ◆ inhalation and dermal contact during manufacture and use
- ◆ ingesting contaminated food
- ◆ broiled beef

REGULATORY STATUS/INFORMATION
- ◆ Clean Air Act Amendment, Title III: hazardous air pollutants
- ◆ Resource Conservation and Recovery Act
  - • list of hazardous constituents. 40CFR261, 40CFR261.11
  - • U waste #: U404
- ◆ CERCLA and Superfund
  - • designation of hazardous substances. 40CFR302.4
  - • list of chemicals. 40CFR372.65
- ◆ Clean Water Act
  - • designation of hazardous substances. 40CFR4116.4
  - • determination of reportable quantities. 40CFR4117.3
- ◆ OSHA 29CFR1910.100 Table Z1
- ◆ chemical regulated by the State of California

ADDITIONAL/OTHER COMMENTS
- ◆ Symptoms
  - • vapors: irritate nose, throat, and lungs, coughing, choking, difficult breathing
  - • eyes: severe burns
  - • contact to wet clothing: skin burns
- ◆ First Aid
  - • wash eyes and skin immediately with large amounts of water
  - • remove to fresh air immediately or provide respiratory apparatus if necessary
  - • wash out mouth with water
- ◆ fire fighting: use carbon dioxide, dry chemical powder, or appropriate foam
- ◆ personal protection: protective clothing, goggles, and respiratory apparatus
- ◆ storage: keep tightly closed away from heat, sparks, and open flame in a cool, dry place

SOURCES
- ◆ *Health Effects Notebook for Hazardous Air Pollutants*, U.S. Environmental Protection Agency, December 1994
- ◆ *Material Safety Data Sheets (MSDS)*
- ◆ *Sax's Dangerous Properties of Industrial Materials*, 9th edition, 1996
- ◆ *Hawley's Condensed Chemical Dictionary*, 12th edition, 1993
- ◆ *Aldrich Chemical Company Catalog*, 1996–1997
- ◆ *Environmental Law Index to Chemicals*, 1996 edition

◆ *Environmental Contaminant Reference Databook*, volume 1, 1995
◆ *Merck Index*, 12th edition, 1996

TRIFLURALIN

CAS #: 1582-09-8
MOLECULAR FORMULA: $F_3C(NO_2)_2C_6H_2N(C_3H_7)_2$
FORMULA WEIGHT: 335.32
SYNONYMS: agreflan, agriflan 24, crisalin, digermin, 2,6-dintro-n,n-di-n-propyl-alpha-alpha-alpha-trifluro-p-toluidine, 2,6-Dinitro-4-trifluormethyl-n,n-dipropylanilin (German), 4-(di-n-propylamino)-3,5-dinitro-1-trifluoromethylbenzene, n,n-di-n-propyl-2,6-dinitro-4-trifluoromethylaniline, n,n-dipropyl-4-trifluoromethyl-2,6-dinitroaniline, elancolan, L-36352, Lilly 36,352, M.T.F., NCI-C00442, nitran, olitref, super-treflan, su seguro carpidor, synfloran, trefanocide, treficon, treflam, treflan, treflanocide elancolan, tri-4, trifloran, trifluoralin (USDA), alpha-alpha-alpha,trifluoro-2,6-dinitro-n,n-dipropylaniline, trifluralina 600, trifluraline, trifurex, trikepin, trim, tristar

USES
◆ Used as a Herbicide
  • control annual grasses
  • broadleaf annual weeds
  • crops
  • shrubs
  • flowers
  • cotton
  • soybeans
  • fruits and vegetables

PHYSICAL PROPERTIES
◆ yellow-orange crystalline solid or crystals
◆ melting point 48.5–49°C
◆ boiling point 139–140°C
◆ vapor pressure 1.99×10^{-4} mm Hg at 29.5°C
◆ slightly soluble in water
◆ very soluble in Me_2CO and xylene

CHEMICAL PROPERTIES
◆ stable

HEALTH RISK
◆ Toxic Exposure Guidelines
  • LOAEL (dog) 3.75 mg/kg/d
  • NOAEL (dog) 0.75 mg/kg/d
  • RfD (0.0075 mg/kg/d)
◆ Toxicity Data
  • mma-sat 1 mg/plate
  • mrc-asn 100 µg/plate
  • cyt-human:lym 2 ppm
  • sce-human:lym 1 mg/L
  • cyt-mouse-intraperitoneal 200 mg/kg
  • oral-mouse TDLo 10 mg/kg
  • intraperitoneal-mouse TDLo 200 mg/kg
  • oral-mouse TDLo 180 g/kg/78W-C:CAR
  • intraperitoneal-mouse TDLo 2600 µg/kg/39D-I:ETA
  • subcutaneous-mouse TDLo 2600 µg/kg/39D-I:ETA
  • oral-mouse TD 340 g/kg/78W-C:CAR

- oral-rat LD_{50} 1930 mg/kg
- inhalation-rat LC_{50} 2800 mg/kg
- skin-rat LD_{50} >5 g/kg
- oral-mouse LD_{50} 3197 mg/kg
- intraperitoneal-mouse LDLo 1500 mg/kg
- oral-dog LD_{50} >2 g/kg
- oral-rabbit LD_{50} >2 g/kg

◆ Acute Risks
- irritation of skin and eyes
- irritation of upper respiratory tract
- burning sensation
- weakness
- effects on central nervous system
- redness of eyes

◆ Chronic Risks
- no information available on the chronic effects in humans
- decreased weight gain in dogs
- changes in hematological parameters in dogs
- increased liver weight in dogs
- skeletal abnormalities in offspring of mice
- depressed fetal weight in rats and rabbits
- fetotoxic and teratogenic effects in rodents
- increased incidences of urinary tract tumors in rats
- increases in thyroid tumors in rats
- urinary metabolite in rats
- nonneoplastic renal pathology
- EPA Group C: possible human carcinogen

HAZARD RISK
◆ Decomposition emits toxic fumes of F^- and NO_x.

EXPOSURE ROUTES
◆ inhalation or dermal contact
◆ production, manufacture, or application as a herbicide
◆ dermal exposure by farmworkers
◆ adsorbed to clothing even after many washings
◆ fugitive emissions during production
◆ fugitive emissions in wastewater effluent
◆ ambient air through application as a herbicide
◆ released to surface water because of agricultural runoff
◆ lawn products
◆ ingestion of contaminated agricultural products
◆ ingestion of fish in contaminated waters

REGULATORY STATUS/INFORMATION
◆ Clean Air Act Amendment, Title III: hazardous air pollutants
◆ Safe Drinking Water Act: priority list of drinking water contaminants. 55FR1470
◆ Federal Insecticide, Fungicide, and Rodenticide Act
  - tolerances and exemptions from tolerances for pesticide chemicals in or on raw agricultural commodities. 40CFR180.102-1147
  - tolerances for pesticides in food. 40CFR185
◆ CERCLA and Superfund

- list of chemicals. 40CFR372.65
◆ chemical regulated by the State of California

ADDITIONAL/OTHER COMMENTS
◆ Symptoms
  - inhalation: irritation of skin, eyes, and upper respiratory tract
  - ingestion: burning sensation, weakness
  - eye contact: lacrimation, redness
◆ First Aid
  - wash eyes and skin with water
◆ personal protection: compressed air/oxygen apparatus, gas-tight suit
◆ storage: keep upwind

SOURCES
◆ *Health Effects Notebook for Hazardous Air Pollutants*, U.S. Environmental Protection Agency, December 1994
◆ *Material Safety Data Sheets (MSDS)*
◆ *Sax's Dangerous Properties of Industrial Materials*, 9th edition, 1996
◆ *Hawley's Condensed Chemical Dictionary*, 12th edition, 1993
◆ *Environmental Law Index to Chemicals*, 1996 edition
◆ *Merck Index*, 12th edition, 1996

2,2,4-TRIMETHYLPENTANE

CAS #: 540-84-1
MOLECULAR FORMULA: $(CH_3)_2CHCH_2C(CH_3)_3$
FORMULA WEIGHT: 114.26
SYNONYMS: isobutyl-trimethylmethane, isooctane, trimethylpentane(2,2,4-)

USES
◆ Used as a Component and an Intermediate
  - azeotropic distillation entrainer
  - determination of octane numbers of fuels
  - spectrophotometric analysis
  - solvent and thinner
  - organic synthesis
  - antiknock agent
  - solvent for rubber processing and for other applications

PHYSICAL PROPERTIES
◆ clear liquid
◆ odor of gasoline
◆ melting point −116°C
◆ boiling point 99.2°C
◆ density 0.692 g/cm³ at 20°/4°C
◆ vapor pressure 40.6 mm at 21°C
◆ vapor density 3.93
◆ critical temperature 270.8°C
◆ critical pressure 25.34 atm
◆ insoluble in water

CHEMICAL PROPERTIES
◆ flash point 10°F
◆ autoignition temperature 779°F
◆ flammability limits 1.1 lower and 6.0 upper
◆ highly flammable

◆ reacts with oxidizing and reducing agents

HEALTH RISK

- ◆ Toxic Exposure Guidelines
  - • NIOSH REL TWA (Alkanes) 350 mg/m^3
- ◆ Toxicity Data
  - • There is no toxicity data available for 2,2,4-trimethylpen-
tane.
- ◆ Acute Risks
  - • necrosis of skin and tissue
  - • irritation of lungs
  - • edema
  - • hemorrhage
  - • central nervous system depression
  - • labored breath
  - • unconsciousness
  - • vomiting
  - • dizziness
  - • headache
  - • nausea
  - • weakness
  - • prickling of skin and eyes
  - • redness of skin and eyes
  - • blisters
  - • asphyxiating
- ◆ Chronic Risks
  - • no information on chronic
effects in humans
  - • dermatitis
  - • central nervous system depression
  - • kidney and liver effects in
rats

HAZARD RISK

- ◆ very dangerous fire hazard when exposed to heat, flame, and
oxidizers
- ◆ reacts vigorously with reducing materials
- ◆ explosive with air in the form of vapor when exposed to heat
or flame
- ◆ Decomposition emits acrid smoke and irritating fumes.
- ◆ may generate electrostatic charges
- ◆ formation of corrosive vapors

EXPOSURE ROUTES

- ◆ manufacture, use, and disposal of products in petroleum and
gasoline field
- ◆ automotive exhaust
- ◆ evaporative emissions
- ◆ inhalation of ambient air
- ◆ inhalation during the refining of petroleum
- ◆ during use and disposal of petroleum products and gasoline

REGULATORY STATUS/INFORMATION

- ◆ Clean Air Act Amendment, Title III: hazardous air pollutants
- ◆ chemical regulated by the State of California

ADDITIONAL/OTHER COMMENTS

- ◆ Symptoms
  - • inhalation: labored breath, unconsciousness, vomiting, dizziness, headache, nausea, asphyxiating
  - • skin absorption: prickling, redness, blisters
- ◆ First Aid

- wash skin and eyes immediately with large amounts of water
  - provide respiratory support
- fire fighting: use dry chemical, foam, or carbon dioxide
- personal protection: wear compressed air/oxygen apparatus
- storage: flammable liquids storage room

SOURCES
- *Health Effects Notebook for Hazardous Air Pollutants*, U.S. Environmental Protection Agency, December 1994
- *Material Safety Data Sheets (MSDS)*
- *Sax's Dangerous Properties of Industrial Materials*, 9th edition, 1996
- *Hawley's Condensed Chemical Dictionary*, 12th edition, 1993
- *Aldrich Chemical Company Catalog*, 1996–1997
- *Environmental Law Index to Chemicals*, 1996 edition

VINYL ACETATE

CAS #: 108-05-4
MOLECULAR FORMULA: $CH_3COOCH:CH_2$
FORMULA WEIGHT: 86.10
SYNONYMS: acetic acid, ethenyl ester, acetic acid, ethylene ester, acetic acid, vinyl ester, acetoxy-ethylene, acetoxy-ethylene(1-), ethenyl-acetate, ethenyl-ethanoate, vinyl-a-monomer, vinyl-acetate-monomer, vinyl-ethanoate

USES
- Used in Polymerization Form
  - plastic masses
  - films
  - lacquers
- Used in the Production Process
  - emulsion paint substances
  - finishing
  - impregnation materials
  - glue
  - polyvinyl acetate
  - polyvinyl alcohol
  - safety glass interlayers
  - water based paints
  - adhesives
  - paper coatings
  - nonwoven binders
  - inks
  - films
  - chewing gum base
  - hair sprays
- Used as an Intermediate
  - resins
  - polymers
  - emulsions

PHYSICAL PROPERTIES
- clear, colorless liquid
- mobile
- sweet, pleasant, fruity odor
- odor threshold 0.5 ppm
- melting point −93.2°C
- boiling point 72.2°C
- density 0.932 at 20°C
- surface tension 23.95 dynes/cm
- viscosity 0.42 mPa at 20°C
- vapor pressure 115 mm Hg at 25°C

- ◆ vapor density 3
- ◆ miscible in alcohol and ether

CHEMICAL PROPERTIES

- ◆ flammable
- ◆ flash point −6°C
- ◆ autoignition temperature 426°C
- ◆ explosion limits in air 2.6% lower and 13.4% upper
- ◆ polymerizes to solid on exposure to light

- ◆ slightly soluble in water
- ◆ log K_{ow} 0.21–0.73

- ◆ heat of combustion −9754 Btu/lb
- ◆ heat of vaporization 163 Btu/lb
- ◆ incompatible with acids, bases, oxidizing agents, peroxides, and heat

HEALTH RISK

- ◆ Toxic Exposure Guidelines
  - OSHA PEL TWA 10 ppm
  - OSHA PEL STEL 20 ppm
  - ACGIH TLV TWA 10 ppm
  - ACGIH TLV STEL 20 ppm
  - DFG MAK 10 ppm
  - NIOSH REL CL 15 mg/m³/15M
  - MSHA standard 30 mg/m³
- ◆ Toxicity Data
  - eye-human 22 ppm
  - skin-rabbit 10 mg/24H open
  - eye-rabbit 500 mg/24H MLD
  - cyt-human:lym 250 μmol/L
  - sce-hamster:ovr 125 μmol/L
  - oral-rat TDLo 500 mg/kg/D
  - oral-rat TDLo 100 g/kg/2Y-C:CAR
  - inhalation-rat TCLo 600 ppm/6H/5D/2Y-I:ETA
  - oral-rat LD_{50} 2920 mg/kg
  - inhalation-rat LC_{50} 4000 ppm/2H
  - oral-mouse LD_{50} 1613 mg/kg
  - inhalation-mouse LC_{50} 1550 ppm/4H
  - inhalation-rabbit LC_{50} 2500 ppm/4H
  - skin-rabbit LD_{50} 2335 mg/kg
  - inhalation-guinea pig LC_{50} 6215 ppm/4H
  - intraperitoneal-guinea pig LDLo 500 mg/kg
- ◆ Acute Risks
  - eye irritation
  - upper respiratory tract irritation
  - nasal irritation
  - labored breathing
  - lung damage
  - convulsions
  - narcosis
- ◆ Chronic Risks
  - upper respiratory tract irritation
  - cough
  - hoarseness
  - nasal epithelial lesions in mice and rats
  - inflammation and irritation of upper respiratory tract in mice and rats
  - reduced body weight gain in rats
  - fetal growth retardation in rats

- increased incidence of nasal cavity tumors in rats
- neoplastic nodules of liver in rats
- adenocarcinomas of the uterus in rats
- C-cell adenomas or carcinomas of the thyroid in rats
- EPA Group C: possible human carcinogen

HAZARD RISK
- highly dangerous fire hazard when exposed to heat, flame, or oxidizers
- storage hazard: may undergo spontaneous exothermic polymerization
- reaction with air or water to form peroxides that form explosions
- Reaction with hydrogen peroxide forms the explosive peracetic acid.
- reacts with oxygen above 50°C to form an unstable explosive peroxide
- reacts with ozone to form explosive vinyl acetate ozonide
- Solution polymerization of the acetate dissolved in toluene produces explosions.
- Polymerization reaction with dibenzol peroxide + ethyl acetate may release ignitable and explosive vapors.
- Vapor may react vigorously with desiccants.
- incompatible with 2-amino ethanol, chlorosulfonic acid, ethylenediamine, ethyleneimine, HCl, HF, HNO_3, oleum, peroxides, and H_2SO_4

EXPOSURE ROUTES
- workplace during manufacture and use
- inhalation of ambient air near facilities manufacturing this compound

REGULATORY STATUS/INFORMATION
- Clean Air Act Amendment, Title III: hazardous air pollutants
- Resource Conservation and Recovery Act
  - constituents for detection monitoring. 40CFR258
  - list of hazardous constituents. 40CFR258
- CERCLA and Superfund
  - designation of hazardous substances. 40CFR302.4
  - list of chemicals. 40CFR372.65
- Clean Water Act
  - designation of hazardous substances. 40CFR116.4
  - determination of reportable quantities. 40CFR117.3
- Toxic Substance Control Act
  - list of substances. 40CFR716.120.a
- OSHA 29CFR1910.1000 Table Z1
- chemical regulated by the State of California

ADDITIONAL/OTHER COMMENTS
- Symptoms
  - exposure: irritation of eyes, skin, and upper respiratory tract

◆ First Aid
  • wash eyes immediately with large amounts of water
  • flush skin with plenty of soap and water
  • remove to fresh air immediately or provide respiratory support if necessary
◆ fire fighting: use carbon dioxide, dry chemical powder, or appropriate foam
◆ personal protection: wear self-contained breathing apparatus, goggles, and protective clothing
◆ storage: keep tightly closed away from heat, sparks, and open flame; refrigerate

SOURCES
◆ *Health Effects Notebook for Hazardous Air Pollutants*, U.S. Environmental Protection Agency, December 1994
◆ *Material Safety Data Sheets (MSDS)*
◆ *Sax's Dangerous Properties of Industrial Materials*, 9th edition, 1996
◆ *Hawley's Condensed Chemical Dictionary*, 12th edition, 1993
◆ *Aldrich Chemical Company Catalog*, 1996–1997
◆ *Environmental Law Index to Chemicals*, 1996 edition
◆ *Merck Index*, 12th edition, 1996
◆ *Environmental Contaminant Reference Databook*, volume 1, 1995

VINYL BROMIDE

CAS #: 593-60-2
MOLECULAR FORMULA: CH_2CHBr
FORMULA WEIGHT: 106.96
SYNONYMS: bromoethene, bromoethylene, bromure de vinyle (French), vinile (bromuro di) (Italian), Vinyl-bromid (German), vinyl bromide, inhibited (DOT), vinyle (bromure de) (French)

USES
◆ Used in the Manufacturing Process
  • flame retardant synthetic fibers
  • preparing films
  • laminating fibers
  • rubber substitutes
  • modacrylic fibers for carpet-backing material
  • comonomer with acrylonitrile in production of fabrics
  • sleepwear
  • home furnishings
  • leather
  • fabricated metal products
  • intermediate in organic synthesis
  • preparation of plastics

PHYSICAL PROPERTIES
◆ colorless gas
◆ colorless liquid under pressure
◆ characteristic pungent odor
◆ melting point −139.5°C
◆ boiling point 15.8°C at 760 mm Hg
◆ density 1.4738 at 25°C
◆ surface tension 22.54 mN/m at 20°C

- ◆ vapor pressure 1033 mm Hg at 25°C
- ◆ vapor density 3.8 at 15°C
- ◆ insoluble in water
- ◆ miscible in alcohol and ether
- ◆ log K_{ow} 1.57

CHEMICAL PROPERTIES
- ◆ autoignition temperature 986°C
- ◆ explosion limits in air 6% lower and 15% upper
- ◆ incompatible with strong oxidizing agents, peroxides, copper and alloys, and plastics

HEALTH RISK
- ◆ Toxic Exposure Guidelines
  - OSHA PEL TWA 5 ppm
  - ACGIH TLV TWA 5 ppm
  - DFG MAK human carcinogen
  - NIOSH REL lowest detectable level
  - MSHA standard 1100 mg/m^3
- ◆ Toxicity Data
  - mma-sat 2 pph/16H
  - mmo-sat 2 pph/16H
  - otr-rat-inhalation 2000 ppm/14W-I
  - inhalation-rat TCLo 10 ppm/6H/2Y-I:CAR
  - inhalation-rat TC 10 ppm:ETA
  - inhalation-rat TCLo 250 ppm/1Y:NEO
  - inhalation-rat TC 52 ppm/6H/2Y-I:CAR
  - oral-rat LD$_{50}$ 500 mg/kg
- ◆ Acute Risks
  - primary target organ: liver
  - dizziness
  - disorientation
  - sleepiness
  - skin irritation
  - liver and kidney damage in rats
  - neurological effects in rats
  - irritating to eyes of rabbits
- ◆ Chronic Risks
  - No information is available on the chronic effects in humans.
  - damage of the liver in rats
  - foci of the liver in rats
  - hematological effects in rats
  - elevated kidney and liver weights in rats
  - potent carcinogen in rats
  - EPA Group B2: probable human carcinogen

HAZARD RISK
- ◆ very dangerous fire hazard when exposed to heat or flame
- ◆ can react violently with oxidizing materials
- ◆ may polymerize in sunlight
- ◆ incompatible with strong oxidizing agents, peroxides, copper and alloys, and plastics
- ◆ hazardous decomposition products: carbon monoxide, carbon dioxide, and hydrogen bromide gas
- ◆ Decomposition emits toxic fumes of Br$^-$.

EXPOSURE ROUTES
- ◆ inhalation during manufacture or use
- ◆ vicinity of facilities that manufacture vinyl bromide

REGULATORY STATUS/INFORMATION
- ◆ Clean Air Act Amendment, Title III: hazardous air pollutants
- ◆ OSHA 29CFR1910.1000 Table Z1
- ◆ chemical regulated by the State of California

ADDITIONAL/OTHER COMMENTS
- ◆ Symptoms
  - • vapor: irritating to eyes, mucous membranes, and upper respiratory tract, skin irritation
- ◆ First Aid
  - • wash eyes and skin immediately with large amounts of water
  - • remove to fresh air immediately or provide respiratory apparatus if necessary
  - • wash out mouth with water
- ◆ fire fighting: use carbon dioxide, water spray, or dry chemical
- ◆ personal protection: wear self-contained breathing apparatus, protective clothing, and goggles
- ◆ storage: keep tightly closed away from heat, sparks, or flame

SOURCES
- ◆ *Health Effects Notebook for Hazardous Air Pollutants*, U.S. Enviromental Protection Agency, December 1994
- ◆ *Material Safety Data Sheets (MSDS)*
- ◆ *Sax's Dangerous Properties of Industrial Materials*, 9th edition, 1996
- ◆ *Hawley's Condensed Chemical Dictionary*, 12th edition, 1993
- ◆ *Aldrich Chemical Company Catalog*, 1996–1997
- ◆ *Environmental Law Index to Chemicals*, 1996 edition
- ◆ *Environmental Contaminant Reference Databook*, volume 1, 1995

VINYL CHLORIDE

CAS #: 75-01-4
MOLECULAR FORMULA: $CH_2:CHCl$
FORMULA WEIGHT: 62.50
SYNONYMS: chlorethene, chlorethylene, chloro-ethene, chloro-ethylene, ethene, chloro-, ethylene, chloro-, ethylene-monochloride, mono-chloroethene, mono-chloroethylene, vinyl-C-monomer, vinyl-chloride-monomer

USES
- ◆ Used in the Manufacturing Process
  - • polyvinyl chloride
  - • plastics
  - • vinyl products
  - • refrigerant gas
  - • manufacture of other chemicals

PHYSICAL PROPERTIES
- ◆ colorless liquid or gas
- ◆ faintly sweet odor
- ◆ odor threshold 3000 ppm
- ◆ melting point −160°C
- ◆ boiling point −13.9°C
- ◆ density (liquid) 0.9195 at 15°/4°
- ◆ vapor pressure 2600 mm Hg @ 25°C
- ◆ vapor density 2.15

- ◆ slightly soluble in water
- ◆ soluble in alcohol and ether

CHEMICAL PROPERTIES
- ◆ highly flammable
- ◆ flash point 25°C
- ◆ autoignition temperature 882°F
- ◆ explosion limits in air 4% lower and 22% upper
- ◆ reacts with chemically active metals, aluminum and alloys, copper, and nitrogen oxides

◆ $\log K_{ow}$ 1.36

HEALTH RISK
- ◆ Toxic Exposure Guidelines
  - • OSHA PEL cancer suspect agent
  - • OSHA (15 min) 12.8 mg/m³
  - • OSHA (8 h) 2.6 mg/m³
  - • ACGIH TLV TWA 5 ppm
  - • ACGIH STEL 25.6 mg/m³
  - • NIOSH REL lowest detectable level
- ◆ Toxicity Data
  - • mma-sat 1 pph
  - • cyt-human:hla 10 mmol/L
  - • inhalation-man TCLo 30 mg/m³
  - • inhalation-rat TCLo 500 ppm/7H
  - • inhalation-man TCLo 200 ppm/14Y-I:CAR,LIV
  - • oral-rat TDLo 3463 mg/kg/52W-I:CAR
  - • inhalation-rat TCLo 1 ppm/4H/52W-I:CAR
  - • inhalation-rat TCLo 10,000 ppm/4H
  - • intraperitoneal-rat TDLo 21 mg/kg/65W-I:ETA
  - • inhalation-human TC 300 mg/m³/W-C:CAR,BLD
  - • oral-rat LD_{50} 500 mg/kg
- ◆ Acute Risks
  - • effects on the central nervous system
  - • dizziness
  - • headaches
  - • giddiness
  - • irritation to eyes
  - • irritation to upper respiratory tract
  - • skin burns
- ◆ Chronic Risks
  - • liver damage
  - • "vinyl chloride disease"
  - • effects on the lung
  - • poor circulation in fingers
  - • changes in the bone at the end of fingers
  - • thickening of skin
  - • changes in blood
  - • increased incidence of birth defects
  - • decreased fetal weight in animals
  - • testicular damage to rats
  - • angiosarcoma of the liver: special liver cancer
  - • cancer of the brain
  - • cancer of the lung
  - • cancer of the digestive tract
  - • EPA Group A: human carcinogen

HAZARD RISK
- ◆ very dangerous fire hazard when exposed to heat, flame, or oxidizers
- ◆ Large fires of this material are practically inextinguishable.
- ◆ severe explosion hazard in form of vapor when exposed to heat or flame
- ◆ Long-term exposure to air forms peroxides which initiate explosive polymerization of the chloride.
- ◆ can react vigorously with oxidizing materials
- ◆ can explode on contact with oxides of nitrogen
- ◆ Decomposition emits highly toxic fumes of Cl^-.

EXPOSURE ROUTES
- ◆ ambient air
- ◆ discharge of exhaust gases
- ◆ evaporation where chemical wastes are stored
- ◆ air inside new cars
- ◆ drinking water
- ◆ water in contact with polyvinyl pipes
- ◆ production, use, transport, storage, and disposal of the chemical
- ◆ microbial degradation product of trichloroethylene in groundwater

REGULATORY STATUS/INFORMATION
- ◆ Clean Air Act Amendment, Title III: hazardous air pollutants
- ◆ Safe Drinking Water Act
  - • 83 contaminants required to be regulated under the SDWA of 1986. 47FR9352
  - • public notification. 40CFR141.32
  - • maximum contaminant level goal for organic contaminants. 40CFR141.50(a)
  - • maximum contaminant level for organic chemicals. 40CFR141.61
  - • variances and exemptions from the maximum contaminant levels for organic and inorganic chemicals. 40CFR142.62
- ◆ Resource Conservation and Recovery Act
  - • design criteria (for municipal solid waste landfill under RCRA). 40CFR258.40
  - • constituents for detection monitoring (for municipal solid waste landfill). 40CFR258
  - • list of hazardous constituents. 40CFR258, 40CFR261, 40CFR261.11
  - • groundwater monitoring list. 40CFR264
  - • risk specific doses. 40CFR266
  - • health based limits for exclusion of waste-derived residues. 40CFR266
  - • D waste #: DO43
  - • U waste #: U043
- ◆ CERCLA and Superfund
  - • designation of hazardous substances. 40CFR302.4
  - • list of chemicals. 40CFR372.65

- ◆ Clean Water Act
  - toxic pollutants. 40CFR401.15
  - total toxic organics (TTOs). 40CFR413.02
  - 126 priority pollutants. 40CFR423
- ◆ OSHA 29CFR1910.1000 Table Z1 and specially regulated substances
- ◆ chemical regulated by the State of California

ADDITIONAL/OTHER COMMENTS

- ◆ Symptoms
  - exposure: dizziness, headaches, giddiness, irritation to eyes, irritation to upper respiratory tract, and skin burns
- ◆ First Aid
  - wash eyes and skin immediately with large amounts of water
  - remove to fresh air immediately or provide oxygen if necessary
- ◆ fire fighting: stop flow of gas
- ◆ personal protection: wear self-contained breathing apparatus, goggles, and protective clothing
- ◆ storage: keep tightly closed away from heat, sparks, and open flame in a cool, dry place

SOURCES

- ◆ *Health Effects Notebook for Hazardous Air Pollutants*, U.S. Environmental Protection Agency, December 1994
- ◆ *Material Safety Data Sheets (MSDS)*
- ◆ *Sax's Dangerous Properties of Industrial Materials*, 9th edition, 1994
- ◆ *Hawley's Condensed Chemical Dictionary*, 12th edition, 1993
- ◆ *Aldrich Chemical Company Catalog*, 1996–1997
- ◆ *Environmental Law Index to Chemicals*, 1996 edition
- ◆ *Merck Index*, 12th edition, 1996

VINYLIDENE CHLORIDE

CAS #: 75-35-4
MOLECULAR FORMULA: $CH_2=CCl_2$
FORMULA WEIGHT: 96.94
SYNONYMS: chlorure de vinylidene (French), 1-1-DCE, 1,1-dichloroethene, 1,1-dichloroethylene, NCI-C54262, sconatex, VDC, vinylidene chloride, vinylidene chloride (II), vinylidene dichloride

USES

- ◆ Used in the Production Process
  - polyvinylidene chloride copolymers
  - flexible films for food packing (Saran@ and Velon@ wraps)
  - packing materials
  - flame retardant coatings for fiber and carpet backing in piping
  - coating for steel pipes
  - adhesive applications
- ◆ used as an intermediate for organic chemical synthesis

PHYSICAL PROPERTIES

- colorless liquid
- mild, sweet odor
- odor threshold 190 ppm
- melting point −122°C
- boiling point 31.6°C
- specific gravity 1.213
- vapor pressure 591 mm Hg at 25°C
- vapor density 3.46
- practically insoluble in water
- soluble in organic solvents
- $\log K_{ow}$ 2.13

CHEMICAL PROPERTIES

- ◆ flash point −15°C
- ◆ autoignition temperature 519°C
- ◆ explosion limits 6.5% lower and 15.5% upper
- ◆ polymerizes to a plastic
- ◆ incompatible with oxidizing agents, copper, aluminum, and peroxides

HEALTH RISK

- ◆ Toxic Exposure Guidelines
  - OSHA PEL TWA 1 ppm
  - ACGIH TLV TWA 5 ppm
  - ACGIH TLV STEL 20 ppm
  - DFG MAK 2 ppm
  - NIOSH REL TWA reduce to lowest detectable level
- ◆ Toxicity Data
  - mmo-sat 5 pph
  - dns-mouse-inhalation 50 ppm
  - oral-rat TDLo 200 mg/kg
  - inhalation-rat TCLo 55 ppm/6H
  - inhalation-mouse TCLo 25 ppm/4H/52W-I:CAR
  - skin-mouse TDLo 4840 mg/kg:NEO
  - inhalation-human TCLo 25 ppm:CNS,LIV,KID
  - oral-rat LD_{50} 200 mg/kg
  - inhalation-rat LC_{50} 6350 ppm/4H
  - oral-mouse LD_{50} 194 mg/kg
  - oral-dog LDLo 5750 mg/kg
  - intravenous-dog LDLo 225 mg/kg
  - subcutaneous-rabbit LDLo 3700 mg/kg
- ◆ Acute Risks
  - adverse neurological effects
  - central nervous system depression
  - inebriation
  - convulsions
  - spasms
  - unconsciousness
  - respiratory effects
  - inflammation of mucous membranes
  - irritation to skin
- ◆ Chronic Risks
  - liver and kidney toxicity
  - effects on the lungs
  - gastrointestinal effects
  - cardiovascular effects
  - neurological system effects
  - birth defects in offspring of rats
  - maternal toxicity in rats
  - increase in kidney and mammary tumors in mice
  - increase in adrenal tumors in rats
  - EPA Group C: possible human carcinogen

HAZARD RISK

- very dangerous fire hazard when exposed to heat or flame
- moderately explosive in the form of gas when exposed to heat or flame
- forms explosive peroxides when exposed to air
- potentially explosive reaction with chlorotrifluoroethylene at 180°C
- Reaction with ozone forms dangerous products.
- explosive reaction with perchloryl fluoride when heated above 100°C
- can explode spontaneously
- reacts violently with oxidizing materials
- Decomposition emits toxic fumes of Cl^-.

EXPOSURE ROUTES

- air releases
- emissions from polymer synthesis and fabrication industries
- inhalation or dermal contact in workplace
- drinking water

REGULATORY STATUS/INFORMATION

- Clean Air Act Amendment, Title III: hazardous air pollutants
- Resource Conservation and Recovery Act
  - constituents for detection monitoring (for municipal solid waste landfill). 40CFR258
  - list of hazardous constituents. 40CFR258
- CERCLA and Superfund
  - designation of hazardous substances. 40CFR302.4
  - list of chemicals. 40CFR372.65
- Clean Water Act
  - designation of hazardous substances. 40CFR116.4
  - determination of reportable quantities. 40CFR117.3
- chemical regulated by the State of California

ADDITIONAL/OTHER COMMENTS

- Symptoms
  - exposure: inebriation, convulsions, spasms, unconsciousness, skin irritation
- First Aid
  - wash eyes immediately with large amounts of water
  - flush skin with plenty of soap and water
  - remove to fresh air immediately or provide respiratory support if necessary
- fire fighting: use carbon dioxide, dry chemical powder, or appropriate foam
- personal protection: wear self-contained breathing apparatus, goggles, and protective clothing
- storage: keep tightly closed away from heat, sparks, and open flame; handle with care

SOURCES

- *Health Effects Notebook for Hazardous Air Pollutants*, U.S. Enviromental Protection Agency, December 1994
- *Material Safety Data Sheets (MSDS)*

◆ *Sax's Dangerous Properties of Industrial Materials*, 9th edition, 1996
◆ *Hawley's Condensed Chemical Dictionary*, 12th edition, 1993
◆ *Aldrich Chemical Company Catalog*, 1996–1997
◆ *Environmental Law Index to Chemicals*, 1996 edition
◆ *Merck Index*, 12 edition, 1996

XYLENES (isomers and mixture) (rtc)

CAS #: 1330-20-7
MOLECULAR FORMULA: C_8H_{10}
FORMULA WEIGHT: 106.18
SYNONYMS: benzene (dimethyl-), dimethylbenzene, methyl toluene, violet-3, xylene methyl toluene, xylol, others

USES
◆ Manufacture of Other Chemicals
  • dyes
  • varnishes
  • resins
  • ethylbenzene
  • benzoic acid
  • terephthalic acids
  • synthesis of organic chemicals
  • hydrogen peroxide
◆ Used as a Component or in the Manufacturing Process
  • paint
  • polyester fibers
  • solvent for adhesives
  • sterilizing catgut
  • aviation gasoline
  • solvent
  • quartz crystal
  • perfumes
  • insect repellents
  • pharmaceuticals
  • leather industry

PHYSICAL PROPERTIES
◆ clear, colorless liquid
◆ boiling point 140°C
◆ density 0.864 at 20°C
◆ sweet odor
◆ specific gravity 0.860
◆ miscible with absolute alcohol, ether, and many other organic liquids
◆ vapor pressure 18 mm Hg at 37.7°C
◆ insoluble in water
◆ vapor density 3.7
◆ odor threshold 5.00 E-5 ppm

CHEMICAL PROPERTIES
◆ easily chlorinated, nitrated, and sulfonated
◆ reacts with oxidizing agents and rubber
◆ flash point 29°C
◆ autoignition temperature 464°C
◆ water partition coefficient $\log K_{ow} = 3.15$
◆ flammability limits lower 1.1% and upper 7%
◆ corrosive to some forms of plastic and rubber

HEALTH RISK
◆ Toxic Exposure Guidelines
  • ACGIH TLV TWA 100 ppm
  • NIOSH REL to xylenes TWA 100 ppm
  • OSHA TWA 100 ppm

◆ Toxicity Data
  • oral-human LDLo 50 mg/kg
  • eye-human 200ppm
  • ingestion-rat LD_{50} 4.3 g/kg/4H
  • inhalation-rat LC_{50} 6330 ppm/4H
  • inhalation-mouse LC_{50} 3907 ppm/6H
  • goldfish LD_{50} 13 mg/L/24H
  • rainbow trout LC_{50} 13.5 mg/L/96H
◆ Acute Risks
  • irritation of eyes, mucous membranes, lungs, and upper respiratory tract
  • pulmonary edema
  • Exposure can cause a narcotic effect.
  • depression of the central nervous system
  • chest pain
  • gastrointestinal difficulties
  • dermatitis
◆ Chronic Risks
  • liver, kidney, and nerve damage
  • memory loss
  • chronic bronchitis
  • blood effects
  • ringing in the ears
  • excessive fatigue

HAZARD RISK

◆ dangerous fire hazard
◆ Vapor can travel long distances to the source of ignition and flash back.
◆ Container may explode in a fire.
◆ reacts strongly with oxidizing agents and rubber
◆ forms explosive mixtures in the air

EXPOSURE ROUTES

◆ inhalation
◆ ingestion
◆ absorption through the skin and eyes
◆ inhalation of air near filling stations
◆ ingestion of contaminated well water
◆ occupational exposure in industries that use xylene in manufacturing process
◆ emissions from heavy traffic

REGULATORY STATUS/INFORMATION

◆ Clean Air Act
  • title 3: list of hazardous air pollutants 42USC7412
◆ Resource Conservation Recovery Act
  • discarded commercial chemical products, off-specification species, container residues, and spill residues thereof. 40CFR261.33
  • appendix 1: constituents for detection monitoring 40CFR258
  • appendix 4: reference air concentrations 40CFR266
  • appendix 9: groundwater monitoring chemicals list 40CFR264
◆ CERCLA and Superfund
  • designation of hazardous substances. 40CFR302.4
  • appendix A: sequential CAS registry # list of CERCLA hazardous substances

- appendix B: radionuclides
- chemical and chemical categories to which this part applies 40CFR372.65
◆ Safe Drinking Water Act
  - 83 contaminants required to be regulated under the SDWA of 1986 47FR9352
  - public notification 40CFR141.32
  - waste specific prohibitions—solvent wastes 40CFR148.10
◆ Clean Water Act
  - designation of hazardous substances 40CFR116.4
  - table 116.4A list of hazardous substances
  - table 116.4B list of hazardous substances by CAS #
  - determination of reportable quantities
    - table 117.3 reportable quantities of hazardous substances designated pursuant to sec. 311 of CWA
◆ Occupational Safety and Health Act
  - limits for air contaminants under the OSHA 29CFR1910.1000
◆ chemical regulated by the State of California

ADDITIONAL/OTHER COMMENTS
◆ Symptoms
  - headaches
  - fatigue
  - irritation of eyes, nose, and throat
  - irritability
  - sleeplessness
◆ First Aid
  - flush eyes or skin with large amounts of water
  - remove contaminated clothes
  - remove to fresh air
  - if not breathing give artificial respiration; if breathing is difficult give oxygen
  - if ingested, wash mouth with water providing person is conscious
◆ fire fighting
  - extinguish with dry chemical powder, carbon dioxide, or appropriate foam
◆ personal protection: wear approved respirator, chemical-resistant gloves, and safety goggles
◆ Storage
  - keep containers tightly closed and store in a cool, dry area
  - keep away from heat, sparks, and open flame

SOURCES
◆ *Environmental Reference Contaminant Databook*, volume 1, 1995
◆ *Environmental Law Index to Chemicals*, 1996 edition
◆ *Health Effects Notebook for Hazardous Air Pollutants*, U.S. Environmental Protection Agency, December 1994
◆ *MSDS*, Aldrich Chemical Company, Inc., 1996–1997
◆ *MSDS*, NIOSH, 1997
◆ *MSDS*, University of Kentucky, 1996–1997

◆ *Sax's Dangerous Properties of Industrial Materials*, 9th edition, 1996

m-XYLENES. *See* p. 543 following o-xylenes.

o-XYLENES (rs)

CAS #: 95-47-6
MOLECULAR FORMULA: $C_8H_{10}(C_6H_4(CH_3)_2)$
FORMULA WEIGHT: 106.18
SYNONYMS: benzene, 1,2-dimethyl, o-dimethylbenzene, 1,2-dimethylebenzene, o-methyltoluene, 1,2-xylene, ortho-xylene, o-xylol

USES
◆ Manufacture of Other Chemicals
  • diphthalic anhydride
  • phthalic anhydride
  • phthalonitrile
  • benzoic acid
  • ethylbenzene
◆ Used as a Component or in the Manufacturing Process
  • high performance polymers
  • plasticizers
  • alkyd resins
  • glass reinforced polyesters
  • vitamin and pharmaceutical syntheses
  • dyes
  • insecticides
  • motor fuels

PHYSICAL PROPERTIES
◆ clear, colorless liquid
◆ melting point -25 to $-23°C$ (-13 to $-9.4°F$)
◆ boiling point 143–145°C (289.4–293°F) at 760 mm Hg
◆ flash point 32°C (90°F)
◆ specific gravity 0.88 at 68°F (water at 39.2°F)
◆ ionization potential 8.56 eV
◆ dielectric constant 2.568 at 25°C
◆ solubility 0.02% by weight (in water at 68°F, g/100 mL)
◆ vapor pressure 7 mm Hg at 68°F
◆ density 0.810 g/cm³ at 20°C
◆ vapor density 3.7 g/cm³
◆ viscosity 0.810 cP at 20°C
◆ surface tension 30.10 dynes/cm (in contact with air)
◆ heat of vaporization 9998.5 gcal/gmol
◆ soluble in acetone, benzene, alcohol, ether; insoluble in water
◆ sweet, aromatic odor
◆ odor threshold in air 0.05 −1.8 ppm
◆ vapors heavier than air
◆ Liquid floats on water.

CHEMICAL PROPERTIES
- lower explosive limit (lel) 0.9% (by volume in room temperature air)
- upper explosive limit (uel) 6.7% (by volume in room temperature air)
- autoignition temperature 463°C (865.4°F)
- heat of combustion 1091.7 kg/cal
- slight explosive potential in vapor form when exposed to heat or flame
- class IC flammable liquid
- reacts with oxidizing materials, strong acids

HEALTH RISK
- Toxic Exposure Guidelines
  - OSHA PEL 8H TWA 100 ppm (435 mg/m^3)
  - OSHA STEL TWA 150 ppm (653 mg/m^3)
  - NIOSH REL 10H TWA 100 ppm (435 mg/m^3)
  - NIOSH CL TWA 200 ppm for 10 minutes
  - IDLH 900 ppm
  - ACGIH THL TWA 100 ppm
  - ACGIH STEL TWA 150 ppm
  - ACGIH BEI methyl hippuric acids in urine at end of shift 1.5 g/g creatinine
- Toxicity Data
  - inhalation-rat TCLo 150 mg/m^3/24H (7–14 days pregnant): nontransmissible changes in offspring
  - intraperitoneal-rat TDLo 500 mg/kg (male 2 days prior to mating): reproductive effects
  - inhalation-human LCLo 6125 ppm/12H
  - oral-rat LDLo 5 g/kg
  - inhalation-rat LCLo 6125 ppm/12H
  - inhalation-mouse LCLo 30 g/m^3
  - intraperitoneal-mouse LC$_{50}$ 1364 mg/kg
  - EPA RfD (reference dose) 2 mg/kg/d
- Target Organs
  - nerves
  - liver
  - kidneys
  - eyes
  - gastrointestinal tract
- Acute Risks
  - moderate health hazard
  - harmful if absorbed through skin, swallowed, or inhaled
  - irritating to skin, eyes, and respiratory system
  - narcotic in high concentrations
  - Combustion may produce irritants and toxic gases.
  - central nervous system depression
  - dermatitis
  - gastrointestinal disturbances
- Chronic Risks
  - pulmonary damage

- hemorrhage
- may impair fertility
- liver damage
- kidney damage

HAZARD RISK
- ◆ very dangerous fire hazard when exposed to heat or flame
- ◆ explosive in the form of vapor when exposed to heat or flame
- ◆ incompatible with oxidizing materials
- ◆ Decomposition emits acrid smoke and irritating fumes.

EXPOSURE ROUTES
- ◆ inhalation
- ◆ oral
- ◆ dermal

REGULATORY STATUS/INFORMATION
- ◆ Clean Air Act
  - Title III: regulated hazardous air pollutant
- ◆ Superfund
  - 40CFR372.65, designation as a specific toxic chemical
- ◆ Toxic Substance Control Act
  - 40CFR716.120.a, list of substances regulated under TSCA
- ◆ Safe Drinking Water Act
  - as mixed isomers (xylene), regulated under the SDWA 1986
  - as mixed isomers (xylene), 40CFR141.32.52, national drinking water standards
  - as mixed isomers (xylene), 40CFR143.10, hazardous waste injection restrictions
- ◆ Federal Insecticide, Fungicide, and Rodenticide Act
  - as mixed isomers (xylene), 40CFR180, tolerances and exemptions from tolerances for pesticide chemicals in or on raw agricultural commodities
- ◆ CERCLA and Superfund
  - as mixed isomers (xylene), 40CFR302.4, designation of hazardous substances
  - as mixed isomers (xylene), table 302.4, list of substances and reportable quantities
- ◆ Clean Water Act
  - as mixed isomers (xylene), 40CFR116.4, designation of hazardous substances
  - as mixed isomers (xylene), table 116.4A, list of hazardous substances
  - as mixed isomers (xylene), table 116.4B, list of hazardous substances by CAS
  - as mixed isomers (xylene), 40CFR117.3, determination of reportable quantities
  - as mixed isomers (xylene), table 117.3, reportable quantities of hazardous substances designated pursuant to section 311 of CWA
- ◆ Occupational Safety and Health Act
  - as mixed isomers (xylene), 29CFR1910.1000 table Z1, limits for air contaminants under OSHA

- ◆ California Environmental Protection Agency
  - • as mixed isomers (xylene), chemicals regulated by the State of California
- ◆ DOT Class 3, flammable and combustible liquid

ADDITIONAL/OTHER COMMENTS
- ◆ Symptoms
  - • dizziness
  - • excitement
  - • drowsiness
  - • staggering gait
  - • nausea
  - • vomiting
  - • dermatitis
  - • flushing and reddening of face
  - • disturbed vision
  - • feeling of increased heat
- ◆ First Aid
  - • eyes: rinse immediately with plenty of water; get to physician
  - • inhalation: fresh air (artificial respiration if needed)
  - • respiratory distress: oxygen inhalation
  - • skin: remove all contaminated clothing; wash with soap and water
  - • ingestion: wash out mouth with water (if person is conscious)
- ◆ Fire Fighting
  - • use dry chemical, foam, or carbon dioxide; use water spray on exposed containers to control temperature
  - • flammable vapor emitted from fire
- ◆ personal protection: full protective clothing and positive pressure self-contained breathing apparatus
- ◆ Spill or Leak
  - • eliminate all ignition sources
  - • stop or control discharge; avoid contact with liquid and vapor
  - • use appropriate foam to blanket release and suppress vapors
  - • control runoff and isolate discharged material for proper disposal
  - • notify local health and pollution control agencies
- ◆ storage: store under nitrogen
- ◆ preventative measures: avoid breathing vapors; keep upwind.
- ◆ NFPA diamond: Health 2, Flammability 3, Reactivity 0
- ◆ DOT: UN 1307
- ◆ Electrical Equipment: Class I, Group D

SOURCES
- ◆ *Aldrich Chemical Company Catalog*, 1996–1997
- ◆ *Environmental Contaminant Reference Databook*, volume 1, 1995
- ◆ *Environmental Law Index to Chemicals*, 1996 edition

- *Fire Protection Guide to Hazardous Materials*, 11th edition, 1994
- *MSDS*, Aldrich Chemical Company, Inc., 1996–1997
- *Merck Index*, 12th edition, 1996
- *NIOSH Pocket Guide to Chemical Hazards*, June 1994
- *Sax's Dangerous Properties of Industrial Materials*, 9th edition, volumes 1 and 2, 1996
- *Hawley's Condensed Chemical Dictionary*, 12th edition, 1993
- *EPA Health Effects Notebook for Hazardous Air Pollutants*, U.S. Environmental Protection Agency, December 1994

m-XYLENES

CAS #: 108-38-3
MOLECULAR FORMULA: $C_6H_4(CH_3)_2$
FORMULA WEIGHT: 106.18
SYNONYMS: benzene, 1,3-dimethyl-, benzene-dimethyl(m-), dimethyl-benzene(1,3-), dimethyl-benzene(m-), methyl-toluene(3-), methyl-toluene(m-), xylene(1,3-), xylene(3-), xylol(m-)

USES
- Used in Production Process
  - benzoic acid
  - phthalic anhydride
  - isophthalic and terphthalic acids
  - polyester fibers
  - solvents
  - dyes
  - sterilizing catgut
  - oil-immersion in microscopy
  - clearing agent in microscopic technique
  - paints and coatings
  - blended into gasoline

PHYSICAL PROPERTIES
- colorless, mobile liquid
- sweet odor
- odor threshold 1.1 ppm
- melting point −47.9°C
- boiling point 139°C
- specific gravity 0.868
- vapor pressure 10 mm Hg at 28.3°C
- vapor density 3.7
- insoluble in water
- soluble in alcohol and ether
- $\log K_{ow}$ 3.12–3.20

CHEMICAL PROPERTIES
- flash point 24°C
- autoignition temperature 527°C
- explosion limits in air 1.1% lower and 7% upper
- incompatible with oxidizing agents

HEALTH RISK
- Toxic Exposure Guidelines
  - OSHA PEL TWA 100 ppm
  - OSHA PEL STEL 150 ppm
  - ACGIH TLV TWA 100 ppm
  - ACGIH TLV STEL 150 ppm
  - NIOSH REL (xylene) TWA 100 ppm
- Toxicity Data
  - skin-rabbit 10 μg/24H open SEV
  - inhalation-rabbit TCLo 500 mg/m³/24H

- oral-mouse TDLo 30 mg/kg
- inhalation-man TCLo 424 mg/m^3/6H/6D:CNS
- inhalation-man TCLo 870 mg/m^3/4H-I:CNS
- oral-rat LD$_{50}$ 5 g/kg
- inhalation-rat LCLo 8000 ppm/4H
- inhalation-mouse LCLo 2010 ppm/24H
- intraperitoneal-mouse LD$_{50}$ 1739 mg/kg
- skin-rabbit LD$_{50}$ 14,100 mg/kg

◆ Acute Risks
- dyspnea
- irritation of nose and throat
- gastrointestinal effects
- nausea
- vomiting
- gastric discomfort
- mild transient eye irritation
- performance decrements in numerical ability
- alterations in equilibrium and body balance
- transient skin irritation
- dryness and scaling of skin
- respiratory toxicity
- neurological toxicity
- motor activity changes
- ataxia
- effects on cardiovascular system in animals
- effects on central nervous system and kidneys in animals

◆ Chronic Risks
- neurological effects
- headache
- dizziness
- fatigue
- tremors
- incoordination
- labored breathing
- impaired pulmonary function
- increased heart palpitation
- severe chest pain
- abnormal EKG
- effects on blood and kidneys
- effects on liver in animals
- effects on central nervous system in animals
- developmental effects in animals
- EPA Group D: not classifiable as to human carcinogenicity

HAZARD RISK
- ◆ very dangerous fire hazard when exposed to heat or flame
- ◆ can react with oxidizing materials
- ◆ explosive in the form of vapor when exposed to heat or flame
- ◆ Decomposition emits acrid smoke and irritating fumes.

EXPOSURE ROUTES
- ◆ air
- ◆ rainwater
- ◆ soils
- ◆ surface water
- ◆ sediments
- ◆ drinking water
- ◆ aquatic organisms
- ◆ home use products such as paints
- ◆ production and use of mixed xylenes

REGULATORY STATUS/INFORMATION
- ◆ Clean Air Act Amendment, Title III: hazardous air pollutants

- ◆ CERCLA and Superfund
  - • designation of hazardous substances. 40CFR302.4
  - • list of chemicals. 40CFR372.65
- ◆ Toxic Substance Control Act
  - • list of substances. 40CFR716.120.a
- ◆ chemical regulated by the State of California

ADDITIONAL/OTHER COMMENTS

- ◆ Symptoms
  - • exposure: burning sensation, coughing, wheezing, laryngitis, shortness of breath, headache, nausea, vomiting, narcotic effect, lung irritation, chest pain, edema
- ◆ First Aid
  - • wash eyes and skin immediately with large amounts of water
  - • remove to fresh air immediately or provide respiratory apparatus if necessary
  - • wash out mouth with water
- ◆ fire fighting: use foam, carbon dioxide, and dry chemical
- ◆ personal protection: self-contained breathing apparatus, protective clothing, and goggles
- ◆ storage: keep tightly closed away from heat, sparks, and open flame in a cool, dry place

SOURCES

- ◆ *Health Effects Notebook for Hazardous Air Pollutants*, U.S. Environmental Protection Agency, December 1994
- ◆ *Material Safety Data Sheets (MSDS)*
- ◆ *Sax's Dangerous Properties of Industrial Materials*, 9th edition, 1996
- ◆ *Hawley's Condensed Chemical Dictionary*, 12th edition, 1993
- ◆ *Aldrich Chemical Company Catalog*, 1996–1997
- ◆ *Environmental Law Index to Chemicals*, 1996 edition
- ◆ *Merck Index*, 12 edition, 1996

p-XYLENES

CAS #: 106-42-3
MOLECULAR FORMULA: $C_6H_4(CH_3)_2$
FORMULA WEIGHT: 106.18
SYNONYMS: benzene, 1,4-dimethyl-, benzene(p-), dimethyl-, benzene-dimethyl(p-), dimethyl-benzene(1,4-), dimethyl-benzene(p-), methyl-toluene(4-), methyl-toluene(p-), xylene(1,4-), xylol(p-)

USES

- ◆ Used in Production Process
  - • benzoic acid
  - • phthalic anhydride
  - • isophthalic and terphthalic acids
  - • polyester fibers
  - • solvents
  - • dyes
  - • sterilizing catgut
  - • oil-immersion in microscopy
  - • clearing agent in microscopic technique
  - • paints and coatings
  - • blended into gasoline
  - • pharmaceutical synthesis
  - • insecticides
  - • preparing tissues for light microscopic examination

- Used as a Chemical Intermediate
  - dimethyl terephthalate
  - terephthalic acid
  - dimethyl tetrachloroterephthalate-herbicide

PHYSICAL PROPERTIES

- colorless plates or prisms at low temperature or liquid
- sweet odor
- odor threshold 0.05 ppm
- melting point 13.3°C
- boiling point 138.37°C
- density 0.86104 at 20°C
- surface tension 28.3 dynes/cm
- viscosity 0.648 cP at 20°C
- insoluble in water
- soluble in alcohol, ether, acetone, and benzene
- vapor pressure 10 mm Hg at 27.3°C
- vapor density 3.66
- log K_{ow} 3.15

CHEMICAL PROPERTIES

- flash point 27.2°C
- autoignition temperature 528°C
- heat of combustion 1089.1 kg/cal
- heat of vaporization 9809.9 cal/mol
- flammability limits 1.1% by volume lower and 7.0% by volume upper
- incompatible with strong oxidizers, strong acids, dichlorohydantoin
- attacks some plastics and rubber compounds
- Hazardous polymerization will not occur.

HEALTH RISK

- Toxic Exposure Guidelines
  - OSHA PEL TWA 100 ppm
  - OSHA PEL STEL 150 ppm
  - ACGIH TLV TWA 100 ppm
  - ACGIH TLV STEL 150 ppm
  - NIOSH REL (xylene) TWA 100 ppm
- Toxicity Data
  - inhalation-rabbit TCLo 1 g/m³/24H
  - oral-rat LD_{50} 5 g/kg
  - inhalation-rat LC_{50} 4550 ppm/4H
  - intraperitoneal-rat LD_{50} 3810 mg/kg
  - inhalation-mouse LCLo 15 g/m³
  - intraperitoneal-mouse LD_{50} 2110 mg/kg
- Acute Risks
  - dyspnea
  - irritation of nose and throat
  - gastrointestinal effects
  - nausea
  - vomiting
  - gastric discomfort
  - mild transient eye irritation
  - performance decrements in numerical ability
  - alterations in equilibrium and body balance
  - transient skin irritation
  - dryness and scaling of skin
  - respiratory toxicity
  - neurological toxicity

- ◆ Chronic Risks
  - neurological effects
  - headache
  - dizziness
  - fatigue
  - tremors
  - incoordination
  - labored breathing
  - impaired pulmonary function
  - increased heart palpitation
  - severe chest pain
  - abnormal EKG
  - effects on blood and kidneys
  - effects on liver in animals
  - effects on central nervous system in animals
  - developmental effects in animals
  - EPA Group D: not classifiable as to human carcinogenicity

HAZARD RISK
- ◆ very dangerous fire hazard when exposed to heat or flame
- ◆ can react with oxidizing materials
- ◆ explosive in the form of vapor when exposed to heat or flame
- ◆ potentially explosive reaction with acetic acid + air, 1,3-dichloro-5,5-dimethyl-2,4-imidazolidindione, and nitric acid + pressure
- ◆ Decomposition emits acrid smoke and irritating fumes.

EXPOSURE ROUTES
- ◆ air
- ◆ rainwater
- ◆ soils
- ◆ surface water
- ◆ sediments
- ◆ drinking water
- ◆ aquatic organisms
- ◆ home use products such as paints
- ◆ production and use of mixed xylenes

REGULATORY STATUS/INFORMATION
- ◆ Clean Air Act Amendment, Title III: hazardous air pollutants
- ◆ CERCLA and Superfund
  - designation of hazardous substances. 40CFR302.4
  - list of chemicals. 40CFR372.65
- ◆ Toxic Substance Control Act
  - OSHA chemicals in need of dermal absorption testing. 40CFR712.30.e10, 40CFR716.120.d10
  - list of substances. 40CFR716.120.a
- ◆ chemical regulated by the State of California

ADDITIONAL/OTHER COMMENTS
- ◆ Symptoms
  - exposure: headache, dizziness, eye and skin irritation, severe coughing, distress, pulmonary edema
  - ingestion: nausea, vomiting, cramps, headache, coma
- ◆ First Aid
  - wash eyes immediately with large amounts of water
  - wash skin immediately with soap and water
  - remove to fresh air immediately or provide respiratory apparatus if necessary
  - wash out mouth with water or milk

- fire fighting: use foam, carbon dioxide, and dry chemical
- personal protection: self-contained breathing apparatus, protective clothing, and goggles
- storage: keep in tightly closed container away from heat, flame, and sources of ignition

SOURCES

- *Health Effects Notebook for Hazardous Air Pollutants*, U.S. Environmental Protection Agency, December 1994
- *Material Safety Data Sheets (MSDS)*
- *Sax's Dangerous Properties of Industrial Materials*, 9th edition, 1996
- *Hawley's Condensed Chemical Dictionary*, 12th edition, 1993
- *Aldrich Chemical Company Catalog*, 1996–1997
- *Environmental Law Index to Chemicals*, 1996 edition
- *Environmental Contaminant Reference Databook*, volume 1, 1995
- *Merck Index*, 12th edition, 1996

The following seventeen hazardous air pollutants, which are categories or groups of chemicals for which no CAS number is available, have been integrated into the alphabetical order of the entries rather than being grouped at the end as they are in the 1990 Amendments to the Clean Air Act.

Antimony Compounds
Arsenic Compounds (inorganic including arsine)
Beryllium Compounds
Cadmium Compounds
Chromium Compounds
Cobalt Compounds
Coke Oven Emissions
Cyanide Compounds
Glycol ethers
Lead Compounds
Manganese Compounds
Mercury Compounds
Fine mineral fibers
Nickel Compounds
Polycyclic Organic Matter
Radionuclides (including radon)
Selenium Compounds

APPENDIX I
ALPHABETICAL LIST

The hazardous air pollutants entries are listed in the alphabetical order used in section 112 of the 1990 Amendments to the Clean Air Act, with one exception. The seventeen hazardous air pollutants, such as antimony compounds and fine mineral fibers, which are categories or groups of chemicals for which no CAS number is available, have been integrated into the alphabetical order rather than being grouped at the end of the list as they are in the 1990 Amendments to the Clean Air Act. Also, numbers, letters (such as N or n), Greek letters, and (except for bis) prefixes such as sym, tris, ortho, meta, and para that precede the name do not affect the alphabetical order.

It is important to check the alphabetical list carefully as the use of two words for a name rather than one word could affect its location on the list. Names consisting of two or more words come before one word names. For example, methyl hydrazine would be found in a different location than methylhydrazine. The most common one word/two word variations of the names of the hazardous air pollutants, as well as many of their most common synonyms, are listed in this appendix. If the name of the chemical being sought is not found in this appendix, it may be one of the many other possible synonyms. In that case, the CAS Number Cross-Reference List (Appendix II) may be used to help locate the chemical. If the reader is not aware of the CAS number, two good sources for locating synonyms and their corresponding CAS numbers are:

1. *Sax's Dangerous Properties of Industrial Materials*, Richard J. Lewis, Van Nostrand Reinhold, 9th edition, 1996 (in 3 volumes), and
2. *Suspect Chemicals Sourcebook*, Roytech Publications, Bethesda, Maryland, 1996 edition.

The names used for the hazardous air pollutant entries are in bold print. The order of the bold print names is the order in which the entries appear. If the name of the chemical being sought is found on this list but is not in bold print, use the CAS Number Cross-Reference List (Appendix II) to locate the name under which the entry may be found.

Chemical or Trade Name	CAS No.

A

Acetaldehyde	**75-07-0**
Acetamide	**60-35-5**
2-Acetamidofluorene	53-96-3
Acetic acid amide	60-35-5
Acetic acid ethylene ester	108-05-4
Acetic aldehyde	75-07-0
Acetoaminofluorene	53-96-3
Acetonitrile	**75-05-8**
Acetophenone	**98-86-2**
2-Acetylaminofluorene	**53-96-3**
N-Acetyl-2-aminofluorene	53-96-3
Acetylbenzene	98-86-2
Acetylene tetrachloride	79-34-5
Acraldohyde	107-02-8
Acroleic acid	79-10-7
Acrolein	**107-02-8**
Acrylaldehyde	107-02-8
Acrylamide	**79-06-1**
Acrylic acid	**79-10-7**
Acrylic acid, ethyl ester	140-88-5
Acrylic aldehyde	107-02-8
Acrylic amide	79-06-1
Acrylonitrile	**107-13-1**
Aldifen	51-28-5
Allyl chloride	**107-05-1**
Amianthus	1332-21-4
p-Aminoaniline	106-50-3
2-Aminoanilsole	90-04-0
Aminobenzene	62-53-3
4-(4-Aminobenzyl)aniline	101-77-9
4-Aminobiphenyl	**92-67-1**
6-Aminocaproic acid lactam	105-60-2
3-Amino-2,5-dichlorobenzoic acid	133-90-4
p-Aminodiphenyl	92-67-1
Aminoethylene	151-56-4
1-Amino-2-methylbenzene	95-53-4
2-Amino-1-methylbenzene	95-53-4
Aminophen	62-53-3
3-Amino-p-toluidine	95-80-7

Chemical or Trade Name	CAS No.
5-Amino-*o*-toluidine	95-80-7
Amphibole	1332-21-4
Aniline	**62-53-3**
2-Anisidine	90-04-0
o-Anisidine	90-04-0
Antimony Compounds	**Not available**
Aprocarb	114-26-1
Aroclor	1336-36-3
Aroclors	1336-36-3
Aroclor 1221, 1232, 1242, 1248, 1254, 1260, 1262, 1268, 2565, 4465, 5442	1336-36-3
Arsenic Compounds (inorganic including arsine)	Not available
Asbestos	**1332-21-4**
Azacyclopropane	151-56-4
1-Azanaphthalene	91-22-5
Azimethylene	334-88-3
Aziridine	151-56-4
1H-Azirine, dihydro	151-56-4

B

Baygon	114-26-1
BBC 12	96-12-8
BCME	542-88-1
1-Benzazine	91-22-5
Benzenamine	62-53-3
Benzenamine, 2,6-dinitro-*N,N*-dipropyl-4-(trifluoromethyl)	1582-09-8
Benzene (including benzene from gasoline)	**71-43-2**
Benzene chloride	108-90-7
Benzene, 1-methyl-2,4-dinitro	121-14-2
Benzeneacetic acid, 4-chloro-α-(4-chlorophenyl)-α-hydroxy-ethyl ester	510-15-6
Benzeneamine, *N,N*-dimethyl	121-69-7
p-Benzenediamine	106-50-3
1,2-Benzenedicarboxylic acid	84-74-2
1,2-Benzenedicarboxylic acid anhydride	85-44-9
1,2-Benzenedicarboxylic acid, bis(2-ethylhexyl) ester	117-81-7
1,2-Benzenedicarboxylic acid, dimethyl ester	131-11-3
1,2-Benzenediol	120-80-9

Chemical or Trade Name	CAS No.
1,4-Benzenediol	123-31-9
Benzenol	108-95-2
Benzenyl chloride	98-07-7
Benzidine	**92-87-5**
Benzine	71-43-2
Benzoic trichloride	98-07-7
Benzol	71-43-2
Benzo(*b*)pyridine	91-22-5
Benzoquinone	106-51-4
p-Benzoquinone	106-51-4
Benzotrichloride	**98-07-7**
Benzoyl methide	98-86-2
Benzyl chloride	**100-44-7**
Benzylidyne chloride	98-07-7
BEPH	117-81-7
Bertholite	7782-50-5
Beryllium Compounds	**Not available**
Betaprone	57-57-8
4,4'-Bi-*o*-toluidine	119-93-7
N,N'-Bianiline	122-66-7
p,p-Bianiline	92-87-5
Bianisidine	119-93-7
Bibenzene	92-52-4
1,2-Bichloroethane	107-06-2
Biethylene	106-99-0
Biocide	107-02-8
Biphenyl	**92-52-4**
1,1'-Biphenyl	92-52-4
(1,1'-Biphenyl)-4-amine	92-67-1
(1,1'-Biphenyl)-4,4'-diamine	92-87-5
4-Biphenylamine	92-67-1
2,2'-Biphenylene oxide	132-64-9
Bis amine	101-14-4
Bis(4-amino-3-chlorophenyl) methane	101-14-4
Bis(*p*-aminophenyl)methane	101-77-9
Bis(2-chloroethyl)ether	111-44-4
Bis(chloromethyl)ether	542-88-1
Bis-CME	542-88-1
Bis(2-ethylhexyl)-1,2-benzenedicarboxylate	117-81-7
Bis(2-ethylhexyl)phthalate (DEHP)	**117-81-7**
Bis(chloromethyl)ether	**542-88-1**

Chemical or Trade Name	CAS No.
Bis(4-isocyanatophenyl)methane	101-68-8
Bivinyl	106-99-0
BPL	57-57-8
Bromo ethylene	593-60-2
Bromo-*o*-gas	74-83-9
Bromoethene	593-60-2
Bromoform	**75-25-2**
Bromofume	106-93-4
Bromomethane	74-83-9
1,3-Butadiene	**106-99-0**
1,3-Butadiene, 2-chloro-	126-99-8
α,γ-Butadiene	106-99-0
Butanone	78-93-3
2-Butanone	78-93-3
1,2-Butene oxide	106-88-7
cis-Butenedioic anhydride	108-31-6
Butter yellow	60-11-7
n-Butyl phthalate	84-74-9
1,2-Butylene oxide	106-88-7

C

Cadmium Compounds	**Not available**
Calcium carbimide	156-62-7
Calcium cyanamide	**156-62-7**
Camphechlor	8001-35-2
Camphor tar	91-20-3
Caprolactam	**105-60-2**
Captan	**133-06-2**
Carabatox	63-25-2
Carbacryl	107-13-1
Carbamic acid, methyl-, 2-(1-methylethoxy) phenyl ester	114-26-1
Carbamic acid, methyl-, 1-naphthyl ester	63-25-2
Carbaryl	**63-25-2**
Carbinol	67-56-1
Carbolic acid	108-95-2
Carbon bisulfide	75-15-0
Carbon chloride	56-23-5
Carbon dichloride oxide	75-44-5
Carbon disulfide	**75-15-0**
Carbon monoxide monosulfide	463-58-1

Chemical or Trade Name	CAS No.
Carbon oxide sulfide	463-58-1
Carbon oxychloride	75-44-5
Carbon sulfide	75-15-0
Carbon tet	56-23-5
Carbon tetrachloride	**56-23-5**
Carbona	56-23-5
Carbonic chloride	75-44-5
Carbonyl chloride	75-44-5
Carbonyl dichloride	75-44-5
Carbonyl sulfide	**463-58-1**
Catechin	120-80-9
Catechol	**120-80-9**
α-Chloracetophenone	532-27-4
Chloramben	**133-90-4**
Chlordane	**57-74-9**
Chlorene	75-00-3
Chlorethyl	75-00-3
Chlorethylene	75-01-4
Chlorex	111-44-4
Chlorinated biphenyl	1336-36-3
Chlorinated camphene	8001-35-2
Chlorinated camphene 60%	8001-35-2
Chlorinated diphenylene	1336-36-3
Chlorinated hydrochloric ether	75-34-3
Chlorine	**7782-50-5**
Chloro (chlormethoxy) methane	542-88-1
2-Chloro-1,3-butadiene	126-99-8
1-Chloro-2,3-dibromopropane	96-12-8
3-Chloro-1,2-dibromopropane	96-12-8
1-Chloro-2,3-epoxy-propane	106-89-8
2-Chloro-1-phenylethanone	532-27-4
3-Chloro-1,2-propylene oxide	106-89-8
Chloroacetic acid	**79-11-8**
α-Chloroacetic acid	79-11-8
2-Chloroacetophenone	**532-27-4**
α-Chloroallyl chloride	542-75-6
γ-Chloroallyl chloride	542-75-6
3-Chloroallyl chloride	542-75-6
3-Chloroallylene	107-05-1
Chlorobenzene	**108-90-7**
Chlorobenzilate	**510-15-6**

Chemical or Trade Name	CAS No.
Chlorobenzol	108-90-7
Chlorobutadiene	126-99-8
3-Chlorochlordene	76-44-8
Chloroethane	71-55-6
Chloroethane	75-00-3
Chloroethene	75-01-4
Chloroform	**67-66-3**
Chloroformic acid dimethylamide	79-44-7
Chlorohydric acid	7647-01-0
Chloromethane	74-87-3
(Chloromethyl) ethylene oxide	106-89-8
Chloromethyl methyl ether	**107-30-2**
2-(Chloromethyl) oxirane	106-89-8
Chloromethylbenzene	100-44-7
Chlorophen	1336-36-3
p-Chlorophenyl chloride	106-46-7
Chloroprene	**126-99-8**
β-Chloroprene	126-99-8
3-Chloroprene	107-05-1
3-Chloropropene	107-05-1
3-Chloropropylene	107-05-1
Chlorothene	71-55-6
α-Chlorotoluene	100-44-7
Chromium Compounds	**Not available**
Cinnamene	100-42-5
CMME	107-30-2
Cobalt Compounds	**Not available**
Coke oven emissions	**Not available**
Cresols/Cresylic acid (isomers and mixture)	**1319-77-3**
o-Cresol	95-48-7
m-Cresol	108-39-4
p-Cresol	106-44-5
Cresorcinol diisocyanate	584-84-9
Cresylic acid	1319-77-3
Cresylol	1319-77-3
Cumene	**98-82-8**
Cumol	98-82-8
Curene 442	101-14-4
Cyanamide	156-62-7
Cyanide Compounds[1]	**Not available**
iso-Cyanatomethane	624-83-9

Chemical or Trade Name	CAS No.
Cyanoethylene	107-13-1
Cyanomethane	75-05-8
Cyanophos	62-73-7
Cyclohexadienedione	106-51-4
1,4-Cyclohexadienedione	106-51-4
Cyclohexanone *iso*-oxime	105-60-2
Cyclohexatriene	71-43-2
4-Cyclohexene-1,2-dicarboximide	133-06-2

D

2,4-D, salts and esters	**94-75-7**
DAB	60-11-7
DBCP	96-12-8
DBE	106-93-4
DBP	84-74-2
DCB	91-94-1
1,1-DCE	75-35-4
1,2-DCE	107-06-2
DCEE	111-44-4
DCM	75-09-2
DDC	79-44-7
DDE	**72-55-9**
DDM	101-77-9
DDVF	62-73-7
DEA	111-42-2
DEHP	117-81-7
Diamide	302-01-2
Diamine	302-01-2
4,4'-Diamino-3,3'-dichlorobiphenyl	91-94-1
4,4'-Diamino-3,3'-dimethoxybiphenyl	119-90-4
1,3-Diamino-4-methylbenzene	95-80-7
2,4-Diamino-1-methylbenzene	95-80-7
1,4-Diaminobenzene	106-50-3
4,4'-Diaminobiphenyl	92-87-5
2,4-Diaminotoluene	95-80-7
2,4-Diaminotoluol	95-80-7
Dianisidine	119-90-4
o,o'-Diansidine	119-90-4
Dianisyltrichloroethane	72-43-5
Diazirine	334-88-3
Diazomethane	**334-88-3**

Chemical or Trade Name	CAS No.
Dibenzo(*b,d*)furan	132-64-9
Dibenzofurans	**132-64-9**
1,2-Dibromo-3-chloropropane	**96-12-8**
Dibromoethane	106-93-4
1,2-Dibromoethane	106-93-4
sym-Dibromoethane	106-93-4
Dibutyl ester	84-74-9
Di-*n*-butyl phthalate	84-74-2
Dibutylphthalate	**84-74-2**
1,4-Dichlorobenzene	**106-46-7**
p-Dichlorobenzene	106-46-7
Dichlorobenzidine base	91-94-1
Dichlorobenzidine	91-94-1
3,3'-Dichlorobenzidine	**91-94-1**
4,4'-Dichlorobenzilic acid ethyl ester	510-15-6
p-Dichlorobenzol	106-46-7
3,3'-Dichloro-1,1'-biphenyl-4,4'-diamine	91-94-1
Dichlorodene	57-74-9
3,3'-Dichloro-4,4'-diaminodiphenylmethane	101-14-4
sym-Dichloro-dimethyl ether	542-88-1
p,p'-Dichlorodiphenyldichloroethylene	72-55-9
α, β-Dichloroethane	107-06-2
1,1-Dichloroethane	75-34-3
1,2-Dichloroethane	107-06-2
sym-Dichloroethane	107-06-2
1,1-Dichloroethene	75-35-4
2,2-Dichloroethenyl ester phosphoric acid	62-73-7
Dichloroethyl ether	**111-44-4**
β,β'-Dichloroethyl ether	111-44-4
sym-Dichloroethyl ether	111-44-4
1,1-Dichloroethylene	75-35-4
Dichloromethane	75-09-2
1,2-Dichloropropane	78-87-5
1,3-Dichloropropene	**542-75-6**
2,2-Dichlorovinyl dimethyl ester phosphoric acid	62-73-7
Dichlorvos	**62-73-7**
Diethanol amine	111-42-2
Diethanolamine	**111-42-2**
N,N-Diethyl aniline	**121-69-7**
Diethyl ester sulfuric acid	64-67-5

Chemical or Trade Name	CAS No.
Di-2-ethylhexylphthalate	117-81-7
Diethyl para-nitrophenyl thiophosphate	56-38-2
Diethyl sulfate	**64-67-5**
Diethyl tetraoxosulfate	64-67-5
(Diethylamino) ethane, TEN	121-44-8
Diethylane deopide Di(ethylene oxide)	123-91-1
1,4-Diethylene dioxide	123-91-1
1,4-Diethylene oxide	123-91-1
N,N-Diethylethanamine	121-44-8
O,O-Diethyl-*O*-(*p*-nitrophenyl) ester phosphorothioic acid	56-38-2
Diethylparathion	56-38-2
4,5-Dihydroimidazole-2(3H)-thione	96-45-7
Dihydroxybenzene	123-31-9
p-Dihydroxybenzene	123-31-9
1,2-Dihydroxybenzene	120-80-9
2,2′-Dihydroxydiethylamine	111-42-2
1,2-Dihydroxyethane	107-21-1
4,4′-Diisocyanatodiphenylmethane	101-68-8
1,1-Diisocyanatohexane	822-06-0
2,4-Diisocyanato-1-methylbenzene	584-84-9
Dimazin, *N,N*-dimethylhydrazine	57-14-7
Dimethoxy-DDT	72-43-5
3,3′-Dimethoxybenzidine	**119-90-4**
Dimethyl aminoazobenzene	**60-11-7**
Dimethyl benzene	1330-20-7
3,3′-Dimethyl benzidine	**119-93-7**
Dimethyl carbamic acid	79-44-7
Dimethyl carbamoyl chloride	**79-44-7**
Dimethyl ester sulfuric acid	77-78-1
Dimethyl formamide	**68-12-2**
Dimethyl hydrazine	57-14-7
1,1-Dimethyl hydrazine	**57-14-7**
Dimethyl monosulfate	77-78-1
Dimethyl phthalate	**131-11-3**
Dimethyl sulfate	**77-78-1**
Dimethylaminoazobenzene	60-11-7
4-Dimethylaminoazobenzene	60-11-7
p-Dimethylamino-azobenzene	60-11-7
(Dimethylamino)benzene	121-69-7
(Dimethylamino) carbonyl chloride	79-44-7

Chemical or Trade Name	CAS No.
tris (Dimethylamino) phosphorus oxide	680-31-9
N,N-Dimethylaniline	121-69-7
N,N-Dimethyl-*p*-azoniline	60-11-7
3,3'-Dimethylbenzidine	119-93-7
Dimethylcarbamoyl chloride	79-44-7
N,N-Dimethylcarbamoyl chloride	79-44-7
Dimethylchloroether	107-30-2
Dimethylchloroformamide	79-44-7
3,3'-Dimethyl-4,4'-diamine-1,1'-biphenyl	119-93-7
Dimethylene oxide	75-21-8
Dimethyleneimine	151-56-4
Dimethylformamide	68-12-2
N,N-Dimethylformamide	68-12-2
1,1-Dimethylhydrazine	57-14-7
Dimethylnitromethane	79-46-9
Dimethylnitrosamine	62-75-9
N,N-Dimethylnitrosamine	62-75-9
Dimethylphenylamine	121-69-7
N,N-Dimethyl-4-(phenylazo)-benzenamine	60-11-7
Dinitro-*o*-cresol	534-52-1
4,6-Dinitro-o-cresol, and salts	**534-52-1**
Dinitrodendtroxal	534-52-1
Dinitromethyl cyclohexyltrienol	534-52-1
2,4-Dinitro-6-methylphenol	534-52-1
α-Dinitrophenol	51-28-5
2,4-Dinitrophenol	**51-28-5**
2,4-Dinitrotoluene	**121-14-2**
Dinitrotoluol	121-14-2
Dioctyl phthalate	117-81-7
1,4-Dioxane	**123-91-1**
1,3-Dioxophthalan	85-44-9
1,4-Dioxybenzene	106-51-4
o-Diphenol	120-80-9
Diphenyl	92-52-4
Diphenylene oxide	132-64-9
sym-Diphenylhydrazine	122-66-7
1,2-Diphenylhydrazine	**122-66-7**
4,4'-Diphenylmethanediamine	101-77-9
Dithiocarbonic anhydride	75-15-0
DMCC	79-44-7
DMDT	72-43-5

Chemical or Trade Name	CAS No.
DMF	68-12-2
DMFA	68-12-2
DMH	57-14-7
DMN	62-75-9
DMNA	62-75-9
DMP, methyl phthalate	131-11-3
DMS	77-78-1
2,4-DNP	51-28-5
DNT	121-14-2
2,4-DNT	121-14-2
Dowchlor	57-74-9
Dowfume	74-83-9
Dowicide 2	95-95-4
Dowicide 2S	88-06-2
Dowicide B	95-95-4
DS	64-67-5
E	
EB	100-41-4
ECH	106-89-8
EDB	106-93-4
EDC	107-06-2
EI ethylimine	151-56-4
E.O. oxifume	75-21-8
Epichlorohydrin	**106-89-8**
1,2-Epoxybutane	**106-88-7**
Epoxyethylbenzene	96-09-3
1,2-Epoxyethylbenzene	96-09-3
1,2-Epoxypropane	75-56-9
Epoxystyrene	96-09-3
Erythrene	106-99-0
Ethanal	75-07-0
Ethanamide	60-35-5
Ethane dichloride	107-06-2
Ethane hexachloride	67-72-1
Ethane trichloride	79-00-5
1,2-Ethanediol	107-21-1
Ethanoic acid ethenyl ester	108-05-4
Ethene oxide	75-21-8
Ethenylbenzene	100-42-5
Ethinyl trichloride	79-01-6

Chemical or Trade Name	CAS No.
Ethoxycarbonylethylene	140-88-5
Ethyl acrylate	**140-88-5**
Ethyl aldehyde	75-07-0
Ethyl benzene	**100-41-4**
Ethyl carbamate	**51-79-6**
Ethyl chloride	**75-00-3**
Ethyl methyl ketone	78-93-3
Ethyl nitrile	75-05-8
Ethyl parathion	56-38-2
Ethyl propenoate	140-88-5
Ethyl sulfate	64-67-5
Ethyl urethane	51-79-6
Ethylbenzene	100-41-4
Ethylbenzol	100-41-4
Ethylene chloride	75-01-4
Ethylene dibromide	**106-93-4**
Ethylene dichloride	**107-06-2**
1,2-Ethylene dichloride	107-06-2
Ethylene dihydrate	107-21-1
Ethylene glycol	**107-21-1**
Ethylene hexachloride	67-72-1
Ethylene imine	**151-56-4**
Ethylene oxide	**75-21-8**
Ethylene thiourea	**96-45-7**
Ethylenecarboxamide	79-06-1
Ethylenecarboxylic acid	79-10-7
Ethyleneimine	151-56-4
1,3-Ethylenethiourea	96-45-7
Ethyl-ethylene oxide	106-88-7
Ethylidene chloride	75-34-3
Ethylidene dichloride	**75-34-3**
o-Ethylurethane	51-79-6
ETU	96-45-7

F

FAA	53-96-3
Fine mineral fibers[2]	**Not available**
2-Fluorenylacetamide	53-96-3
N-9H-Fluoren-2-yl-acetamide	53-96-3
Fluoric acid	7664-39-3
Folbex	510-15-6

Chemical or Trade Name	CAS No.
Formaldehyde	**50-00-0**
Formalin	50-00-0
Formic aldehyde	50-00-0
N-Formyldimethylamine	68-12-2
Formyl trichloride	67-66-3
Freon 30	75-09-2
Fumagon	96-12-8
2,5-Furandione	108-31-6

G

Glycol	107-21-1
Glycol alcohol	107-21-1
Glycol bromide	106-93-4
Glycol dichloride	107-06-2
Glycol ethers[3]	**Not available**

H

Halon 1001	74-83-9
HCB	118-74-1
HCBD	87-68-3
HCCPD	77-47-4
Heptachlor	**76-44-8**
Heptachlorane	76-44-8
1,4,5,6,7,8,8-Heptachloro-3a,4,7,7a-tetrahy-dro-4,7-methanoindene	76-44-8
Hexa CB	118-74-1
Hexachloro-1,3-butadiene	87-68-3
Hexachloro-1,3-cyclopentadiene	77-47-4
1,2,3,4,5,6-Hexachloro-1,3-cyclopentadiene	77-47-4
Hexachlorobenzene	**118-74-1**
Hexachlorobutadiene	**87-68-3**
Hexachlorocyclopentadiene	**77-47-4**
Hexachloroethane	**67-72-1**
1,1,1,2,2,2-Hexachloroethane	67-72-1
Hexahydro-2H-azepin-2-one	105-60-2
Hexamethyl phosphoramide	680-31-9
Hexamethylene diisocyanate	822-06-0
Hexamethylene-1,6-diisocyanate	**822-06-0**
Hexamethylene ester	822-06-0
Hexamethylphosphoramide	**680-31-9**

Chemical or Trade Name	CAS No.
Hexamethylphosphoric acid triamide	680-31-9
Hexane	**110-54-3**
n-Hexane	110-54-3
Hexone	108-10-1
HMPA	680-31-9
HMPT	680-31-9
HPT	680-31-9
Hydracrylic acid β-lactone	57-57-8
Hydrazine	**302-01-2**
Hydrazobenzene	122-66-7
Hydrazomethane	60-34-4
Hydrochloric acid	**7647-01-0**
Hydrochloride	7647-01-0
Hydrofluoric acid	7664-39-3
Hydrogen chloride	7647-01-0
Hydrogen fluoride	**7664-39-3**
Hydrogen phosphide	7803-51-2
Hydroquinol	123-31-9
Hydroquinone	**123-31-9**
p-Hydroquinone	123-31-9
3-Hydroxy-1-propanesulphonic acid sulfone	1120-71-4
Hydroxybenzene	108-95-2
4-Hydroxynitrobenzene	100-02-7
1-Hydroxypentachlorobenzene	87-86-5

I

2-Imidazolidinethione	96-45-7
2,2'-Iminobisethanol	111-42-2
2,2'-Iminodiethanol	111-42-2
Iodomethane	74-88-4
Isoacetophorone	78-59-1
1,3-Isobenzofurandione	85-44-9
Isobutyl methyl ketone	108-10-1
Isobutyltrimethylmethane	540-84-1
Isocarb	114-26-1
Isocyanic acid	822-06-0
Isocyanic acid, methyl ester	624-83-9
Isocyanic acid, methylphenylene ester	584-84-9
Isoforon	78-59-1
Isonitropropane	79-46-9
Isooctane	540-84-1

Chemical or Trade Name	CAS No.
Isophorone	**78-59-1**
o-Isopropoxyphenyl ester, N-methyl carbamic acid	114-26-1
Isopropyl benzene	98-82-8
Isopropylacetone	108-10-1

L

β-Lactone 3-hydroxy-propionic acid	57-57-8
Lead Compounds	**Not available**
Leucethane	51-79-6
Lime-nitrogen	156-62-7
Lindane (all isomers)	**58-89-9**

M

Maleic acid anhydride	108-31-6
Maleic anhydride	**108-31-6**
Manganese Compounds	**Not available**
MB	74-83-9
MBOCA	101-14-4
MBX	74-83-9
MCA	79-11-8
MCB	108-90-7
MDA	101-77-9
MDI	101-68-8
MEC	107-21-1
MEK	78-93-3
2-Mercaptoimidazoline	96-45-7
Mercury Compounds	**Not available**
Methacide	108-88-3
Methaldehyde	50-00-0
Methanal	50-00-0
Methane dichloride	75-09-2
Methane tetrachloride	56-23-5
Methane trichloride	67-66-3
Methanecarbonitrile	75-05-8
Methanecarboxamide	60-35-5
Methanol	**67-56-1**
Methenyl trichloride	67-66-3
Methogas	74-83-9
2-Methoxy-1-aminobenzene	90-04-0

Chemical or Trade Name	CAS No.
2-Methoxy-benzenamine	90-04-0
2-Methoxy-2-methyl-propane	1634-04-4
2-Methoxyaniline	90-04-0
o-Methoxyaniline	90-04-0
2-Methoxybenzenamine	90-04-0
Methoxychlor	**72-43-5**
Methyl alcohol	67-56-1
2-Methyl aziridine	75-55-8
Methyl benzene	108-88-3
Methyl bromide	**74-83-9**
Methyl chloride	**74-87-3**
Methyl chloroform	**71-55-6**
Methyl cyanide	75-05-8
Methyl ester methacrylic acid	80-62-6
Methyl ethyl ketone	**78-93-3**
Methyl hydrate	67-56-1
Methyl hydrazine	**60-34-4**
Methyl iodide	**74-88-4**
Methyl isobutyl ketone	**108-10-1**
Methyl isocyanate	**624-83-9**
N-Methyl-2-isopropoxyphenylcarbamate	114-26-1
Methyl methacrylate	**80-62-6**
Methyl 2-methyl-2-propenoate	80-62-6
N-Methyl-α-naphthylcarbamate	63-25-2
N-Methyl-*N*-nitrosomethamine	62-75-9
Methyl oxirane	75-56-9
Methyl phenyl ketone	98-86-2
Methyl sulfate	77-78-1
Methyl tert-butyl ether	**1634-04-4**
Methyl toluene	1330-20-7
Methyl tribromide	75-25-2
Methyl trichloride	67-66-3
Methyl-acetone	78-93-3
2-Methyl-acrylic acid methyl ester	80-62-6
2-Methyl-aziridine	75-55-8
2-Methyl-benzenamine	95-53-4
1-Methyl-2,4-dinitrobenzene	121-14-2
1-Methyl-1-nitrosourea	684-93-5
4-Methyl-2-pentanone	108-10-1
4-Methyl-phenylene diisocyanate	584-84-9
2-Methyl-2-propenoic acid methyl ester	80-62-6

Chemical or Trade Name	CAS No.
Methylacetaldehyde	123-38-6
2-Methylaniline	95-53-4
2-Methylazcyclopropane	75-55-8
Methylbenzol	108-88-3
Methylcarbamate 1-naphthalenol	63-25-2
Methylchloromethyl ether	107-30-2
4,4′-Methylene bis(2-chloroaniline)	**101-14-4**
Methylene chloride	**75-09-2**
Methylene dichloride	75-09-2
Methylene diphenyl diisocyanate	**101-68-8**
Methylene glycol	50-00-0
Methylene oxide	50-00-0
4,4′-Methylenebisbenzenamine	101-77-9
4,4′-Methylenebis-benzenamine	101-77-9
1,1′-Methylenebis(4-isocyanato-benzene)	101-68-8
Methylene-bis-orthochloroaniline	101-14-4
4,4′-Methylenedianiline	**101-77-9**
4,4′-Methylenediphenyl isocyanate	101-68-8
4,4′-Methylenediphenylisocyanate	101-68-8
Methylenedi-*p*-phenylene ester isocyanic acid	101-68-8
2-(1-Methylethoxy) phenol methylcarbamate	114-26-1
(1-Methylethyl)-benzene	98-82-8
Methylethylene oxide	75-56-9
Methylethylenimine	75-55-8
Methylhydrazine	60-34-4
1-Methylhydrazine	60-34-4
Methylol	67-56-1
Methyltrichloromethane	71-55-6
MIBK	108-10-1
MIK	108-10-1
MME	80-62-6
MMH	60-34-4
MOCA	101-14-4
Molecular chlorine	7782-50-5
Monochloroacetic acid	79-11-8
Monochlorobenzene	108-90-7
Monochlorodimethyl ether	107-30-2
Monochloroethane	75-00-3
Monochloroethanoic acid	79-11-8
Monochloroethene	75-01-4

Chemical or Trade Name	CAS No.
Monochloromethane	74-87-3
Monomethylhydrazine	60-34-4
MTD	95-80-7
Muriatic acid	7647-01-0
Muriatic ether	75-00-3

N

Naphthalene	**91-20-3**
Naphthalin	91-20-3
NDMA	62-75-9
Nemagon	96-12-8
Nickel Compounds	**Not available**
Nitrobenzene	**98-95-3**
Nitrobenzol	98-95-3
p-Nitrobiphenyl	92-93-3
4-Nitrobiphenyl	**92-93-3**
4-Nitrodiphenyl	92-93-3
Nitroisopropane	79-46-9
Nitropentachloro-benzene	82-68-8
4-Nitrophenol	**100-02-7**
p-Nitrophenol	100-02-7
2-Nitropropane	**79-46-9**
N-Nitroso-N,N-dimethylamine	62-75-9
N-Nitroso-N-methylcarbamide	684-93-5
N-Nitroso-N-methylurea	684-93-5
N-Nitrosodimethylamine	62-75-9
Nitrosomorpholine	59-89-2
N-Nitrosomorpholine	59-89-2
NMH	684-93-5
NMOR	59-89-2
NMU	684-93-5
2-NP	79-46-9

O

Octachlor	57-74-9
1,2,4,5,6,7,8,8-Octachlor-2,3,3a,4,7,7a- hexahydro-4,7-methano-1H-indene	57-74-9
Octachlorcamphene	8001-35-2
1,2,4,5,6,7,8,8-Octachloro-4,7-methano- 3a,4,7,7a-tetra-hydroindane	57-74-9

Chemical or Trade Name	CAS No.
1,2,4,5,6,7,8,8-Octachloro-3a,4,7,7a-tetrahydro-4,7-methanindan	57-74-9
Di-*sec*-Octyl phthalate	117-81-7
Oil of myrbane (mirbane)	98-95-3
Orthocide	133-06-2
Oxane	75-21-8
1,2-Oxathiolane-2,2-dioxide	1120-71-4
Oxidoethane	75-21-8
Oxirane	75-21-8
2-Oxohexamethylenimine	105-60-2
Oxomethane	50-00-0
Oxybis (chloromethane)	542-88-1
1,1'-Oxybis(2-chloro)ethane	111-44-4
Oxycarbon sulfide	463-58-1

P

para-Crystals	106-46-7
para-Dichlorobenzene	106-46-7
Parathion	**56-38-2**
PCBs	1336-36-3
PCE	127-18-4
PCL	77-47-4
PCNB	82-68-8
PCP	87-86-5
PDAB	60-11-7
PDB	106-46-7
PDCB	106-46-7
PENTA	87-86-5
Pentachloronitrobenzene	**82-68-8**
Pentachlorophenate	87-86-5
Pentachlorophenol	**87-86-5**
2,3,4,5,6-Pentachlorophenol	87-86-5
Perchlor	127-18-4
Perchlorobenzene	118-74-1
Perchlorobutadiene	87-68-3
Perchlorocyclopentadiene	77-47-4
Perchloroethane	67-72-1
Perchloroethylene	127-18-4
Perchloromethane	56-23-5
Perclene	127-18-4
Perk	127-18-4

Chemical or Trade Name	CAS No.
PHC	114-26-1
Phenachlor	88-06-2
Phenacyl chloride	532-27-4
Phenic acid	108-95-2
Phenochlor	1336-36-3
Phenol	**108-95-2**
Phenyl benzene	92-52-4
Phenyl chloride	108-90-7
Phenyl hydride	71-43-2
Phenyl methane	108-88-3
Phenyl oxirane	96-09-3
2-Phenyl propane	98-82-8
Phenylamine	62-53-3
p-Phenylaniline	92-67-1
Phenylchloromethylketone, mace	532-27-4
p-Phenylenediamine	**106-50-3**
o-Phenylenediol	120-80-9
Phenylethane	100-41-4
1-Phenyl-ethanone	98-86-2
Phenylethylene	100-42-5
Phenylethylene oxide	96-09-3
Phenylic alcohol	108-95-2
p-Phenyl-nitrobenzene	92-93-3
Phosgene	**75-44-5**
Phosphine	**7803-51-2**
Phosphorus	**7723-14-0**
Phosphorus (white)	7723-14-0
Phosphorus (yellow)	7723-14-0
Phosphorus trihydride	7803-51-2
PHPH	92-52-4
Phthalandione	85-44-9
Phthalic acid dioctyl ester	117-81-7
Phthalic anhydride	**85-44-9**
PNB	92-93-3
Polychlorcamphene	8001-35-2
Polychlorinated biphenyls	**1336-36-3**
Polychloro-biphenyl	1336-36-3
Polycyclic organic matter[4]	**Not available**
PPD	106-50-3
Pracarbamine	51-79-6
Propanal	123-38-6

Chemical or Trade Name	CAS No.
1,3-Propane sultone	**1120-71-4**
Propanolide	57-57-8
3-Propanolide	57-57-8
2-Propenal	107-02-8
Propenamide	79-06-1
2-Propenamide	79-06-1
Propene acid	79-10-7
2-Propenenitrile	107-13-1
Propenoic acid	79-10-7
2-Propenoic acid	79-10-7
2-Propenoic acid, ethyl ester	140-88-5
β-Propiolactone	**57-57-8**
1,3-Propiolactone	57-57-8
Propionaldehyde	**123-38-6**
Propionolactone	57-57-8
Propoxur	**114-26-1**
Propylaldehyde	123-38-6
Propylene chloride	78-87-5
Propylene dichloride	**78-87-5**
Propylene epoxide	75-56-9
Propylene imine	75-55-8
Propylene oxide	**75-56-9**
Propyleneimine	75-55-8
1,2-Propyleneimine	75-55-8
1,2-Propylenimine	**75-55-8**
Pyrocatechin	120-80-9
Pyrocatechol	120-80-9

Q

Quinoline	**91-22-5**
Quinone	**106-51-4**
Quintobenzene	82-68-8
Quintocene	82-68-8
Quintozene	82-68-8

R

Radionuclides (including Radon)[5]	**Not available**
Rotox	74-83-9

Chemical or Trade Name	CAS No.

S

Selenium Compounds	**Not available**
Sevin	63-25-2
Slimicide	107-02-8
Styrene	**100-42-5**
Styrene oxide	**96-09-3**

T

TCE	71-55-6
TCE	79-01-6
TCE	79-34-5
TCM	67-66-3
TDA	95-80-7
TDI	584-84-9
Telone	542-75-6
Tetrachlorocarbon	56-23-5
2,3,7,8-Tetrachlorodibenzo-p-dioxin	**1746-01-6**
1,1,2,2-Tetrachloroethane	**79-34-5**
sym-Tetrachloroethane	79-34-5
Tetrachloroethylene	**127-18-4**
Tetrachloromethane	56-23-5
Tetraform	56-23-5
1,2,3,6-Tetrahydro-*N*-(trichloromethyl-thio)phthalimide	133-06-2
Titanium chloride	7550-45-0
Titanium tetrachloride	**7550-45-0**
Tolidine	119-93-7
o-Tolidine	119-93-7
Toluene	**108-88-3**
2,4-Toluene diamine	**95-80-7**
2,4-Toluene diisocyanate	**584-84-9**
m-Toluenediamine	95-80-7
Toluene-2,4-diamine	95-80-7
Toluene 2,4-diisocyanate	584-84-9
Toluene trichloride	98-07-7
Toluenol	1319-77-3
o-Toluidine	**95-53-4**
Toluol	108-88-3
o-Tolylamine	95-53-4

Chemical or Trade Name	CAS No.
Tolyl chloride	100-44-7
Toxaphene	**8001-35-2**
Treflan	1582-09-8
Tribromomethane	75-25-2
1,2,4-Trichlorobenzene	**120-82-1**
unsym-Trichlorobenzene	120-82-1
1,2,4-Trichlorobenzol	120-82-1
1,1,1-Trichloro-2,2-*bis*(*p*-methoxyphenyl)-ethane	72-43-5
α-Trichloroethane	71-55-6
1,1,1-Trichloroethane	71-55-6
1,1,2-Trichloroethane	**79-00-5**
1,2,2-Trichloroethane	79-00-5
Trichloroethene	79-01-6
Trichloroethylene	**79-01-6**
1,2,2-Trichloroethylene	79-01-6
Trichloromethyl benzene	98-07-7
Trichloromethylbenzene	98-07-7
N-(Trichloromethyl)thio-1H-isoindole-1,3(2H)-dione	133-06-2
2,4,5-Trichlorophenol	**95-95-4**
2,4,6-Trichlorophenol	**88-06-2**
Trichlorophenylmethane	98-07-7
α,α,α,Trichlorotoluene	98-07-7
Tri-clene	79-01-6
Tricresol, hydroxy toluene	1319-77-3
Triethylamine	**121-44-8**
Trifluoralin	1582-09-8
α,α,α-Trifluoro-2,6-dinitro-*N,N*-dipropyl-*p*-toluidine	1582-09-8
Trifuralin	**1582-09-8**
Trifurex	1582-09-8
Trim	1582-09-8
3,5,5-Trimethyl-2-cyclohexen-1-one	78-59-1
1,1,3-Trimethyl-3-cyclohexene-5-one	78-59-1
2,2,4-Trimethylpentane	**540-84-1**

U

Urethan	51-79-6

V

VAC	108-05-4
VC	75-01-4
VCM	75-01-4
VDC	75-35-4
Vinyl acetate	**108-05-4**
Vinyl amide	79-06-1
Vinyl benzene	100-42-5
Vinyl bromide	**593-60-2**
Vinyl chloride	**75-01-4**
Vinyl cyanide	107-13-1
Vinyl ester acetic acid	108-05-4
Vinyl ethanoate	108-05-4
Vinyl trichloride	79-00-5
Vinylbenzol	100-42-5
Vinylformic acid	79-10-7
Vinylidene chloride	**75-35-4**
Vinylidene dichloride	75-35-4
VYAC	108-05-4

W

Weed-B-Gon	94-75-7
Weeviltox	75-15-0
White Phosphorus	7723-14-0
White tar	91-20-3
Wood alcohol	67-56-1
Wood naphtha	67-56-1
Wood spirit	67-56-1

X

Xenylamine	92-67-1
Xylenes (isomers and mixture)	**1330-20-7**
o-Xylenes	**95-47-6**
m-Xylenes	**108-38-3**
p-Xylenes	**106-42-3**
Xylol	1330-20-7

Z

Zytox	74-83-9

Note: For all listings above that contain the word "compounds," and for glycol ethers, the following applies: Unless otherwise specified, these listings are defined as including any unique chemical substance that contains the named chemical (i.e., antimony, arsenic, etc.) as part of that chemical's infrastructure.

[1]$X'CN$ where $X = H'$ or any other group where a formal dissociation may occur. For example KCN or $Ca(CN)_2$.

[2]Includes mineral fiber emissions from facilities manufacturing or processing glass, rock, or slag fibers (or other mineral derived fibers) of average diameter 1 micrometer or less.

[3]Includes mono- and di- ethers of ethylene glycol, diethylene glycol, and triethylene glycol $R-(OCH_2CH_2)_n=cOR'$ where

n= 1, 2, or 3
R = alkyl or aryl groups
R' = R, H, or groups which, when removed, yield glycol ethers with the structure $R-(OCH_2CH)_n=OH$[sic]. Polymers are excluded from the glycol category.

[4]Includes organic compounds with more than one benzene ring and which have a boiling point greater than or equal to 100°C.

[5]A type of atom which spontaneously undergoes radioactive decay.

APPENDIX II
CAS NUMBER CROSS-REFERENCE LIST

The names on this list are the names under which the entries for the hazardous air pollutants may be found. This list is in CAS number order, whereas the text entries are in the alphabetical order used in Section 112 of the 1990 Amendments to the Clean Air Act. For important information about the alphabetical order of the entries, see the comments under ALPHABETICAL ARRANGEMENT in the ABOUT THIS BOOK section and at the beginning of the Alphabetical List in Appendix I. For synonyms, see the Alphabetical List in Appendix I and the synonym field in each entry.

CAS No.	Chemical or Trade Name
50-00-0	Formaldehyde
51-28-5	2,4-Dinitrophenol
51-79-6	Ethyl carbamate
53-96-3	2-Acetylaminofluorene
56-23-5	Carbon tetrachloride
56-38-2	Parathion
57-14-7	1,1-Dimethyl hydrazine
57-57-8	β-Propiolactone
57-74-9	Chlordane
58-89-9	Lindane (all isomers)
59-89-2	N-Nitrosomorpholine
60-11-7	Dimethyl aminoazobenzene
60-34-4	Methyl hydrazine
60-35-5	Acetamide
62-53-3	Aniline
62-73-7	Dichlorvos
62-75-9	N-Nitrosodimethylamine
63-25-2	Carbaryl
64-67-5	Diethyl sulfate
67-56-1	Methanol
67-66-3	Chloroform
67-72-1	Hexachloroethane
68-12-2	Dimethyl formamide

CAS No.	Chemical or Trade Name
71-43-2	Benzene (including benzene from gasoline)
71-55-6	Methyl chloroform
72-43-5	Methoxychlor
72-55-9	DDE
74-83-9	Methyl bromide
74-87-3	Methyl chloride
74-88-4	Methyl iodide
75-00-3	Ethyl chloride
75-01-4	Vinyl chloride
75-05-8	Acetonitrile
75-07-0	Acetaldehyde
75-09-2	Methylene chloride
75-15-0	Carbon disulfide
75-21-8	Ethylene oxide
75-25-2	Bromoform
75-34-3	Ethylidene dichloride
75-35-4	Vinylidene chloride
75-44-5	Phosgene
75-55-8	1,2-Propylenimine
75-56-9	Propylene oxide
76-44-8	Heptachlor
77-47-4	Hexachlorocyclopentadiene
77-78-1	Dimethyl sulfate
78-59-1	Isophorone
78-87-5	Propylene dichloride
78-93-3	Methyl ethyl ketone
79-00-5	1,1,2-Trichloroethane
79-01-6	Trichloroethylene
79-06-1	Acrylamide
79-10-7	Acrylic acid
79-11-8	Chloroacetic acid
79-34-5	1,1,2,2-Tetrachloroethane
79-44-7	Dimethyl carbamoyl chloride
79-46-9	2-Nitropropane
80-62-6	Methyl methacrylate
82-68-8	Pentachloronitrobenzene
84-74-2	Dibutylphthalate
85-44-9	Phthalic anhydride
87-68-3	Hexachlorobutadiene
87-86-5	Pentachlorophenol
88-06-2	2,4,6-Trichlorophenol
90-04-0	o-Anisidine

CAS No.	Chemical or Trade Name
91-20-3	Naphthalene
91-22-5	Quinoline
91-94-1	3,3'-Dichlorobenzidine
92-52-4	Biphenyl
92-67-1	4-Aminobiphenyl
92-87-5	Benzidine
92-93-3	4-Nitrobiphenyl
94-75-7	2,4-D, Salts and Esters
95-47-6	o-Xylenes
95-48-7	o-Cresol
95-53-4	o-Toluidine
95-80-7	2,4-Toluene diamine
95-95-4	2,4,5-Trichlorophenol
96-09-3	Styrene oxide
96-12-8	1,2-Dibromo-3-chloropropane
96-45-7	Ethylene thiourea
98-07-7	Benzotrichloride
98-82-8	Cumene
98-86-2	Acetophenone
98-95-3	Nitrobenzene
100-02-7	4-Nitrophenol
100-41-4	Ethyl benzene
100-42-5	Styrene
100-44-7	Benzyl chloride
101-14-4	4,4'-Methylene bis(2-chloroaniline)
101-68-8	Methylene diphenyl diisocyanate
101-77-9	4,4'-Methylenedianiline
105-60-2	Caprolactam
106-42-3	p-Xylenes
106-44-5	p-Cresol
106-46-7	1,4-Dichlorobenzene
106-50-3	p-Phenylenediamine
106-51-4	Quinone
106-88-7	1,2-Epoxybutane
106-89-8	Epichlorohydrin
106-93-4	Ethylene dibromide
106-99-0	1,3-Butadiene
107-02-8	Acrolein
107-05-1	Allyl chloride
107-06-2	Ethylene dichloride
107-13-1	Acrylonitrile
107-21-1	Ethylene glycol
107-30-2	Chloromethyl methyl ether

CAS No.	Chemical or Trade Name
108-05-4	Vinyl acetate
108-10-1	Methyl isobutyl ketone
108-31-6	Maleic anhydride
108-38-3	m-Xylenes
108-39-4	m-Cresol
108-88-3	Toluene
108-90-7	Chlorobenzene
108-95-2	Phenol
110-54-3	Hexane
111-42-2	Diethanolamine
111-44-4	Dichloroethyl ether
114-26-1	Propoxur
117-81-7	Bis(2-ethylhexyl)phthalate (DEHP)
118-74-1	Hexachlorobenzene
119-90-4	3,3'-Dimethoxybenzidine
119-93-7	3,3'-Dimethyl benzidine
120-80-9	Catechol
120-82-1	1,2,4-Trichlorobenzene
121-14-2	2,4-Dinitrotoluene
121-44-8	Triethylamine
121-69-7	*N,N*-Diethyl aniline
122-66-7	1,2-Diphenylhydrazine
123-31-9	Hydroquinone
123-38-6	Propionaldehyde
123-91-1	1,4-Dioxane
126-99-8	Chloroprene
127-18-4	Tetrachloroethylene
131-11-3	Dimethyl phthalate
132-64-9	Dibenzofurans
133-06-2	Captan
133-90-4	Chloramben
140-88-5	Ethyl acrylate
151-56-4	Ethylene imine
156-62-7	Calcium cyanamide
302-01-2	Hydrazine
334-88-3	Diazomethane
463-58-1	Carbonyl sulfide
510-15-6	Chlorobenzilate
532-27-4	2-Chloroacetophenone
534-52-1	4,6-Dinitro-*o*-cresol, and Salts
540-84-1	2,2,4-Trimethylpentane
542-75-6	1,3-Dichloropropene
542-88-1	Bis(Chloromethyl)ether

CAS No.	Chemical or Trade Name
584-84-9	2,4-Toluene diisocyanate
593-60-2	Vinyl bromide
624-83-9	Methyl isocyanate
680-31-9	Hexamethylphosphoramide
684-93-5	N-Nitroso-N-methylurea
822-06-0	Hexamethylene-1,6-diisocyanate
1120-71-4	1,3-Propane sultone
1319-77-3	Cresols/Cresylic acid (isomers and mixture)
1330-20-7	Xylenes (isomers and mixture)
1332-21-4	Asbestos
1336-36-3	Polychlorinated biphenyls
1582-09-8	Trifuralin
1634-04-4	Methyl *tert*-butyl ether
1746-01-6	2,3,7,8-Tetrachlorodibenzo-p-dioxin
7550-45-0	Titanium tetrachloride
7647-01-0	Hydrochloric acid
7664-39-3	Hydrogen fluoride
7723-14-0	Phosphorus
7782-50-5	Chlorine
7803-51-2	Phosphine
8001-35-2	Toxaphene
Not available	Antimony Compounds
Not available	Arsenic Compounds (inorganic including arsine)
Not available	Beryllium Compounds
Not available	Cadmium Compounds
Not available	Chromium Compounds
Not available	Cobalt Compounds
Not available	Coke oven emissions
Not available	Cyanide Compounds[1]
Not available	Fine mineral fibers[2]
Not available	Glycol ethers[3]
Not available	Lead Compounds
Not available	Manganese Compounds
Not available	Mercury Compounds
Not available	Nickel Compounds
Not available	Polycyclic organic matter[4]
Not available	Radionuclides (including Radon)[5]
Not available	Selenium compounds

Note: For all listings above that contain the word "compounds," and for glycol ethers, the following applies: unless otherwise specified, these listings are defined as including any unique chemical substance that contains the named chemical (i.e., antimony, arsenic, etc.) as part of that chemical's infrastructure.

[1]X'CN where X = H' or any other group where a formal dissociation may occur. For example KCN or Ca(CN)$_2$.

[2]Includes mineral fiber emissions from facilities manufacturing or processing glass, rock, or slag fibers (or other mineral derived fibers) of average diameter 1 micrometer or less.

[3]Includes mono- and di- ethers of ethylene glycol, diethylene glycol, and triethylene glycol R-(OCH$_2$CH$_2$)$_n$=OR' where

   n = 1, 2, or 3

   R = alkyl or aryl groups

   R' = R, H, or groups which, when removed, yield glycol ethers with the structure R-(OCH$_2$CH)$_n$=OH[sic]. Polymers are excluded from the glycol category.

[4]Includes organic compounds with more than one benzene ring and which have a boiling point greater than or equal to 100°C.

[5]A type of atom which spontaneously undergoes radioactive decay.